MICHIGAN AGRICULTURAL COLLEGE

MICHIGAN AGRICULTURAL COLLEGE

The Evolution of a Land-Grant Philosophy, 1855–1925

Keith R. Widder

Michigan State University Press • *East Lansing*

This book was made possible by support from Michigan State University, the Office of the President, Michigan State University, and the Office of the Provost, Michigan State University.

green press INITIATIVE Michigan State University Press is a member of the Green Press Initiative and is committed to developing and encouraging ecologically responsible publishing practices. For more information about the Green Press Initiative and the use of recycled paper in book publishing, please visit *www.greenpressinitiative.org*.

♾ The paper used in this publication meets the minimum requirements of ANSI/NISO Z39.48-1992 (R 1997) (Permanence of Paper).

Michigan State University Press
East Lansing, Michigan 48823-5245

Printed and bound in the United States of America.

11 10 09 08 07 06 05   1 2 3 4 5 6 7 8 9 10

LIBRARY OF CONGRESS CATALOGING-IN-PUBLICATION DATA
Widder, Keith R.
Michigan Agricultural College : the evolution of a land-grant philosophy, 1855–1925 / Keith R. Widder.
p. cm.
Includes bibliographical references and index.
ISBN 0-87013-734-4 (cloth : alk. paper)
1. Michigan Agricultural College—History. I. Title.
S537.M5SW53 2004
630'.71'1774—dc22
2004021674

Book and cover design by Sharp Des!gns, Lansing, Michigan

Visit Michigan State University Press on the World Wide Web at
www.msupress.msu.edu

For the men and women who made the experiment

at Michigan Agricultural College successful,

and for the archivists, curators, librarians,

and scholars who have preserved their record.

Contents

Foreword

THREE GREAT SEISMIC SHIFTS TRANSFORMED AMERICAN HIGHER education in the nineteenth century. First, Protestants, overwhelmed with the spirit of the Second Great Awakening and subsequent waves of revivalism, built new institutions in the first decades of the century to serve as founts of evangelical ardor. From Hamilton in New York through Oberlin in Ohio, Kalamazoo in Michigan, Grinnell in Iowa, and Wake Forest in North Carolina, small-town colleges became the nexus for revivalism and reform. Second, German ideas of the transformed academy began to filter into the country at mid-century, pioneered by Henry Tappan in his brief stay at Michigan in the 1850s, and subsequently led by Charles Eliot at Harvard and Daniel Coit Gilman at Johns Hopkins. Third, and in some ways most revolutionary, a group of reformers in the 1850s began to argue for a more practical cast to higher education. This agitation climaxed in the passage of the Morrill Land Grant Act, but its earliest signal achievement came in the mucklands just to the east of the state capital in Michigan.

Keith Widder's study of the founding and early development of Michigan Agricultural College provides a unique and utterly fresh account not only of the transformation of this institution but of the development of practical higher education in America. It is, indeed, not simply an institutional history, but rather it is the biography of an idea. The conceptualization for an agricultural college began, crucially, not in the world of academics. The celebrated Yale Report of 1828 gave no attention to a practical education. Jeremiah Day stressed "the discipline and furniture of the mind," but his young Elis would not be drawn from Midwestern farm families.[1] It was, instead, the agitation of agricultural reformers throughout the country that brought forward the notion that higher education might consider some of the issues that were filling the pages of the burgeoning number of agricultural journals. The president of the State Agricultural Society of Michigan argued, in an address at the inaugural state fair, that theology, law, medicine, and the military were all well served by American colleges. However, he noted, "poor agriculture, whose hand sows the seed, and whose arm gathers the harvest on which all our earthly comforts and even our very existence depend, as yet has no seminary in which to teach her sons the most valuable of all arts."[2] The next year Massachusetts asked the president of Amherst, Edward Hitchcock, to investigate European ideas of agricultural education.[3]

Michigan proved most ready to act on the dreams of its agricultural reformers. Article 13, Section 11 of the Michigan State Constitution of 1850 declares:

> The Legislature shall encourage the promotion of intellectual, scientific, and agricultural improvements; and shall, as soon as practicable, provide for the establishment of an Agricultural School. The Legislature may appropriate the twenty-two sections of Salt Spring Lands now unappropriated, or the money arising from the sale of the same, where such lands have been already sold, and any Land which may hereafter be granted or appropriated for such purposes, for the support and maintenance of such School, and may make the same a branch of the University, for instruction in agriculture and the natural sciences, connected therewith, and place the same under the supervision of the Regents of the University.[4]

Not surprisingly, the implementation of Section 11 led to a half-decade of squabbles. Alexander Winchell, professor of Geology, Zoology, and Botany at the university in Ann Arbor, and later the Chancellor at Syracuse University, vigorously opposed the new college. Yet, as Widder carefully recounts, the clamor for the creation of a brand-new type of institution, a school that would deal with agricultural education in an innovative and, indeed, modern way, received the blessing of the state government in 1855.

Michigan Agricultural College was at first only a germ of an idea. It was, as Widder so appropriately terms it, an experiment. It would be a place of "practical" education. One president would insist that it was more a center of training than education. Almost from the moment of inception, however, the founders discovered that the word practical meant different things to different people.

To be sure, most everyone understood that "practical" could be defined negatively—it was the opposite of what Tappan had attempted to create in Ann Arbor as he rescued that experiment from imminent collapse. Tappan's German, romantic vision of education, cobbled onto the previous commitment at the University of Michigan to classical training in Greek and Latin, put the institution at the forefront of the transformation of research-based universities as they emerged in the second half of the century. Tappan's goals stood in constant tension with the desire of many trustees and faculties to make the university a center of religious training, and he was soon given his walking papers. Yet he had established the beachhead in Ann Arbor for the notion that if a university were to be great it needed but two ingredients: "scholars and books."[5]

Clearly what was about to take place in East Lansing was different from the Ann Arbor model. Scholarship for its own sake, for the sheer joy of learning, meant less here than sixty miles southeast. Once the agricultural college began, however, it was not sufficient to say that it would be defined largely by what the university in Ann Arbor was not. (Well, perhaps this does suffice on occasional football Saturdays.) Many voices were raised to say what should be meant by this vague idea of practical education. Indeed, as Widder explains in great detail, many farmers remained certain that the faculty in East Lansing constantly got it wrong. It was nevertheless an idea that had a life of its own, which developed in dialogues between the faculty and the students, between the presidents and the alumni, between the extension agents and the citizens of Michigan.

"Practical" might best be understood as having three components in the early years of M.A.C.: pragmatism, politics, and place. It would be several decades before Charles Pierce, William James, and their circle began to see the peculiarly American value of the pragmatic method.[6] Yet the notion of sorting out live propositions from dead ones, the goal of formulating questions that could possibly be answered, held sway in the middle of Michigan almost from the inception of the college there. Truth at M.A.C. came constantly from results. While the ideology of the founding generation perhaps was to create better and more productive farmers, the students quickly proved much less malleable than the faculty might have hoped. These young men wanted to learn much more than the best new ideas in fertilizer. Students insisted that they have a range of learning available to them—they wanted to know about the latest ideas in chemistry and soil sciences, to be sure, but they were also curious about the latest novels. A style of teaching soon emerged that put students in action—experimenting with soils, cutting up plants and examining their structures under microscopes, learning by doing. Soon imaginative young men were attracted to join this great experiment. William James Beal arrived on campus in the early 1870s full of Darwinian ideas, and he immediately began to develop teaching gardens, sending his students out into the research laboratory of wildflowers and grasses he created for them.

Even the institution itself constantly checked its progress against reality. By the end of the century, some, Beal included, wondered whether the experimental college in East Lansing had run its course. Expanding to include women, to teach a wider range of classes (including engineering, even though that infuriated some in the agriculture lobby), to attract a wider range of students seemed the only logical approach to a college that never stood still.

The seat-of-the-pants pragmatism described in the following pages required a second kind of practical skill: a talent for politics. Only a short trolley ride from the seat of state government, the college always had to pay attention to the needs of constituents. Even its beginning had an element of the political: the founding of the farmer-oriented Republican party in Jackson in 1854, and the election of a Republican majority within the state government that year meant that there was a powerful base of support in 1855 for the new experiment in practical education. Eighteen fifty-five stands as a peculiarly uneventful year in a tragically eventful decade. No great wars began. No great operas debuted. Dickens had an off-year, producing the little-admired serialization of *Little Dorrit*. Even tempestuous American politics had its most calm year of the decade; Kansas continued to roil, but the death of the Whigs and the emergence of the Republicans made this an organizational rather than confrontational year politically. Perhaps the agricultural college was their first attempt, therefore, in consolidating power.

During the Civil War, when Republicans came to have the sole voice in national government, the Morrill Land Grant Act helped develop the East Lansing experiment further. Yet at the same time, Lewis Fisk had to fend off the attempt to reduce the new college to a minor role as a two-year technical training facility. Throughout its history, indeed, Michigan State has been unusually adept at balancing its role as the land-grant college with its place in the long shadow of state office buildings. Long before John Hannah perfected the art of the politician as president, a series of leaders made use of the pragmatic, expansive idea of the college in East Lansing to try out a variety of roles: experimental gardens, agricultural extension agents, co-educational and multi-racial learning, international studies. Without shrewd leaders who could speak to the state legislature, the excitement and vigor of the college might have been snuffed out. Unlike other colleges and universities that tried to remain insulated from public and political battles, our history is engagement with these forces, for good and sometimes for ill.

Perhaps the most challenging political chore, and one that continued throughout the history of Michigan State, was the constant need to deal with farmers and, more generally, agricultural interests. Although agrarian support was essential for the creation of the college, farmers proved to be the font of regular criticism for activities in East Lansing. What was practical education? Should it not focus on how to farm effectively? Why were these eggheads up at the ag school insisting on all these experiments? Most important, why were the sons of Michigan farmers going to East Lansing, filling their heads with modern ideas, and not coming back to work the land? To be sure, the administration found some brilliant political solutions to these troubling questions, which frankly could not be answered satisfactorily in a direct manner. Instead, the practical, politically savvy leaders created the agricultural experiment station and, even more crucial, the system of agricultural extension agents, who could represent college interests throughout the counties of the state.

The third element of practical education is the place itself. As the pages that follow make amply clear, East Lansing became a central place for generations of young men and, just before the turn of the twentieth century, young women as well. It was a peculiar but inspired choice. Michigan's first two institutions of higher learning, the university at Ann Arbor and the normal school in Ypsilanti, both sat close to the stagecoach road between Detroit and Chicago and the first railroad, built to carry passengers from St. Joseph to Detroit. Both lay on the outskirts of the emerging urban mass in Wayne County, where Detroit served as the state capital until 1847. In contrast, the agricultural college was placed in the middle of swampy land just to the east of the new capital, Lansing. The choice of Lansing as the new government center made rough geographic sense, although the legislator who proposed it may have intended the suggestion as a joke. As every child in Michigan learns soon after birth, you explain where you live (in the Lower Peninsula, that is) by holding up your right hand. Lansing sits more or less in the center of the palm. Yet when the legislators first found their way there, Lansing had but a single log house and a sawmill, with oak and pine woods covering most of the land at the confluence of the Red Cedar and the Grand Rivers. For years the railroad stopped forty miles east, in Howell, and the rest of the journey was a bumpy ride on a plank road.

In the late 1850s, as the college began to be built, students found themselves enlisted in the project of making the place livable. The Burr farm, purchased because it was cheap and in a natural opening in the oak forest, required constant work, and the students earned their keep as lumbermen and log haulers. Quickly built in 1856, College Hall would collapse in 1918, probably because this wetland was given to shifting soils. The history department today occupies the third floor of Morrill Hall, a handsome brick and sandstone building dedicated in 1900 as the Women's building. (I tell my students that my office was once the shower room, based upon some rumor that has floated from generation to generation.) We're told that the main supports for Morrill are wood beams, which have long since deteriorated, and that, like College Hall, Morrill will one day return to the earth from whence it came.

Yet there was an important element to choosing a place (like Ezra Cornell's institution in Ithaca, New York) that was so centrally inconvenient. Its very isolation meant that the great experiment had room to breathe and grow. Drawn from farms throughout the state (and, soon, from as far away as Massachusetts and the Deep South), students came to the middle of Michigan, to the increasingly beautiful parkland of the campus, and for a year or more they joined the mystical union that is a college.

When I first flew from California to the middle of Michigan in 1979, I was startled by the pattern of lights. In California, at night, huge blocks of darkness alternate with enormous seas of light. Flying into Michigan from the west, the view from above was of small islands of light, each separated by about sixty miles of apparent emptiness. Upon landing, I found that the airport was bordered on the north by farmland, and that, even on the drive to campus, we regularly passed signs of agriculture. The next morning, we looked out of the Kellogg Center window on a particularly beautiful but extraordinarily cold and snow-covered winter scene, a pretty little ice-covered stream (the Red Cedar, I learned) cutting through a vast field of white. We had come for my wife's job visit—I would only be hired later as an afterthought—so I had the opportunity to wander around the place. East to west, it seemed to be about half an hour's walk. North to south, I couldn't seem to capture any sense of the distance. Later I discovered that the buildings give way after a few miles heading south to thousands of acres of barns, alfalfa fields, and woods. Unlike most universities in the country, this remains a meeting place between nature and society, where the rural past and the post-modern future still meet.

One last and important observation: Widder is only beginning to tell a complex tale. This history of the creation and development of Michigan State University should be the source of a raft of research projects. Why the history of Michigan institutions has lagged so far behind other states is a source of constant puzzlement. To be sure, a fire in the state archives made some projects more challenging. Yet the rich heritage of native culture, conquest, settlement, state formation, and institution building have remained largely unexamined by scholars. Popping out of every page of this story are more stories to be told. What became of so many of the products of the university—where did the students go, how did they serve to transform other institutions? Widder gives us a first cut in understanding the human impact of the university. How did the ideas generated by this place, the myriad transformations of lives and meaning, expand beyond its boundaries? It is the hallmark of this fine book, in the end, that Widder entices future scholars to pursue issues he has only begun to explore.

DAVID T. BAILEY
Associate Professor of History
Michigan State University

Acknowledgments

THE EFFORTS OF MANY PEOPLE HAVE MADE THE WRITING, PRE-
paration, and publication of this work possible. Michigan State
University is fortunate to have remarkable collections in its
archives, museum, and library from which the history of the
institution can be learned. As they go about their work, administrators,
faculty, students, alumni, staff, and others produce the "stuff" of these
collections. But the collections themselves are built only by highly
skilled and dedicated people who make researching and interpreting
Michigan State's history possible. Most collecting, preservation, and
organizing of historical documents, objects, images, tapes, and other
materials is invisible to the public. Even researchers and students are
usually too focused on their particular topic to be aware of the work
required to make that folder of correspondence, set of photographs,
objects used many years ago, type collections of archaeological arti-
facts, or plant and animal specimens available to them. The rich and
significant history of Michigan State University would remain
unknown and untold without the work of librarians, archivists, curators, and scholars who have built the university's research collections.
Great collections are a hallmark of a great university.

Director Fred Honhart and the staff of the Michigan State University Archives and Historical Collections oversee a remarkable collection, which holds far more stories than the one told in this book.
I am indebted to Fred, Whitney Miller, Carl Lee, Jeanine Mazak,
Sarah Roberts, and Portia Vescio for their faithful assistance and
advice throughout the research and writing of this book. The collections of the Michigan State University Museum contain another gold
mine. Val Berryman knows these collections inside out, and his willingness, along with Ilene Schechter's, to help me discover and

capture important pieces of the university's history, contained in the museum's collections, have enriched this book's interpretation of MSU's land-grant mission. More chapters of the institution's history are found in Special Collections of the Michigan State University Libraries. I am grateful for the efforts made by Peter Berg, Jerry Paulins, Randy Scott, and Kris Baclawski to locate everything I needed as I pushed forward in my research.

The history of agriculture in Michigan forms an essential corollary to the establishment and growth of Michigan Agricultural College. Anita Ezzo, of the MSU Libraries, has compiled an indispensable on-line tool for studying and researching Michigan agriculture: *Agriculture and Rural Life in Michigan: A Bibliography of State and Local Literature, 1820–1945*. This work is part of the United States Agricultural Information Network's National Preservation Project.

I am obliged to authors of earlier histories of Michigan State. William J. Beal's *History of the Michigan Agricultural College and Biographical Sketches of Trustees and Professors* (1915) and Madison Kuhn's *Michigan State: The First Hundred Years, 1855–1955* (1955) are invaluable sources for the institution's history. Kuhn, a professor of history at Michigan State also served as university historian, creating a significant collection of materials pertaining to MSU's past that form part of the university archives. In 2002, Linda O. Stanford and C. Kurt Dewhurst published *MSU Campus—Buildings, Places, Spaces: Architecture and the Campus Park of Michigan State University*. This work is a wonderful source for the history of the campus's architecture and use of space; this book lessened my need to tell Michigan State's architectural history.

Interpreting the evolution of the land-grant philosophy at Michigan Agricultural College has been a complicated, often uncertain undertaking. Many people have critiqued my thinking, writing, and factual presentation throughout the preparation of the manuscript. Stephanie Perentesis, Ilene Schechter, Fred Honhart, Peter Berg, Stacey Bieler, Ben Kilpela, Terry Shaffer, and Clarence Suelter read all or parts of the manuscript. Mary Black Junttonen, Lisa Fine, Frank Dennis, Carolyn Lewis, Curtis Stokes, and Austin Jackson helped to shape my thinking during conversations or by providing me with information or sources to consult. Drawing on his vast knowledge of the history of Michigan State University, David Thomas passed along to me important occurrences that influenced the institution's growth. Dorothy Frye made available to me her research files on the experience of African-American students at MSU, and she commented on part of the manuscript. Mary McMillen of the Archives and Records Management Division, City of Boston, Office of the City Clerk, and Cynthia Beavers Wilson of the Archives and Museums, Tuskegee University, answered critical questions relative to my research regarding early African-American graduates of Michigan Agricultural College.

Fred Bohm, director of Michigan State University Press, served as a sounding board for many of the ideas and interpretations expressed in this book. As he shared his insights into and knowledge of land-grant institutions, he helped me to stay on course. Likewise, his critical reading of the manuscript led to suggestions that enabled me to present my story with greater clarity and precision. I am especially grateful to Kristine Blakeslee for her expert copyediting of the manuscript. Annette Tanner and Julie Loehr, also of the Press, performed their usual miracles to bring the manuscript into print. The Press helps Michigan State fulfill its mission as a land-grant university by publishing scholarship that interprets research done at the university and elsewhere for the public.

David Bailey's continuous and growing interest in my research and the interpretations that emerged from it provided me with encouragement at times when I most needed it. His comments and suggestions raised questions that I needed to address, challenged me to expand the historical context of the creation and development of Michigan Agricultural College, and reinforced my belief that the college's history has a significance far greater than simply being the first agricultural college to succeed in the United States.

Finally, I could not have completed this project without the encouragement, patience, and love of my wife, Agnes. For several years she listened to my reports of new findings and my new interpretation for the day—only to hear the "real" significance of a piece of the college's history the next day. Agnes faithfully read every chapter and subsequent revisions, offered insightful comments, and asked a seemingly never ending list of relevant questions. Writing a book such as this one can at times be all consuming, but her enthusiasm for what we were learning, and what it meant, spurred me along.

The Names of
the Institution

During its history, Michigan State University has had six official names:

AGRICULTURAL COLLEGE OF THE STATE OF MICHIGAN
(1855–1861)

THE STATE AGRICULTURAL COLLEGE
(1861–1909)

MICHIGAN AGRICULTURAL COLLEGE
(1909–1925)

MICHIGAN STATE COLLEGE OF AGRICULTURE AND APPLIED SCIENCE
(1925–1955)

MICHIGAN STATE UNIVERSITY OF AGRICULTURE AND APPLIED SCIENCE
(1955–1964)

MICHIGAN STATE UNIVERSITY
(1964–)

Long before the college's name changed to Michigan Agricultural College in 1909, that name was commonly used. Consequently, I have used it interchangeably with the name State Agricultural College throughout the narrative. I use the names Michigan State College and Michigan State University only when it is chronologically appropriate.

Introduction

CHARTERED IN 1855, THE AGRICULTURAL COLLEGE OF THE STATE OF Michigan welcomed its first students two years later. The institution had been created in response to the demands of the state's agricultural leaders for a college to teach both practical farming and theories of the emerging scientific agriculture that were coming out of Europe. In 1855 no one knew what an agricultural college would look like or if it could succeed. Proponents of agricultural education in Michigan could not agree on whether the institution should be a department in the University of Michigan in Ann Arbor or a separate entity. Although the Michigan State Agricultural Society and editors of farm publications, notably the *Michigan Farmer,* vehemently called for an agricultural school, most Michigan farmers either had no interest in such an institution or were downright hostile to it. When the Michigan legislature passed an act creating the agricultural college, there was no certainty that the school would thrive or even survive. With no precedent for a four-year agricultural college in the United States, Michigan's attempt to create such a school was, as Governor Moses Wisner said at his inauguration in 1859, "an experiment in the agricultural history of our country."[1] In that same year, Alonzo Sessions, a member of the Michigan House of Representatives from Ionia County, called the college's formation "an experiment, wholly new and untried in this country"[2] in a report of the Joint Committee on Education and Agriculture. One hundred and fifty years later, Michigan State University stands as living proof that the work begun in a wooded swamp three miles east of Lansing has made a difference in the lives of millions of people. The dreamers of the 1850s understood their changing world quite well.

Seven years after the agricultural college was founded, Congress passed the Morrill Act, which did much to shape the institution's growth and development. A close look at the creation of Michigan's agricultural school reveals, however, that the die for its success—with many rough edges—had been cast before 1862. The Morrill Act called for the formation of colleges that would "promote the liberal and practical education of the industrial classes in the several pursuits and professions of life."[3] By 1862 the State Agricultural College already had fashioned a curriculum that embodied the core and the spirit of what became known as the land-grant philosophy of higher education in America.

The provisions of the Morrill Act also put in place the principle that both state and federal government should provide the financial support needed by land-grant schools. Congress allotted public lands to each state that, when sold, yielded an endowment for the operation of each state's college. (The practice of the federal government reserving the proceeds of land sales to support public education started with the Northwest Ordinances of 1785 and 1787 that set aside section 16 in each township to finance common schools.[4]) The Morrill Act also, however, required that construction costs be borne by the states. The federal government had, in effect, created a national system of higher education, but in the process it allowed each state to determine the programs offered by its college so long as the institution offered courses of study in agriculture, the mechanic arts, and military tactics. In the years that followed, Congress increased federal support for the additional responsibilities assumed by land-grant colleges. In particular, the Hatch Act of 1887 supplied federal funding for agricultural experiment stations, and the Smith-Lever Act of 1914 authorized and funded, in part, a cooperative extension service in each state. Furthermore, land-grant colleges were linked together by their work with the United States Department of Agriculture, which was created in 1862 by Congress in response to agitation by state and local agricultural interests and societies.

To understand the origins of Michigan State University, we need to be cognizant of the perceptions and the place of agriculture in the United States at the time. In the 1850s most Americans were engaged in some form of agriculture. At that point few people thought of farming as anything more than planting, cultivating, and harvesting crops, and raising livestock, intended first to provide for their families and then to sell or trade as surpluses on the market. In the South, African-American slaves labored on small farms and large plantations at the direction of slaveholders with little hope that their rewards would provide them with anything beyond subsistence. The availability of vast tracts of fertile land encouraged free farmers of the North, South, and West to uproot their families in order to start anew in states— including Michigan—and territories farther west. By 1890 the federal government had pushed Native Americans off most of the lands desired by European Americans for settlement. The availability of fresh land often resulted in farmers exhausting their fields without trying to improve their methods of production. Still, there were some who recognized that science, especially the infant discipline of agricultural chemistry, could unlock the secrets of soil composition, plant growth, and proper nourishment of farm animals. Editors of farm journals, scientists, and visionary agriculturalists argued that this newly discovered knowledge could be used to improve yields, restore soil, and reduce labor on American farms. These people provided the leadership that led to the creation of institutions of agricultural education in Michigan and across the United States.

Even as American farmers cultivated more land and increased production after the Civil War, the relative importance of agriculture to the overall American economy shrunk due to the rapid expansion of industry and manufacturing. The number of people working on farms relative to those employed in industry diminished as well. In 1850 over 84 percent of all Americans lived in rural areas, but by 1930 the proportion slipped to less than 44 percent. This shift was even greater in Michigan, where the rural population fell from 93 percent in 1850 to under 32 percent in 1930. The proportion of American workers engaged in agriculture declined from 54.8 percent in 1850 to 21.6 percent in 1930. Between 1880 and 1930 the proportion of Americans living on farms dropped from 44 percent to 25 percent.[5] Agricultural colleges contributed to the dramatic change reflected in these statistics by increasing the farmers' productivity—put simply, thanks to the work of these institutions, fewer people were needed on farms. Furthermore, land-grant institutions offered courses of study that trained young people to create, operate, and manage the new manufacturing plants and businesses that drove the American economy— including the agricultural sector.

The first students and some of the early leaders of Michigan's agricultural college, notably president pro tem Lewis R. Fisk, recognized that the key to the success of their school depended upon a paradox: if their institution, born of the desire to educate young men from the farm to farm more effectively, was to survive and grow, it also had to educate those young men (and later young women) to find places in a rapidly changing world *off* the farm as well. When a crisis over the curriculum arose in 1859, the early students made it known that if the college only taught them to be better farmers, they would *not* attend, for many intended to live in a world away from cows and wheat fields. They expected a broader education, and the college moved quickly to meet this demand.

The United States stood on the brink of enormous change in 1855. Secession by Southern states in 1861 triggered the start of the Civil War. After the war ended in 1865, the nation took twelve years to put itself back together—a time known as Reconstruction. Over the next fifty years, settlers carved out thousands of farms on western lands, entrepreneurs built great industries, and millions of immigrants swelled the populations of America's cities. Farmers traded their produce and manufacturers sold their wares in a national market linked by a newly constructed railroad system that connected the east coast with the west coast and the Gulf of Mexico with the Great Lakes. The telegraph, the telephone, and the radio created rapid communication between most parts of the land by 1925. Electricity and the gasoline engine also played key roles in changing the way Americans lived and conducted their business.

In the midst of change, hundreds of thousands of people living on isolated farms and in crowded cities suffered enormous hardships that created massive challenges for American institutions. Lack of education, the presence of disease, dangerous workplaces, economic depression, and the demands of the market led those who suffered to cry out for help and motivated those who heard their laments to initiate reforms to improve the lives of the masses. Consequently, Americans expended considerable resources to create public schools, to build water and sewage systems, to find ways to treat the sick and prevent disease, to improve diets, to conserve natural resources, to extend the vote to women, to end the use of alcoholic beverages, and to help lonely, often poor, people and immigrants to connect with the larger community around them. Churchgoers challenged organized religious bodies to look after the physical well-being of people as well as their spiritual state. Agrarians formed organizations such as the Grange and the Farmers' Alliance, and industrial workers joined unions to redress their ills. Increasingly, reformers, including the Populists and the Progressives, looked to government to alleviate the excesses and corruption brought about by rapid economic expansion in the United States after the Civil War.

Ignited by imperialism, the world experienced much upheaval and conflict during the late nineteenth and early twentieth centuries. Great Britain, Germany, France, Japan, Russia, Austria, and Italy, among other nations, competed with each other for colonial markets and spheres of influence, especially in Africa and Asia. A seemingly endless series of wars sharpened hostilities between people around the globe. Included among these conflicts were the Franco-Prussian War, 1870–71; the Sino-Japanese War, 1894–95; the Boer War, 1899–1902; the Boxer Rebellion, 1900; and the Russo-Japanese War, 1904–5. The United States engaged in its own imperialistic initiative when it went to war with Spain in 1898. As a result, the country found itself with an overseas empire composed of the Philippine Islands, Guam, and Puerto Rico, in addition to exercising considerable influence in Cuba. All of this international conflict led to the outbreak of World War I in 1914, which marked the beginning of three decades of turmoil and suffering that saw many of the same countries join World War II between 1939 and 1945.

If the United States hoped to meet the challenges and opportunities that it faced both at home and abroad, it was imperative that the nation's youth receive education that prepared them for this task. This required an extension of the democratization of higher education envisioned by Thomas Jefferson when he worked for the establishment of the University of Virginia in 1825. The intellectual abilities of young people from all classes in society needed to be cultivated, nourished, and unleashed if Americans intended to build a strong, democratic country. The Morrill Act's recognition that the "industrial classes" should receive a "liberal and practical education" was a giant step in that direction. America needed men and women who could apply scientific principles to agriculture, industry, public health, teaching, domestic life, and construction. But a vital democracy also relies on men and women educated in literature, history, philosophy, political economy, and the arts to provide leadership in government, journalism, medicine, law, religion, and community development. Since 1862, land-grant colleges have played a major role in the democratization of the American people.

The history of Michigan State University is just one chapter in a much larger story of higher education in Michigan. The Agricultural College of the State of Michigan was the third state-supported institution of higher learning founded in Michigan. By the time it was established, the University of Michigan, in Ann Arbor, was well on its way to becoming the leader among state universities in the country, and the State Normal School, in Ypsilanti, was training teachers for the public schools. State normal schools were established in Mount Pleasant (1895), Marquette (1899), and Kalamazoo (1903). Each of these institutions evolved into a regional state university. In 1886 the Michigan School of Mines offered its first classes in Houghton, and in 1893 construction began in Big Rapids for Ferris Institute. Throughout the twentieth century public higher education in the state grew at a fast pace. In addition, private institutions affiliated with Christian denominations, including Kalamazoo, Albion, Hope, and Adrian colleges, also offered college educations.

Although the achievements and contributions of Michigan Agricultural College form a unique story, we must remember that other land-grant institutions across the country have similar histories. Michigan Agricultural College played a significant role in linking together land-grant schools by helping to form the Association of Land-Grant Colleges and Universities in 1887. M.A.C. was a part of the national success story generated by land-grant colleges, which in turn, was just one part of the nation's larger system of higher education.

One key to M.A.C.'s survival, success, and acceptance was its refusal to surrender to demands—even from its friends and supporters—to deviate from the mandate it had received from the public through the Michigan legislature and the U.S. Congress. Many forces, including farmers' attitudes, economic cycles, politics, and law, influenced the growth and development of the school. But administrators, faculty, students, and alumni never lost sight of the purpose of their institution spelled out in its enabling legislation enacted in 1855 and 1861 and the Morrill Act of 1862. The people who made up the college community, not the public or the legislature, defined the meaning of land-grant, and the institution worked tirelessly to have the public understand its relevance for their lives. During most of its first seventy years Michigan State University struggled to build a strong and unbreakable equilateral triangle that joined together teaching, research, and outreach. The emerging land-grant philosophy or way required that the scope of the study and practice of agriculture and other disciplines keep growing. Consequently, M.A.C. perceived agriculture to be more than planting corn, milking cows, and picking apples; the college's view of agriculture embraced, among other things, scientific research, marketing agricultural products, and educating farmers and their families to live fuller lives. Michigan farmers did not welcome the introduction of the mechanical or engineering course in the mid 1880s, but the college fulfilled that provision of the Morrill Act nonetheless. Students attending Michigan Agricultural College received a well-rounded education that prepared them to assume positions of leadership in the professions, commerce, industry, education, and civic life, as well as agriculture. It must be kept in mind that M.A.C. was a small college—its enrollment did not exceed three thousand until 1926.

Three interpretive themes explain the evolution of the land-grant philosophy at Michigan Agricultural College and how this new, experimental, even revolutionary, way of approaching higher education came to play a vital role in the history of Michigan, the United States, and the world. First, the conflict between Michigan Agricultural College and many Michigan farmers over the role of the college generated the creative tension that shaped the land-grant philosophy at M.A.C. The historian Richard Hofstadter identified one of the forces at work in Michigan and across the United States that had serious implications for the development of the college during its first sixty years when he noted that a fundamental change occurred in American agriculture during the second half of the nineteenth century: "The triumph of commercial agriculture not only rendered obsolete the objective conditions that had given to the agrarian myth so much of its original force, but also showed that the ideal implicit in the myth was contesting the ground with another, even stronger ideal—the notion of opportunity, of career, of the self-made man."[6] Hofstadter tells us that as new generations grew up on farms, they looked beyond their family farms for their future. This led to stress between the generations that had implications for M.A.C. To prepare young men and women coming from farms, small towns, and cities to pursue careers in industry, education, business, government, and other professions as well as agriculture, Michigan Agricultural College fashioned a curriculum based on science, experimentation, and liberal arts that formed the core of the agricultural course. This did not sit well with many Michigan farmers.

Believing that the college's primary, and even only purpose was to train students to become better practical farmers, many farmers questioned the value of scientific agriculture or felt that the college's program subverted their desire to have their children return to the farm. Two examples illustrate the nature and complexity of farmer's attitudes toward M.A.C. Late in life, Clark Brody, '04, recalled his desire to attend college and his belief that doing so would "open up unusual opportunities" for him, just as it had for some of his friends and relatives. When he asked his father if he would allow him to go to college, the elder Brody gave his consent on the condition that Clark "study a profession," but he told him that if he wanted to be a farmer, he should "stay on the farm." Clark's father, like many of his peers had little regard for the scientific agriculture being taught at M.A.C.[7] Clark Brody enrolled at M.A.C. intending to become a mechanical engineer, but the potential of scientific agriculture appealed to him so much that he took the agricultural course. He became a farmer, and he worked for the Michigan Farm Bureau for thirty-five years. During that time he showed many other farmers how to use the technology developed and methods learned through scientific experimentation at M.A.C. and other institutions on their own farms.

In contrast to Clark Brody's father, who encouraged his son to study mechanical engineering, if anything, at M.A.C., other farmers denounced the college for having an engineering program at all, even though the provisions of the Morrill Act of 1862 required land-grant schools to offer such a course. In May 1914 the Michigan Supreme Court overturned legislation passed in 1913 that limited the institution's ability to expend federal funds for its engineering program. In response to this episode, the *Adrian Daily Telegram* published a scathing editorial denouncing M.A.C. for teaching engineering. The editorial oozes with frustration, bitterness, and anger over what is per-

ceived to be the intent of the college to entice students to leave the
farm for other professions.

> In one sense this effort is unfair. It is almost underhand[ed], because
> the M.A.C. uses the name of "agriculture" and the prestige of the
> state "agricultural" college to induce farmers' boys to quit the farm.
> Unquestionably the fact that the engineering school is attached to
> the Agricultural college results in many boys leaving the farm for-
> ever, who otherwise would have become farmers of the most prac-
> tical type. Even when a boy goes to East Lansing to study farming,
> he is immediately subjected to temptation to become an engineer.
> This is not because engineering is better than farming, but because
> it is novel and interesting, and many boys are fascinated by machin-
> ery and scientific experiments. We cannot imagine a more effective
> "away-from-the-farm" influence than an engineering school mas-
> querading as part of an agricultural college.[8]

While the *Telegram* articulated the perception that M.A.C.
behaved in a subversive manner, the editorial raises the question of
whether there was at the time a significant number of disgruntled
farmers who felt betrayed by their agricultural college. Two months
before the *Telegram*'s blast, the *Detroit Tribune* discussed the meaning
of the attack on M.A.C. over the existence of the engineering program.
Although the *Tribune* did not necessarily agree with farmers who
argued that M.A.C. existed only to train practical farmers, it did note:
"There is clear evidence it is the agriculturalists of the state who are
most displeased with them [President Snyder and the State Board of
Agriculture]; that it is for a very considerable constituency back on
the farms that the makers of laws have to make laws that clip the
wings of the college."[9] What makes these harsh words so striking is
that they appeared fifty-nine years after the founding of the college,
and they originated with the group of people that the college was
intended to help the most.

But was the college was subversive? The answer to this question
requires a look at the meaning of subversive and the manner in
which it can enhance our understanding of the history of Michigan
Agricultural College. According to the *Oxford English Dictionary* (second edition), subvert can mean: "To bring about the overthrow or
ruin of (a person, people, or country, a dynasty, etc.)" or "To disturb
(the mind, soul): to overturn, overthrow (a condition or order of

things, a principle, law, etc.)." Clearly, the farmers were not accusing the college of trying to overthrow or ruin either the country or themselves. Many farmers, however, believed that the college willfully disturbed their children's—particularly their sons'—minds to motivate them to turn their backs on farming and the way of life so revered by generations of agriculturalists.

The college developed courses of study that were clearly intended to educate young men and women to be well-informed citizens and to pursue evolving professions that served commercial agriculture in new ways or were independent of agriculture. Though the college never despaired of training its students to be better farmers, nor lost sight of its purpose to serve agriculture, critics of the college found the number of alumni who actually tilled the fields to be far too few relative to the number of students who attended the college. The *Telegram* stated their views quite well: the college was underhanded in that it created programs (engineering in particular) to masquerade as agricultural, when in fact they were designed to encourage young men to leave the farm and never return. From this perspective, the college's administration, faculty, and even many of its students and alumni were subversives. In addition, as the experience of Clark Brody illustrates, many farmers had great difficulty appreciating the potential value of scientific agriculture for their own enterprises—which, in turn, generated much misunderstanding of work being done at M.A.C. and much hostility toward it. The world that surrounded Michigan farms was in a state of rapid change that many people did not comprehend or like. Since the college was an agent of change, it became an easy target for agrarian discontent.

On the other hand, M.A.C. worked tirelessly to build up and improve the practice of agriculture and the quality of life in rural Michigan. In order for the college to accomplish its mission, it needed to challenge, circumvent, undermine, and overcome the negative attitudes that many of its constituents directed toward the institution. This Herculean effort required M.A.C. to fulfill the agrarian demand that it train practical farmers. To this end, the creation and development of the Cooperative Extension after the passage of the Smith-Lever Act in 1914 eventually enabled the college to educate farmers in their own communities or on their own farms. County agricultural agents and home demonstration agents working with local organizations and drawing upon the resources of the college created a system that met the needs of rural men and women where they lived. At long last, in the minds of many farmers, the college ceased to be a subversive agent when it thus began fulfilling its promise to train practical farmers. Together the college and the farmers came to recognize the obvious: the person who needed to be turned into a better practical farmer was the farmer himself, and not necessarily his son.

The second theme of this evolution is reflected in how the students, the sons and daughters of the "industrial classes," helped to shape the college's curriculum and fulfilled the mission of land-grant education through their work in the wider world they entered after they left the institution. Administrators, professors, instructors, and researchers set the college's direction through their teaching, research, and public service. The faculty worked tirelessly to take their hard-won knowledge to the public through both the spoken and the written word, but even more significantly their students (some of whom served on M.A.C.'s faculty and governing board) touched the lives of thousands of people each day following the completion of their studies. M.A.C. alumni labored on farms, conducted experiments in laboratories, designed and built bridges and machines, treated the sick, served in government at all levels, marketed fruit, edited newspapers and journals, taught in schools and colleges across the country, and traveled the world, leaving with people and their communities some of what they had learned on the banks of the Red Cedar.

Finally, the third theme of this evolution is demonstrated in how the land-grant philosophy that evolved at Michigan Agricultural College expanded and strengthened American democracy. The college provided a place, a curriculum, and an atmosphere that enabled thousands of young men and women to receive a higher education in a "shirt sleeve" rather than a "dress-suit" environment. Many of these young people were from poor and "ordinary" families, and would not have been able to find places in more elite colleges and universities.[10] M.A.C. served as a point of entry into higher learning. A large percentage of the students who enrolled in M.A.C. stayed for only a term or a year or two before transferring to the University of Michigan or another school to pursue a course of study not available in East Lansing. And Michigan Agricultural College worked hard to ensure that knowledge gained on its campus was made relevant to the general public. The land-grant way demanded that the benefits of scholarly research be disseminated to the public and not reserved for the edification of a small elite. M.A.C. alumni and faculty directed much of the knowledge gained at M.A.C. toward the ordinary things and occurrences in daily life. In effect, the professors and their students at M.A.C. helped to break down barriers between "learned" and "ordinary" people. Not everyone, however, participated equally in the growth of democracy at M.A.C. The college was slow to incorporate women into its full program, although that changed dramatically after the creation of the women's course in 1896. African Americans found few opportunities in the college before 1925. International students did not come to study agriculture until the 1870s.

In 1945, reflecting upon his long life in the small town of Amherst, Massachusetts, Ray Stannard Baker, '89, captured the essence of M.A.C.'s democratic ideal when he wrote in his autobiography: "But it is not enough to live in a country town: one must also live with it. Little towns are shy, not easily taken by assault. Their acceptances are often slow; but they are genuine. Nothing much is said about democracy; but we manage to get a good deal of it into our common affairs. We somehow know that the core of democracy lies not in being able to make something out of the town, but of contributing something to the town."[11] Baker's life was characterized by "contributing something" to the world. He and his fellow students learned at Michigan Agricultural College that democracy was more dependent upon what they gave to their community than upon what they received from it. A healthy democracy thrives when people actively participate in community affairs, rather than sit by passively and absorb its benefits.

In a sense, the story of Michigan State University is a re-creation of an old Greek story. Joseph R. Williams, John C. Holmes, Lewis R. Fisk, and others built a Trojan horse that looked like, felt like, and smelled like an agricultural college—and it was an agricultural college. But inside the college much more went on than learning about farming. Non-agricultural courses guaranteed that students were equipped to take on responsibilities in many vocations and positions of public trust. M.A.C. first took the nickname "Aggies," but it is perhaps apropos that when the institution's name changed to Michigan

State College of Agriculture and Applied Science, it drew upon Greek history to adopt "Spartans" as its new nickname. Together Aggies and Spartans have served the people of Michigan, the nation, and the world. Unlike the Greek soldiers at Troy, they have not sought to conquer; instead they have tried to liberate. They have improved people's diets and health, designed roads and factories, and made agriculture more productive. They have taught, built, managed, led, and more important, they have served people throughout the world.

It is fitting that this land-grant institution that eschewed the study of Greek can be understood in terms of a story drawn from Greek antiquity, for herein lies the essence of the land-grant philosophy that has been taking form for a century and a half. It is a philosophy that will not be bound by the past, yet that is informed by the past. It seeks out new ways to educate new generations to meet the challenges of their world, not just the needs of their parents' and grandparents' generations. The land-grant philosophy is optimistic, enthusiastic, innovative and, above all, courageous and anticipatory. Professors, administrators, and students alike have not been afraid to try one thing after another in the pursuit of knowledge and to apply the fruits of their research to real life in the real world. After all, that is how you carry out "an experiment, wholly new and untried."

1

The Beginning

O N 13 MAY 1857 JOSEPH R. WILLIAMS, THE RECENTLY APPOINTED president of the Agricultural College of the State of Michigan, told those assembled at the school's dedication: "We have no guides, no precedents. We have to mark out the Course of Studies, and the whole discipline and policy to be followed in the administration of the Institution."[1] Williams welcomed this challenge as he launched America's first four-year college devoted to the study and teaching of agriculture. For Williams, and for the other proponents of an agricultural college for Michigan, the opening of the school represented both a culmination and a beginning. The long struggle to gain a college had ended, but now began the even more arduous task of making the institution work. Its founders and supporters were well aware that they had undertaken a pioneering effort to make higher education available to the sons of farmers and others educated in the common schools, not only in Michigan but throughout the United States. The ceremonies three miles east of Michigan's Capitol on the second Wednesday in May 1857 inaugurated what soon would come to be known as America's land-grant system of colleges and universities.

The setting foreshadowed the difficult journey facing the Agricultural College of the State of Michigan. Robert F. Johnstone, editor of the *Michigan Farmer,* offered to his readers an account of the fledgling institution in the newspaper's May 1857 issue. His description of the "The Farm" revealed the enormous amount of work that stood before the college.

This land is like all in that section, heavily timbered and has as yet undergone but little improvement. Some twenty or thirty acres had been chopped before the purchase was made, and during the past

Joseph Rickelson Williams, first president of Agricultural College of the State of Michigan. A classmate of Wendell Phillips at Harvard, Williams opposed the expansion of slavery in Kansas and played a key role in the establishment of the Republican Party in Ohio. He served as editor of the *Toledo Blade* before accepting the position as president of the recently established Agricultural College of the State of Michigan. Williams laid much of the groundwork for the creation of the land-grant system of higher education as he worked diligently for the passage of first Morrill bill in 1859, only to see President Buchanan veto it. A year after Williams's death, President Lincoln signed the Morrill Act of 1862 into law. Courtesy of Michigan State University Archives and Historical Collections (People, Williams, Joseph R., no. 2962).

summer and winter, there has been cleared about one hundred acres now lying principally around the buildings, and between the buildings and the plank road leading to Detroit. Nearly all this land was wild, unimproved, and unenclosed. There were no fences, even along the highway, and the rails have had to be split, and the fences built, during the past winter and present spring, which were absolutely necessary for enclosure. The land upon which the timber has been cut, looks and is as rough and wild, at present, as it is possible for a new piece of land to appear. The cut timber and brush is piled in heaps, and ready as soon as the season will permit, to be set on fire. The stumps are all on hand, looking as green and as sturdy, and as obstinate, in their determination to retain possession of the land, as green hard wood stumps generally appear. The surface in many places was covered with water, especially on the flat places, where there was little or no declivity to drain off the water. What our farming community would call "cat holes," had evidently been numerous. These places are well known as the habitation of frogs and mosquitoes.[2]

Farmers easily understood a setting like this as the start of a farm, but they had no comprehension of this place as an institution of higher learning. And, it took them a long time to learn the difference between a farm and an agricultural college.

The primitive state of the college's environment fit in well with the frontier appearance of Lansing and Ingham County. Native Americans still frequented the Lansing area, which stood near the northern edge of land being settled by European-American and European immigrant farmers in Lower Michigan in the 1850s.[3] The Ottawa Chief Okemos, a famous leader among local Indians, died in 1858.[4] Lansing's population reached 3,085 in 1860, and Ingham County claimed 17,456 residents at that time. Seventy-five folks called the village of Okemos home, and Meridian and Lansing townships had populations of 825 and 497, respectively. Michigan had a total of 758,252 inhabitants with 45,387 of them residing in Detroit.[5] Built in 1847, the Capitol appeared before adequate public accommodations had been constructed to serve it. But within a decade of being designated the state capital, Lansing took on the appearance of a growing community with churches, schools, and hotels serving the needs of residents and visitors—including politicians.[6] The Amboy, Lansing & Traverse Bay Railroad reached Lansing in 1862, giving the city its first direct rail link to the outside world.[7] No longer did travelers to the state

capital have to endure an unpleasant ride on a stage coach over ruts or through mud from Jackson or Owosso.

Michigan farmers, like their counterparts elsewhere in the United States, found themselves in an expanding capitalist market that was transforming the economic, political, and social life of the country.[8] The demands of the market, over time, compelled agriculturalists to change the way they planted, cared for, harvested, and sold their crops. In fact, the agricultural college came into existence in part to help landowners adapt to the fluidity and unpredictability of market forces. Growers needed to figure out how to apply the results of scientific research to their fields, orchards, gardens, vineyards, and barnyards. The founders of the agricultural college and their supporters believed that their new institution would play a critical, decisive role in the metamorphosis of a farming community rooted in the past into one driven by scientific discovery and the pursuit of profits.[9]

The assembled group took part in the opening phase of another struggle to enlarge American democracy by giving more people the

Lake Shore and Michigan Southern Railroad in North Lansing, Franklin Street (Grand River Avenue), circa 1870. Courtesy of Michigan State University Archives and Historical Collections (Michigan, Lansing).

This 1873 photograph shows the "Sacred Space" in front of the three buildings that launched and sustained M.A.C. in its early years. "Saints' Rest," the first dormitory stands to the left of Williams Hall, the second boarding hall, built in 1869. College Hall, the first classroom building, chemistry laboratory, and library is on the right.

Courtesy of Michigan State University Archives and Historical Collections (Michigan State University, Buildings).

freedom to realize their potential through education. Antebellum America witnessed a series of movements to empower more people, especially women, African Americans, and children, to participate more fully in public life in the United States. Many forward-looking people believed that a strong system of education, starting with the common schools all the way through the university level, was necessary for a thriving democracy.[10] The historian Eric Foner has noted that the meanings of freedom have emerged out of "debates and struggles," and the "understandings of freedom are shaped by, and in turn help to shape, social movement and political and economic events."[11] This insight helps to illuminate the significance of the land-grant philosophy of higher education launched by Williams and those assembled at the dedication of the Agricultural College of the State of Michigan. Sons and daughters of farmers and laborers educated at Michigan's land-grant college assumed responsibilities put upon them by American democracy and worked to extend democracy's benefits to more people.

This extension of American democracy rested upon two interrelated aspirations, one held by the founders of the Agricultural College of the State of Michigan and another by the students who attended it. First, Williams and his allies intended to lead a new, revolutionary approach to higher education in the United States, not just Michigan. They understood that their success depended in no small part on creating a curriculum that was not centered on the classical languages of Greek and Latin if young people who did not have the chance to learn these "dead" languages were to be given the opportunity to continue their formal education beyond the common schools and high schools.[12] The proponents of the agricultural college triggered a welcoming response from young men and women living on farms and in towns throughout Michigan who desired to receive the education that the new school promised to make available to them. They were prepared to read works of English literature, write compositions, and study science in order to pursue agricultural and non-agricultural vocations and careers in the changing world where they would be living. Since that world often extended beyond their parents' farms, many of their mothers and fathers reacted negatively toward the college.

The Genesis of Agricultural Education in the United States

From the birth of the United States, Americans believed that education should prepare young people to break with the past and meet the intellectual and practical challenges that would come their way. The American historian Bernard Bailyn has shown that as America emerged from its colonial experience, education "released rather than impeded the restless energies and ambitions of groups and individuals."[13] When, in 1782, J. Hector St. John de Crèvecoeur replied to the rhetorical question "What is an American?" part of his answer was: "The American is a new man, who acts upon new principles; he must therefore entertain new ideas and form new opinions."[14] Horace Mann argued in 1839 that the country's young needed to receive a thorough education in order for the republic to thrive. Education enabled people "to act, *formatively,* upon the crude substance of nature,—to turn a wilderness into cultivated fields, forests into ships, or quarries and clay-pits into villages and cities."[15] Thus, Crèvecoeur and Mann helped to pave the way for an approach to higher education that went beyond the classical curriculums that focused on improving a student's "moral character" and "mental faculties."[16]

Informed by this evolving philosophy that education could be, even should be, designed to teach people to do practical things, American agriculturalists cited two developments in Europe that appeared to hold great promise for farmers in the United States. First, the rise of the study of agricultural chemistry raised an awareness that chemical analysis of soils, fertilizers, and manures could

Lewis Ransom Fisk started the tradition of excellence in chemistry at M.A.C. when he outfitted the college's first laboratory in College Hall as part of his duties as the institution's first professor of chemistry. His greatest contribution to M.A.C., which has gone largely unnoticed, came when he guided the college through some very uncertain days while serving as president pro tem from 1859 until 1862. A Methodist minister, Fisk pastored several churches and held the presidency of Albion College for many years after he left M.A.C. Courtesy of Michigan State University Archives and Historical Collections (People, Fisk, Lewis R., no. 992).

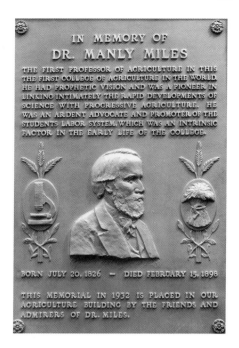

Manly Miles Memorial. Born in Homer, New York, Miles grew up on a farm near Flint, Michigan. A physician by training, he practiced medicine for only nine years after receiving his M.D. from Rush Medical College in Chicago in 1850. His interest in geology, zoology, and farming led him to M.A.C., where he was professor of zoology and animal physiology from 1861 until 1865 and professor of practical agriculture from 1865 until 1875. Courtesy of Michigan State University Archives and Historical Collections (People, Miles, Manly, no. 1983).

lead to more productive farming. The works of the German scientist Justus Liebig attracted many avid readers in the United States in the 1840s after he published *Organic Chemistry in Its Applications to Agriculture and Physiology* (1840) and *Animal Chemistry* (1842). Although he contributed much to the birth of agricultural chemistry, Liebig oversimplified what proved to be much more complex chemical problems related to agriculture. In 1842 Samuel L. Dana challenged some of Liebig's work and generated debate within farming circles with the publication of *Muck Manual*. Such controversies showed that much needed to be learned before agricultural science could produce results that would make significant, practical differences in the fields of American farms. By the mid 1850s scientists "began to stress the need for long-term experimentation into agricultural problems as the only hope for improvement."[17]

The existence of agricultural schools in Europe motivated Americans to build similar institutions in the United States. To see this inspiration at work, we need look no further than the executive meeting of the State Agricultural Society of Michigan in 1850. Committee members noted that the "three kingdoms of Great Britain" and the "leading countries of Europe" all had agricultural schools and colleges. They believed that these institutions, along with their model farms, provided the "most important aid to the successful progress of agriculture." Distinguished scientists conducted experiments and gave lectures on natural science at these institutions. Furthermore, they disseminated their findings through publications and trained young men from the United States to be professors of applied chemistry at Harvard and Yale. The leaders of the society saw the real value of agricultural education to be in the training of "an army of practical men" in expert farm management, arming them with the knowledge that the application of scientific principles could improve farming. They envisioned these young men becoming, in essence, "normal schools of agriculture," sharing their knowledge and wisdom with other farmers.[18]

The birth of agricultural institutions of higher learning in the United States, and in Michigan, in particular, followed, not surprisingly, an uncertain, tortuous path. The rise of scientific research and the question of the place of science, particularly agricultural science, in curriculums that grew out of the study of the classical languages of Greek and Latin posed a real challenge to many. In 1846 Yale took the lead by setting up professorships of agricultural chemistry, animal and vegetable physiology, and practical chemistry. At Harvard, Abbott Lawrence endowed the Lawrence Scientific School. But inroads such as these at established institutions of higher learning did not lead to large-scale efforts to create applied scientific agricultural curriculums.[19]

A brief survey of a few early attempts to create agricultural education programs illustrates that many people contributed to this effort before the Agricultural College of the State of Michigan opened its doors. In the 1790s King's College (Columbia University) recognized a need for a professor of agriculture and the New York legislature granted funds for the position, but the study of agriculture did not take hold. In 1823 the Gardiner Lyceum in Maine started its ten-year run as an agricultural school. Some people considered James Smithson's bequest to the United States, the foundation of the Smithsonian Institute, as a potential source of funding for agricultural education in the late 1830s and 1840s.[20]

In 1855 the Pennsylvania legislature authorized the establishment of the Farmer's High School. Construction of the school did not begin until 1857, and it welcomed its first class in early 1859. Later that year Evan Pugh assumed the presidency of the young institution, and he skillfully guided its early growth. In 1862 it was renamed the Agricultural College of Pennsylvania. The Massachusetts legislature set up a board of trustees for the Massachusetts School of Agriculture in 1856, but the institution did not come into existence until after the passage of the Morrill Act in 1862. The Maryland Agricultural College, founded in 1856, enrolled its first students in 1859. In the same year, the Iowa legislature set aside $10,000 to start the Iowa State Agricultural College and Farm, but it would be ten years before any students received instruction there.[21]

THE ESTABLISHMENT OF THE AGRICULTURAL COLLEGE OF THE STATE OF MICHIGAN

On 12 February 1855 Governor Kinsley S. Bingham signed into law the legislation establishing the Agricultural College of the State of Michigan. Michigan had relied upon funds derived from the sale of federal land to support another institution of higher learning that included agricultural education since long before 1855. Soon after Michigan attained statehood in 1837 the legislature passed an act organizing the University of Michigan, which became the sole beneficiary of the sale of seventy-two sections of land reserved by Congress in 1826 for the support of "a seminary of learning." Sales from the federal land were invested to provide earnings to support the university's operations. The act of 1837 also required the university to set up branches throughout the state (no more than one per county) to function as preparatory schools for the university. Section 20 of the act provided that: "In each of the branches of the University, there shall be a department of agriculture, with competent instructors in the theory of agriculture, including vegetable physiology and agricultural chemistry, and experimental and practical farming and agriculture."[22] Within ten years the university stopped supporting the branches that had been established in Detroit, Pontiac, Monroe, Tecumseh, Kalamazoo, Niles, Romeo, and White Pigeon.[23] Although the university's first effort at agricultural education had been at the preparatory level, it would fight hard to become the institution for agricultural education at the university level. The real significance of Section 20 was that the people of Michigan served notice at the beginning of statehood that they expected state government, relying in part upon federal resources, to provide agricultural education at an institution of higher learning.

Voices throughout Michigan in the 1840s and early 1850s articulated the need for formal agricultural education in the state and spread their message far and wide through agricultural periodicals and reports. The *Michigan Farmer* became a leading advocate for reform in agriculture and for the creation of an agricultural college. In 1848 Warren Isham, editor of the *Michigan Farmer,* stated bluntly to his readers on Michigan farms that "it is the duty of Government to institute Normal Schools for the promotion of agriculture."[24] A year later E. H. Lothrop addressed the Michigan State Agricultural Society and lamented that while both the people and government liberally supported "literary and other public institutions" (theological seminaries, law and medical schools, and military academies), "poor agriculture, whose hand sows the seed and whose arm gathers the harvest, on which all our earthly comforts and even our very existence depends, as yet has no seminary in which to teach her sons the most valuable of all arts."[25] Concerned over the failure of the national government to support institutions devoted to agricultural education, the Michigan legislature petitioned Congress in 1850, through its senators and representatives, to donate 350,000 acres of federal land to the state to enable it to create and operate agricultural schools.[26] Although nothing came of this memorial, Michigan continued to provide leadership at the national level for the passage of federal legislation to support the establishment of institutions of higher education for agriculture.

In 1850 Bela Hubbard, who had surveyed much of the state earlier in his career and knew the state's geography and geology intimately, outlined a plan for the legislature. Before flickering candlelight in farm kitchens and offices of small businesses, readers of *Michigan Farmer* could nod in agreement with Hubbard's challenge to the legislature that it establish "a State Central Agricultural Office, with which shall be connected a Museum of Agricultural Products and Implements, and an Agricultural Library, and as soon as practicable, an Agricultural College and a Model Farm."[27] He saw the college and its model farm as "the most important aid to the successful progress of agriculture." Although Hubbard wished to see the school affiliated with the University of Michigan, he had identified the first components of what would become a separate agricultural college.

The seeds sown by Isham, Lathrop, Hubbard, and many others bore fruit at the state constitutional convention that rewrote the Michigan Constitution in 1850. Article 13, Section 11 put into law the wishes of the leaders of the Michigan agricultural establishment, but it did not settle the question of whether the agricultural school would be an independent institution or a part of the University of Michigan. Section 11 reads as follows:

> The Legislature shall encourage the promotion of intellectual, scientific, and agricultural improvements; and shall, as soon as practicable, provide for the establishment of an Agricultural School. The Legislature may appropriate the twenty-two sections of Salt Spring Lands now unappropriated, or the money arising from the sale of the same, where such lands have been already sold, and any Land which may hereafter be granted or appropriated for such purposes, for the support and maintenance of such School, and may make the same a branch of the University, for instruction in agriculture and

the natural sciences, connected therewith, and place the same under the supervision of the Regents of the University.[28]

A fierce battle ensued over the next four years to determine the disposition of Michigan's agricultural college.

When examining the separate institution–university debate, it must be recognized that at the core of this argument lie two different—and not necessarily incompatible—visions of schools of higher learning. J. W. Scott, who lived in Adrian, articulated the human side of what is often seen as being primarily an intellectual and political fracas. Scott wrote in the *Michigan Farmer* that he wanted his son to be educated as a "mechanic." Scott desired that the boy learn some French and German and hear "lectures on scientific subjects, and on moral and political philosophy." He wanted him "to have a practical education" and "to participate in the benefits of an institution that *should* be adapted to the wants of the chief occupations of our people." Scott went on to reject the value of dead languages (Greek and Latin). He then raised a series of questions for his audience to answer for themselves: "Well, our University—what does it do to educate the farmer's son, for a farmer?—what to make scientific mechanics?—what to form business men?"[29] For Scott it was of paramount importance that his son receive an education that prepared him to make a living in the professions created by industrial growth that were emerging in the middle of the nineteenth century. While Scott expressed no opposition to the university training young men to become lawyers, doctors, and professors, he questioned whether the Ann Arbor school could meet the educational needs of *his* son.

In Ann Arbor, Henry P. Tappan assumed the presidency of the University of Michigan in 1852. For the next three years he worked hard to have the agricultural college established as a department in his institution. And he was not alone in this campaign. For example, in December 1852 Titus Dort, a member of the Executive Committee of the Michigan State Agricultural Society, asked the society to urge the legislature to create a school of agriculture as a branch of the university.[30] Proponents of this view argued that the university already had professors, scientific apparatus, a library, and other facilities in place. It would be much more economical to expand the university than to build an entirely new institution.[31]

Tappan shot down the suggestion that the Michigan State Normal School, which had opened its doors in Ypsilanti in 1852, should host the agricultural school. Even though the enabling legislation for the normal school seemed to allow for the possibility of an agricultural curriculum in Ypsilanti, Tappan contended that the law could not be interpreted in this manner. The law directed the normal school to train women and men to teach subjects, including "the Arts of husbandry and agricultural Chemistry," that were appropriate for the common schools. According to Tappan, this provision did not allow the normal school to set up a separate entity designed to teach agriculture: agricultural schools were of a "higher grade of education" than normal schools.[32] Nonetheless, the normal school tried to assume the state's responsibility to provide agricultural education when Professor Lewis R. Fisk offered lectures on agricultural chemistry.

On 3 December 1852 Tappan told John C. Holmes, secretary of the Michigan State Agricultural Society, that the university was preparing a scientific course that would be distinct from the classical course. This resulted in several afternoon lectures during each week in spring 1853 and a summer course in 1855 teaching practical analysis and chemical manipulation.[33] It is noteworthy that while Tappan observed and commented on model farms in Europe, he does not seem to have envisioned establishing one at his university.[34]

It is with the issue of the model farm where Tappan parted company with many agriculturalists, because his vision for the University of Michigan differed fundamentally from that of most supporters of an agricultural college. Tappan told the state agricultural society in September 1853 that he intended "to make the University one of the first in our country, and if we can, second to none in the world; and therefore there is no branch of knowledge that we can lawfully omit."[35] He added that it was "better to have one great institution than half a dozen abortions"; consequently, "an Agricultural Department belongs to the University." In Tappan's university there certainly was room for an agricultural department, but not a school, much less a model farm. His observations of the model farm at Hohenheim and that scientific agriculture was taught wherever the other sciences were taught in Germany led him to conclude that "the model farm of Hohenheim is of great practical value considered independently of the school."[36]

John C. Holmes, January 1883. Driven by an unyielding determination, Holmes led the successful campaign to establish the Agricultural College of the State of Michigan in 1855 as an institution separate from the University of Michigan. Holmes provided strong leadership in agricultural circles in Michigan. He served as the secretary of the Michigan State Agricultural Society from its inception in 1849 until 1857. He also was president of the Detroit Horticultural Society and held membership in a wide range of organizations including the Detroit Scientific Society and the Michigan Pioneer and Historical Society. Courtesy of Michigan State University Archives and Historical Collections (People, Holmes, John C., no. 1491).

Backers of agricultural education could not be satisfied with an arrangement where their interests would have to compete with the growing influence of the legal and medical professions and the university's commitment to the literary course built upon a study of the classical languages. Science also would make up an important part of the university's curriculum, and theoretical scientific and technological research became one of its hallmarks. Although there was nothing inherently wrong with Tappan's vision for the university, it did not coincide with most agriculturalists' visions of higher education for their children, which required a model farm. Tappan's proposed agricultural department at the University of Michigan could not provide enough to satisfy Michigan's agricultural establishment.

In 1857 Tappan told the Society for the Advancement of Education in Albany, New York, that normal schools and agricultural colleges should fall between primary schools and universities in the hierarchy of American education. Later in his speech he did not even acknowledge Michigan's new agricultural college as being part of the state's "promising educational organization."[37]

A critical look at the model farm also exposes the differences that arose between the agricultural college's perception that the farm was to be an experimental farm and the farmers' expectation that it would be a practical farm. This fundamental dispute over the role of the farm led to heated controversy between the college and the farmers that lasted for decades and served as the genesis of the land-grant philosophy of higher education in Michigan. Since most Americans lived and worked on farms or plantations in the mid 1850s, an individually owned farm fulfilled a key promise of democracy for free men and women and a hope for enslaved men and women. Many agriculturalists wanted their college of agriculture to be in effect a farm. Farmers envisioned the model farm to be a place at the college where young men would learn to operate and manage a farm in the most efficient, profitable manner possible. Working under the supervision of their professors, students would clear land, plant and harvest crops, tend to livestock, plant trees, fertilize land, and keep records of their labors in order to improve the management of farms as businesses. Fathers, politicians, and clerics, among others, extolled the virtues of manual labor for the well-being of youth, and thus the model farm became a cornerstone of their view of agricultural education. On the farm students would learn to apply what they learned in the classroom and professors would practically apply the results of their scientific research. Young men educated on the college farm could show their younger brothers and neighbors back home how they could become better farmers. On this point, Henry Tappan waxed prophetic (although not in the way he intended) when he told his Albany audience in 1857: "Universities alone can multiply universities; universities alone can properly form and order all the subordinate institutions."[38] Michigan Agricultural College sent scores of graduates to teach at other agricultural colleges and high schools in the decades that followed, thereby helping to multiply institutions that taught agriculture.

The model farm as envisioned by the Agricultural College of the State of Michigan was, however, more than a physical entity. It symbolized the potential for a new kind of learning that broke out of the bounds of the classical curriculum held so dearly by Henry Tappan, and it portended to change the meaning of farming as well. As this symbol was transformed into reality at the college it generated the creative tensions between the theoretical (or the experimental) and the practical that shaped the curriculum and the role that the institution would play in the state, the nation, and ultimately the world. In a very real way, it was on this "model farm" where the children of farmers were transformed into citizens equipped to provide leadership in many professions and walks of life essential to a growing democracy.

Advocates of scientific agricultural research realized that true experimentation could not take place on a farm that was designed only to teach good agricultural practices. Consequently, the symbolic model farm evolved into the experimental farm. When John C. Holmes recollected his opposition to the creation of an agricultural college as a department in the university, he succinctly articulated the importance of the farm as a place "upon which to teach practical Agriculture; apply science to practice; test theories; try experiments; test new plants and implements."[39] Without a farm an agricultural school could not hope to achieve the objectives that Holmes held so dear. In the 1850s agricultural science was in its infancy, and it took several decades to create and refine research objectives and techniques. At the heart of this process stood the need for controlled experiments designed to isolate specific chemicals, determine the effects of additions to and removal of substances from the soil, and to ascertain which crops grew best in different types of soil—and this was only the beginning. This work could only take place in the laboratory and on the experimental farm, where over time the symbol became reality. Most farmers, however, only understood the practical farm. Even though they disputed with scientific researchers over the purpose of the farm, the argument took place on the farm—a place understood and valued, although differently, by both sides.

Much of the argument over whether a separate college or a new department in the University of Michigan should be formed took place after the adoption of the Constitution of 1850. Correspondents to the *Michigan Farmer* voiced strong opinions that a separate college devoted to agriculture should be set up. In February 1854 Philo Cultus argued for a "new college" that offered more than "a few third or fourth rate lectures on the science and art of agriculture." He captured the need for and the essence of the model farm, even though no one yet fully understood what it meant. Cultus wrote: "Besides this you want your farm attached, so that much of the instruction may be given in the field—not exactly a *model farm,* but an experimental farm, where all sorts of slovenly as well as neat farming may be exhibited and the results demonstrated."[40] Later in the year J. S. Tibbits, writing from Plymouth, Michigan, urged that an agricultural college be established "where a thorough knowledge of agricultural chemistry, together with a practical knowledge of the analysis of soils, grains and manures may be obtained."[41] These men provided clear evidence that the value of the farm at the college resided in the experimental farm, even though most farmers failed to see beyond the potential of the practical farm.

When Warren Isham reported on his visits to agricultural schools in Europe in the *American Citizen,* a newspaper published in Jackson, he exposed another blind spot in Tappan's approach to agricultural education. Isham found that, with the exception of the University of Geisen, where Justus Liebig held forth, European universities placed little emphasis on agricultural education. European institutions devoted specifically to agriculture were much more effective. Isham quoted Professor James Johnston of Edinburgh as saying:

> It is necessary that such a school should be in the charge of men who understand agriculture, and the wants and wishes of Agriculturalists,

and who know what should be done to improve both. However profoundly learned a Professor may be, if he is not acquainted with
practical agriculture, he will be apt to take up crude notions, and
inculcate them, and thus do harm rather than good.[42]

In order to teach agriculture to young farmers, the professors needed
to know how to farm. The lack of first-hand experience with farming
would hamper some of the college's early instructors. The farm community in Michigan rightly demanded that the agricultural college get
to know farmers and farming—a learning process that knows no end.

The intellectual debate and the accompanying political struggle
over the creation of an agricultural college in Michigan culminated
when the legislature passed an act in 1855 calling for the organization
of the Agricultural College of the State of Michigan. The Michigan
State Agricultural Society was instructed to select a site of no less than
five hundred acres located within ten miles of Lansing "for the purpose of an experimental farm and site for such Agricultural
School. . . . the chief purpose and design of which shall be to improve
and teach the science and practice of agriculture." The course of study
was to include both "an English and Scientific Course."[43] The college
was placed under the authority of the State Board of Education.

National Interest in Michigan's New Agricultural College

Before the first student had set foot on campus, President Williams
placed announcements in newspapers and agricultural periodicals
across the country proudly announcing the establishment of the Agricultural College of the State of Michigan. Several examples illustrate
that young men from around the country, either at the instigation of
their fathers or on their own, wanted to participate in the Michigan
experiment. Responding to a notice in the *Chicago Tribune,* J. Wilkins
from Vandalia, Illinois, told Williams that he had dreamed of such an
institution for a long time and wished it "crowned with complete success." He also hoped that his two sons, ages 10 and 18, could be educated there.[44] Four days after the college opened, C. Beckington, 21,
from Lyons, Iowa wrote to Williams asking for the admission requirements.[45] In July 1857 H. K. Moss wrote Williams from Yazoo, Mississippi, that he had by chance seen a reference in an agricultural paper

to the fact that Michigan had put into successful operation a "State Agricultural College." He said that for twenty years he had wished to educate one of his sons "as a Farmer in Such an Institution." Now, he hoped that his seventeen-year-old son would be accepted into the college.[46] Other inquiries poured in from New York, Vermont, Ohio, Wisconsin, Kentucky, and other states. Taken in their entirety, these applications reveal a deep and lingering hunger and thirst for the new kind of education that was to be offered at the new college near Lansing, Michigan.

Williams's correspondents, living across the country, made it clear that Michigan's undertaking had national significance and that its success would help to lead the way to the creation of more institutions like the Agricultural College of the State of Michigan. In San Francisco, the editor of the *California Farmer* gladly promised to send Williams copies of his journal for the college's library. Furthermore, he told Williams, "We rejoice with you Dear Sir, and with the Trustees of your noble Institution at the prospect before you for yours is a noble work & you have it in your power to accomplish great good not only for Michigan but the whole Union."[47] Closer to home, John W. Hoyt, the editor of the *Wisconsin Farmer,* saw the developments in Michigan as an example for Wisconsin. Hoyt summed up the feelings of many people across the country when he wrote, "America has no material or educational need so great as the multiplication of Agricultural schools or Colleges, & it is my earnest hope & determination that Wisconsin shall not be slow to follow your noble example."[48] Further east, the editor of the *Vermont Stock Journal* expressed his unqualified support: "I consider the Success of your own & similar institutions to be of most importance to the Agricultural interests of our Country & I esteem it the duty of every good citizen whether he be devoted to agricultural pursuits or not to lend them a cheerful & ready support."[49]

Not surprisingly the Agricultural College of the State of Michigan served as both an inspiration and a model for other states. Just two months after the college opened its doors, Theodore Brown, who lived in Jefferson County, Kentucky, asked Williams for information related to the "Education of Farmers" at his institution, since there was great interest in creating such schools in Kentucky. Though Brown also looked to other states for models, he was drawn to Michigan and "the Course pursued, its connection with Literary Education, with the practice of Agriculture (I believe you have an Experimental Farm)."[50] He no doubt found the college's catalog to be of great interest. In his report on the college to the Michigan State Board of Education in December 1858, Williams boasted: "The institution continues to attract interest in other States. It should be a subject of honorable pride in Michigan, that her example in taking the lead in a great movement, indicative of educational progress, is so generally applauded, and in fact, imitated." Iowa and Minnesota had "almost literally copied" the college's enabling legislation. Alabama, South Carolina, Wisconsin, and Massachusetts gave strong indications that they, too, would establish similar institutions.[51]

THE COLLEGE GETS STARTED

John C. Holmes, who had championed the cause of a separate college during the political battles with Tappan and the university forces, now faced the challenge of turning the concept of an agricultural college into a physical reality.[52] Holmes, the secretary of the Michigan State Agricultural Society, and the society's executive committee chose a 677-acre site three miles east of Lansing, in part because on its northern boundary it was adjacent to the plank road that ran from Detroit to Lansing. Anyone bouncing in a coach as it rambled over the boards in 1855 could scarcely have imagined that a great university would grow up out of the marshes, swales, oak openings, and heavily timbered spreads of land before them. Bear, deer, turkeys, and especially mosquitoes called this area home. Many soil types, ranging from clays to rich river bottoms, offered great potential for agricultural experimentation, but a lot of stumps would need to be pulled before serious scientific work, not to mention the planting of the first turnips, could begin.

Holmes and college officials had to construct buildings upon this inhospitable land. Students, faculty, and livestock all needed a place to live, and professors required facilities where they could give instruction. S. M. Bartlett of Monroe oversaw the erection of College Hall, the dormitory dubbed "Saints' Rest" by the students, and a small barn. Holmes designed College Hall as a place to teach scientific agriculture, and there was no other building like it in the country. The

Faculty Row, circa 1896. Houses built for faculty stood along "Faculty Row" on West Circle Drive. Looking left to right the houses were reserved for Professor of English, Professor of Agriculture, Professor of Horticulture, Secretary to State Board of Agriculture, and Howard Terrace (faculty apartments). Courtesy of Michigan State University Archives and Historical Collections (Michigan State University, Buildings, Faculty Row).

hall contained a lecture room, classrooms, a library, a museum, and laboratories. The dormitory was divided into sleeping rooms, kitchen, parlor, laundry room, dining room, and an apartment for the steward. A small stable housed the college's first livestock. Faulty construction coupled with insufficient supervision of the work led to serious structural problems that plagued College Hall and the dormitory long after they opened.[53]

The State Board of Education appointed Joseph R. Williams as president in 1857, as well as a faculty who conscientiously fashioned a curriculum and prepared the college for the arrival of students. Williams brought to his position an impressive set of credentials. He had graduated from Harvard in 1831, had served as editor of the *Toledo Blade,* and had been closely involved with Michigan agriculture as a miller and merchant in Constantine. John C. Holmes joined him as professor of horticulture and treasurer of the college. His long service in the state agricultural society (1849– 57) and tireless efforts to have the college created as a separate entity had prepared him well to help turn a concept into bricks and mortar. Calvin Tracy, who had grown up on a farm near Norwich, Connecticut, became professor of mathematics. Tracy previously had taught at Norwich Academy and had written materials used to teach arithmetic. Robert D. Weeks, a printer by trade who also had taught and farmed, assumed the professorship of English literature and farm economy. Perhaps the most impressive member of this first faculty was Lewis R. Fisk. He had graduated from the University of Michigan, had studied chemistry at Harvard, and had been professor of natural science at the Michigan State Normal School in Ypsilanti. After his appointment, he traveled extensively, acquiring the apparatus and chemicals needed for his laboratory.

Several other faculty played important roles in the first few years of the college. Henry Goadby, a medical doctor who had lectured at the Royal College of Surgeons in London, taught animal and vegetable physiology and entomology in 1859. In 1858 he had published *A Text-Book of Vegetable and Animal Physiology.* George Thurber, who held an M.D. from the University Medical College of New York, joined the faculty in 1860 as professor of botany and horticulture, offering courses in vegetable physiology, botany, and horticulture. Theophilus Capen Abbot, a native of Maine, brought an impressive classical education to his professorship of English literature. He had graduated from Colby University at the age of nineteen, had studied for two years at Bangor Theological School, and had traveled in Europe. After Manly Miles received his medical training at Rush Medical College, he had returned to his native Flint, Michigan. His interest in the natural world motivated him to assemble a collection of zoological specimens. In 1861 Miles became professor of zoology and animal physiology.[54] Although two or three of the first faculty members had grown up on farms, none of them had been trained to teach agriculture—a circumstance that did not go unnoticed by the college's critics.

President Williams and the faculty designed the college's first curriculum, which brought together science, an English course, and the farm. Potential students learned from an announcement published on 10 December 1856 that the first course of studies at the college entailed the following:

- An ample Chemical Laboratory has been purchased by the Professor of Chemistry, inferior to few in the country, and instruction in that Science will be thorough and practical.
- Ample instruction will be given in the Natural Sciences.
- The Course in Mathematics will be comprehensive.

- The application of Science to the business and arts of life will be practically illustrated in the field and the Lecture Room, especially where it bears upon Agriculture.
- Instruction in Ancient and Modern Languages is not included as an object of the Institution.
- A thorough English education is deemed indispensable, including Rhetoric, History, Moral and Intellectual Philosophy, Political Economy, the elements of Constitutional Law, etc., etc.
- The Farm being almost entirely in a state of nature, a very large amount of the labor of Students must at first be bestowed where it will yield little immediate profit. Had the Institution possessed a large tract of arable land, at the commencement, the earlier results would be far more profitable than they can now prove.[55]

This description of the college's first curriculum foreshadowed the institution's success, for it melded together a practical education rooted in the study of literature, history, philosophy, and other subjects of the mind, with chemistry, a science dependent upon research. Every student was also expected "to devote a portion of each day to manual labor," for which he would "be entitled to receive an equitable remuneration." As unrefined as this course description may be, it reveals the founding faculty's aim that the agricultural college educate its students to be much more than just better farmers. Young men who followed this regimen stepped from their fathers' farms, shops, factories, or stores into the dynamic and expanding world of the second half of the nineteenth century prepared to fulfill whatever duties and responsibilities fell their way.

Seventy-three young men from throughout Michigan appeared before the faculty for examination to determine their eligibility for entrance to the college during the second week in May 1857. Professors Tracy, Weeks, and Fisk tested their potential students' capabilities and knowledge in spelling, geography, arithmetic, composition, and penmanship. In their judgment, the common schools had adequately prepared fifty-nine for admission. They and those who followed during the next several years played a pivotal role in shaping the college's direction, thereby ensuring its survival.

The first students at the agricultural college demanded that they receive a liberal education that encompassed more than learning about agriculture. In 1861 "a farmer's wife" articulated a prevalent attitude among farm boys: "There is a growing dissatisfaction among farmer's sons, a dislike to farming as a vocation or desire to try something else, that is much to be deplored."[56] While this woman viewed this trend with alarm, undoubtedly the young men at the college envisioned a brighter future away from the never-ending labor required to operate their family businesses. The sweat of their youth had led them to reject idealistic views held by many of the virtues of working the land. They parted company with those people who shared Thomas Jefferson's glorification of the yeoman farmer.

Some youth in Michigan, farmers and non-farmers, took great delight in learning, and they welcomed the opportunity to pursue an education at the new college that did not require any knowledge or study of Greek or Latin. Edward G. Granger, '58–'59, came from Detroit, apparently not from a farm, and never graduated. His diary gives significant insights into the curiosity and intellectual interests of students, their study habits, and social life at the Agricultural College of the State of Michigan in its infancy. While we must be careful not to draw unwarranted generalizations from the words and experiences of one person, Granger's revelations tell us much about the assumptions and expectations that he had for his own intellectual development and how his studies at the institution could help him grow. References that Granger makes to other students suggest that his outlook was shared by others.

Granger and his fellow students had deep interest in literature and history. They read such literary giants as John Milton, Sir Walter Scott, William Shakespeare, Oliver Wendell Holmes, and Harriet Beecher Stowe. Granger added to his knowledge of history as he made his way through "Abbot's history of Hannibal," "Mrs. Sheldon's history of Michigan," and "Life of Wellington by an old soldier." He welcomed the arrival of the *Atlantic Monthly* that came to his classmate Harvey T. Bush, '57–'59. Several of the boys rejoiced when the mail carrier delivered the books they had ordered from Philadelphia. Most noteworthy of their new acquisitions was William Tredick's ('57–'60) copy of "Spenser's 'Faerie Queene.'" Granger, to his consternation, had his reading of *Paradise Lost* interrupted when Harvey Bush, in a huff, locked the book in his book case. On 8 January 1859 Granger, Albert J. Cook, '62, and William A. Thomas, '58–'60, made their way into

Lansing to observe the legislature in session. Before entering the chamber, the boys read at the state library. On another occasion, Granger enjoyed "a large volume of illustrations from William Hogarth's paintings and two volumes of Shakespeare" at the state library. There is no mention of Justus Liebig, Samuel Dana, or any other authors of agricultural works. Such lively intellectual interests prepared Granger for the instruction he received from his professors.

Granger and his classmates expected their professors to challenge their intellects, but they did not tolerate what they deemed to be violations of their rights as American citizens. Granger made a strong statement demanding that the college not interfere with their exercise of democracy in an entry in his diary dated 17 December 1858:

> Went to Lyceum. The President tried to have an amendment made to the constitution of Our Lyceum but we voted it down. It was, that there should be nothing said in the Lyceum or in the Paper, disrespectful to the Faculty or ridiculing any of their sayings or doings. *Free Country*. President Williams requested us to pass the amendment but fortunately the[re] was more of the spirit of liberty in the society than to pass such an abomination.

Young men enrolled in the college to learn to think and speak critically, not to have a muzzle placed over their mouths.

Granger's formal studies show how the faculty presented a curriculum that integrated liberal studies, science, and practical agriculture, which at the same time required students to perform the manual labor needed to clear land, look after animals, and grow foodstuffs for the college family. We observe both the excitement and the pain experienced by a young man as his mind is being stretched. In the mornings Granger recited chemistry before Fisk, geometry before Tracy, and literature before Abbot. He marveled at the "very pretty experiments" Fisk demonstrated with arsenic in the laboratory. When Professor Abbot asked Granger to write a composition, he chose the topic "Sir Walter Raleigh and his times." Granger put considerable effort into his writing, for he knew that Abbot would criticize it carefully and demand that the final work be of high quality. As Granger fashioned his composition, he proved the truth of his earlier prognosis that "literature is the hardest thing, I ever Studied, but I guess it will be interesting after a while."[57] Granger was not the only student who kept his nose to the grindstone. A couple of years later, Charles Jewell, '62, reported that he worked outside each afternoon in addition to spending two and one-half hours doing chemical analysis in the morning. The rest of the time he engaged in "reading writing and study."[58]

Edward Granger's intellectual development took a break for three hours each afternoon when he performed the mandatory labor section of his curriculum that brought him into contact with the soil. He and his friends literally hacked a farm out of the wooded, swampy terrain that comprised the college lands. His record of his labors reveals that he got a broad experience in practical matters, but little in experimental farming. Among other tasks, Granger planted trees and gardens, laid brick, weeded, hoed the peach trees, sowed turnips, planted potatoes, cleaned the barn, spread manure, logged, picked brush, split wood, and ground corn. He spent more time chopping and splitting wood than doing anything else.[59]

Tree removal could be dangerous work, with each stroke of an ax exposing the handler to potential injury, as was demonstrated by a serious accident that happened ten years after the college opened. While chopping down a tree on 12 March 1867, Henry Graham Reynolds, '70, hit his foot with an errant swing of his ax. A $1\frac{1}{2}$-inch slice over his instep that went to the bone fortunately "cut just between the cords." President T. C. Abbot, Professor Albert J. Cook (Granger's classmate), and students brought Reynolds books, meals, and good cheer while he was confined to his room for a few days while his wound healed. Reynolds became homesick as he wrote letters to family and friends during his newly acquired free time. No doubt he appreciated the attention given to him by his college family, but Henry, who was only fifteen years old, longed for his mother's care while he endured pain and loneliness brought on by his injury.[60]

But even as Granger's blisters turned to calluses, Professor Holmes instructed him in rudimentary scientific farming. In a small section of his scrapbook Granger recorded scientific observations gleaned from several sources. In an entry that foreshadows the emergence of a scientific, experimental farm at the agricultural college Granger touched upon some of the larger problems facing farmers: pest control, the chemistry of soil, and the preparation of seeds for planting. Granger's "Notes on Agriculture" note, in full:

Carrots consume 199 pounds of lime to the acre; Turnips but 90 pounds. News Paper. Three parts of flour to one of pepper will protect vines from bugs. Cook. The first mess of green peas which we had at the Agricultural College of Mich. this year (1858) was on the 23d of July. Dragging in the spring is good for Winter Wheat. Morse. Fredick, & Hastings. To sustain a healthy prolific vegetation, soils must contain, silex, alumina, carbonate of lime, sulphate of lime potash, soda, magnesia, sulphur, phosphorus, oxide of iron, manganese, chlorine, and probably iodine. In addition to these they also contain carbon, oxygen, nitrogen, hydrogen. It is best to soak beet seed in soft water for one or two days before planting; as it requires a large share of moisture to produce germination. Sulphur is good to keep bugs off of vines. Ibid. Rural New Yorker. Prof. Holmes says that the College potatoes have been hoed so much that they will not amount to any thing. The Turnips are all going to seed because there is not enough potash in the soil, which is almost all muck.[61]

This entry also demonstrates how the farming community received information from popular periodicals. As the college's scientific agricultural research program evolved, faculty made known their findings and their practical applications on the farm through both popular and scholarly publications. Granger also notes information learned from his classmates.

Like students at many of the classical colleges, Granger and his friends enjoyed the intellectual exercises offered by the Lyceum and regularly participated in religious services. On 10 February 1859 James Birney addressed the Lyceum on culture. Birney challenged his audience to consider his argument that human capabilities were no different in the present than in the past and that the arts in the nineteenth century had not advanced much beyond that of the Greeks and Romans. Each morning Granger attended chapel services conducted by faculty members in the lecture hall. In order not to favor any denomination, the college called upon the faculty, rather than local clergy, to lead worship. On Sundays students went either to the college chapel or to a church in Lansing. Religious belief was clearly important to Granger, who read through the entire Bible in 1858.[62]

Before the college became coeducational, the men had to find female companionship in the community, which helped to integrate the institution into the growing city of Lansing. Miss Abigail Rogers, who operated the Michigan Female College on the north side of town, invited Theodore Garvin, '59–'60, and his friends to a husking bee in October 1859. Founded in 1855, the college had an enrollment of about twenty-five boarding students and another fifty day students who lived at home.[63] While "Thede" enjoyed the company of "some pretty nice girls," he appeared to be most impressed by the coffee, pie, gingerbread, and doughnuts served to him to replenish his strength after a busy afternoon husking corn. These were the best treats that he had had since he left his mother's home.[64] The next month Professor Tracy welcomed members of the community and the students to a social gathering at the college. Sidney Sessions, '58–'59, was pleased that Tracy had invited "the *Young Ladies* of the Female College; who are very essential to make things go off right *you* know."[65] A year or two later Charles Jewell and a couple of classmates each asked a young woman to ride with them to Pine Lake (Lake Lansing) on their rented livery. The party had a pleasant time boating, walking along the shore, and eating a picnic dinner before returning home.[66]

Granger took his meals in the dormitory's dining room, but he and his friends especially enjoyed treats sent from home. He implied that the regular diet was rather nondescript with this comment at Christmas, 1858: "Had a fine Christmas dinner considering that it was in the Agricultural College." Later he picked up a Christmas box sent by his mother from Mrs. John Holmes, who gave him popcorn and "some very nice apples." The package contained peaches and "other good things of this world." Evidently some of those items found their way to the table in his room the next evening, when he and some of his friends had a "feast" of "chickens and peaches, brown bread and ginger-snaps."[67]

A FORK IN THE ROAD

Between 1859 and 1861 the college came to a fork in the road where a wrong turn could have meant ruin while another turn led to growth and success. President Williams, who had been accused of spending too much money and having personal political ambitions, resigned in March 1859. His departure deprived the institution of his forceful leadership and his advocacy for a well-rounded four-year curriculum that incorporated both an English course and practical agriculture.

His commitment to a broad curriculum had led to criticism by farmers who believed that a purely practical course provided their sons with all of the education they needed. The faculty's lack of practical experience in farming contributed to the public's negative perception of Williams's leadership and the work of the college. Clearly the college struggled as it tried to create a course of study to meet the needs of its students and to build a faculty able to bring together practical agriculture, scientific agriculture, and liberal studies.[68]

Without Williams at the helm, John M. Gregory, state superintendent of public instruction, took a much more active role in the affairs of the college. Gregory instituted very tight spending policies in an effort to stretch the institution's small budget as far as he could. A large part of the college's funds were needed to put a new roof on the dormitory and replace much of the plaster inside the building. Gregory managed so stringently that he failed to expend about one-fourth of the budget. As a result, the legislature not only took back these dollars, but it also reduced the college's appropriation proportionately for the next two years.

Of even greater significance, Gregory proposed and the State Board of Education approved a change in the college's curriculum from a four-year course to a two-year program designed to teach young men to manage a farm without receiving the benefits of the broader general education. This elicited a negative response from the faculty and, especially, the students, who demanded that the college continue the curriculum that they had come to study. Gregory envisioned professors of agricultural chemistry, botany, zoology, and engineering each offering a two-year sequence of systematic science and applied science. Gone were courses in literature, history, mathematics, and philosophy. Applicants could make up deficiencies in algebra, geometry, trigonometry, and general chemistry in a one-year preparatory course before embarking on the regular two-year program. Perhaps nothing epitomized the change more than the reassignment of Professor Abbot, formerly the professor of English literature, as the new professor of civil and rural engineering.[69] It is doubtful that the college would have survived long had it had followed the path laid out by Gregory, but fortunately others fought for an alternative.

Professor Lewis R. Fisk, who served as president pro tem after Williams's exit, championed the four-year curriculum. His report to the newly formed State Board of Agriculture dated 4 April 1861 is a remarkable document. He spent many pages informing the board of the condition of the institution that they now governed. When Fisk gets to the "Course of Study" section, he quotes extensively from his 1859 report to the State Board of Education. It is here that we see the crux of the argument for a four-year program. Fisk reaffirms his commitment to the principle that "this institution should be built upon an agricultural basis, . . . that no department of practical farming should be left unexplained; . . . It should . . . be a place where science and practice shall be beautifully combined." But he did not stop here when he continued on to say: "The only way then to teach agriculture here, is to teach literature also." Fisk stated emphatically that the faculty supported this position, and added that they believed that a four-year "comprehensive course of instruction" was essential, if for no other reason than because the students demanded it.[70] T. C. Abbot recorded the students' sentiment: "March 1 [1860]. . . . We met students and told them we should organize the old academic studies for those who wished—the purely agricultural for those who wished. All voted for the academic."[71] The battle for a four-year comprehensive academic and agricultural course of study united the students and the faculty.

The controversy between the students, the faculty, and Fisk on one side and Gregory and the State Board of Education on the other shines as a wonderful democratic moment. The students would have no part of the two-year curriculum, and they played the decisive role in shaping the agricultural college's future direction. Young men such as Edward Granger expected to learn more at college than better farming. They embodied the idealistic rhetoric of the institution's proponents that the sons of farmers, mechanics, and shopkeepers needed an education that would prepare them for life's duties and responsibilities. In essence their whole person—mind and body—deserved to be challenged and nourished. Bluntly, Fisk first told the State Board of Education and then the State Board of Agriculture that "There is probably not one young man that has come here for the sole purpose of studying the science of agriculture." Fisk's strong words were not simply an expression of anger directed towards the board, who seemed oblivious to the students' needs. Fisk and the rest of the faculty knew their students well, and as professional educators they cared deeply

The 1861 Act of the Michigan Legislature that created the State Board of Agriculture named Hezekiah G. Wells as one of its members. For nearly all of the next twenty-two years Judge Wells held a seat on the board and was its president. In the 1860s Wells blocked attempts in the legislature to move the college to Ann Arbor as a department in the University of Michigan. President Abbot trusted and valued Wells's advice and contributions to the college's growth. Born in Ohio in 1812, Wells moved to Kalamazoo County in 1833, where he lived until his death on 4 April 1885. Courtesy of Michigan State University Archives and Historical Collections (People, Wells, Hezekiah G., no. 2911).

about how well they prepared their students "to act in any capacity in society which they may be called upon to occupy."[72]

Student desire for "a good general education" flew in the face of the apathy or hostility many farmers felt towards the college, but it reaffirmed the founding fathers' vision for the institution.[73] The students' unwillingness to attend a strictly agricultural school of higher learning showed that Williams, Holmes, Fisk, and others spoke for a class of young people who were frustrated by their inability to get the intellectual training that they wanted in order to develop their abilities in ways different from their parents. Edward Granger, Harvey T. Bush, Albert J. Cook, and William A. Thomas willingly presented themselves as elements for the academic test tube that was being fashioned in the woods near Lansing. The students expected that, when mixed together with their mentors in the laboratory, in the lecture hall, in the library, in the fields, or at the dinner table, they would take on new characteristics—not unlike those of a compound that the chemist produced when he mixed together two or more unlike substances. Youth from Michigan and elsewhere enrolled in the agricultural college because they wanted to be transformed into enlightened citizens, not just better farmers.

Former president Williams provided the leadership to strengthen the agricultural college after his election to the Michigan Senate in 1860. Convinced that the State Board of Education did not understand the college's purpose, Williams prodded the legislature to pass an act to change radically the governance of the college. An analysis of "An Act to Reorganize the Agricultural College of the State of Michigan, and to Establish a State Board of Agriculture" reveals that the members and friends of the college had come a long way in six years. The legislature prescribed that the course of study include both an English and an agricultural curriculum. The students had been heard. Section 33 defined the role of the farm, and the advocates of scientific agriculture scored big. The words of the opening part of this provision succinctly bring together scientific research, the teaching of students, and the application of knowledge gained on the college's farm: "All agricultural operations on the farm shall be carried on experimentally, and for the instruction of the students, and with a view to the improvement of the science of agriculture in the State of Michigan."[74] The act also required the State Board of Agriculture to publish an annual report that would make available to farmers accounts detailing virtually every aspect of growing crops. Agricultural researchers were to report on such topics as the number of fields and acres planted, methods of planting, cultivation and harvesting, types of seeds used, methods of soil preparation, kinds of manure used and for which crops, time expended working each field, and yields from each acre planted.

Two other provisions of the act ensured the existence of the college as an independent entity and bound the institution closely to Michigan's agricultural community. The State Board of Agriculture came into being for the express purpose of governing the State Agricultural College (the new name specified in Section 13) and its farm. The governor was empowered to appoint future members from a list of nominees presented by county agricultural societies. Men who were experienced in and knowledgeable about farming would now oversee the school's operation and future. No longer would the State Board of Education have to agonize over an institution whose mission they did not fully understand.

The act did not stop here, for Section 6 ordered the new board to appoint a secretary as a nonvoting member of the board. The secretary was to carry out duties that would link the college to different constituencies throughout the state, thereby applying findings on the farm and in the laboratory to the lives of people living in rural and urban Michigan. He was to send circulars and reports to local societies on topics pertaining to agriculture and domestic life. Farmers could expect to learn about advances in raising cattle, harvesting crops, dairying, the development of new seeds, and the understanding of different soil types, among other topics. The secretary was to publish agricultural information in newspapers throughout the state. He was to function as the college's business manager, and he was expected always to be current in his knowledge of the school's affairs. He was to make reports to the legislature and the governor regarding both the state of the college and agriculture in Michigan.[75] In effect, the secretary was to function as a master cylinder who pumped information through a pipeline to the organizations and people who were essential players in Michigan agriculture.

Justin Smith Morrill was elected to the U.S. House of Representatives in 1854 and served his district in Vermont for twelve years. In 1867 Morrill became a U.S. Senator from Vermont—an office he held until his death in 1898. Courtesy of Michigan State University Archives and Historical Collections (People, Morrill, Justin S.).

THE 1862 MORRILL ACT

While the formation of the State Agricultural College of Michigan was a very important occurrence in the history of higher education in the United States, the enactment of the Morrill Act or Land-Grant College Act of 1862 was even more significant. Scientific agricultural research and schools that taught agriculture in Europe led Americans to want to establish institutions in their country devoted to both of these activities. Michigan's successful inauguration of its experiment in 1857 and efforts in Pennsylvania, Maryland, Massachusetts, Iowa, and other states to start similar schools demonstrated that the demand for agricultural colleges was widespread. By the late 1850s the time had come for the federal government to act.[76] On 17 December 1857, Justin Morrill, Representative from Vermont, introduced a bill in the House of Representatives that called for the establishment of what came to be known as the land-grant colleges.

On 20 April 1858 Justin Morrill told his colleagues on the floor of the House of Representatives that there was great need for agricultural education in the United States and that the federal government ought to provide the means to bring this about. Farmers were exhausting soil at an alarming rate, which led to declining fertility and crop yields. This could not continue, for agriculture's "products, the aggregate, are not only of greater value than those of any other branch of industry, but greater than all others together." Fortunately, farmers had begun to make their needs known through the circulation of a large number of periodicals, the creation of agricultural societies, and annual fairs. Clearly, American farmers craved more information in order to improve their way of doing business. But Morrill argued that this was not enough. He pointed out that farmers could only "snatch their education, such as it is, from the crevices between labor and sleep." Pushing further, Morrill lamented that the country had colleges, such as the military academy at West Point, to teach how to make war but not "to teach men the way to feed, clothe, and enlighten the brotherhood of man." Farmers and mechanics required "special schools and appropriate literature quite as much as any one of the so-called learned professions." He reminded his listeners that agricultural colleges and schools were a common feature throughout much of Europe, while in the United States "plentiful lack of funds" had prevented most attempts to create agricultural schools from succeeding. He noted that there were "some examples, like that of Michigan, liberally supported by the States, in the full tide of successful experiment," but, he contended, if agricultural colleges were to succeed, Congress would have to pass his bill, which would set aside large grants of land for each state. The proceeds from the sale of such lands would be used to endow at least one college devoted to an agricultural curriculum that was open to the sons of farmers and mechanics in each state. The economic health of the nation depended upon the education that would be provided at these institutions. Morrill believed that the work done by agricultural colleges would increase the value of land across the country, thereby adding to the country's wealth.[77]

Michigan's own Joseph R. Williams played an important role in generating support for Morrill's bill. While in Washington, D.C., during the winter of 1857–58, Williams lobbied congressmen to support the measure.[78] With much satisfaction, he told Morrill that segments of the Vermonter's speech in the U.S. House had appeared in many agricultural papers and journals issued across the country. Morrill had incorporated many of Williams's views into his address, and he had credited Michigan for showing leadership in the agricultural college movement. In the South, Williams tried to insert articles favorable to Morrill's bill only in agricultural periodicals, since the political papers concerned themselves with hotter topics related to the deepening crisis with the North.[79] In October 1858 Williams urged an audience at the New York State Fair in Syracuse to rally support for Morrill's bill. Williams reminded his listeners that the drive for the establishment of "a new order" of "Agricultural and Industrial Colleges" was taking place on the national level. He identified Michigan, New York, Pennsylvania, Maryland, Iowa, Minnesota, Ohio, and Massachusetts as having already established agricultural colleges and predicted that Virginia, South Carolina, Alabama, and Wisconsin would soon follow.[80] As 1858 drew to a close, Williams kept up the pressure on representatives and senators to pass the agricultural college bill.[81]

Congress approved Morrill's bill in 1859, but President James Buchanan vetoed it. The passage of this bill appeared to cap the long battle to enact federal legislation to establish land-grant colleges.[82] But the sectional politics that were leading to the Civil War cast a large shadow over Morrill's bill. Congressmen and senators from the South

and the West opposed the measure on a number of grounds. South-
erners, fearing that the North would use this measure to impose
dominance over their region, argued that the measure was unconsti-
tutional. Westerners did not want to create a mechanism whereby
speculators could get their hands on large chunks of land below the
minimum price set by the government. Although western states
wanted agricultural colleges, some of their leaders feared that this legislation would give more money to wealthier eastern states than to
their own.[83]

Buchanan's veto did not stop the people who had worked so diligently to bring the land-grant legislation before Congress from continuing their campaign until they saw it signed into law in 1862.
Throughout the 1850s leaders living throughout America had shared
their ideas for agricultural education with friends and foes alike. The
veto neither changed their minds nor prevented them from bringing
their vision into reality. During this decade the concept of making
higher education available to the growing "industrial classes" became
implanted in American public life. Horace Greeley used his *New York
Tribune* to drum up support. Ezra Cornell and Thomas Green Clem-
son championed the cause in New York and South Carolina respectively, and the land-grant schools in their states would be named after
them. Beginning in the early 1850s, Jonathan B. Turner worked for the
creation of an agricultural school in Illinois, and his ideas and efforts
were instrumental in the formulation of Morrill's bill.[84] In 1852
Turner first proposed the sale of public lands by the federal government to fund "a general system of popular Industrial Education" that
would make money available for each state to finance an institution
committed to teaching agriculture and the mechanic arts.[85] Evan
Pugh, president of the Farmers' High School in Pennsylvania, also
threw his energy behind the Morrill bill.[86]

On 2 July 1862, President Abraham Lincoln signed the Morrill bill
into law. It is worth noting that although Lincoln does not appear to
have participated in the movement that culminated with the Morrill
Act, he had supported education for the laboring classes before he
entered the White House. In an address that Lincoln gave before the
Wisconsin State Agricultural Society on 30 September 1859 he tied
together labor and education. He set on its head the notion that
educated people should not do physical work by suggesting just the

opposite. He posed the crucial question: "How can labor and education be the most satisfactorily combined?" Lincoln articulated his conviction that agriculture provided an unlimited potential for learning and application when he told the Wisconsin State Agricultural Society:

> Every blade of grass is a study; and to produce two, where there was but one, is both a profit and a pleasure. And not grass alone; but soils, and seasons—hedges, ditches, and fences, draining, droughts, and irrigation—plowing, hoeing, and harrowing—reaping, mowing, and threshing—saving crops, pests of crops, diseases of crops, and what will prevent or cure them—implements, utensils, and machines, their relative merit, and [how] to improve them—hogs, horses, and cattle—sheep, goats, and poultry—trees, shrubs, fruits, plants, and flowers—the thousand things of which these are specimens—each a world of study within itself.[87]

Lincoln also emphasized that "Free Labor insists on universal education."[88] The Morrill Act promised to help realize this ideal through funding a new type of college in each state that would be accessible for young men and women from all social classes. It was fortunate that the hope offered by the Land-Grant College Act should become law at the same time that the Civil War appeared to be destroying the nation.

The Morrill Act generated support for a system of higher education that dramatically expanded American democracy, which helped to lead the way for economic growth in the United States. Land-grant universities trained engineers, agriculturalists, and scientists who made possible the "managerial revolution" that built the American economy in the late nineteenth and twentieth centuries. Men and women applied skills learned in their study of science, mathematics, agriculture, domestic science, economics, literature, and history to provide leadership in all realms of life. As graduates of land-grant institutions played increasingly larger roles in the growing American economy, other colleges and universities made changes in their curriculums to enable their students to participate in this managerial revolution.[89]

A look at the act itself reveals that the Agricultural College of the State of Michigan influenced this truly revolutionary legislation. Section 4 required the states to use the funds from the sale of their land grants to set up a perpetual fund:

While the Civil War raged, President Abraham Lincoln signed the Morrill Bill into law on 2 July 1862. On 1 July the Battle of Malvern Hill had brought to an end General George B. McClellan's unsuccessful Seven Days' Campaign on the Virginia Peninsula to capture Richmond, the Confederate capital. Courtesy of Michigan State University Archives and Historical Collections (People, Lincoln, Abraham).

to the endowment, support, and maintenance of at least one college where the leading object shall be, without excluding other scientific and classical studies and including military tactics, to teach such branches of learning as are related to agriculture and the mechanic arts, in such manner as the legislatures of the States may respectively prescribe, in order to promote the liberal and practical education of the industrial classes in the several pursuits and professions in life.[90]

This provision encapsulates the essence of the Michigan experiment: a curriculum designed to teach practical and scientific agriculture, an English course, and most important, a curriculum open to young people from farm and working-class families. No longer would higher education be limited primarily to the children of the nation's social and economic elite. The act also fulfilled the dream of Williams and others that the federal government support education of this sort in each state without dictating its terms.[91] The State Agricultural College and the nation would be the beneficiaries of the Morrill Act.

THE STATE AGRICULTURAL COLLEGE DURING THE CIVIL WAR

The Civil War caused disruptions in the development of the college, but it did not prevent it from moving forward. Soon after hostilities broke out, Professor George Thurber introduced military drills for students who formed a unit known as the "Plow-Boy Guards." In September 1861 members of the college's first class of degree candidates were allowed to leave school before they formally graduated, to serve with Captain E. P. Howland's topographical engineers, who were with General John Fremont's troops in Missouri. The State Board of Agriculture awarded its first seven bachelor of science degrees in November, even though none of the recipients were present. Two members of the class of 1861 gave their lives in the service of the Union. Gilbert Dickey, 2nd lieutenant, Company G, Michigan 24th Infantry, died at Gettysburg, Pennsylvania, on 1 July 1863, and Henry D. Benham, 1st lieutenant, Company B, First Colored Infantry, died at Beaufort, South Carolina, on 3 July 1864.[92] Even though more students enlisted in the army, the college maintained a small but steady enrollment, with sixty-five attending in 1861, seventy-four in 1862, sixty in 1863, sixty-two in 1864, and eighty-eight in 1865.

The State Agricultural College continued to resist attempts by supporters of the University of Michigan to move it to Ann Arbor when the legislature considered the college's appropriation bill in 1863. At the same time, friends of the university tried to have the funds generated by the Morrill Act directed to it. The proponents of this view were encouraged by events in Massachusetts, where the governor had recommended that the land grant be given to Harvard. The Reverend Dr. Erastus O. Haven helped to lead the successful opposition to this measure in the legislature, which ensured the establishment of a separate land-grant college in Massachusetts. Before 1863 ended, Haven also assumed the presidency of the University of Michigan.[93]

In the midst of war and uncertainty, the State Board of Agriculture appointed a new president who, in conjunction with the faculty, carried out the provisions of Sections 15 and 16 of the 1861 State Agricultural College Reorganization Act. The board chose not to make Lewis R. Fisk permanent president. Instead, they appointed Theophilus Capen Abbot to the post in February 1862, and he led the college with great skill and sensitivity until 1885. The faculty fashioned a four-year course of study, as called for in Section 16, that met student demands for both an English and an agricultural curriculum as required in Section 15:

> Sec. 15. The course of instruction shall embrace the English language and literature, mathematics, civil engineering, agricultural chemistry, animal and vegetable anatomy and physiology, the veterinary art, entomology, geology, and such other natural sciences as may be prescribed, technology, political, rural and household economy, horticulture, moral philosophy, history, book-keeping, and especially the application of science and the mechanic arts to practical agriculture in the field.[94]

Simultaneously the faculty designed the college farm, gardens, and orchards as laboratories for experimentation and for the teaching of practical and scientific agriculture.[95] By 1865 the program put in place to carry out the college's mission seemed clear.[96] But this proved not to be the case.

2

Defining the College's Mission

DURING MICHIGAN AGRICULTURAL COLLEGE'S FIRST SIXTY YEARS, farmers, students, and faculty had conflicting expectations of the college that led to serious misunderstandings over its role in the higher education of Michigan's young men and women. The central point of contention was whether or not the State Agricultural College should provide more than practical educations for farmers' sons—in other words whether it should offer a curriculum that included engineering, science, and English. Some, both outside and inside the college, recommended turning over some parts of the curriculum to the University of Michigan, and some went so far as to advocate moving the entire institution to Ann Arbor. Many Michigan farmers saw little of value deriving to them from the classrooms, laboratories, and fields east of Lansing. They charged that M.A.C. failed to send enough of its graduates back to the farm. Their views shaped both the public's perception of the school and its response to its critics. M.A.C. administrators countered by arguing that a majority of the school's graduates were engaged in farming or in a career related to agriculture—as its name proclaimed, M.A.C. was an agricultural college. Both sides remained committed to the idea of an agricultural college even though they differed over the means to achieve their objective and the method for operating it.

M.A.C.'s necessarily vigorous defense of its agricultural credentials, however, often hid the scientific and literary segments of its curriculum from the view of potential students, and the noisy debate over the college's commitment to agriculture stymied its growth. Enrollments started to increase after President Jonathan L. Snyder launched an aggressive campaign in 1896 to inform Michigan citizens that M.A.C. prepared its students for careers in agriculture *and* engineering, domestic science, and other fields—M.A.C. was an agricultural

Theophilus Capen Abbot came to the Agricultural College of the State of Michigan as professor of English literature in 1858. The State Board of Agriculture appointed him president of the college in 1862, a position he held until 1885. Abbot never wavered from his belief that scientific agriculture should be at the core of the college's curriculum, research, and outreach. Abbot inspired many students to stretch their minds in order to fulfill their intellectual potential, and as a result they held him in very high esteem. He was born in Vassalboro, Maine, in 1826, and married Sarah Merrylees in 1860. She, too, earned the love and respect of many students for her sensitivity and attention to their needs. Abbot died in 1892. Courtesy of Michigan State University Archives and Historical Collections (People, Abbot, Theophilus, no. 8).

college and more. Tensions generated by opposing viewpoints energized college administrators, faculty, alumni, supporters, and students to build an institution that successfully answered and defended its purpose and program. And in the process they forged land-grant education in Michigan.

Behind the controversy over the State Agricultural College lurked the meaning of "liberal and practical education" as this phrase was used in the Morrill Act of 1862, the place of science in the institution, and the definition of agriculture. Friction between the scientific and the practical also inflamed tensions between generations. Many parents rooted their notions of present and future in their vision of the past, while their children looked to a land-grant education as a means to liberate them from their family's past. Over several decades, M.A.C.'s detractors and supporters negotiated a functional understanding of the relationship between scientific research and its practical application to agriculture in Michigan. Farmers and professors collaborated in ways that allowed each to carry out their work more productively. Scientists continued to do research, but they committed themselves to taking the results of their experiments directly to farmers and the general public. Although publications formed a vital part of this outreach, the real breakthrough occurred after professors presented their ideas face-to-face with farmers in their communities at farmers' institutes beginning in 1876. Grudgingly, the farmers became receptive to suggestions made by professors to apply new methods in their fields, orchards, and barns. This process, often slow and frustrating, became integral to the land-grant philosophy at Michigan Agricultural College, which in turn, expanded the horizons of agriculture.

AGRICULTURE IN TURMOIL

Throughout the nineteenth century, an expanding economy and a growing population were transforming the United States from a country being driven by the farm to a nation being driven by the farm and the factory. In the late eighteenth and early nineteenth centuries the life of the yeoman farmer was extolled as the ideal existence in America. In 1787 Thomas Jefferson, who would become the third president of the United States in 1801, argued that men who worked the ground were "the chosen people of God, if ever he had a chosen people."[1] The future well-being of the democracy rested on their shoulders, which were free from the evils of the factory and the city. As the nineteenth century wore on, other writers such as Nathaniel Hawthorne, Herman Melville, and Ralph Waldo Emerson agonized over the impact of the machine or mechanization on the "pastoral ideal" that had been engrained in the American psyche.[2] Tension between the farm and the factory kept rising as the nineteenth century dissolved into the twentieth century.

During the second half of the nineteenth century, several factors that were reshaping the world of Michigan farmers converged on them; this, in turn, influenced their perceptions of the purpose of their State Agricultural College. Farmers looked to M.A.C. to educate their youth to sustain the traditional role of farmers in American society. But their children and the college both viewed the education offered by the institution as a means to prepare young people to take an active role in the America of the future—not the America of Thomas Jefferson. Although many uncertainties clouded the future, the growth of manufacturing, the spread of commercialism, and the rise of cities all promised to change the role played by agriculture in American life. If this was not enough, science and scientific research discovered new knowledge and developed new technology that affected the way people ate, treated disease, constructed buildings, communicated, traveled, and even cared for their fruit trees or planted and harvested their crops. Farmers living across the United States reacted to the power wielded by people and special interests that they considered to be hostile to them by trying to curb the excesses of industry and business through political action. Discontent abounded on American farms in the late nineteenth and early twentieth centuries. The State Agricultural College went through its adolescence during these troubled times for American farmers.

Agriculturalists did not yield to the transformation of American life gracefully, and they organized themselves to take political action to advance their interests. Having accepted, reluctantly, the fact that commercialized industry had become a permanent feature of their world, agrarians fought for "a more equitable distribution of wealth."[3] Throughout the country, farmers worked for reform of the currency system, the regulation of railroads, the cleaning up of corrupt government, and the lowering of taxes, haulage, and mortgage rates. Between 1867 and 1896 the Patrons of Husbandry, the Grange, the Farmers' Alliances, and finally the radical Populists all marched across the political stage in Midwestern and Southern states supporting candidates committed to improving life on the farm. Since many farmers believed that their enemies had formed conspiracies against them, the political struggles were intense and bitter.[4] The reformers saw money lenders, mortgage holders, railroad barons, corrupt politicians, and powerful industrialists as the leading conspirators. Bad weather coupled with economic downturns, especially during the depression of 1893, increased rural suffering and spurred on even louder cries for reform. The nomination of William Jennings Bryan to be the Democratic Party's candidate for president in 1896 marked the culmination of this agrarian political protest.

Michigan farmers appeared to fare better than their counterparts in other Midwestern states in the mid 1890s, but they nevertheless experienced some of the pain being felt on farms across the nation.[5] Farmers complained bitterly that tariffs drove up artificially the prices of goods that they purchased, while the commodities that they sold on the world market enjoyed no such protection. When the prices farmers received for their products decreased, many of them had great difficulty paying off debts that they had incurred during times of inflation. Farmers also groaned under the burden of a taxation system that placed a disproportionate share of taxes on their shoulders.[6] Late in life, Clark Brody, who grew up on a farm in Michigan during the 1880s and 1890s, remembered agrarian discontent during his youth: "But the farmer believed that he was exploited by everyone. It was a rare farmer who did not resent his inferior political and economic status. The organization of farmers associations—granges, farm bureaus, and alliances—sprang from this felt disparity."[7] The Michigan State Grange, founded in 1873, gave farmers and farm families an organization through which they could work to bring about social, economic, and political reform. In addition, the Grange consistently supported the State Agricultural College and lobbied the legislature for appropriations needed to create and maintain solid programs.[8]

Liberty Hyde Bailey, '82, a son of a Michigan farmer, and a scientist educated at M.A.C., explained how the different perceptions of agriculture held by producers and scientists led to vastly different definitions of agriculture itself when in 1898 he wrote:

> The greatest difficulty in the teaching of agriculture is to tell what agriculture is. To the scientist, agriculture has been largely an application of the teachings of agricultural chemistry; to the stockman, it is chiefly the raising of animals; to the horticulturist, it may be fruit-growing, flower-growing, or nursery business; and everyone, since the establishment of the agricultural colleges and experiment stations, is certain that it is a science. The fact is, however, that agriculture is pursued primarily for the gaining of a livelihood, not for the extension of knowledge: it is, therefore, a business, not a science. But at every point, a knowledge of science aids the business. It is on the science side that the experimenter is able to help the farmer. On the business side the farmer must rely upon himself; for the person who is not a good business man cannot be a good farmer, however much he may know of science. These statements are no disparagement of science, for, in these days, facts of science and scientific habits of thought are essential to the best farming; but they are intended to emphasize the fact that business method is the master, and that teachings of science are the helpmates.[9]

Small wonder that practical farmers had difficulty connecting with M.A.C., where science formed the basis of research and teaching. The two did not come to a meeting of the minds until both entities learned how to turn the "teachings of science" into "helpmates" on the farm. In effect, the scientist and the grower expanded the definition of agriculture to include both experimentation and the application of new knowledge in the fields, orchards, barns, and homes of the farmer. It took more than half a century to achieve a mutual understanding of the college's place in Michigan agriculture.

FARMERS' DISSATISFACTION WITH M.A.C.

Michigan farmers ridiculed the college's early efforts at farming. James Gunnison, '57–'60, recalled an incident that created a long lasting negative impression. In swampland alongside the Lansing to Detroit plank road (behind Morrill Hall) Oscar Palmer, '57–'58, had sown too much turnip seed to produce a harvestable crop. As farmers

Michigan Avenue, which went through farm land, connected M.A.C. to downtown Lansing and the Capitol. It was still a dirt road at the turn of the twentieth century. Courtesy of Michigan State University Archives and Historical Collections (Michigan, Lansing).

traveled past the field of useless turnip tops, they were aghast that the men who had supervised the planting of this crop would have the audacity to tell them how to improve their methods.[10] Farmers were practical people who survived by planting the correct amount of seed, at the right time of year, and in the right kind of soil. They had learned this from their fathers and grandfathers and from experience. Proponents of the agricultural college argued that experiments performed on the college farm would provide information that would enable practical farmers to improve their yields, while farmers were incredulous that any good thing could come out of this enterprise if the professors did not even know how much seed to plant to produce an edible turnip. The Agricultural College of the State of Michigan was off to a bad start with its prime constituency, and its enemies tried to drive it out of existence.

Throughout the 1860s opponents of the State Agricultural College made unsuccessful attempts to close it down and reincarnate it as part of the University of Michigan in Ann Arbor.[11] Critics charged that the college's location was too remote, its faculty was ill-suited to the institution's mission, and it cost too much.[12] The matter came to a head in the legislature in 1869. In January, President Theophilus C. Abbot worked hard to convince the Senate Committee for Agriculture and Education and the House Committee for the Agricultural College to pass an annual appropriation of $20,000 for operational costs for 1869 and 1870 and to provide $30,000 to construct a new dormitory.[13] In February, the *Detroit Free Press* demanded that the legislature no longer fund the college, and along with the *Detroit Post* the paper argued that the college ought to be made a branch of the university in Ann Arbor. The *Free Press* charged that "not one in ten [farmers] would vote to continue the separate existence of the Agricultural College another day." At the same time, Erastus O. Haven, president of the

"The Forest of Arden" became the "Sacred Space" on the Michigan State University campus, where open spaces and their oak trees were preserved free of buildings and intrusions. Williams Hall is visible through the trees in this photograph taken in 1886. Courtesy of Michigan State University Archives and Historical Collections (Michigan State University, Buildings, Williams Hall [1]).

University of Michigan, assured college officials that he did not agree with the newspaper's proposals.[14] Richard Haigh, assistant to Sanford Howard, secretary to the State Board of Agriculture, recorded the drama of 16 March as he and his colleagues awaited the Senate's vote on the college's appropriation bill.

> Extremely quiet, internally considerable excitement. Every one almost holding his breath for tidings of the result.
> Afternoon. Bill passed the Senate by 22 to 8[,] a fair majority. A thrill like an electric shock seems to pass all through the machinery and the motion is increased as if by magic. Everything runs smoothly. Everything looks bright and every body is as happy as a small boy with his first pair of boots. Plans for a new dormitory looked at but nothing decided upon.[15]

The State Agricultural College survived. Within a week, stones arrived for the new dormitory which came to be known as Williams Hall.

During the 1869 session of the legislature, friends of the State Agricultural College also had to refute a petition presented to the House by fifty citizens who laid out what they considered to be the institution's failures. This document expressed many of the farmers' criticisms of the college as well as their understanding of its reason for being. The memorialists charged that the college was not "promoting the cause of agriculture and the mechanic arts," the institution's

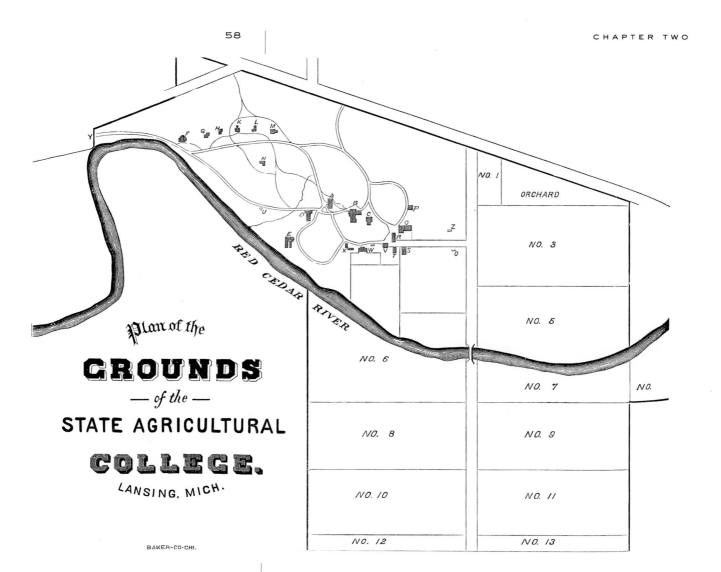

Plan of the
GROUNDS
— of the —
STATE AGRICULTURAL
COLLEGE.
LANSING, MICH.

BAKER-CO-CHI.

very purpose for existence, and that no useful information on either of these enterprises had resulted even though $300,000 had been spent. The petitioners alleged that only a minority of the thirty-four graduates had become "model agriculturalists and mechanics." Among the others, one had become a minister, another a doctor, two were professors at universities, and one had the audacity to engage in photography. The critics condemned President Abbot and the "principal professors" as being "strictly literary and professional men, who can have little or no sympathy with labor or the laborer." Furthermore, the faculty ran the farm through a foreman, when it should have been operated by "an intelligent and efficient superintendent" who was not under the thumb of the professors. The signatories concluded by expressing their desire "to require a pledge of every student entering the College, that he intends to make the calling of agriculture or the mechanic arts his business, and to graduate no student unless he will so pledge himself."[16]

The House Committee on the Agriculture College rejected the complaint as being "wholly unjust and unfounded," retorting that "indeed, the contrary is strictly true," and the committee stated bluntly that "the trouble with the College is, it is not understood." The representatives expressed confidence in President Abbot, under whose leadership the school had "attained the very first rank among similar institutions in the United States." They defended Manly Miles, the professor of agriculture, pointing out that he had spent most of his life as "a practical agriculturalist." The committee found it incredible that anyone could fault graduates who taught "the science of agriculture" at other colleges (William W. Daniells, '64, at the University of Wisconsin, and Albert N. Prentiss, '61, at Cornell University). Of the first twenty-four graduates, the committee argued, fifteen were engaged in agricultural or mechanical work. They stated emphatically that students came to the college "to receive a thorough English education" in addition to studying scientific agriculture and performing manual labor in the fields and shops. They also claimed that results of experiments carried out on the farm were "extensively copied and published." Although the college had made mistakes in its curriculum and management, it was on sound footing. The committee gave Abbot, the faculty, and the students a resounding vote of confidence as they concluded in their report "that the time will surely come when the State Agricultural College will rank beside the University, in the pride and affection of the people of the State."[17]

The citizen's petition encapsulated the issues that challenged the State Agricultural College to find its place as an institution of higher learning and to determine its role in Michigan's growth and development. For decades, many people articulated these views in one form or another: the college did not sufficiently devote itself to agriculture, it did not make the results of its experiments accessible to the public, the professors were not practical enough (they were more interested in theory or literary pursuits), the school was poorly managed, and too few graduates returned to the farm. The college did not take such attacks lying down. It worked tirelessly to overcome criticisms and to

OPPOSITE: "Plan of the Grounds of the State Agricultural College. Lansing, Mich." This diagram was published in the *Twentieth Annual Catalogue* for the State Agricultural College in 1876.

KEY:

A   College Hall
B   Boarding Hall (Williams Hall)
C   Boarding Hall (Saints' Rest)
D   Chemistry Laboratory
E   Greenhouse
F   Dwelling of President
G   Dwelling of Professor of Agriculture
H   Dwelling of Professor of Entomology
K   Dwelling of Secretary for State Board of Agriculture
L   Dwelling of Professor of Chemistry
M   Dwelling of Professor of Literature
N   Dwelling of Professor of Botany and Horticulture
O   Dwelling of Herdsman
P   Farmhouse
Q   Cattle barn
R   Horse barn
S   Sheep barn
T   Tool shed
U   Apiary
V   Shop
W   Piggery
X   Garden barn and tool shed
Y   Principal entrance with self-opening gate
Z   Windmill and tank for supplying water to the yards and barns

Fields range in size from 12 to 27 acres. Crops for the current year are designated by field number:
No. 1 hay
No. 2 summer fallow for wheat
No. 3 oats
No. 4 forage
No. 5 clover
No. 6 corn
No. 7 wood-pasture
No. 8 hay
No. 9 wheat
No. 10 roots
No. 11 corn
Nos. 12 and 13 rough pasture
The remainder of the farm is woodland.

Courtesy of Michigan State University Archives and Historical Collections (*Twentieth Annual Catalogue . . . State Agricultural College of Michigan*, 1876).

convince the petitioners and Michigan farmers that it had their best interests at heart.

President Abbot, who was aware of the House committee's report, corrected misunderstandings of the State Agricultural College's operations in a letter to the politically influential State Agricultural Society, one of the institution's staunch friends. Abbot assured the society that Professor Miles knew farming well, directed two competent farmers on the college farm, and oversaw the labor of students who worked under his supervision in the gardens and orchards each day. While Abbot acknowledged that the bloodlines of the livestock could be improved, he pointed out that the state had never made any appropriations for this purpose. He defended his professors of agriculture, chemistry, and horticulture, and other staff as men who willingly dirtied their hands in the fields. Abbot seemed pleased to note that more than half of the first twenty-four graduates were engaged in agricultural work, including Albert J. Cook, who taught zoology and entomology at his alma mater.[18]

Remembering his student years on campus, Daniel Strange, '67, recorded useful insights into academic life during some of the early, difficult years in the 1860s when he critiqued the faculty:

It has always been a matter of great wonder to me that so feeble a college should have been manned by such able and strong men. Pres[ident] Abbot was a man of marvelous scholastic attainments. America has known very few who were in this respect his equal. Dr. [Robert C.] Kedzie's praises have been sung so long, so loud, so repeatedly or so constantly that I need not add a word. If indeed he was, as is so often claimed, the back-bone of the college for many years, then was Pres[ident] Abbot the brain and the heart and Dr. Miles was the active limbs. [Oscar] Clute, [Albert N.] Prentiss, [George T.] Fairchild, and [Albert J.] Cook all achieved later a national reputation as eminent educators. Not one of these men had any practical knowledge of farming except possibly Dr. Miles and his knowledge was certainly very limited when he became Prof[essor] of Agriculture, but he was a man of eminently scientific trend of thought, a faithful student and an indefatigable worker. His early collegiate instruction had been quite limited and however little or much he may have known of Agriculture when appointed, he lived until those best qualified to know said that if there is any science of agriculture, Dr. Miles knew more of it than any other man living.[19]

Strange's recollections in 1911 put the 1869 debate into perspective. Most farmers were practical people who had "but slight appreciation of scientific research" or little patience with experiments that appeared to produce nothing of value to them.[20] Their criticisms had merit and must not be dismissed as an uninformed diatribe. Nevertheless, the college's defenders were right to refute harsh attacks questioning the institution's purpose, the faculty's abilities, and the occupations of the graduates. As talented instructors and pupils struggled to carry out their experiment in higher education, they made mistakes, but they lacked neither commitment nor integrity. All parties interested in the State Agricultural College had years of toil in front of them before they could work out a mutual understanding of what a land-grant institution should be.

The Michigan State Grange recognized M.A.C.'s good work while always asking it to do more. The Grange appreciated the value of farmers' institutes, which started in 1876, where "intelligent farmers" learned about the latest research in agriculture, entomology, horticulture, and livestock. Speakers at institutes told their audiences about soil analyses, methods of fertilization, pests, seeds, and animal feed. But the Grange never let its agricultural college rest on its laurels, as it urged the institution to establish a veterinary medicine course, create a women's department, and start a program for "technical mechanical instruction."[21] Although often critical, the Grange proved to be an able ally of M.A.C., and it helped to gain legislative appropriations for improvements at the college through political agitation.

On some occasions, professors challenged farmers to open their minds in order to understand the value of scientific agriculture. On 9 July 1879, Professor William J. Beal gave an address on campus discussing the relationship between botany and farmers. Beal revealed how faculty perceived the chasm between their work and many a farmer's attitude towards experimental agricultural research. On his first visit to the campus in 1870 Beal learned firsthand that many Lansing area agriculturalists saw little value in the "state farm," as they called the college.[22] Beal defined a true farmer not as a man who simply raises produce to sell but as one who has "head and heart, loves truth as well as food, and eats to live, not lives to eat." He chided farmers "to escape that intellectual littleness that asks only what is the use?" He thought that farmers should be interested in learning the

LEFT: William J. Beal pioneered the science of botany during his storied career at Michigan Agricultural College from 1870 until 1910. He was born in Adrian, Michigan, in 1833 and received his undergraduate education at the University of Michigan. Beal studied under the noted scientists Asa Gray and Louis Agassiz at Harvard College, where under Gray's tutelage his commitment to Charles Darwin's theory of evolution matured. True to his Quaker heritage, Beal lived a frugal life and was opposed to the use of alcohol and tobacco. Included among his contributions to botany at M.A.C. were his starting a botanic garden, building a laboratory and collections for a botanic museum, collecting over 100,000 specimens for a herbarium, doing pioneer research into grasses, carrying out experiments leading to hybrid corn, and conducting experiments with tree plantings in northern Michigan. Beal also planted grass and weed gardens, an arboretum, and a pinetum, and labeled trees and shrubs around the campus. An active member of the Grange, he never tired in his efforts to communicate his knowledge to farmers in the field. But Beal was a teacher above all else. His daughter, Jessie, and her husband, Ray Stannard Baker, recognized this in their biography of Beal, published in 1925, the year after his death: "It is probable, indeed, that notable as were Dr. Beal's services as a scientist and to science, it was as a man and a teacher that his life was truly most valuable" (Ray Stannard Baker and Jessie Beal Baker, *An American Pioneer in Science: The Life and Service of William James Beal* [Amherst, Mass.: Privately printed, 1925], 14).
Courtesy of Michigan State University Archives and Historical Collections (People, Beal, William J., no. 243).

RIGHT: Hannah A. (Proud) Beal made her presence felt at M.A.C. during her thirty-nine years on campus. She married William J. Beal in 1863 and had two children, Jessie Irene, born in 1870, and a son, born in 1873, who died in infancy. Mrs. Beal touched the lives of many students in ways that are invisible today but that certainly helped them through low points during their days on campus. In his history of the college W. J. Beal paid tribute to his wife: "Cut off by three miles of rough road, the College was forced to live much to itself. Its life was that of a large family and of that family many of the students remember Mrs. Beal, truly, as the mother. She cheered many a homesick boy; she watched faithfully at the bedside of more than one that was sick. Her home was always open and at times of reunion, always crowded" (W. J. Beal, *History of the Michigan Agricultural College and Biographical Sketches of Trustees and Professors* [East Lansing: The Agricultural College, 1915], 415). Hannah Beal died on 22 December 1909. Courtesy of Michigan State University Archives and Historical Collections (People, Beal, Mrs. William J., no. 276).

structure and characteristics of plants. Beal argued that, since a field of wheat or apple trees in an orchard constituted a large number of plants of the same type, what botanists discovered about one plant applied to the entire field or orchard.[23] He implied that what was true for botany and the farmer was also true for other sciences as well. If farmers hoped to understand the value of Beal's and his colleagues' work, they had to expand their concept of being practical to include the application of newly discovered information to their crops and livestock. As the 1870s drew to a close, much still needed to be done to convince the men who plowed fields and picked apples that the agricultural college was relevant to them.

Beal's pronouncements did not put an end to the demand that the State Agricultural College's course of study be designed to be more practical. In 1891 the *Michigan Farmer* expressed dismay that the college's appropriation bill was running into trouble in the legislature because "agricultural instruction, such as will fit young men for the practical duties of farm life, is being more and more ignored." The periodical's position was significant, for editor Robert Gibbons was probably closer to Michigan farmers than any other person in the state. Gibbons frequently castigated M.A.C. for not training enough men to be practical farmers.[24] Earlier in the year, State Board of Agriculture member Ira H. Butterfield had complained to President Oscar Clute that people, including elected officials, uncritically believed journalists' allegations (especially those who wrote for the *Michigan Farmer*) that M.A.C. was wandering from its commitment to agriculture.[25] Butterfield argued that the antidote to this problem required the college to get its message to both legislators and the public.[26]

While serving as editor of the *Grange Visitor* in 1894 and 1895, Kenyon L. Butterfield, '91, (Ira's son) sent out questionnaires to alumni around the state asking them to ascertain the opinions of farmers, living in their counties, of their agricultural college. The responses clearly show that M.A.C., as it reached its fortieth year, still faced an uphill struggle to change the attitude of Michigan's farmers. (Butterfield's surveys provide a good context for the reforms that Jonathan L. Snyder implemented after he assumed the presidency in 1896. These measures are discussed in depth in chapters 4 and 5.)

Farmers in Ionia County held the college in high esteem and expressed a desire to receive bulletins issued by the experiment

Born in Lapeer, Michigan, in 1868, Kenyon Leech Butterfield graduated from M.A.C. in 1891 and went on to become one the college's most distinguished sons. He edited the *Grange Visitor* for five years, superintended the farmers' institutes at M.A.C. for four years, earned a master's degree in economics and rural sociology at the University of Michigan, and taught rural sociology at Michigan for a year before becoming president of Rhode Island College of Agriculture and Mechanic Arts. In 1906 he assumed the presidency of Massachusetts Agricultural College in Amherst. Butterfield believed that land-grant colleges should teach agriculture in such a way that students learned to understand the entirety of the business that encompassed the production, marketing, distribution, and use of agricultural products. He toured China in 1921 to study its educational system, showing a particular interest in agricultural instruction. In 1924 he left Massachusetts to return to Michigan Agricultural College as president, a post he occupied for four years. He died in 1935. Courtesy of Michigan State University Archives and Historical Collections (People, Butterfield, Kenyon L., no. 497).

station.[27] M.A.C. also fared well in places where growing fruit made up a large part of its agriculture. In Berrien County farmers thought highly of the institution, although only a few relied upon results of its experiments for their work.[28] E. O. Ladd, '78, reported that in Grand Traverse County, people attending farmers' institutes and grange meetings considered M.A.C. to be one of "the very best of our State educational institutions" and took interest in its research.[29] It appears that fruit growers noticed the professors' work in horticulture.

But farmers in other counties found M.A.C. to be of little relevance to them. In Shiawassee County a majority of farmers did not regard it with much favor. Although some appreciated the work of the experiment station and the farmers' institutes, many criticized the institutes as consisting "of too much talk and too little teaching."[30] In January 1894 President Lewis Gorton and Professors Robert C. Kedzie and Fred B. Mumford, '90, addressed farmers gathered at the institute held in Highland, Oakland County. This meeting helped those in attendance to overcome their suspicions of M.A.C.[31] "Prejudice among farmers against book-farming which tells against the college" led to little interest in the school in Kalamazoo County.[32] Elisha D. A. True, '78, articulated a widely held attitude when he said that Macomb County farmers thought the agricultural college was a "useless piece of expense" that "must cost something to run . . . and consequently must add to there [sic] taxes."[33] Perhaps the most damning report came from F. N. Clark, '89, when he told Butterfield that farmers living around Milford, Oakland County, did not speak much of the college "except for ridicule."[34] Men who thought so little of M.A.C. did not encourage their sons to attend the school.

In February 1896, the Michigan State Grange's Special Committee on Agricultural College reinforced the results of Kenyon Butterfield's survey in its report to the readers of the *Visitor*. An excerpt from the committee's review of the relationship between the farmers of Michigan and M.A.C. confirmed that the gulf between the two was wide:

> There is and has been for a number of years a lack of students from farm homes who intend to become farmers after graduation. While this is no doubt due partly to the depressed condition of agriculture it is also largely due to a lack among farmers of confidence in the college itself.

We regret to say in many sections the farmers seem to feel that the influence of the college tends to draw students away from the farm instead of attracting them toward it. That this sentiment exists is proved abundantly by evidence placed before your committee. While it is only justice to say that this feeling is owing partly to misrepresentation and unfair criticism on the enemies of the institution, still it is due largely to the fact that only a small proportion of its graduates become practical farmers.

Their [sic] is also a feeling of disappointment in many sections in the results of experimental work and a wish for different and more practical lines of investigation, and an idea is prevalent among many of our more progressive farmers that in some respects the college is falling behind the times. These conditions can not be remedied by attacks on the college, neither will defense and explanation win the much needed support of the great masses of farmers in the state.[35]

The Edwards, Smith, and Kedzie Report, 1896

Troubled by the lack of students coming from the farms to attend M.A.C., the State Board of Agriculture appointed, in September 1895, a committee of three faculty to investigate why enrollment had stagnated and to make recommendations to revitalize the institution. The implementation of the suggestions made by Professors Howard Edwards, Clinton D. Smith, and Frank S. Kedzie, '77, would profoundly change the course of the college. They found that flat enrollments from agricultural families over the past twenty years revealed a stationary interest in the college by farmers; public opinion expressed in newspapers generated unrealistic expectations; and costs for "industrial education" were higher than anticipated. Edwards, Smith, and Kedzie captured the conflicting philosophies of land-grant education at M.A.C. when they reported:

This original disappointment and consequent distrust still survives and is kept alive by the continued disagreement in the theory and policy among those who guide the destinies of our agricultural colleges. There are those who contend that the land grants were intended to aid in establishing colleges where a general scientific education under farm auspices, so to speak, might be obtained; on the other extreme are those who think that the science-training should be of the crudest sort, and the main part of the instruction should be in empirical methods of tillage and manual training in the ordinary farm operations. There is one class of men who would make all the teaching in an agricultural college technical; there is another who, with the smallest possible concession to the name agricultural in the way of technical instruction, would build a literary college on an agricultural foundation. Between these extremes are all grades and shades of opinion; until the name agricultural college has, in the minds of the people, no definite meaning, and agricultural education has been pronounced by many a sham and a fraud.[36]

The committee's explanation for the lack of growth embodied many of the points raised in Kenyon L. Butterfield's surveys over the previous two years. They found that farmers believed the professors did not teach their sons "how to make a business success of farming," and that the course of study drove young men away from the farm. Reality, however, differed from perception, for the report pointed out that while only 11 percent of entering students wished to become farmers, 33 percent either farmed or taught agriculture after graduation. Other factors contributing to M.A.C.'s problems included the general depression in agriculture, widespread ignorance of the college's accomplishments, lack of feeder schools to prepare students to enroll in its curriculum, hostility from the press, and the requirement for students to pay between $75 and $100 for room and board before they could begin their studies.

Edwards, Smith, and Kedzie spelled out ten recommendations, beginning with what they called "a clear and authoritative definition of the charter of our agricultural and mechanical courses":

> The Michigan Agricultural College is a school for technical and professional training in farming and engineering. Its aim is to develop all its pupils into broadminded men, good citizens and ideal farmers or mechanical engineers. Its methods are, science applied to all duties and labors, united action of brain and hand and eye until skill is attained, development of character through "the blessed companionship of wise thoughts and right feelings."[37]

The committee called for more teaching of technical or practical subjects during the freshmen and sophomore years, offering at least six

six-week short courses in winter, changing the long vacation from winter to summer, and providing agricultural training to teachers in country schools. They also wanted to establish a six-month preparatory course for students who could not meet entrance requirements, to reduce student costs, and to abolish the dormitory system. Edwards, Smith, and Kedzie echoed the voices of women in the state, as communicated through the Grange and the alumni, when they called for the creation of a "ladies" course with a curriculum featuring domestic science and arts. Almost immediately, Snyder and the board implemented this recommendation.

The committee's analysis of the student body over the years offers some useful insights into the college's relationship with the agricultural community. In 1895 proponents of practical agriculture certainly were dismayed to learn that only 160 of 676 graduates (23.7 percent) were active farmers and that another 71 (10.8 percent) taught agriculture. And the proportion of graduates who either farmed or entered a career related to agriculture had dropped from 53 percent between 1861 and 1865 to 30 percent between 1890 and 1895. The proportion of students coming from farms dropped from an average of 55.6 percent between 1883 and 1886 to just under 40 percent between 1891 and 1894. This followed the national trend, and M.A.C.'s 167 enrollees in the agricultural course compared favorably with agricultural colleges in other states. Student demand for nonagricultural college education was attested to by increasing attendance at other colleges in Michigan, while the number of agricultural students remained relatively constant at M.A.C. For example, between 1880 and 1895 the approximate enrollment at the University of Michigan jumped from about 1,450 to about 2,850; at the State Normal School, from about 450 to about 930; and at Albion College, from about 180 to about 650. In 1895 M.A.C. had 117 students studying engineering and 29 women taking subjects primarily from the agricultural course.[38]

NOT ENOUGH PRACTICAL FARMERS

After the turn of the century, much criticism of Michigan Agricultural College still came from supporters who believed that its primary purpose ought to be to train young men to farm. In 1905 A. H. Zenner, speaking for himself and the Michigan Improved Live Stock Breeders Association, put it bluntly to Governor Fred Warner: "The first object of the agricultural college should be, to my mind, the training of young men as practical agriculturalists, in all the varied lines of production that makes agriculture the basis of national wealth and prosperity, and returning them to the farm to put this knowledge into practice and profit."[39] Zenner argued that the domestic science program should be transferred to the State Normal School at Ypsilanti and that "higher mechanics, including electrical engineering," forestry, and other departments that did not apply directly to agriculture, should be taught at the University of Michigan. He also advocated that M.A.C. introduce a two-year course in practical agriculture "omitting studies of a literary and scientific character" that could be offered to young men who did not wish to take the time to earn a four-year degree. After fifty years, well-meaning people still could not grasp that the land-grant philosophy was predicated upon a curriculum that taught much more than practical agriculture. One can almost hear the ghosts of Joseph R. Williams, John C. Holmes, Lewis R. Fisk, and T. C. Abbot groaning on the banks of the Red Cedar.

The Michigan State Grange continued to insist that graduates take up farming. Leo M. Geismar, who was designated chairman of the Committee of Education for the Grange's convention in 1911, intended to help promote M.A.C. by insisting "that education does not educate away from the farm." Trying to gather support for his position, he asked President Snyder for a tally of the percentage of students who returned to the farm after they completed their studies.[40] Although Snyder's response is unknown, it appears that it did not satisfy disgruntled grangers. Four years later some members of the Grange protested an increase in state appropriations for M.A.C. because they felt that too few graduates actually engaged in farming.[41]

Geismar's position reveals that differences existed within the staff of M.A.C. Geismar, himself, served as superintendent of the Upper Peninsula station of the Michigan Experiment Station, and Clinton D. Smith, who was appointed professor of agriculture in 1893, expressed a view similar to Geismar's when he told the board that M.A.C.'s reason for being was "to graduate successful farmers rather than expert scientists," and that he thought it was essential that any student of his "know how to rig a plow."[42] Smith, who directed the experiment station from 1900 to 1908 before going to Brazil, still held in 1912 that M.A.C. "ought to remain an agricultural college pure and simple."[43]

In 1912 Smith, the Grange and the farmers' lobby fought still another battle to preserve the myth that the land-grant college of Michigan should primarily train practical farmers when they opposed efforts by engineering faculty and alumni to change the institution's name to Michigan State College.[44] This controversy clearly shows that Michigan Agricultural College had evolved into an institution that offered courses of study that prepared young men and women to contribute to society in practical ways both on and off the farm, just as the Morrill Act of 1862 had intended. In order to preserve the integrity of the college's agricultural mission, Smith proposed that "the courses in science and letters" and engineering be taught only at the University of Michigan.[45] This view ignored the legal requirement set forth by the Morrill Act that clearly called for instruction in the "mechanic arts." Defenders of the engineering course offered a simple answer to their critics: when the legislature accepted the land-grant funds, it obligated M.A.C. to create what became the engineering course.[46] The name of the college remained the same, but the number of nonagricultural subjects grew.

Ever mindful that some people continued to challenge the extent of studies offered at M.A.C., Snyder restated positions argued in the 1850s and 1860s that justified its status as a separate entity. As he defended M.A.C.'s existence, he spelled out important characteristics of land-grant colleges that set them apart from the state universities. Snyder pointed out that the Morrill Act broadened the scope of higher education to include practical training. But even after the passage of the Morrill Act, established colleges and universities saw little or no value in "industrial training," as they educated young people from the elite for the "learned professions, such as law, medicine, and ministry." In states where agricultural schools were created as part of state universities, agriculture made slow headway until late in the nineteenth century. Separate schools provided greater benefits to agriculture and to the students, many of whom came from poor families. Snyder made the distinction clear: "A large university is controlled by persons who are in sympathy with the other type of education. They draw from a different class of students."[47] Furthermore, the academic culture at the "great universities did not nurture agricultural and technical

President Jonathan LeMoyne Snyder led Michigan Agricultural College from 1896 until 1915. In this photograph Snyder is flanked by his son, Robert, and wife, Clara, in the library of their home on campus (circa 1897). Snyder was born in 1859 near Slippery Rock, Pennsylvania, and he received his bachelors and doctoral degrees from Westminster College. During his student days he excelled in athletics and debating. Before coming to M.A.C. he worked as the principal of the Fifth Ward Schools of Allegheny (North Side Pittsburgh), where he introduced, among other things, home economics into the district's curriculum. Under Snyder's leadership, M.A.C. underwent numerous changes that enabled the institution to adapt successfully to the changing world of the early twentieth century. He died in 1918. Courtesy of Michigan State University Archives and Historical Collections (People, Snyder, Jonathan L., no. 2545).

courses."[48] At the land-grant college, the study and teaching of agriculture and engineering trained practically minded people to serve society in both the country and the city.

In 1914, W. J. Beal shocked President Snyder and the State Board of Agriculture when he proposed that M.A.C. be merged with the University of Michigan. While doing research for his history of Michigan Agricultural College, Beal was dismayed to learn that only a few institutions of higher learning still looked to M.A.C. as a training ground for professors. Beal formed his opinion by drawing upon his forty years in East Lansing and on a body of correspondence with more than thirty-five alumni who taught agriculture in universities or separate land-grant colleges across the United States. He suggested that combining the two institutions would spare the university from duplicating courses offered at M.A.C. Students from both schools would be able to study under professors on both faculties, some of whom would teach on both campuses. Beal thought that M.A.C., on its own, would not be able to either acquire sufficient scientific equipment or attract enough expert faculty on a level to match their counterparts in Ann Arbor. He also wanted engineering students to attend the university, believed that both institutions should teach forestry, and encouraged students to take classes in both East Lansing and Ann Arbor.[49]

Beal sent his initiative to Snyder and President Harry B. Hutchins of the University of Michigan, both of whom threw "cold water" on his proposal. Snyder castigated Beal for not consulting with the State Board of Agriculture before sending out his scheme, and he expressed fear that the college's opponents would use Beal's argument to seek a reduction in M.A.C.'s appropriation from the legislature. Snyder told Beal that his own career showed that outstanding faculty could be retained at M.A.C., and that procuring enough equipment only required having the money to pay for it.[50] Hutchins replied to Beal that he would "seriously question the wisdom" of the movement that he suggested.[51] Hutchins also promised Snyder that he would not share Beal's ideas outside of his office, especially with the newspapers. Then Hutchins reassured Snyder that he held dear "the friendly relations" between the two schools, that he hoped to stand with Snyder before the next legislature, and that they would "have to hold back as far as we can, some of our injudicious friends."[52]

A final word on this episode belongs to Jason Woodman, '81, a member of the State Board of Agriculture, a former master of the Michigan State Grange, and a prominent farmer from Kalamazoo. He felt that Michigan farmers would react so strongly against any "serious effort" to merge M.A.C. with the University of Michigan that it would take ten years to calm them down. Woodman summed up the situation:

> The trouble with Doctor Beal is that he seems to think that the taxpayers of the state of Michigan are maintaining our agricultural college for the sole purpose of educating agricultural scientists. Primarily our college exists for the purpose of training farmers for farming. If it is to continue to do this work, it will have to be a separate institution. If it is ever combined with the university of this state, it will mean that the efforts of those who have endeavored to give agricul-

tural education in this state less, in order that the university might have more of the public funds, will have been successful.[53]

Many farmers still clung to the idea that M.A.C. existed primarily to educate their sons and daughters to return to the farm. Ironically, both Beal and Woodman were out of touch with the real debate over the curriculum at M.A.C., which no longer revolved around teaching practical farming or scientific agriculture. Rather, at the demand of younger generations, the college had restructured its courses of study to enable students to choose from an expanding range of subjects. The Great War would soon turn everyone's world upside down, and after the war the college would begin to implement new programs in applied science and liberal arts that would prepare the way for Michigan Agricultural College to become Michigan State College of Agriculture and Applied Science and, finally, Michigan State University.

TEACHING STUDENTS TO MAKE PRACTICAL APPLICATION OF KNOWLEDGE

As M.A.C. struggled to make itself relevant to men and women who actually toiled on the farm and, at the same time, to convince farmers that it was pertinent to them, a new definition of who the target student was evolved. This process both formed and expanded the core of the land-grant philosophy of higher education at M.A.C., and it is essential to an understanding of the institution's history. The different concepts of M.A.C.'s mission were demonstrated by the college's response to never-ending charges that it was not teaching enough practical agriculture. College officials argued forcefully that many of its students did go into careers devoted to all phases of agriculture, not just farming. As it made its case to distrustful farmers, the college never wavered from its commitment to train young men and women to do other things, and it took great pains to point out alumni and institutional contributions to agriculture. The faculty and administration created programs that brought them into direct contact with real farmers: the farmers' institutes began in 1876, the winter short courses started on campus in 1894, and the Extension Service sent out its first agent in 1908. Through such mechanisms, many farmers themselves became M.A.C. students, even though most of them never set foot on campus. The reasons given by prospective students for enrolling in M.A.C. provide another set of insights into the public's understanding of its mission.

After his inauguration in 1896, President Jonathan L. Snyder articulated a strong, aggressive definition of M.A.C.'s mission and its role in the state and beyond.[54] Snyder understood the land-grant philosophy of education to be a vital, innovative force in society that required faculty, students, and alumni to see Michigan, the United States, and the world as a whole. If work at M.A.C. was intended simply to enable Michigan farmers to squeeze more milk out of their cows or put more wheat into their granaries, the college would be failing. Michigan's land-grant institution would do all it could to empower the sons and daughters of the state's "industrial classes."

Snyder received outside help in his effort to promote a clearer understanding of the meaning of land-grant education. On 15 June 1900, Booker T. Washington, principal of the Tuskegee Normal and Industrial Institute in Alabama, identified basic tenets of the land-grant philosophy of education in his commencement address at M.A.C. Washington had founded Tuskegee in 1881, as an institution that taught agriculture and practical skills to African-American students.[55] He believed that agriculture had "constituted the main foundation upon which all races have grown useful and strong." He stressed that black youth needed to be educated in such a way that they could return to their "father's farm," rather than try to live in a city where they knew neither the culture nor the people. As Washington laid out his hopes for educating poor black youth in the South, he spelled out a philosophy that applied to Michigan's young people, black and white, as well.

Washington argued that intellectual and vocational education worked together to improve the human condition:

> No race can be lifted till its mind is awakened and strengthened. By the side of industrial training should always go mental and moral training. But the mere pushing of abstract knowledge into the head means little. We want more than the mere performance of mental gymnastics. Our knowledge must be harnessed to things of real life.

As Washington continued to make his case for educating the African-American children and grandchildren of former slaves, he said: "I

Booker T. Washington (1856–1915) became the leading spokesman for the education of African Americans in the late nineteenth and early twentieth centuries. Born into slavery and poverty, he worked hard to gain an elementary education. Washington studied for three years at Hampton Institute, in Hampton, Virginia, where he learned brickmasonry. After he left Hampton he devoted the rest of his life to teaching African Americans. In 1901 he published his autobiography, *Up From Slavery*, which was read by many. Washington's emphasis on industrial training caused his critics to accuse him of not paying enough attention to the intellectual training needed by African Americans to achieve racial equality in the United States. In 1895 Washington spelled out in the "Atlanta Compromise," an approach to race relations that called for African Americans to make accommodations to racial views held by most whites. In essence, Washington asked African Americans not to agitate for the right to vote and to look to sympathetic whites to help them improve their lives. While white Americans embraced Washington's ideas, important black leaders disagreed with him. W. E. B. Du Bois advocated that the right to vote was essential to the African Americans' struggle to gain equality. Du Bois charged that "Mr. Washington's programme practically accepts the alleged inferiority of the Negro races." Du Bois did not stop here: "But so far as Mr. Washington apologizes for injustice, North or South, does not rightly value the privilege and duty of voting, belittles the emasculating effects of caste distinctions, and opposes the higher training and ambitions of our brighter minds,—so far as he, the South, or the Nation, does this,—we must increasingly and firmly oppose them" (W. E. B. Du Bois, "Of Mr. Booker T. Washington and Others," in *The Souls of Black Folk* [New York: Dover, 1994], 30). Courtesy of Michigan State University Archives and Historical Collections (People, Washington, Booker T.).

plead for industrial developments, not because I want to cramp the Negro, but because I want to free him. I want to see him enter the great and all-powerful business and commercial world."[56] Washington's students suffered from slavery's evil effects, which had denied them formal education and relegated them to a life of grinding poverty. Although many M.A.C. students also came from poor families, they had not experienced the brutality of racism directed against America's African-American people. But Washington's remarks were nevertheless applicable to the educational needs of M.A.C.'s almost all white student body, for M.A.C. students too needed to have their minds stimulated intellectually, as well as to learn practical skills that would enable them to find a place in America's emerging economy.

Seven years later, President Theodore Roosevelt outlined his thoughts on agricultural and industrial education in his commencement address when M.A.C. celebrated its Semi-Centennial on Friday, 31 May 1907. He told his audience that "industrial training is one of the most potent factors in national development," that with it, individuals work together effectively to strengthen the nation's economy. He expressed concern that the social fabric of America's rural communities was suffering as the coherence between families, churches, schools, and other institutions was slipping with declining farm populations, and said that this state of affairs needed to be reversed. The president noted with pride that agricultural science had done much to improve farming. He urged farmers to use new knowledge discovered at agricultural colleges and to insist that their children's schools develop "a constantly more practical curriculum." He encouraged farmers to cooperate with the United States Department of

Michigan Agricultural College chose 1907 to mark its fiftieth birthday because it commemorated the college's opening—not its founding. The highlight of the Semi-Centennial occurred on the afternoon of Friday, 31 May, when President Theodore Roosevelt presented his address "The Man Who Works with His Hands" to the graduates, a large number of dignitaries, and a crowd estimated by the *New York Times* to number 25,000. Roosevelt's appearance brought the Class of 1907 to their feet with shouts of "Rah, rah, rah," to which he responded with a rousing shout "Touchdown," to the enthusiastic delight of all the students.

People who attended the Semi-Centennial celebration could share the occasion with friends and family back home by sending them this postcard. Courtesy of Michigan State University Archives and Historical Collections (Michigan State University, Special Events, Semi-Centennial, 1907, Folder 1).

Ransom E. Olds won a flip of a coin in President Snyder's office that allowed him to drive President Roosevelt to campus from the Capitol in one of his Reo automobiles. After the festivities ended, Roosevelt went to the train station in Lansing in the Oldsmobile of the men who had lost the coin toss. In this well-known picture, Roosevelt and Snyder ride in the back seat as Olds, flanked by William Loeb Jr., Roosevelt's secretary, steers his car down Michigan Avenue. Roosevelt had spoken at the Capitol, where he had enjoyed the hospitality of Governor Fred Warner, who had attended the college during the 1880–81 school year. Courtesy of Michigan State University Archives and Historical Collections (Michigan State University, Special Events, Semi-Centennial, 1907, Folder 1).

President Roosevelt addressing the Class of 1907, seated in front of him. Though there were no loudspeakers, the president's strong voice projected well, which enabled most people to hear his lengthy speech. Myrtle Craig, the first African-American woman to graduate from M.A.C., and her ninety-five classmates, walked across this stage to receive their diplomas from Roosevelt. Following this, Snyder conferred honorary degrees on sixteen men, including James Burrill Angell, president of the University of Michigan; James Wilson, secretary of the Department of Agriculture; and Gifford Pinchot, chief forester of the Department of Agriculture.

Enoch Albert Bryan, President of the Agricultural College of the State of Washington was among those who received the honorary degree of Doctor of Laws (L.L.D.). He, like the other recipients of honorary degrees and representatives of other institutions, offered his congratulations and sense of gratitude to M.A.C. for its leadership in higher education in the United States. Bryan brought this message:

> The New Northwest sends greetings to the child of the Old Northwest. The State College of Washington presents its congratulations and felicitations to the Michigan Agricultural College on occasion of the fiftieth anniversary of its establishment. The distinguished service to the commonwealth and to humanity rendered by the College during the past fifty years will forever continue to be an inspiration to men and to states.
>
> May the oak tree, emblematic of long life, strength, beauty, and usefulness henceforth be inscribed on your coat of arms.

Courtesy Michigan State University Archives and Historical Collections (Michigan State University, Special Events, Semi-Centennial, 1907, Folder 2).

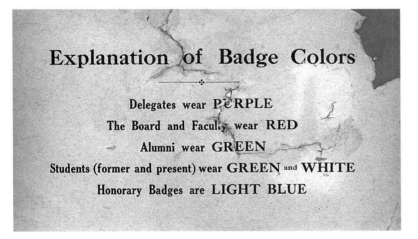

People attending the Semi-Centennial celebration wore color-coded badges to identify their relationship to the college. Faculty and board members wore red badges like the ones belonging to Professor Frank S. Kedzie and Registrar Elida Yakeley. Francis Hodgman, '62, and other alumni sported green badges. Courtesy of Michigan State University Museum Collections.

Agriculture, state governments, and with each other through associations. Roosevelt laid out his conviction that education at agricultural colleges "should create as intimate relationship as is possible between the theory of learning and the facts of actual life." Education needed to get beyond book learning. It had to "train executive power" and shape public opinion to enable people to best solve the country's social and political problems.[57] The president of the United States challenged Michigan Agricultural College and its sister land-grant colleges to play an important role in the nation's development during the twentieth century.

As Snyder made his case for M.A.C.'s mission, he left no doubt that the institution and its land-grant philosophy served the larger general public, not just the agricultural community. Responding to questions about the cost of educating a young person at M.A.C., he argued that the cost was "insignificant when compared with the great service rendered." Men and women trained in scientific agriculture devised ways that benefited more people than the farmer. If researchers found a method to enable hens to lay one more egg each year, it would yield more earnings to the farmer and lower the cost to consumers. Snyder summed up the value to society of land-grant educated young people: "The man who has had a college training lives on a higher level; he gets more out of the world; he is able to bring up a better family and in every way contribute a great deal more to society."[58] Snyder's M.A.C. had a far wider mission than simply to educate a few men to be better practical farmers.

Even though Snyder's vision stretched beyond the farm, he expended much time trying to convince farmers that M.A.C. served them well and to create a better public understanding of its purpose and achievements. In the process, thousands of Michigan farmers became students of the agricultural college. By 1907 more than 40,000 farmers were receiving free copies of bulletins (twelve to fifteen annually) informing them of the results of research done at the experiment station. Men trained at M.A.C. contributed many articles to agricultural papers, provided scientific knowledge for the beet-sugar industry, and worked to improve the state's dairy farms. Snyder emphasized that the six- or eight-week winter short courses that drew almost two hundred young men from farms to study specific agricultural subjects made them more informed practical farmers. He took great pride in farmers' institutes, which were held in every county. At these meetings speakers from the faculty and staff and from the local community made presentations on a wide range of farm-related topics to thousands of people.[59] Over the years, the institutes demonstrated to farmers that M.A.C.'s researchers did more than read books and carry out useless experiments. The professors' work provided farmers with better fertilizers, effective methods of pest control, stronger breeds of livestock, and improved ways of making butter and cheese. The professors also inspected animals, fields, and orchards, looking for ways to advise owners on how to increase production.[60] In 1908 M.A.C. began to reach farm families through its extension service, answering questions relative to matters of the home, the field, and the barn. Whether farm operators realized it or not, they looked to M.A.C.'s faculty, staff, and alumni to teach them, just as resident students did on campus. They had become students of Michigan Agricultural College. In 1911 Snyder boasted that Dean Robert S. Shaw had "gotten close to the farmers of the state" and that M.A.C. now had "the farmers with us."[61]

The Snyder years (1896–1915) ensured that M.A.C. would never be just an agricultural college. Snyder's successor, Frank S. Kedzie, led it through the difficult years of World War I and into the changed world that followed in the war's wake. In 1917 and 1918 most of the college's male students joined the military, many faculty and staff left campus to lend their knowledge and skills to the war effort, and female students and alumni contributed to food and health needs on the home front and on the battlefield. Consistent with the land-grant commitment to service, M.A.C. did its part to support the United States as it joined a coalition that was waging a brutal war against Germany and its allies. (The war years at M.A.C. are discussed in more detail in chapter 10.)

In the 1920s, Presidents Frank S. Kedzie, David Friday, Kenyon L. Butterfield, and Robert S. Shaw initiated changes that made it clear that M.A.C. intended to educate young men and women to earn their living in a wide range of occupations. The applied science course, which was introduced in 1921, made it possible for students to select subjects that prepared them to pursue an expanding list of opportunities created by science, business, industry, education, government, and agriculture.[62] The adoption of this program fulfilled Snyder's belief, expressed in 1912, that "the Agricultural College should be an institution of applied science."[63] The 1920s also saw the creation of the liberal arts course, the use of radio to communicate to large, distant audiences, and the introduction of studies in hotel and restaurant management. When the name of the institution was changed to Michigan State College of Agriculture and Applied Science in 1925, the people of Michigan finally recognized that M.A.C.'s mission was broader than the promotion of practical agriculture.

ATTRACTING STUDENTS TO ATTEND M.A.C.

All of the debate over the purpose of M.A.C. would have gone for naught had not enough students thought it worth their time to attend. Prior to 1896, criticism directed against the institution by political enemies and agricultural faultfinders, such as the *Michigan Farmer,* and the college's failure to make its programs widely known discouraged young people from coming to the campus. Even after the mechanical program was established in 1885, the size of the student body grew by only 98 students in the following ten years, with 128 out of a total of 393 taking that course. The addition of the women's program in 1896 and the increase in the number of short courses spurred faster growth. By 1901, 158 students in the women's course and 94 in short courses had helped to boost enrollment to 689. Five years later 1,001 students signed up for programs in agriculture (252), engineering (379), the women's course (161), forestry (11), and short courses (203). By the time Snyder retired in 1915, the total enrollment had swelled to 1,993, which was divided into programs in agriculture (827), engineering (408), home economics (368), forestry (47), veterinary (67), and short courses (387).[64]

A look into the motivations behind some of these students' desire to study at the agricultural college reveals how they and, in some cases, their parents understood its mission. In February 1872, E. S. Ingalls, who lived in Menominee, wanted his nineteen-year-old son, Charles, to take subjects that would prepare him for a career in the mines located in the Upper Peninsula. The elder Ingalls believed that if Charles studied civil engineering, geology, mineralogy, and chemistry, while continuing to learn German and receive instruction in the piano or organ for two years, he would be in a position to make long-term plans for his life. Parenthetically, Charles's father thought that "such agricultural knowledge as he may acquire will always be useful."[65] Charles enrolled in M.A.C. for the spring term. While he seemed "well satisfied with his studies," Charles expressed concern that the full four-year course would cause too long a delay before he could take up engineering. Still he wished to study the subjects that would "directly or incidentally aid or become useful to an engineer." His father concluded that the agricultural college was the right place for Charles to receive the practical education he needed to prepare to work in the mining industry.[66] Unfortunately, Charles never completed his course, for he died on 25 August 1875; but his experience shows that for some students the study of agriculture was not central to an education designed to prepare them for a wide range of opportunities.

Students studying at Michigan Agricultural College in the 1880s recognized their need for a broad education, and they understood the college's mission far better than the larger public did. In 1884 the *College Speculum* (which became the *Speculum* in September 1888), the student newspaper, called attention to the fact that the school was "not properly known and appreciated." Many people never looked beyond the institution's name and thereby assumed that its students studied only agriculture. The editorialist claimed that agriculture was "not even the distinctive characteristic of the College, but it is indisputably scientific." He went on to say that if the public understood the opportunities for a practical education offered at M.A.C., many more young people would fill its classrooms. The *Speculum* pointed out that efforts to promote M.A.C.'s agricultural mission to the public obscured the scope of its curriculum, preventing students from looking to M.A.C. as a place to study chemistry, mathematics, veterinary science, and natural sciences.[67]

Students believed that an education at M.A.C. promoted equality of opportunity, a vital organ of democracy in American society. Frank F. Rogers, '83, suggested that in the United States, young men "of average ability" could get a college education if they really wanted one, and that it prepared them to deal with "life's great struggle." The study of mathematics, natural science, and "the best thoughts of the best men of all ages" disciplined the minds of both the rich and the poor. Colleges produced men, from all social classes, "who understand human nature, who become disheartened at nothing, but persevere to the end" as they found their places in the business world and the professions.[68]

Allen C. Redding, '83, Rogers's classmate, added that many farmers in Michigan viewed a college education as a time of "elegant-do-nothingism" rather than as an experience that stretched their sons' and daughters' minds. As a result, many "well-to-do farmers" neither allowed nor encouraged their children to receive "the lasting benefit of a real, true, practical, scientific education" offered at M.A.C. Since the sons of the rich were more likely to go to college, children from

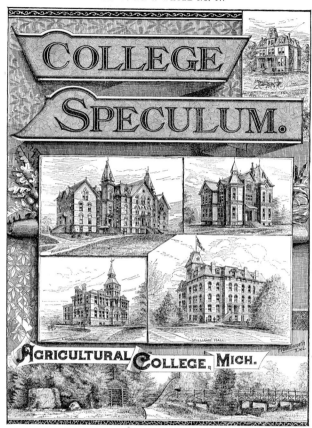

VOL. V: NO. 1.—WHOLE NO. 17.

W. S. GEORGE & CO., PRINTERS & BINDERS, LANSING, MICH.

In the 1880s and 1890s M.A.C. students sharpened their literary skills by writing articles for the *College Speculum*. This cover for the issue that appeared on 1 August 1885 features five key buildings on campus: (*upper right hand corner*) President's house (1873); (*center section, clockwise, starting upper left*) Wells Hall (1877), Botanical Laboratory (1880), Williams Hall (1869), and Library-Museum (1881). Courtesy of Michigan State University Archives and Historical Collections (*College Speculum* 5, no. 1 [1 August 1885]).

the farm needed an education at M.A.C. in order to compete with the elite. Redding reiterated student demands that scientific colleges teach both science and literary subjects to produce balanced minds.[69]

Even though M.A.C. students repeatedly called for a broad education, they also insisted that agriculture constitute a significant place in the college's curriculum. In his commencement address in 1888, George L. Teller, '88, told his classmates and their guests that scientific agriculture uncovered "the ultimate principles which underlie the producing of farm commodities," but cautioned that theories must be made practical to be useful.[70] The next year John W. White, '90, argued that farmers needed to learn how to increase their yields as the availability of inexpensive land diminished. Farmers who acquired a liberal education were best suited to apply the results of scientific research to their own fields, orchards, and livestock, and to serve as teachers for others.[71] In 1895, Orson Phelps West, '96, told readers of the *Speculum* that farmers faced a great challenge to keep up with the explosion of new knowledge in agriculture. This made it imperative that farm youth have "free access to all the knowledge which bears upon agriculture."[72] M.A.C. students had not lost sight of the place occupied by agriculture in their school, even though they demanded an education that introduced them to many other scholarly disciplines.

When Harry J. Eustace, '01, made inquiry about coming to M.A.C. from Rochester, New York, he asked a series of questions that illuminated his and undoubtedly many others' reasons for going to an agricultural college. Eustace wanted to know what graduates of M.A.C. did after they finished their studies. More specifically, he asked about "the prospect of securing a position in some experiment station." He expressed his interest in pursuing a course in horticulture, but he wanted President Lewis G. Gorton's opinion whether or not this was "a good study to make a specialty of." Eustace considered his college education as his entree into a career in some branch of agriculture away from the farm.[73]

Discouraged by the difficult economic times faced by American agriculture in the mid 1890s, many potential M.A.C. students had no hope of making a living on the farm. Hudson Shelton, a concerned citizen, summed this up quite well in 1894: "Farming seems unprofitable. The majority of those who seek education seek it with a view of escaping from the farm. It is rare that you can find a young man of 17 or 18 who is doing well in school and who returns to the farm on his own accord."[74] Since M.A.C. was seen as an agricultural college, many young people went to the University of Michigan or elsewhere to find their way to a more promising career than that presented by their family's occupation. Students at M.A.C. understood Shelton's lament. James S. Mitchell, '95, detailed the dire straits of American farmers in the *Speculum*'s 15 May 1895 issue. Farmers suffered under a burden of large debts, low prices for their products, high prices for goods they purchased, and high taxes.[75] Farm youth were quite aware of the economic health of their families' businesses, and that could be a major factor in their choice of college and vocation.

Some students considered the nature of learning more philosophically. In 1897, sophomore Katherine McCurdy, '96–'98, expressed a view of education that exposed the heart of the land-grant philosophy at M.A.C. She identified two kinds of students. On one hand were many "who depend upon the memory rather than the thought powers." These students might "retain the materials of knowledge, but these materials, since they have not been brought into proper relations with those already in the mind, will not be converted into real knowledge." On the other hand were students "who endeavor to understand their subjects thoroughly, every idea or image voluntarily received is rapidly woven into the fabric of the mind, adding to its strength, usefulness, and beauty." McCurdy went on to say: "What is true of the student in natural history and the student in literature is true of all real students in every line. He examines objects of study at first hand, thinks independently upon them, and assimilates what is acquired thereby for genuine, practical improvements of his mind and his life."[76] T. C. Abbot would have given his hearty approval to McCurdy's appreciation for an education that taught her and her classmates how to think. At the turn of the twentieth century, M.A.C. students understood that their college was preparing them well for life's challenges and opportunities as opposed to forcing them into a mold that suited idealized yeoman farmers.

With the introduction of the women's course in 1896, young women were able to look to M.A.C. as a college where they, too, could receive a practical and scientific education that would prepare them for a career off or on the farm. No longer would women have to

Harry J. Eustace was professor of horticulture and the horticulturalist for the Experiment Station from 1908 until 1919. His students took very responsible positions across the country in research, teaching, orchard management, and marketing. Eustace had a national reputation, and Herbert Hoover tapped him to join the Food Administration during World War I. Soon after Eustace returned to M.A.C. he took a job, in 1919, as the advertising manager for the Pacific Coast division of the Curtis Publishing Company in San Francisco, California. Michigan State University Archives and Historical Collections (People, Eustace, Harry J., no. 925).

follow a curriculum designed to educate young men. Female students brought together the theoretical and the practical in their study of domestic art and science, and many of them linked this to their desire to teach in the public schools. Blanche Clark, a special student in 1910, informed readers of the student newspaper, *Holcad,* that as women followed careers outside the home, their knowledge of cooking, sewing, cleaning, etc. had diminished to the point where many could no longer teach their daughters the skills required to keep up a home. Consequently, young women trained in domestic science and art needed to teach girls the basics of homemaking in the public schools. The teacher should have studied physics, anatomy, chemistry, biology, and bacteriology in order to explain such things as how a sewing machine works, how cloth takes color during dyeing, and the reasons for certain methods of food preparation. Scientific training would also enable the teacher to understand bulletins and literature that showed how to put theoretical discoveries to practical use. Clark, her classmates, and her instructors endeavored to extend the reach of the land-grant way into the kitchens, bedrooms, and parlors of Michigan homes through the public schools.[77]

The introduction of the short courses in the mid 1890s opened up opportunities for a new group of students to study agriculture at M.A.C. that established positive connections with the agricultural community and increased farmers' understanding of the college's potential benefit to them. For many farm youth, the demands of farm life prevented them from attending college for a full academic year, much less for four years. But many expressed keen interest in coming to campus for six or eight weeks in the winter to learn new methods to make themselves better farmers or to pursue an agricultural trade. Practical courses that taught students how to make butter, cheese, and ice cream drew much attention. Creameries dispatched employees to the college to improve their skills. In 1911, E. V. O'Rourke, the proprietor of the Bear Lake Creamery, sent Elmer Alto to take the butter maker's course.[78] The next year the Port Huron Creamery intended to have one of their workers study the management of dairy farming so that he could become "a first class herdsman."[79] George W. Murrow, of Bitely, found the short course to be just right for his son, Harold, but the father still clung to that agrarian mistrust of higher education: "This seems to be my only course, however, as the boy must have

school at once, of some sort, and other schools with which I have corresponded make such a strong point of 'increased salary,' with education, that they are to be avoided by the boy who is to remain on the
farm if possible."[80] George Murrow held out hope that M.A.C. might
teach his son useful things without interesting him in other careers.

The short-course students formed an integral part of the student
body, and by 1905, 184 men, over 19 percent of the total number of students, enrolled in the winter classes. Over the next ten years short-
course students made up about 20 percent of the student population on
campus. M.A.C. placed advertisements for each year's offerings in the
newspapers read by Michigan farmers and sent circulars to granges,
farmers' clubs, institutes, creameries, and cheese factories. Prospective
students learned that they could receive special training in livestock
and general farming, creamery management and dairy works, fruit culture, floriculture, cheese making, and beet sugar production. Specialists
taught each class, and students learned "the most recent methods" and
came away inspired to apply their new knowledge on their farms or in
their creameries.[81] Alumni of the short courses, along with former students of the degree programs, became ambassadors for M.A.C. in their
communities scattered throughout Michigan.

Even as Snyder and Michigan Agricultural College made extensive efforts to recruit more young men and women to study on campus, they felt the need to demonstrate that many of the college's students came from farms and chose careers in agriculture. This
constituted an important piece of their argument to farmers that
M.A.C. was, in fact, an agricultural college. Nevertheless, while in
1897 60 percent of entering freshmen came from farms, this proportion shrank to 39.3 percent by 1905, when 114 (90 men and 24
women) out of 290 first-year students had living fathers who were
farmers.[82] One explanation given for this decline is that some students whose fathers engaged in businesses in addition to operating
farms identified their fathers' occupations as lawyer, banker, merchant, etc. Snyder believed that three-quarters of M.A.C.'s students
had some connection to agriculture at the time. In 1905 about equal
numbers of boys from off the farm chose to study agriculture and
engineering.[83] Most alumni of the agricultural course went into some
line of work pertaining to agriculture, often taking jobs in teaching, at
experiment stations, or in the U.S. Department of Agriculture.[84]

Ten years later Snyder unabashedly defended the college's entire curriculum when he reported to the Michigan State Grange that about 60 percent of the 744 agricultural students enrolled in the long course were farm kids. Snyder said that although he could not predict how many would actually farm, "the entire number" intended "to follow some agricultural pursuit," such as teaching and research, that "contributed to the up-bringing of Agriculture." In addition, approximately 400 were studying engineering, 300 home economics, 100 forestry, and 50 veterinary medicine. These students were committed to "some form of industrial work." M.A.C., according to Snyder was "the one institution in the state which is turning out producers." M.A.C. educated its charges, most of whose families were poor, to participate in the growing, industrial economy of Michigan.[85] But agriculture was "just beginning really to develop in a scientific way."[86]

During the Snyder presidency M.A.C. mounted an extensive recruitment program in order to attract more students who would prepare themselves to be producers. Advertisements and contacts made by the college with individuals and other institutions led to a better understanding of M.A.C.'s land-grant mission. Snyder's advertising campaign went on all year. Before New Year's Day he sent thousands of calendars graced with photographs of campus scenes to legislators, agricultural leaders, newspaper editors, and others whom he wished to remind that M.A.C. was growing and in need of support. Four times a year, bulletins targeted to specific audiences filled many mail bags. The February bulletin informed 5,000 farmers what to expect at the current year's institutes. The May issue went to as many as 8,000 potential students on a list of names provided by schools, ministers, and agricultural organizations. These students also received a copy of the *M.A.C. Record*'s special issue, designed to convince them that Michigan Agricultural College offered a course of study to meet their needs. In August, Michigan citizens learned about one-day excursions to the campus during the third week of the month. Annually, as many as 8,000 farmers responded to the invitation to come to see their agricultural college in person. The fourth bulletin, in November, relayed to farmers the content of the upcoming winter short courses.[87]

Snyder and the faculty personally explained to the public M.A.C.'s mission and programs by answering requests for information and encouraging people to send students their way. For example, in 1896 Snyder sent a personal letter and a circular describing the mechanical department to W. E. Weatherly of the Fifth Ward Manual Training School in Allegheny, Pennsylvania. Snyder asked Weatherly to share the information with some of his students and to urge them to come to M.A.C. In particular, Snyder hoped that Weatherly would talk to one of his students, Walter Huber, with whose mother Snyder had corresponded about the college.[88] In 1897, Snyder sent materials explaining the domestic science curriculum to Mrs. James F. Hancock, of Grand Haven, who served as chair of household economy of the Women's Clubs of Michigan. He asked her to involve the clubs in an effort to include instruction in cooking and sewing for girls in the public schools.[89] Ernest E. Bogue, professor of forestry, thanked Gifford Pinchot, chief of the Forestry Service in the U.S. Department of Agriculture, for encouraging two students to come to M.A.C. and telling them that it "was the best place for men of their preparation."[90]

Many inquiries for information about M.A.C. came from people who answered advertisements placed in periodicals that circulated throughout Michigan and across the country. In August 1911, *Cosmopolitan Magazine,* published in New York, informed Snyder that Charles Davis of Tahoe Tavern, California, wanted to know about the forestry program, and the *Chicago Tribune*'s Bureau of School Information passed along S. G. Salvesen's interest in M.A.C.[91] These young men had read announcements like the one below that informed them that:

THE MICHIGAN STATE AGRICULTURAL COLLEGE
offers the best opportunities for those desiring a practical education based upon scientific principles. Four and five year courses in Agricultural and allied sciences, Mechanical Engineering and Domestic Economy. All lead to B.S. degree. New $100,000 Women's Building. Athletics and Gymnasium. Military Drill.

EXPENSES LOW
College supported by State and National Governments. No tuition to residents of Michigan. Rooms rented at cost of repairs, lighting and heating. Board and text-books supplied at cost through co-operative plan. Send for circulars and catalog, mentioning this paper to
J. L. Snyder, President
Michigan Agricultural College[92]

It should be noted that M.A.C., ever mindful of its suspicious agrarian constituents, almost always listed agriculture first in its promotional literature and often made no mention of literary or liberal arts courses.

College administrators challenged alumni to entice young people to attend M.A.C. In 1908 Robert S. Shaw, dean of agriculture, sent a letter to men who had taken the agricultural course asking them to recruit at least one person for the next year's class. He anticipated that this action would double the college's enrollment.[93] The *M.A.C. Record* asked its readers, who were in large part alumni, to let parents and potential students know about the educational opportunities that the East Lansing campus made available. In 1911 Snyder gave another round of ammunition to recruiters when he ordered each division to make up a twelve-page brochure describing its curriculum and facilities.[94]

Snyder was not content to simply promote Michigan Agricultural College; he wanted to know which methods of advertising were most effective. In 1913, 400 freshmen returned a questionnaire in which they indicated who or what helped them to decide to enroll at M.A.C. (see figure 1).[95] A pattern emerged from this data. The catalog caught prospective students' attention, but good words from alumni, teachers, and citizens who were knowledgeable about M.A.C.'s programs assured young men and women that the school in East Lansing was right for them. A visit to campus also helped. The 1,993 students who attended M.A.C. in 1915 gave evidence that efforts to promote the public's understanding of the institution were bearing fruit.

Mina Jarrard, '98–'01, who had enrolled in the women's course in 1898, held a series of ten meetings in the basement of the Presbyterian church in Yale, Michigan, in the late summer and early autumn of 1903. A look at her work serves as a fitting conclusion to this chapter. The ladies of the church had invited Jarrard to share with them her knowledge of domestic science. They paid her $15 and reimbursed her for train fare to and from Lansing. Jarrard showed her audience how to determine the nutritional value of different foods and how to cook them without diminishing their nourishment or flavor. As word spread of her work, women from other communities came to Yale to hear this M.A.C.-trained young woman tell them how to improve their cooking. As a result, Jarrard received two invitations to come to other towns to repeat her presentations. At the conclusions of the meetings,

Figure 1. Snyder's Questionnaire

Three questions asked in the survey were:

1. "Which of these causes led you to first become interested in the College?"

2. "Which aided you most in keeping in touch with the College?"

3. "Which were the most influential in helping to make the decision?

The answers submitted are keyed to the columns below.

INFLUENCED BY	RESPONSES	1	2	3
A graduate	144	83	66	57
Former student	99	53	34	38
Present student	18	19	14	15
School teacher	110	48	40	39
Citizen of the state	93	48	28	34
College circulars	66	31	29	11
Catalog of the college	218	71	104	80
Newspaper advertisements	6	3	0	1
College paper	13	2	6	1
Visit to college	121	25	37	68
Grange or Farmers' Club	14	4	5	0
Attending Farmers' Institute	24	4	8	4
Interview with M.A.C. professor	28	4	10	19
Letters from college	44	5	9	29
General reputation		7	1	9
Self				4
Last year of preparatory course				5

Jarrard explained how the agricultural college benefited the public. The Reverend Charles D. Ellis commented on her effort: "I can hardly imagine anything that could be done in a community that would do more to advertise and commend the college to a people than just such work as Miss Jarrard has done here." He anticipated the creation of the extension service when he added: "And perhaps her work here might be a hint as to one way of widening the influence of the college."[96]

Snyder, Bogue, and others could answer letters, send out bulletins, calendars, and slick brochures, and give speeches exalting the virtues of M.A.C., but all that meant nothing if the women and men educated on the campus did not serve the public as the college had promised. The students ultimately provided the answer to the House Committee on the Agricultural College's concern that "the trouble with the College is, it is not understood." After learning from Mina Jarrard, the citizens of Yale understood the Michigan Agricultural College. Its alumni and students became its best and most significant ambassadors.

3

Transforming Knowledge

THE ULTIMATE SUCCESS OR FAILURE OF MICHIGAN AGRICULTURAL College depended upon its capacity to transform knowledge into courses of study that gave substance to its evolving land-grant philosophy. Professors and students discovered ways to create new understandings that went beyond existing knowledge and employed new ways to diffuse information to scholars and to the general public. Central to the land-grant philosophy was teaching—professors instructing students who, in turn, learned how to impart their newly acquired knowledge to others. Research and teaching in the sciences that made up what became known as agricultural science distinguished M.A.C.—and other land-grant colleges—from the University of Michigan and most other institutions of higher learning in the United States. M.A.C. became simultaneously the servant of and an agent of change for science, technology, and society. The land-grant way demanded that knowledge not be kept inside the heads of a few, but that M.A.C. and the men and women educated there should show other people how to transform understandings generated on the campus to meet their everyday needs. As the curriculum evolved at M.A.C., it directed teachers and researchers to address—progressively—the needs of the field, the barn, the shop, the home, and the community.

A look at the curriculum that had developed at the University of Michigan by 1890 shows the ramifications of the agricultural college's establishment as a separate institution rather than as a department in the University of Michigan. When comparing the courses of study offered at the two institutions, several things stand out that demonstrate how the work done at both schools served the people of Michigan well and would continue to benefit the state. The university's *Calendar* for 1890–91 contains a concise statement of its purpose.

Edwin Willits received his bachelor of arts and master of arts degrees from the University of Michigan. He had been a successful lawyer and an active Republican who had served twelve years (1861–73) on the State Board of Education before he became president of the college in 1885. He had also been elected to three terms in the United State House of Representatives. Willits was very popular with students, who lamented his departure after President Benjamin Harrison appointed him to be the first assistant secretary of agriculture in 1889. Willits died in 1896 in Washington, D.C. Courtesy of Michigan State University Archives (People, Willits, Edwin, no. 2970).

> The University aims to complete and crown the work that is begun in the public schools, by furnishing ample facilities for liberal education in literature, science, and the arts, and for thorough professional study of medicine, pharmacy, law, and dentistry.[1]

The University of Michigan's curriculum confirmed the institution's unfailing commitment to this mission.

In 1890 the 2,420 students enrolled in the University of Michigan found nearly four hundred classes listed in the catalog. The Department of Literature, Science, and the Arts had an impressive array of offerings distributed throughout forty "Courses of Instruction" that fulfilled the university's intent to provide a thorough liberal education. In history, for example, the university's faculty taught several courses each in English, American, and European history, and the history of civilization, among other subjects. At M.A.C. students took only the required courses in ancient history and constitutional law and history of the United States—enough to receive an overview of ancient and American history. People who wanted to study history in depth went to the university, where they could enroll in classes that reached deeper into the subject. They could also learn there the ancient and modern languages required to carry out primary research and to read contemporary works.[2]

It is also significant that the University of Michigan had committed itself to work for the improvement of the state's public schools. In 1886 Professor John Dewey and other educators formed the Michigan Schoolmasters' Club, which brought together secondary school teachers and university professors to talk about matters pertaining to teaching in high schools, particularly preparing students for college.[3] Courses in "The Science and Art of Teaching" taught the practical aspects of teaching and the history and philosophy of education. It is noteworthy that Michigan taxpayers supported innovations in teaching at their state university and, as will be discussed later, at their agricultural college, as well. University of Michigan professor John Dewey's evolving educational philosophy blended theory and practice as it pertained to teaching in the public schools.[4] Although there is no evidence that educators at M.A.C. looked to Dewey for guidance or inspiration, his thinking helps to put M.A.C.'s emphasis on transforming knowledge into useful or practical application beyond the classroom into the overall story of teaching and learning in late nineteenth-

and early twentieth-century America. Dewey may or may not have been aware of M.A.C.'s program, but he articulated some of what had been happening at the college for a long time when on 18 January 1899 he told his students at the University of Chicago that "the possibility of having knowledge become something more than the mere accumulation of facts and laws, of becoming actually operative in character and conduct, is dependent on the extent to which that information is evolved out of some need in the child's own experience and to which it receives application to that experience."[5] Young men and women attending M.A.C. found a learning environment where simply mastering information was not enough. Professors devised methods that enabled their students to transform facts and results of experiments into actions that met the needs of both the student and the public, even before Dewey advocated using similar techniques in elementary and secondary schools.

The University of Michigan was realizing Henry Tappan's vision that it become a leading public institution of higher learning in the United States and the world. By 1891 it had awarded more than 10,700 degrees, had a faculty of more than one hundred, and had a student body of which 52 percent came from outside of Michigan.[6] The list of courses of instruction in the Department of Literature, Science, and the Arts for 1890 (figure 2) gives ample evidence that Tappan's university had for the most part faithfully followed his desire that it omit "no branch of knowledge" as it had grown and matured. There was one exception to this, however, that was forced upon the university by the establishment of the Agricultural College of the State of Michigan in 1855.

The people of Michigan had mandated that agriculture, including sciences related to it, be taught at the new institution rather than at the University of Michigan. Administrators, professors, and students in East Lansing forged courses of study that looked quite different than those offered in Ann Arbor. Although English, history, philosophy, modern languages, and other liberal subjects were taught at M.A.C., they did not occupy the place of prominence that they enjoyed at the university. M.A.C.'s commitment to the practical application of scientific knowledge to the everyday life of people, especially those who lived in rural areas, took the college's scientific courses in different directions than those at the university. M.A.C.'s

Figure 2. "Courses of Instruction" in the Department of Literature, Science, and the Arts at the University of Michigan, 1890

Greek	Bibliography
Latin	Mathematics
Sanskrit and Comparative Philology	Physics
Hebrew	General Chemistry
Assyrian	Analytical Chemistry and Organic Chemistry
French	Hygiene and Physiological Chemistry
Italian	Astronomy
Spanish	Mineralogy
German	Geology
Gothic	General Biology
Swedish	Zoology
Danish-Norwegian	Botany
English and Rhetoric	Physiology
History	Surveying
Philosophy	Civil Engineering
The Science and Art of Teaching	Mechanical Engineering
Political Economy	Marine Engineering
International Law	Mining Engineering
Music	Metallurgy

Source: *Calendar of the University of Michigan for 1890–91* (Ann Arbor: University of Michigan, 1891), 43–77.

land-grant philosophy directed scientific inquiry and knowledge to the home through domestic science and arts, to the health of animals through veterinary medicine, to the improvement of the physical facilities of the farm through agricultural engineering, and to the fruit industry through horticulture and entomology. After the Cooperative Extension was formed in 1914, it ensured that M.A.C. took its message to every county in Michigan. The manner of acquiring knowledge and transforming it into forms that were useful to the public at Michigan Agricultural College and the University of Michigan proceeded along different paths that were both compatible and

President Oscar Clute working at his desk in the northwest corner of the Library-Museum building, circa 1891. Clute graduated from M.A.C. in 1862 and held appointments as instructor (1863–65) and professor (1865–67) of pure and applied mathematics at the college. In 1868 he married Mary Merrylees, sister of President Abbot's wife. The couple had six children. For many years he pastored Unitarian churches in New Jersey, Iowa, and California. Clute assumed the presidency of M.A.C. in 1889; he held the post until 1893, when he accepted the presidency of Florida Agricultural College. Upon his retirement in 1897 he moved to California where he died in 1902. Courtesy of Michigan State University Archives and Historical Collections (People, Clute, Oscar, no. 654).

complementary, as well as essential to the growth and development of Michigan.

Agricultural Education Takes Shape

Mastery of the physical sciences, teaching practical applications of scientific discoveries, and the education of well-rounded citizens formed the underlying assumptions of the State Agricultural College's programs. The 1866 catalog set forth the five "objects of the institution," which articulated the nascent land-grant educational philosophy and comprised an emphatic statement of the basic principles that shaped and drove learning and teaching at M.A.C.:

1st. To impart a knowledge of Science, and its application to the arts of life. . . .

2d. To afford to its students the privilege of daily manual labor. . . .

3d. To prosecute experiments for the promotion of Agriculture. . . .

4th. The organic law of the College, as well as the act of Congress donating lands for Agricultural Colleges, contemplates courses of instruction in the military art, and in the applications of Science to the various arts of life. . . .

5th. To afford the means of a general education to the farming class. . . . Added to this are the branches of study which help to make an intelligent and useful citizen, which cultivate his taste, and enable him to give expression to his knowledge and opinions.[7]

Michigan Agricultural College linked together science, practicality, and liberal studies to create a curriculum that prepared young people to live successfully in their ever-changing world.

The officers of the college and faculty numbered eight in 1866, and they worked tirelessly to put together courses of study that enabled the State Agricultural College to achieve its objectives. Members of the faculty were:

- T. C. Abbot, A.M., President; Professor of Mental Philosophy and Logic.
- Manly Miles, M.D., Professor of Animal Physiology and Practical Agriculture, and Superintendent of the Farm.
- R. C. Kedzie, A.M., M.D., Professor of Chemistry.
- Albert N. Prentiss, M.S., Professor of Botany and Horticulture, and Superintendent of the Gardens.
- Oscar Clute, M.S., Professor of Pure and Applied Mathematics.
- Sanford Howard, Secretary.
- George T. Fairchild, A.M., Professor of English Literature.
- W. W. Daniells, B.S., Assistant in Chemistry.

The seven instructors all had distinguished careers at M.A.C. and elsewhere. Abbot retired as president in 1885 and stayed on the M.A.C. faculty until 1889. After Clute left M.A.C. in 1867, he became a Unitarian minister. He returned to the college to assume the presidency in 1889. Following his departure, Prentiss, '61, spent many years teaching botany and horticulture at Cornell University. Daniells, '64, went from M.A.C. to the Lawrence Scientific School at Harvard for three years of study. In 1868 he moved on to the University of Wisconsin where he pioneered the teaching of chemistry.[8] Fairchild served as acting president at M.A.C. during 1872–73 and became president of Kansas State Agricultural College in 1879. Miles won the respect and admiration of his students as he gave definition to the new professorship in practical agriculture, and Kedzie built a top-notch course in chemistry.

The 1866 catalog revealed that the number of subjects offered at the agricultural college had expanded since 1857. Chemistry had grown to three half-year courses in elementary, analytical, and agricultural fields. Natural science was taught through courses in geology, structural botany, vegetable physiology, entomology, systematic botany, horticulture, physics, animal physiology, zoology, and astronomy. Mathematics now included classes in both surveying and civil engineering to go along with algebra, geometry, and trigonometry. Freshmen and seniors took full-year courses in practical agriculture. Students also had semester courses in bookkeeping and landscape gardening. Liberal studies brought students into the disciplines of history, English literature, inductive logic, rhetoric, mental and moral philosophy, political economy, and French.

The course descriptions told potential students of the methods used by their professors to teach them. Young men, just off the farm or out of their fathers' shops, read textbooks, listened to lectures, gave recitations of their lessons, and wrote compositions. And newer teaching techniques, often in their infancy, enabled young scholars to gain knowledge through their own observations and manual exercises. Students dissected plants and animals to learn the structure of stems and the arrangement of internal organs. Peering through a microscope they could see features of tissue and fiber that were invisible to the naked eye and then make drawings of their observations. Collections of insects aided entomology students as they looked for ways to identify and eradicate pests. Sophomores met their manual labor requirement by working in the gardens and on the grounds. This gave them an opportunity to put into practice what they had heard in their lectures. Chemistry students learned the properties of different chemicals, soils, manures, and other substances during their daily three-hour sessions in the laboratory. The English professor challenged them to read critically some of the great works of English literature, and he required that they write original compositions.

Knowledge and how it was taught all had a purpose that extended beyond learning just for the sake of learning. The 1866 catalog spelled it out: "It is believed that students who complete the course will be qualified to follow agricultural pursuits with intelligence and success; or should some other profession seem more congenial, they will have the discipline and scientific acquirements that will enable them to pursue with profit the studies preparatory to entering it."[9] With this sentence M.A.C. told the world that agricultural pursuits went beyond practical farming, and the college intended to prepare students to go elsewhere, if they deemed it necessary, to continue studying for their desired occupations. Michigan Agricultural College was not a provincial institution.

Persons seeking admission to the State Agricultural College needed to be at least fourteen years old, had to pass an examination in arithmetic, geography, grammar, reading, spelling, and penmanship, and be willing to spend nine months on campus. The institution also offered a one-year preparatory course designed to teach basic skills in the above subjects to students not yet ready to enter the regular course. From 1861 until 1867 and again in 1870 young men who failed to pass their entrance examinations studied the required subjects in the preparatory class. The highest number of preparatory students was 51 out of a total of 108 students in 1866; in some years these students made up more than half of the student body—in 1865 the ratio of preparatory students was 45 out of 88.[10] After enrollment in the regular course grew to more than 100 in 1871, the college eliminated the preparatory class. The school year in 1866 was divided into two terms. The first ran from 28 February to 27 June, and the second from 10 July to 28 November. The long vacation occurred in the winter in part because it enabled students to earn money for college by teaching in the common schools.

President Theophilus C. Abbot never wavered from the vision put in place by the faculty in 1857, the mandate of the Agricultural Reorganization Act of 1861, and its re-statement in the college's annual catalogs.[11] Abbot and others strengthened the philosophical basis for M.A.C.'s curriculum that enabled President Jonathan L. Snyder, who led M.A.C. from 1896 until 1915, and his successors to extend the land-grant way in new directions. Abbot's educational philosophy, which encapsulated the key elements of land-grant education, gave impetus for the growth of M.A.C.'s efforts to transform knowledge into a curriculum that taught students how to show others to apply it in their lives. Politically astute, Abbot placed the education of the farmer at the center of his institution's program. He supposed that only education could lift farmers and artisans to a social level equal to that occupied by lawyers, doctors, and people in other "liberal professions." Educated farmers could then exert more influence in governmental affairs. Abbot believed that women needed to contribute to the social elevation of the industrial classes, and he called for the establishment of a women's department devoted to the study of "common household arts."

On 4 March 1875, Abbot, who had been at M.A.C. for seventeen years, spelled out his views of an "Agricultural Education" before the Michigan House of Representatives. In this oration he argued that the pursuit and diffusion of knowledge at M.A.C. had been and would continue to be guided by thoughtful purpose. Abbot's vision for the college was one of a vital, forward-looking, institution focused on preparing students to make a difference in the world. Hence he deemed it essential that the agricultural college teach literature, history, and political economy to enable students to interact intelligently with people around them. Abbot thought it important that a student have a thorough knowledge of English, that it "should be a flexible instrument at his command, which he should speak and write with ease and vigor, that he may instruct and impress others, avert mischief, or inculcate truth."[12] Liberal studies educated the whole person and elevated him beyond the mass of ignorant people who were incapable of providing leadership to their own social and economic class.

Nevertheless, scientific studies formed the core of Abbot's curriculum and the basis for experimental research and its practical application to agriculture. Students wanted "to know what to do, and how to do it, but also why?"[13] This "why" formed the nub of agricultural studies at M.A.C. in 1875, and it has driven scientific research at the institution ever since. Chemistry, botany, horticulture, entomology, animal and vegetable physiology, meteorology, and mechanics all provided answers to problems facing farmers. In order to conduct research that would help farmers improve crop yields, fatten livestock, and fertilize soil, the researchers had to be in touch with practicing farmers. Professors needed to visit the fields, barns, orchards, and grange halls of Michigan to find out firsthand about the producers' needs and to offer them advice on how to improve their methods. The faculty also had to respond to readers' requests for information in newspapers and journals, and to write articles for these publications in addition to bulletins sharing the results of experiments. Abbot also believed that students had to contribute manual labor in the fields, laboratories, barns, and shops as a means of learning how to conduct research and to make practical applications of its findings.

Experimentation and original investigation provided instruction to students in both theoretical and practical matters as well as generated new knowledge. Abbot divided experiments into two categories—"rough" and "exact." Rough experiments could be conducted by farmers themselves, for example, by observing the results after planting different varieties of seeds to determine which were suitable for their soils and climate. Exact experimentation proceeded along a more demanding path and could only be performed at the agricultural college. The inquisitor sought to answer only one question at a time, under controlled conditions, recording his findings exactly as they occurred "adding no inference of his own," at least to the extent that this was possible. The challenge faced by the researcher was to keep asking the right question. Such experiments were expensive, but they were essential if agriculture was to "change from an undeveloped art into the standing of a science."[14] Even though science advanced slowly, the agricultural college had an obligation to conduct costly, exact experimentation. Students taught by their professorial investigators would be able to interpret experimental findings for others and make practical applications of scientific findings in the real world.

In order for land-grant education to achieve its purposes, Michigan Agricultural College had to open its doors to women. With Abbot's blessing, in the 1870s and 1880s women advocated for an expanded curriculum to educate young women in domestic science and art in the same way that young men were taught to be practical farmers and well-informed citizens. Women had sought admission to the college even before it had opened its doors in 1857, and in 1870 ten women enrolled, taking the agricultural course or a special course. In 1879 Eva D. Coryell, '79, the first woman to graduate from M.A.C., challenged her alma mater to "substitute in place of agriculture some study to a girl's education," which she said would result in "an excellent lady's course." Coryell believed that the agricultural course did not teach young women the practical skills they needed to properly manage their households. She argued that this was necessary because throughout history "the charge of the housework has devolved upon the woman."[15] Coryell, like Abbot, believed that the ladies should also study French, psychology, moral philosophy,

English, and natural sciences. Eva Coryell envisioned an M.A.C. where men and women would follow courses of study suited to equip them for what society perceived as the role of each gender, but where both sexes read the works of great literary figures, performed some of the same laboratory experiments, and heard the same lectures in history and philosophy.

The daughters of the industrial classes found their champion in Mary Mayo, an activist in the Michigan State Grange. Mayo vigorously advanced the idea that land-grant education at Michigan Agricultural College also should be for women. In 1881 she told her neighbors at the Battle Creek Farmers' Institute that so long as a person had "not learned to apply to the practical concerns of life his acquired knowledge, he or she is a useless mortal to themselves and a burden to society."[16] Mayo argued that a land-grant education equipped both women and men to work for reforms that could overcome the evils of poverty, illness, and ignorance. She believed that education should make women and men ready to be mothers and fathers and to teach them how best to raise their children. According to Mayo: "If our young women are thoroughly skilled in cooking, physiology, hygiene, the proper care of children and the home, many of the social and poor problems would be solved."[17] She stressed that education goes beyond the absorption of knowledge; it enables people to develop and apply their abilities.[18] Mayo worked tirelessly to convince the Michigan legislature to appropriate funds for the construction of a dormitory needed for women so that a women's course could be offered at M.A.C.[19] Mayo's optimistic and progressive outlook resonated with others in the late nineteenth century who assumed that educated women could not only manage their households more effectively, but also lead the way to women's suffrage, prohibition of alcoholic beverages, and greater sexual freedom for women. When M.A.C. adopted the women's program in 1896, it finally committed itself to fully integrating women into the land-grant way. In order for Mayo and Abbot to see their vision for M.A.C. transformed into reality, the institution had to build its courses of study on the emerging sciences that had led to the practice of scientific agriculture.

Mary Bryant Mayo was born near Battle Creek, Michigan, in 1845. She and her husband, Perry, knew the trials and hardships of farm people from working long hours on their own farm. Mary Mayo spent thirty years trying to improve the lives of rural women and children through her extensive career with the Grange. At her instigation, the Grange instituted an annual children's day in 1886. Mayo made numerous speeches promoting the welfare of women and children before audiences across Michigan and wrote extensively for *The Grange Visitor*. A deeply religious woman, she served as the state chaplain for the Grange for thirteen years. She died in 1903. *Courtesy of Michigan State University Archives and Historical Collections (People, Mayo, Mary, no. 1900).*

SCIENTIFIC AGRICULTURE

Over the course of Michigan Agricultural College's first seventy-five years, scientists in Europe and the United States produced a body of knowledge that made up the basis of scientific agriculture. A review of some of their work puts into context the discovery of new knowledge at M.A.C. In Michigan, the State Agricultural College distilled the results of scientific experiments and found ways to teach farmers how to use scientific research beneficially in their work.[20] Farmers, whether they realized it or not, looked to agricultural science to increase their productivity, to strengthen their family life, and to help them keep their farms during the difficult economic times. Most of the knowledge that led to innovations on the farm came from experiments performed at agricultural colleges and experiment stations, and by scientists working for the United States Department of Agriculture.[21]

After Congress established the United States Department of Agriculture (USDA) in 1862, its first commissioner, Isaac Newton, spelled out the department's objectives. Newton wanted to acquire and publish scientific information relative to agriculture, provide seeds for planters, and respond to specific questions posed by farmers. He employed chemists to analyze crops, soils, fertilizers, and relationships between them. He looked beyond the nation's borders and hoped to introduce plants and processes used in other countries into American agriculture. If agricultural colleges were to carry out the work envisioned by the department, they needed to employ professors of entomology and botany, and establish museums and libraries.[22] Major early research efforts supported by the USDA included looking for the causes of hog cholera, fire blight, and Texas fever. The department's early vision of agricultural research was reflected in the development of the curriculum at the State Agricultural College.

Tension between pure science and applied science still prevailed in the 1920s, and scientific agriculture never crystallized as a separate academic discipline. The contentious relationship between the pure and the applied led to growing support for scientific research. Scientists understood that practical applications of scientific discoveries could only come about if pure research made the discoveries first.

Farmers in the field, however, had little or no understanding of the scientific process; they wanted answers to their problems—now. Researchers found that teaching, answering correspondence, giving public lectures, and visiting farms and orchards often left little time to carry out time-consuming and exacting experiments. In addition, younger scientists, who often were better trained than their older colleagues, demanded that they be able to do more pure research. Two pieces of federal legislation that provided funds to support research at experiment stations reflected the increased demand for pure research and tensions within the scientific community. The Hatch Act of 1887 sent $15,000 from the U.S. Treasury to each station to support "agricultural research, both pure and applied."[23] Under terms of the Adams Act in 1906, Congress appropriated another $15,000 for "original investigations" at each station. Since scientists at liberal arts institutions and European universities tended to look down on the practically oriented research being conducted at agricultural colleges, this measure helped to give academic respectability to researchers at experiment stations.[24]

More specialization in scientific research led to the strengthening of existing disciplines and the growth of new fields of study. One result of this trend was that scientific agriculture did not become a distinct academic discipline. Instead, the agricultural courses on college campuses involved students and professors studying a wide range of scientific subjects, including horticulture, botany, zoology, animal and plant pathology, bacteriology, veterinary science, entomology, chemistry, and new disciplines such as rural sociology and agricultural economics. The challenges presented by American agriculture were too numerous to be sufficiently addressed by a single discipline.[25] The boundaries of pure science could not be defined by the practical needs of agriculture alone.

A brief survey of several scientific fields demonstrates that as experimental science grew more sophisticated, its practitioners at agricultural colleges expanded the applications of their findings to agricultural concerns. Chemistry played a fundamental role in the development of agricultural science. Research done by European (especially German) chemists and American chemists trained in Europe provided much of the impetus for chemical research in the United States before 1900.[26] They studied a wide range of problems, including the chemistry of nitrogen, calcium, sugar, fertilizers, bacteria, plant and animal nutrition, and soils, as well as ways to determine the best conditions and methodologies for their experiments. Applications of chemical research, such as the test for determining the fat content in milk developed in the early 1890s by Stephen M. Babcock at the University of Wisconsin (another land-grant institution), had a profound impact on dairy farming. Babcock's test enabled farmers to receive a price for their milk based on fat content and to build herds of cows that produced milk with a high content of butterfat.[27] At the turn of the century, agricultural chemistry led to improvements in the production of starch, glucose, sugar, fertilizer, wine, beer, whiskey, and hides.[28] The growth of chemical research relative to agriculture during the first quarter of the twentieth century saw the number of people employed by the USDA's Bureau of Soils jump from 10 in 1895 to 218 in 1927.[29] Further evidence of the importance of chemical investigation is illustrated in that in 1923 hundreds of projects involving chemistry were underway at agricultural experiment stations across the country. Researchers had turned their attention to a myriad of questions pertaining to soil, fertilizer, animal nutrition, insecticides, fungicides, dairy products, plant nutrition, and "various miscellaneous studies of food, etc."[30]

The applied science of economic entomology proved to be another area of great value to agriculturalists during the first seventy years of Michigan Agricultural College's existence. The relative isolation of American farms in the eighteenth and early nineteenth centuries shielded crops from major insect attacks. As European settlement spread across the country, insect damage to farm crops rose dramatically. Dr. Thaddeus William Harris, a pioneering entomologist in Massachusetts, stressed that it was "necessary to know the life history of the insects before suggesting remedies," which served as a basic tenet of entomology. Between the Civil War and the end of the century entomologists promoted the use of arsenicals, kerosene emulsions, and hydrocyanic acid gas as major chemical weapons against insect pests. They also developed equipment and methods to apply insecticides.[31] Land-grant colleges, state experiment stations, and the USDA all spurred the growth of entomology.[32] More researchers and more experimentation led to important advances, including the introduction of foreign species to destroy pests. This was first demonstrated in

California where USDA scientists introduced the Coccinellid from Australia to eradicate the cottony cushion scale that had done much damage to the citrus crop. Entomologists waged other wars against such nemeses as the gypsy moth, cotton boll weevil, and San Jose scale.[33] By 1920 thousands of men and women were working as entomologists in the United States and Canada, compared to a relative few in 1860. Many of them labored in the powerful USDA's Bureau of Entomology. Besides killing bugs, they were producing an extensive literature, sharing research, and offering commentary on problems faced by agriculturalists.[34]

Plant pathology studies the cause, diagnosis, treatment, and prevention of disease, and by 1900 it had started to bloom as both a pure and an applied science.[35] This science required a thorough knowledge of plant physiology and the environment where plants live. Research done in the 1870s and published in 1881 by Professor Thomas J. Burrill of the Industrial University of Illinois, a land-grant institution, made one of the first contributions to the identification of plant disease by an American botanist. He showed that bacteria caused the highly destructive "fire blight of the pear and the twig blight of the apple."[36] During the last third of the nineteenth century scientists studied fungi, bacteria, and viruses as causes of plant diseases.[37] J. C. Arthur undertook the first "systematic attempt" to study plant disease at an experiment station after he became botanist at the New York State Station in Geneva in 1884. When the USDA placed F. Lamson-Scribner in charge of building a plant pathology program, it set in motion a work that expanded in the decades ahead. He linked laboratory experiments together with treatment in his early work with grape disease.[38] By 1900 researchers working at land-grant colleges, experiment stations, and the Department of Agriculture had created the framework for plant pathology to make profound contributions to the growth of American agriculture.[39]

Soil science, too, emerged as "an independent natural science" during the first decades of the twentieth century. Agricultural chemistry, particularly Liebig's work in the 1840s, had given impetus to the study of soil, its composition, and fertility. As the nineteenth century progressed, scientists devoted attention to fertilizers, the role played by climate and natural vegetation in soil formation, and the geological attributes of soil. Disappointments over practical applications of experiments in the chemistry laboratory led to experimentation with the soil itself. This led to the conclusion in the late nineteenth century that though "chemical analysis of soils might be instructive for scientific research," they had "no use in practical agriculture."[40] Soil scientists expanded their research interests into "the physical conditions and peculiarities of soils," addressing such concerns as drainage, irrigation, temperature, moisture content, and texture, all of which led to making soil maps of agricultural districts designed to assist farmers in determining which crops to plant in their fields.[41] By 1930 soil scientists expressed such optimism in the potential of their discipline that they believed "in the possibility of resuscitating even a dead soil," and perhaps even being able to "treat dead soils and start them on a brand-new evolutionary course from youth to old age."[42]

Other researchers worked to determine causes of animal diseases and to find treatment for contagions that could destroy large numbers of cattle, swine, sheep, or poultry. As was usually the case, Europeans already had done much work before Americans turned their attention to this scientific question. Scientists in France had expended considerable energy trying to diagnose and treat glanders in the second half of the eighteenth century. Edward Jenner's work with cow pox led to the discovery of the serum used to vaccinate humans against smallpox. Research into the cause and spread of anthrax helped to bring about acceptance of the "germ theory" of disease, which enabled scientists to better understand animal diseases.[43]

After the United States Department of Agriculture created the Bureau of Animal Industry in 1884, American researchers focused their efforts to control and eradicate animal disease. Supported by "the necessary regulatory powers to enforce ordinances based upon its scientific findings" the bureau's work proved effective.[44] It achieved a spectacular success when it discovered that cattle ticks carried the parasite that caused Texas Fever to uninfected animals, which led to the use of quarantine and dipping to control the disease.[45] As the nineteenth century ended, the bureau had eliminated pleuropneumonia and had reined in the effects of hog cholera.[46] By 1924 the bureau employed almost 4,500 people to carry out a long list of duties that included "scientific research, education, relieving animal suffering, guarding the public health, promoting the livestock industry, protecting property, stimulating domestic commerce in animals and their

products, and preserving and extending export trade."[47] One important task for the bureau was to inspect meat. Animal science included
a wide range of disciplines including physiology, medicine, pathology,
and food technology.[48]

Veterinary medicine evolved in conjunction with the emergence
of animal science. The first veterinarians in the United States, at the
beginning of the nineteenth century, were graduates of European
schools. By 1850 private schools of veterinary medicine had opened
in Boston, New York, and Philadelphia. These institutions and other
urban veterinary schools educated men to care for the large number
of horses employed to transport people and goods in growing American cities. At the close of World War I, ten agricultural colleges had
started schools of veterinary medicine, training people to meet the
needs of agriculture. By the same time many of the private schools
had closed as machine power replaced horse power.[49] In 1879 Iowa
Agricultural College became the first land-grant institution to begin
operating a veterinary school. Michigan Agricultural College launched
its school in 1909, though Edward A. A. Grange had become the first
professor of veterinary science at the college years before, in 1883.[50]
The Hatch Act of 1887 promoted the growth of veterinary science,
because many of the experiment stations hired veterinarians. Further
impetus for the growth of veterinary medicine came from the organization of the United States Veterinary Medical Association in 1863
(which became the American Veterinary Medical Association in 1898)
and state and local organizations, all of which enabled veterinarians
to exert political influence to create boards that regulated the profession. The invention of the refrigerated rail car revolutionized the meat
industry, as it allowed packers to ship dressed beef across the country.
This, in turn, created the need for adequate inspection of meat and
processing facilities. Federal legislation in 1890 and 1906 committed
the U.S. government to a program of meat inspection, a task that initially fell to veterinarians.[51] By 1920 veterinarians worked to make
sure the nation's milk, meat, and animal foods were pure, built hospitals to care for ailing animals, and continued to battle contagious diseases among farmers' livestock.[52]

The expansion of dairying in the United States demonstrates
how practical concerns and scientific research came together in the
growth and study of livestock industries between the 1860s and

1930. Farmers spent considerable effort to improve the quality of their herds through the introduction of shorthorn breeds, including Ayrshires, Holstein-Friesians, Jerseys, Guernseys, and Brown Swiss. Improved breeding increased the milk and butterfat production of cows, which yielded larger profits for farmers. The invention of the cream separator and the milking machine (in 1902) enabled dairy farmers to sell their products for higher prices and, over time, reduce the amount of labor required to bring milk to the market.[53] Farmers formed cooperative cow testing associations and bull associations in order to determine the productivity of their cows and to pool the use of purebred bulls. Scientific research provided the nutritionally rich hay and developed the commercially produced feeds that made cows healthier and more productive.[54] In 1895 the Dairy Division of the USDA came into being. Although the division did not engage in laboratory research until 1902, by 1929 it was said that the division, which was by then named the Bureau of Dairy Industry, was "95 per cent a research organization."[55] All of these changes, however, had not altered the underlying principle governing the practice of dairying in the United States. In 1926, T. R. Pirtle, wrote: "All dairy herd improvement in the past may be traced to three things: selection, breeding, and feeding. All dairy herd improvement in the future must be based also on selection, breeding, and feeding."[56]

Forestry in America took form as a distinct discipline between 1875 and 1925. In 1876 Franklin B. Hough went to work for the USDA "to study the annual timber consumption, the supply to meet future wants, the influence of forests on climate, and the measures suitable for the preservation and restoration of forests." In the 1890s the U.S. government set aside forty million acres of forest reserves, making it possible to conduct scientific research and to put the findings into operation. Lack of trained foresters, however, prevented much advancement in the practice of scientific forestry at the time. Gifford Pinchot, one of only a few trained foresters in the country, provided dynamic leadership in the forestry movement after he was appointed head of the USDA's Division of Forestry in 1898. To meet the need for competent workers, Cornell University opened a school of forestry in 1898 and Yale followed in 1900. The profession received another boost with the formation of the Society of American Foresters in 1900. Perhaps even more important, President Theodore Roosevelt's commitment to conservation encouraged people to use the nation's natural resources wisely. For example, Roosevelt's establishment of the Inland Waterways Commission in 1907 drew attention to the relationship between forest cover and the flow of water at the heads of navigable streams.[57] As foresters tried to control fire, disease, and insects, they created the major components of forest protection.[58] Only through scientific research could they devise meaningful plans to protect and use the forests. Initially, botanists and horticulturists at state universities and colleges performed forestry related research, but in 1915 the Branch of Research in the U.S. Forest Service came into being. As a result, forestry research gained prestige that stimulated further scientific investigation.[59] Good forest management became a national concern as lumbermen thinned the trees first in the East, then the South, the Great Lakes, and the Northwest. As supplies for the future became more uncertain, private entrepreneurs showed increased interest in scientific forest management. People trained in forestry worked for companies to identify species and grades of wood and to oversee the harvesting of trees and the manufacture of wood products.[60]

THE AGRICULTURAL COURSE

Innovative teaching and scientific experimentation in combination shaped an agricultural course of study at Michigan Agricultural College that equipped students to become competent researchers and effective teachers. The discovery of new knowledge alone was never considered enough to fulfill completely the institution's mission. The faculty fashioned term-length courses that transformed both existing and new knowledge, generated at M.A.C. and elsewhere, into organized units of study. Professors W. J. Beal, A. J. Cook, and R. C. Kedzie and other instructors required students to observe plants, insects, and chemical reactions in the laboratory. Students found themselves looking through microscopes and dissecting plants and animals, and drawing and writing down what they had seen. They learned the substance of horticulture, entomology, botany, and other sciences, and they acquired the tools needed to carry out research themselves. Just as important, they found out how to tell others how to use knowledge

Farm buildings and farm lands made up one of the most visible and important physical features of Michigan's land-grant college. This photograph appeared in "College Farm Buildings," a bulletin of the Experiment Station published in 1908. It shows the reconfiguration of the buildings that occurred after some structures had been moved or remodeled. A new horse barn (the big barn in the middle of the left side) was built and an extension was added to the sheep barn (the long building to the right of the horse barn with two shades of shingles). This cluster of buildings is located south of Agriculture Hall, west of Farm Lane, and north of the Red Cedar River. Courtesy of Michigan State University Archives and Historical Collections (Michigan State University, Buildings, Farm Buildings, Folder 2).

to bring about change—hopefully to improve the way people did things. Innovations in teaching and scientific research at M.A.C. did not occur in a vacuum. M.A.C. professors communicated frequently with colleagues at other institutions. They shared with and borrowed from each other in an effort to make teaching and learning more effective throughout higher education in the United States.

The agricultural course formed the heart of the academic program at Michigan Agricultural College, and putting it together made up a critical phase in the experiment in higher education that was occurring at the institution. The courses of study expanded to meet the changing needs of American agriculture, the growing sophistication of individual scientific disciplines, and the preparation needed by students to take the new jobs being created by the mushrooming American economy.

The fact that new knowledge in the sciences is discovered in the laboratory and the field rather than gleaned from the printed page led professors at M.A.C. to expand and vary their teaching methods even as they relied on traditional technologies. In the 1860s textbooks were the foundation of instruction, especially in natural philosophy (or

The construction of Agriculture Hall in 1909 left no doubt that the study of agriculture was central to the mission of Michigan Agricultural College. "Ag. Hall" dwarfed its neighboring building (Albert J. Cook Hall), built in 1889, which had functioned as the college's first agricultural laboratory. Ag. Hall's massive structure symbolized the college's entrance into the twentieth century and its embrace of an agricultural industry that extended far beyond practical farming. The building also helped to keep M.A.C. in the forefront of agricultural colleges. In November 1910 Eugene Davenport, '78, the dean of the agricultural school at the University of Illinois, took some of his colleagues on a tour of agricultural buildings looking for ideas to help them design a new structure at Illinois. They found Ag. Hall to be the best facility among those they visited. Courtesy of Michigan State University Archives and Historical Collections (Michigan State University, Buildings, Albums).

physics), geology, entomology, and botany. Professors required students to master textbooks and then to tell the class, while standing, what they knew. While this provided little opportunity to communicate knowledge through writing, it did teach students how to think on their feet. Practical experience did augment other courses. Chemistry students carried out experiments in the laboratory after their daily lecture from Professor Robert C. Kedzie or their recitation in the chapel. Students applied the principles learned in their surveying textbook as they used their surveyor's transits, compasses, and levels in the field. Professor Manly Miles's lectures to his zoology students went far beyond the information offered in the textbook. Miles, however, did not make use of the specimens in the museum or direct experiments in the laboratory. Professor Albert N. Prentiss required his botany students to dissect plants and to use microscopes in order to identify their names, and locate specimens in the field. Such was the beginning of laboratory work at M.A.C.[61]

Professors William J. Beal and Eugene Davenport offered insights into the State Agricultural College that summed up challenges faced by M.A.C. during its first two decades and that foreshadowed the future. Both men recalled the college in its infancy and the uncertain place of science in higher education and in agriculture as M.A.C. grew. When he prepared to write his history of the institution published in 1915, Beal reminisced that when he first came to the school in 1870:

The College was in the woods, so to speak, with no model to follow. Nowhere in this broad country were students yet taught advanced stock judging, stock feeding, the examination of dressed meats, soil physics, dairying, plant breeding, plant histology, ecology, plant physiology, farm economics, the growing of forest trees, spraying for insects and fungi. Bacteriology as related to animals, dairying, soils, and plants was a sealed book.[62]

Davenport, writing in 1939, expressed the belief that M.A.C.'s real service was "not so much along the line of what we now call agriculture as it was in the *development of science* which had *no academic standing at the time.* College work meant languages and literature with a modicum of philosophy." He captured the essence of why the college's agricultural course generated so much tension between the scientific and the practical when he said it "was not popular with the academic world who looked with prejudice against the ungodly thing known as science. Nor was it popular with the farmers who had no need for 'book farming' nor confidence in 'college fellers!'"[63]

Beal played a critical role in leading students out of scientific textbooks and into the laboratory at Michigan Agricultural College, thereby influencing the study of science beyond the banks of the Red Cedar. As early as 1872, Professor Beal taught his students how to use microscopes to examine a plant specimen, and to describe and draw what they saw.[64] Ever conscious of the need to expand his own knowledge, Beal kept in touch with leading botanists in the world. He relied upon Harvard's Asa Gray for advice when buying books for his research library, and he corresponded with Charles Darwin, who told him that he was "glad" that he intended "to experimentize."[65] When Beal planned for the botany building to be built in 1880, he sought out the advice of M.A.C. alumni. A. N. Prentiss sent him the plan for his botanical facilities at Cornell. Prentiss emphasized that the laboratory was where students studied and examined things for themselves.[66] Charles Bessey, '69, provided further ammunition in favor of the need of laboratories when he told Beal: "A college which proposes to keep up with the current must provide Botanical and Zoological laboratories."[67] Bessey's admonition to build up-to-date laboratories at his alma mater was an affirmation of the "laboratory method" that he believed M.A.C. should embody.[68] M.A.C. opened the Botanical Laboratory in 1880, the first such facility in the United States. Ten years

later it burned down, and a new structure (now known as Old Botany) took its place in 1892.

In 1880, professor Beal's *The New Botany* was published. In this lecture he articulated the hard kernel of land-grant education's commitment to practical knowledge. He made a distinction between teaching and training: "Teaching communicates ideas; training forms habit." Beal trained his students more than he taught them, which prepared them to become "independent and reliable observers and communicators." Trained students learned how to do things, not just hear about them.[69] The laboratory method was not an easy approach to education; it demanded that students learn to apply their knowledge. It is here where we begin to see why men and women educated at M.A.C. did so well in the world. Professors like Beal stretched their students' minds to get beyond the words and ideas of the printed page to meet the myriad challenges in an America that increasingly relied upon scientific discoveries and their application to everyday life. And Beal's methods influenced others beyond his own campus. For example, Alexander Winchell, professor of geology at the University of Michigan, was so impressed with Beal's teaching methods that he

This photograph, circa 1896, reveals the evolution of some of the scientific disciplines that made major contributions to the development of scientific agriculture. It shows the first three buildings on Laboratory Row and other important structures. (*Left to right*) the Horticultural Laboratory (Eustace-Cole Hall), designed by Liberty Hyde Bailey and constructed in 1888; the Botanical Laboratory (Old Botany), devised by W. J. Beal and built in 1892 to replace the facility that burned down in 1890; a house put up in 1869 and occupied by the foreman of the farm; the first Agricultural Laboratory (Albert J. Cook Hall), erected in 1889; the Library-Museum (Robert S. Linton Hall), constructed in 1881; the first Veterinary Laboratory, built in 1885; and Williams Hall, a dormitory that opened in 1869 and burned in 1919. Courtesy of Michigan State University Archives and Historical Collections (Michigan State University Buildings).

applied them to geology in his short work "Geological Excursions."[70] Beal had taught zoology and physiology in like manner.

Experimentation at M.A.C. received a boost with the establishment of the experiment station following the passage of the Hatch Act in 1887. Professors who had been conducting scientific research at the college for years now conducted their experiments under the auspices of the station. The president of M.A.C. became the director, and some department heads led "corresponding departments of the experiment station," thereby weaving the expertise of the college into the fabric of the station. Professors used their experiments and those of their colleagues to instruct their students and, in some instances, called upon students to assist in the research, which helped them to learn skills needed to conduct scientific investigations.[71]

Students, however, demanded that they be taught as well as be trained, as they called for even more comprehensive methods of instruction. Even as the laboratory method worked to transform the teaching of science, the lecture system prevailed. In 1891 the student newspaper, the *Speculum,* pointed out that most instruction in the agricultural course was by lecture, an approach that had both proponents and critics on campus. The editorialist favored the lecture system because it allowed for "a condensation of subject matter" and "the incorporation of the results of the very latest investigations," which enabled a student to become "familiar with the pith of the subject." On the other hand, the lecture system did not require a student to use any "particular mental *discipline*" or encourage further investigation. To remedy this shortcoming, the *Speculum* suggested that students "search authorities" themselves and make their findings part of the lesson. In the process they would become familiar with the literature and work of authorities in the field in order to supplement what they observed in the laboratory. The columnist pointed out that his position was "in harmony with the spirit of laboratory practice in vogue here."[72]

The laboratory method and the lecture system, even when enhanced as proposed by the *Speculum,* often paid insufficient attention to the practical in the minds of farmers, who never tired of castigating the college for not emphasizing the practical. A severe critic of "purely literary studies," Edwin Phelps, a member of the State Board of Agriculture, told President Oscar Clute in 1892 that many farmers

Professor Levi Rawson Taft lecturing to a horticulture class in the old Agricultural Laboratory (Albert J. Cook Hall). This photograph, among others, appeared in the college exhibit at the World Columbian Exposition held in Chicago in 1893.
Courtesy of Michigan State University Archives and Historical Collections (Horticulture, Classroom Scenes, Folder 2).

felt that M.A.C. was drifting "away from that thorough practical Agricultural training of Students." While he favored offering electives in physics, civil engineering, and agricultural engineering, he opposed having courses in German, logic, and "kindred studies." He suggested that subjects worthy of more attention included "Road making, ventilation, Sewerage, Building material, insect & other enemies to the Farmers crops & animals &c. &c." It is worthy of note that Phelps makes no reference to the sciences in the agricultural course, even though he lamented that "many of the graduates were called to lucrative positions as teachers in other Colleges."[73] Phelps failed to grasp that scientific research provided grist for instruction in practical agriculture.

Farmers' criticism notwithstanding, M.A.C. did not ignore practical agriculture, but the college's efforts to address this issue reveal the ambiguity in the term "practical." The college's concept of practical was informed by science to a much larger degree than desired by many farmers. State law had mandated that students at the agricultural college work three hours a day. The fundamental assumption of the manual labor system was that work in fields, orchards, gardens, and barns would give students practical experience in farming. During the 1880s and 1890s students debated the merits of the system, which they often found lacking. Clearly, most students found little

Alfred G. Gulley, '68, oversees a horti-
culture class in the gardens in 1892.
Courtesy of Michigan State University Archives and
Historical Collections (Horticulture, Field Scenes,
Folder 1).

benefit, except for their pay and physical exercise, when called upon
to perform such mindless tasks as clearing land. Some recognized
value to work that would "yield to study and experiment."[74] But many
found that the faculty did not provide adequate instruction in how to
use tools or how to perform required work. If students were to learn
practical agriculture, they had to do more than shovel dirt or rake
lawns.[75] In 1892 the State Board of Agriculture revamped the system
so that each student in the agricultural course, working under the
guidance of his professor or his assistants, would work on an experiment for which he was responsible and would report the results.[76]
Two years later, Guy L. Stewart, '95, complained that nothing had
really changed, when he charged that he had been assigned to use a
hoe, rake, pruning knife, etc., but that students were never "allowed to
plough, or harrow, to drive a horse except on a cultivator." Stewart
hoped that the day of "elective work" would soon arrive.[77] It is not possible to know how many students shared Stewart's disgruntlement,
but not everyone's experience was as distasteful.

    Samuel Johnson's ten-year tenure as professor of practical agricul-
ture (1879–89) manifested the intertwined tensions between perceived differences in the nature, value, and use of knowledge that
plagued M.A.C. The farmers' awareness that the faculty taught science
over practicality was reinforced by student desire to learn science
rather than practicality, particularly as it was envisioned by Professor

A collection of horse drawn farm equipment was stored in the "Tool Barn." The college included this photograph in its exhibit at the World's Columbian Exposition in Chicago in 1893. Courtesy of Michigan State University Archives and Historical Collections (Michigan State University, Farm Buildings, Interiors).

Johnson. Johnson was a good farmer, but he knew little science. He won the respect of farmers for running the college farm efficiently, for building a herd of "pedigree shorthorns," and for his work with silos and stock feeding. His attitude towards student workers on the farm, however, generated much ill will between them and their professor. Johnson viewed the young men as lazy, unproductive workers, and they, in turn, retaliated by causing disturbances in class, rearranging furniture, and asking embarrassing scientific questions that they knew he could not answer. In 1886 eighteen seniors were suspended for refusing to tattle on classmates who soiled Johnson's classroom by putting hydrogen sulfide in the stove. Three years later, resentment towards Johnson reached new heights when the English professor Elias J. MacEwan lost his job for allegedly belittling Johnson. When students protested MacEwan's forced resignation, Johnson blamed professors Beal, Kedzie, and Cook for his troubles—a charge that could not be substantiated, so Johnson was asked to resign.[78] Johnson's successor, Eugene Davenport, would blend together the practical and the scientific in a ways that Johnson never did.

Students, under the supervision of their professors, could help to create subjects by employing techniques that their instructors used to teach them to generate new knowledge. In May 1895 the *Speculum* carried an editorial entitled "Student Experiments at the Agricultural College" that listed some of the experiments performed by students as part of their manual labor requirement and that demonstrates how students carried out practical experiments, which often grew out of the results of scientific research at the experiment station.

> One continued an experiment in the improvement of corn which had been conducted by different students for a series of years; another tried several experiments in the use of salt as a fertilizer; another reported his success with different methods of killing Canada thistles; another worked all summer trying various ways of killing quack grass; another had obtained some curious results from detasseling corn; two others reported their success in using bisulphide of carbon for killing woodchucks; one performed some interesting experiments on the cross-fertilization of corn; two cultivated several kinds of roots on different soils to see the effect on yield, quality, habit of growth, etc.; one cut repeatedly the grass upon two square rods of meadow and compared its weight when dry with the yield of hay from a similar plot cut but once; twelve students carried out various lines of treatment for wheat smut; another set was similarly engaged upon oat smut, while six others tried various remedies for corn smut; two students had charge of a half acre collection of the less known agricultural plants known as the "curiosity strip"; one reported on the application of various fertilizers to wheat. The students meet occasionally in the class room to talk over their experiments with each other and with the professors in charge. Much enthusiasm is manifested in this work, which is considered a valuable part of the college course.[79]

Many farmers might have had difficulty understanding how such activity would make a young man a better farmer. Student participation in experimentation seemed to practical farmers to contribute to what they saw as the subversive character of the agricultural course. Nevertheless, performing tasks such as these stimulated the imaginations of students in ways that encouraged them to learn even more and to make more significant uses of their skills and knowledge than they could on the family farm. Edwin Phelps had good reason to fear that many graduates were headed away from the farm and towards teaching at other colleges.

Later in 1895 M.A.C. introduced a stronger dose of practical instruction into the agricultural course when it mandated that all manual labor be educational and performed without pay. The freshmen worked in the horticultural department and the sophomores worked

in the farm department. Clinton D. Smith, professor of practical agriculture, acknowledged that many students in the past could not "perform with skill a large part of the ordinary farm operations," but he
intended to improve this situation. During the first term, the sophomores would be divided into four sections, with each group focusing
on each of four areas for three weeks: dairy work, stock judging, farm
machinery, and general farm work. All students were expected to pass
examinations along the way. Farm youth who demonstrated their com-
petence in running farm machinery, caring for horses, and plowing
were not required to enroll in all segments of this program. In subsequent terms, students learned how to manage a farm, "including planting the crops and caring for and harvesting them."[80]

Michigan Agricultural College attempted to attract students who
were not academically prepared for college classes by reinstating the
preparatory course in 1899. The Edwards, Smith, and Kedzie report of
1896 cited reasons for the dearth of students attending M.A.C. and
proposed a remedy. High school teachers trained at the University of
Michigan directed their graduates to Ann Arbor, while district school
teachers educated at the Michigan State Normal School encouraged
their graduates to go to Ypsilanti. Furthermore, students who had only
a district school education were not adequately prepared for M.A.C.'s
academic rigor.[81] Many communities established high schools after
the Civil War that enabled more young men and women to receive
foundational work in English, algebra, geometry, literature, U.S. and
English history, chemistry, botany, zoology, French, German, and even
Latin.[82] Some students who came to M.A.C. with deficiencies tried to
make up work, but few who went to high schools where nothing
related to agriculture was taught ever enrolled at M.A.C. The report
recommended that a preparatory course be established that would
enable M.A.C. to adopt higher entrance requirements and academic
standards.[83] Such a course was offered for engineering students in
1899, and three years later in agriculture and home economics. These
courses attracted many students: 184 (out of 917) in 1903, 140 (out of
1,001) in 1906, and 188 (out of 1,494) in 1909. Until 1909 Michigan
law required that the college admit all students who had finished the
district school. After the law was changed, students had to have completed the tenth grade before they could enter the sub-freshman class,
which was eliminated in 1914. By enrolling students who lacked the

SHORT COURSE MEN, WELCOME.

In winter, young men came from the farms of Michigan to enroll in the short courses designed to improve their practical skills in farming and agricultural industries. *The Holcad* welcomed them in 1913 with this sketch of a timid youth, who just got off the train, about to enter a new world of learning and social opportunities. Courtesy of Michigan State University Archives and Historical Collections (*The Holcad* 5, no. 13 [13 January 1913]: 11).

academic preparation needed for college level classes in the preparatory course, M.A.C. enabled its professors to teach the freshmen, all of whom were high school graduates, at a level that was commensurate with classes taught at other colleges.[84]

By the mid 1890s the State Agricultural College came to the realization that most instruction in practical agriculture as demanded by farmers could not be adequately met by offerings in the regular four-year course. Recitations, lectures, laboratory work, and manual labor all failed to satisfy the practical farmers. Few farmers were able to leave their farms for four years, and when they sent their sons to M.A.C., many of them never returned. The college learned over the years that if it hoped to meet the needs of farmers, as the farmers themselves perceived their needs, it had to take its messages directly to them in their own communities. The farmers' institutes, which started in the 1870s, were the first step in this direction. (The college's outreach programs are discussed in more depth in chapter 4.) The land-grant philosophy also enabled M.A.C. and other agricultural colleges to bring young farmers who couldn't make the four-year commitment to the campus for practical instruction through such programs as the short course.[85]

The genius of the eight-week short course was that its simplicity made it possible for people who had the greatest desire for practical instruction to come to the campus to receive knowledge from the professors and use its facilities. Professor Smith first taught a six-week course in dairying to seventeen students in 1894, and he became the architect and champion of the short course program up until he left M.A.C. in 1908.[86] The Edwards, Smith, and Kedzie report in 1896 called for the establishment of short courses for young farmers "who desire to perfect themselves in a special line of farming, for which they themselves and their farms are best adapted."[87] In the early years, course offerings included fruit culture, livestock, creamery, cheese making, beet sugar production, and dairy husbandry, and a general agriculture course, which grew to include first- and second-year sixteen-week sessions. Short-course offerings were never static, and they kept pace with changing times, as programs appeared in forestry, poultry, housekeeping, trucks and tractors, beekeeping, gardening, cow testing and barn management, and ice cream making. Students learned through lectures, demonstrations, and hands-on activities, but they also "received

Short course students pose in front of a collection of tractors in 1922. Michigan Agricultural College faced a never-ending challenge to keep its curriculum in step with changing technology and to teach people to apply new technology in their work and personal lives. Farmers who used tractors to cultivate their fields were able to increase their productivity, but they needed to learn how to operate and maintain their new machines. In 1917, to help meet this need, M.A.C. introduced a two-week short course in tractors that attracted 130 students. In the background, the agricultural pavilion can be seen extending out at the rear of Agriculture Hall. The Dairy Building (1913) appears in the upper right-hand corner. Courtesy of Michigan State University Museum (Acc. No. 2003:152.7.18.2).

an insight into scientific methods and agricultural training." The description of the creamery course for winter 1909 illustrates the contents of a short course:

> Daily program includes 1 hour Business Methods, 1 hour Bacteriology, 1 hour Chemistry and Physics of Milk, 1 hour Grading Gathered Cream, Butter Judging, Creamery Principles, etc. The entire afternoon is devoted to practical work in the butter room or testing milk and cream or in studying creamery mechanics.[88]

Even in the short course, the professors refused to exclude science from its practical application. The short courses were immensely successful, for nearly seven thousand men and women had taken one by 1925. Michigan Agricultural College educated practical agriculturalists after all.

The adoption of the short courses reflected the trend in the curriculum towards specialization and offering more technical classes. The growth of knowledge in most academic disciplines, especially the scientific, converged with the needs of students to prepare for an expanding list of occupations, which forced the diversification of courses of study available to students.[89] Consequently, M.A.C. established its mechanical course in 1885, the women's course in 1896, forestry in 1909, and veterinary science in 1910. Departments of instruction expanded their subject offerings, professors' interests shaped their research, and students in all courses of study drew upon the classes in the departments to fulfill requirements and choose electives needed to complete their majors.

The level of knowledge in the sciences and liberal subjects grew at a fast pace, and the faculty kept revising and reorganizing the courses of study. A comparison of the 1896–97 catalog with the 1925–26 catalog shows how the agricultural course and the curriculum as a whole became specialized. By 1925 both the general agricultural course and the horticultural course were focused on the practical, heavier concentrations of science fell under the applied science course, and a doctoral program was offered in seven sciences related to agriculture. During the college's early years, the course of study was completely prescribed, and in 1883 seniors were allowed to elect some classes. But by 1896 there was a major departure whereby students in the agricultural course took required subjects until the middle of their junior year, after which they could choose from classes in "stock and grain farming, dairying, vegetable culture, pomology, floriculture and poultry raising—the lines of work likely to be most useful to them."[90] Twenty-eight departments of instruction appear in the 1896–97 catalogue (see figure 3), but the description of courses reveals that not all classes had crystallized into full-term entities. Horticulture, for example, required two and a half terms of practical instruction in vegetable gardening, landscape gardening, and fruit culture. An additional four and a half terms of elective studies in the above subjects plus floriculture were available for "those who expect to make a specialty of horticulture." In addition horticulture students performed manual labor to gain practical experience. By 1925 the horticulture department offered twenty-two full-term courses, including three-term sequences in commercial floriculture, herbaceous crops, and advanced pomology. Students enrolled in the agricultural course began to specialize in their sophomore year in general agriculture, horticulture, forestry, landscape architecture, agricultural economics, or agricultural engineering. They could then specialize even further in dairy husbandry, animal husbandry, poultry husbandry, soils, farm crops, or apiculture.[91] Thirty-five departments offered hundreds of courses in 1925 (see figure 4). (See appendix 1 for a listing of the names of courses offered in each department.)

Figure 3. Departments of Instruction Listed in the 1896–97
Catalogue of Michigan State Agricultural College

Agriculture	German and French
Anatomy and Physiology	History and Political Science
Astronomy	Horticulture
Botany	Mathematics
Chemistry	Meteorology
Domestic Economy	Military Science and Tactics
Drawing and Painting	Music, Vocal
Engineering, Civil	Physics
Engineering, Electrical	Psychology and Ethics
Engineering, Mechanical	Shop Practice
English Language and Literature	Special [Short] Courses
Farmers' Institutes	Veterinary Science
The Farm-Home Reading Circle	Women's Course
Geology	Zoology

Robert Sidey Shaw occupied key leadership positions in the college for four decades. He was a native of Ontario, Canada, and had graduated from Ontario Agricultural College. Shaw came to Michigan Agricultural College in 1902 from Montana State College to become professor of agriculture and superintendent of the farm. In 1907 the State Board of Agriculture appointed him dean of agriculture, and the next year he took on additional responsibilities as director of the Experiment Station. He assumed the presidency of the college in 1928 and guided it through the difficult years of the Great Depression until his retirement in 1941. Courtesy of Michigan State University Archives and Historical Collections (People, Shaw, Robert S., no. 2460).

The ever-changing courses of study required the judicious leadership that Robert Shaw would give to the agricultural program for many years. He joined the faculty as professor of agriculture in 1902, served as dean from 1907 until 1928, acted as president in 1921–22, 1923, and 1928, and in 1928 became president of Michigan State College. Shaw encouraged the growth of the livestock program, made marketing an integral part of the curriculum, and contributed to the organization and growth of extension. He established good working relationships with farmers in the state, especially with livestock breeders' associations, which provided effective lobbyists for the college with the legislature.[92]

The principles that research gives rise to knowledge and that teaching transforms that knowledge into practical application in the world drove the agriculture course. The early study of entomology at Michigan Agricultural College shows these principles in action. Entomology first appeared as a "department" in 1865. The description of the course in 1866 spelled out a purpose and methodology that anticipated the evolution of the importance that entomology would have for agriculture in Michigan and the United States:

Figure 4. Departments of Instruction Listed in the 1925–26
Catalog of the Michigan State College of Agriculture and
Applied Science

Agricultural Engineering	Home Economics
Anatomy	Horticulture
Animal Husbandry	Landscape Architecture
Animal Pathology	Mathematics
Bacteriology and Hygiene	Mechanical Engineering
Botany	Meteorology
Chemistry	Military Science
Civil Engineering	Music
Dairy Husbandry	Physics
Drawing and Design	Physical Education
Economics	Physiology and Pharmacology
Education	Poultry Husbandry
Electrical Engineering	Religious Education
English and Modern Languages	Sociology
Entomology	Soils
Farm Crops	Veterinary Medicine, Surgery, and Clinic
Forestry	Zoology and Geology
History and Political Science	Winter Short Courses

Entomology.—The course in Entomology is illustrated by a valuable collection of native and exotic insects. Particular attention is given to the study of species injurious to vegetation; and the best methods of checking their ravages are discussed. Students by collecting and preserving specimens of our native species, become familiar with their habits in their several stages of development.[93]

A. J. Cook, who served as the professor of entomology from 1868 until 1893, had his students dissecting insects under microscopes, drawing their features, and learning their habits and life cycles. Cook focused on those insects that destroyed plants, and he sought ways to eradicate them.[94]

Cook's research into finding chemical remedies for insect pests received national recognition before the establishment of the experiment station under the provisions of the Hatch Act of 1887. His efforts to determine the potency of the arsenic poison Paris green, when applied to fruit trees to kill the codling moth, contributed to a national debate over the chemical's effectiveness and the dangers that it posed for humans. Cook made significant contributions to the body of research relative to Paris green that resulted in it becoming the "standard remedy against nearly all mandibulate or gnawing insects" by 1887. He also looked for ways to create emulsified kerosene as a treatment for sucking insects.[95] Cook's students knew how to identify and kill insect pests as a result of their professor's research, and they learned how to help producers employ the techniques developed by him.

Chemistry always formed an integral part of the agricultural course, and Professor Robert C. Kedzie made chemical research relevant to Michigan agriculture during his long career at M.A.C., which spanned the years from 1863 until 1902. The first chemistry laboratory in College Hall was state-of-the-art, but the needs of the department soon outgrew its facilities.[96] In 1871 the new Chemical Laboratory building opened with a lecture hall seating 80 and an "analytical room fitted with evaporating hoods and tables for 48 students." Kedzie had his own laboratory and "rooms for researches in higher chemistry."[97] Subsequently, new rooms were added in 1882 and 1911.

Kedzie's experiments benefited Michigan farmers and the public greatly. He devoted much time and energy to studying fertilizers and animal manures. At Kedzie's urging the legislature passed an act that required manufacturers of commercial fertilizers to submit samples to be tested for their composition and that the results be published. This measure helped to prevent fraudulent claims made by producers of fertilizers and enabled farmers to purchase fertilizers that actually increased their crop yields. Kedzie's research gave birth to the sugar beet industry in Michigan and provided useful information regarding the qualities of Clawson wheat for Michigan farmers.[98] Kedzie's work also revealed that arsenic in wallpaper could kill people and that kerosene when improperly mixed with oil could explode. These discoveries led to legislative reforms in the manufacture of wallpaper and kerosene. Kedzie taught his students how to apply practical chemical concepts to agricultural problems faced by farmers in the field and to demonstrate to farmers that

Robert C. Kedzie, professor of chemistry, 20 March 1900. For thirty-nine years (1863–1902) Kedzie, a physician who graduated with the first medical class at the University of Michigan in 1851, set the standard for a professor at the land-grant college. He excelled as a teacher, a researcher, and a public servant. Active in many organizations, Kedzie served as president of the Michigan State Board of Health (1877–81), the Michigan Medical Society (1874), the American Public Health Association (1882), and the Association of Agricultural Colleges and Experiment Stations (1899).

Burt Wermuth, '02, had the good fortune to be a student in Kedzie's last meteorology class in the spring of 1902. In 1929, Wermuth, then the editor of *Michigan Farmer*, remembered the impression that Robert Kedzie had made upon him in a letter to Frank Kedzie: "Besides his remarkable grasp of facts and his wonderful personality, he was a splendid teacher" (Burt Wermuth to Frank S. Kedzie, 20 March 1929, Frank S. Kedzie Papers, UA 2.1.8, Box 894, Folder 18). Scores of others could have expressed similar sentiments.

Courtesy of Michigan State University Archives and Historical Collections (People, Kedzie, Robert C., no. 1625).

The Chemical Laboratory, which opened in 1871, was the first structure built by M.A.C. devoted to a single scientific discipline. Professor Robert C. Kedzie planned it after he and President Abbot traveled throughout the United States and Canada, during fall 1869, to tour similar institutions in order to ensure that it would "combine the conveniences of all." William S. Holdsworth, '78, instructor in drawing, designed an addition that was built in 1882 on the south end (*far end in photo*, circa 1888) of the structure. W. J. Beal observed that because the white brick edifice was of "moderate height" and had a flat roof, people referred to it as the "Chemical Fort." Courtesy of Michigan State University Archives and Historical Collections (Michigan State University, Buildings, Physics and Electrical Engineering, Folder 3).

work done at the State Agricultural College could make a difference in their businesses.[99]

The Michigan Agricultural College devoted considerable energy to the study of the care and management of farm animals and products generated by them. The development of the dairying program, which formed the heart of this segment of the agricultural course, depended upon a quality herd of purebred animals. Manly Miles introduced dairying to the campus when he purchased some Ayrshires 1867, but the early herd was not impressive. It appears that Samuel Johnson, professor of practical agriculture, bought the first registered cow, a Friesien, in 1881. After his arrival in 1893, Clinton Smith built up the purebred herd, especially Holsteins, through experimental feeding and breeding. As the program expanded, so did the need for new space. Initially, dairy classes were taught in the basement of the old Agricultural Laboratory building (now Albert C. Cook Hall). They were moved to a new dairy building in 1900. This structure had been built, in part, because of lobbying of the legislature by the Michigan Dairymen's Association "for the erection and equipment of a suitable dairy building" at the agricultural college.[100] In 1913 the department relocated into the new and larger Dairy Building, located at the north end of Farm Lane.[101]

Practical dairying required students to spend time in the barn, the dairy, and the field, as well as in the classroom, the shop, and the

The note on the back of "Belle Sarcastic's" portrait reads: "Belle Sarcastic, owned by Michigan Agricultural College, made a world's record in 1897 of 23,190 lbs. of milk and 722 pounds of fat. This record stood for eleven years. Her son, Sarcastic Lad, was grand champion at the St. Louis Exposition in 1903 and became one of the noted sires of the breed." M.A.C. professors, staff, and faculty took great pride in prize animals owned by the college. Courtesy of Michigan State University Archives and Historical Collections (Michigan State University, Animals, Cattle, Folder 3).

laboratory. In 1896 the Department of Agriculture's curriculum called for students to gain knowledge of soils, fences, crops, accounts, and planning of farm work in addition to acquiring minimal skills in blacksmithing and carpentry. Students spent two hours a day for ten weeks in the barns learning the characteristics of horses, cattle, sheep, swine, and poultry. A knowledge of chemistry was a prerequisite before studying stock feeding during the first six weeks of winter term of the sophomore year. At mid-term, the students moved to the dairy, where they became familiar with creamers, separators, and churns among other equipment. They also learned how to use the Babcock test and the importance of pasteurization and sterilization when processing milk and cream.[102] Dr. Charles E. Marshall came to M.A.C. in 1896, and his research informed his teaching concerning the role played by bacteria in processing milk into the marketable commodities of bottled milk, condensed milk, cheese, cream, ice cream, and butter.

By 1925 a number of departments in the college focused on different aspects of animal industries. Animal husbandry courses covered the breeds of beef cattle, horses, sheep, and swine, in addition to feeding and judging animals and the production of meat. Dairy husbandry taught the production of milk products, judging of dairy cattle and breeds, barn practice, and farm management. Students learned how to cull, feed, house, judge, market, breed, incubate, and board poultry in courses offered by the poultry husbandry department. The

Researchers conducting experiments on crops depended upon seeds for which they knew the provenance. When agricultural scientists at M.A.C. provided farmers with pure seeds of varieties or strains of grains or grasses that had been shown to produce greater yields, they helped farmers to increase their productivity and profits. Every time this happened, the college fulfilled its land-grant mission. In 1892 the Experiment Station converted its old tool room into a "model seed room." This photograph of the new seed room appeared in M.A.C.'s exhibit at the World's Columbian Exposition in Chicago in 1893. Courtesy of Michigan State University Archives and Historical Collections (Michigan State University, Farm Buildings, Interiors).

soils department dealt with the composition and classification of soils, soil physics, fertility, fertilizers, and the modification of soil through tillage, the application of chemicals, and changes in climate and temperature. Students gained a working knowledge of growing crops in the field, cereals, grading grain, marketing, forage crops, plant breeding, the production and care of seeds, and special crops such as potatoes and sugar beets from instructors in the farm crops department. Agricultural engineering professors taught students how to operate and maintain gasoline and steam engines and farm machinery, how to design farm structures to house cattle, poultry, sheep, horses, and swine, and how to drain swamps and clear land.

Under provisions of the Morrill Act, men enrolled at land-grant institutions were to receive instruction in "military tactics." Organized military activities began in 1885, when the army assigned Lieutenant John A. Lockwood to M.A.C. to train the farm boys who

After military instruction became mandatory for male students in 1888, the corps of cadets became a prominent organization in the college. This photograph shows the corps assembled, in the rain, on the steps of the Michigan Capitol in downtown Lansing. Since the men appear in their Civil War blue uniforms, this photograph was taken before 1892, when they began wearing gray uniforms. Notice how their appearance had changed from the picture of first cadet band. Courtesy of Michigan State University Archives and Historical Collections (Michigan State University, Military Science, 1880–1899, Cadets).

The first cadet band, photographed in 1884 or 1885. Courtesy of Michigan State University Archives and Historical Collections (Michigan State University, Music, Bands, Marching Bands).

A couple strolling by the Armory on a late winter or early spring afternoon. The Armory, which opened in 1886, functioned as a center of campus life. It housed the Military Department and provided space for drills, but many other activities also took place in it. M.A.C. athletes played basketball and competed in indoor track and field events in the building. On many occasions, students built elaborate sets to create the proper atmosphere for dances, including J-Hops. Public lectures and other large group meetings also found a home within the Armory's walls. Courtesy of Michigan State University Archives and Historical Collections (Michigan State University, Buildings, Armory).

had volunteered to drill and do calisthenics twice a week. Through the efforts of Lieutenant W. L. Simpson, military science became a requirement in 1888 for all male students until they entered their senior year. Two hundred and twelve men formed the corps in 1889, and an average of 187 of these showed up for each session. Lack of weapons hampered instruction, for M.A.C. owned only 150 rifles and no field guns.[103] When the weather turned foul, the men assembled in the Armory, which had opened in 1885. Each week the corps, dressed in their dark blue military uniforms, marched to martial music provided by the cadet band. The band had been formed ten years earlier through the efforts of students and the faculty, and it had affiliated with the corps of cadets.[104] It seemed that the corps was always short on facilities and equipment. Upon reviewing the cadets favorably in May 1907, Carl A. Wagner, inspector general of the Michigan National Guard, urged President Snyder to build a 500-yard rifle range to give the men a suitable place to practice shooting.[105]

One more observation relative to the agricultural course casts more light on the evolution of the agricultural curriculum. Out of a total of 2,352 undergraduates in 1925, only 394 were enrolled in the agricultural course. That only 17 percent of the student body chose this course reveals that Michigan State College of Agriculture and Applied Science had by then become much more than an agricultural college. Furthermore, its courses of study and the interests of its students mirrored the declining place of agriculture in the social and economic life of the United States. An analysis of the courses offered in 1925 clearly shows that agricultural education at Michigan State College was geared towards preparing its students for agricultural careers that extended far beyond farming. Landscape architects designed parks, boulevards, playgrounds, gardens, and golf courses in small towns and large cities across America. Horticultural graduates owned or managed truck farms, nurseries, fruit farms, floral shops, and gardens, as well as oversaw privately and publicly owned gardens and arboretums. Agricultural business and marketing majors managed cooperatives and commercial enterprises, worked for commodity exchanges, taught school, took jobs with extension services, and bought and sold agricultural commodities. The general agricultural course prepared students to work as county agricultural extension agents who advised farmers on how to improve their livestock, increase the production of their fields, and design and build good structures for their businesses.

THE MECHANICAL/ENGINEERING COURSE

In 1885 the State Board of Agriculture finally approved a mechanical course as called for in the Morrill Act of 1862. The new course grew out of the agricultural course, the spine of M.A.C.'s curriculum. Agricultural students had taken courses in surveying, civil engineering, and mathematics for years, and now new subjects such as analytical mechanics, mechanical laboratory practice, elements of mechanism, and steam engineering and thermodynamics appeared on class schedules for people enrolled in the mechanical course. Both courses required students to take chemistry, English, rhetoric, public speeches, and United States constitutional history and political economy, but the mechanical students took more mathematics, less literature and agricultural science, a lot of mechanical drawing, physics, and shop practice.[106] Students in the mechanical course were being groomed to contribute to the growing industrialization of the United States—to the consternation of the practical farmers.

Teaching the subjects that comprised the mechanical course became a part of the creation of the subjects themselves because the subject matter—the theoretical and the practical—were constantly changing as scientific research going on around the world made new discoveries. Learning how to understand and apply theories, information, and practices that emerged after a student had graduated became as important as what he had learned while on campus. Formal education in the engineering courses taught students to learn through informal ways, as new knowledge and techniques were communicated in books, articles, news reports, and by word of mouth. M.A.C. professors taught their students how to learn, adopt, and apply new methods and technology where they worked for corporations, utilities, factories, and municipalities located in Michigan, the United States, and the world.

Two members of the mechanical class of 1889 articulated the potential and method of the mechanical course. Just as with the agricultural course, paradox prevailed, as the supposed emphasis on the practical often gave way to the theoretical with the result that the successful student turned out to be a good practitioner. William H. Van Dervoort, '89, shared his ideas with the readers of the *College Speculum* in October 1887. He suggested it was natural that if a young man was interested in the application of physical sciences and mathematics "he should be an engineer." Van Dervoort made a distinction between shop, or practical work, and theoretical study, arguing that one needed to know the practical in order to apply the theory. Schools that neglected teaching such subjects as "pattern making, forging, welding, tempering, and the general manipulating of machine and bench tools" turned out "not mechanics to construct, but engineers, rich in theory to design." He stressed that it was essential that a person who designs a machine or a structure know how the parts are made and fitted together. Van Dervoort briefly surveyed some of the engineering marvels of the nineteenth century, including transcontinental railroads, the Erie Canal, the Brooklyn and Niagara bridges, and the telegraph system, to point out the need for more engineers. The mechanical course at M.A.C. promised to prepare him and his classmates to help meet the increasing demands of the country's "producing and manufacturing interests."[107]

At the seventy-fifth anniversary of Michigan State College in 1932, Edward N. Pagelsen, also of the class 1889, identified the key

element of the mechanical course, as he shared his recollections of his education and how it enabled him and others to pursue successfully careers in engineering. He told his audience that forty-three students entered the new program and only six graduated. Pagelsen provides insights into how engineering was taught in its first years at M.A.C. He credited Professor William F. Durand with laying the foundation of the present engineering school and praised the teaching of the brothers Rolla C. Carpenter and Louis G. Carpenter. These three men taught all of the engineering and mathematics courses. Pagelsen said that his four years at the college "resulted in a broad mental training, although the instructions in practical mechanics were sketchy, to say the least." For example, he said he "learned about steam engines from books, for never during my four years was the interior of the small engine which drove the shop tools shown to us." The other equipment

When M.A.C. implemented the mechanical course in 1885, the institution finally fulfilled the provision of the Morrill Act of 1862 that required land-grant colleges to teach "branches of learning" that applied to the "mechanic arts" as well as to agriculture. Students are shown here working on iron lathes in the machine shop in 1893. M.A.C.–trained engineers found employment across the United States and contributed to the industrial growth that fueled the rise of the United States as a world power. Courtesy of Michigan State University Archives and Historical Collections (Michigan State University, Mechanical Engineering, Machine Shops).

in the shop amounted to only two lathes, a planer, a drill, a black-smith's hand forge, and some hand tools. Mechanical course students studying metallurgy got to see metal founding in the Bement Company's foundry in Lansing. Pagelsen said a high school in 1932 gave better instruction in practical work than he received at the college, "But the theoretical instructions in the mechanics of materials, analytical mechanics and thermo-dynamics were of as high order as could be had at any engineering school, thanks to the painstaking efforts of Professors Durand and Carpenter."

Pagelsen put what appeared to be a very rudimentary training in engineering into historical perspective. "Electricity was an infant," and scientists such as "Edison, DeForrest, Brush, Tesla, and Kelvin were still groping in the dark for the laws controlling it." Few engineers understood how newly invented internal combustion engines operated, and cement was so new that he never saw even a sample while at the college. As an example of the faculty's insights, he noted: "Dr. Robert Kedzie proudly exhibited to each class a piece of aluminum weighing a few ounces and predicted that some day this metal would be used extensively."

Looking back, Pagelsen clearly and emphatically identified the strength of M.A.C.–trained engineers:

> While the new discoveries in electricity, metallurgy, and chemistry have wiped out the theories taught the students of forty, yes, even twenty-five years ago, the mental training received by the older alumni now in engineering lines enabled them to discard the incorrect theories and formulae and replace them with those fitting present day thinking and knowledge, as they came along, and these alumni thus keep abreast of the times, and much of this advance is due to the correct thinking and accurate experimentation of these same older alumni of our old school of mechanic arts. While some of the theories taught in those early days may have been incorrect, the students were trained in the fundamental principles of engineering and the allied sciences and they were trained to think correctly and study accurately and scientifically and they became skillful in applying principles to practical affairs. The training received bred a contempt for sham and a hatred for dishonesty. No other class of men, in a like degree stands for such high principles in the administration of public affairs as well as for private honor.

He then rattled off the names of nearly twenty alumni who had made major contributions to the world of engineering including William H. Van Dervoort.[108]

Pagelsen's understanding of the foundational principle of the mechanical or engineering course was not commonly accepted, and college officials repeatedly had to justify or explain the course among themselves and to the public. In 1907, Herman K. Vedder, who had succeeded Rolla Carpenter as professor of mathematics and civil engineering in 1891, pointed out to President Snyder that M.A.C. was not really "turning out engineers" at graduation. (Vedder said the same was true for all other institutions as well.) Rather, he and his colleagues trained people "who know how to do things, who know how to put into practice some of the simpler theories to the end that they are able to earn a living immediately upon graduation." Over time, experience enabled former students to become engineers. In the meantime, however, it remained critical to "dwell on the eminently utilitarian character of the instruction and its capability of being transformed into wages and breadwinning."[109]

While Vedder stressed the practical, Snyder never wavered from his commitment to the theoretical. It was beyond his comprehension that meaningful instruction in technical aspects of engineering was possible without a thorough grounding in the theoretical underpinnings of the craft. Yes, students worked in the shop, where they operated machines and manufactured objects, but they also studied mathematics, science, and English, among other subjects that formed the heart of their courses of study.[110] Michigan Agricultural College would not become a trade school.

Predictably, over the next thirty years M.A.C. had to justify its engineering courses to farmers and others who questioned the need for them. When called upon to defend the teaching of engineering, the wily Snyder emphasized that M.A.C had kept its "engineering work as far as possible to practical lines," unlike the University of Michigan, which offered a much greater diversity of subjects. Whether people liked it or not, the college taught engineering because when the legislature accepted land-grant funds derived from the Morrill Act, it mandated that the State Agricultural College offer a course in mechanic arts.[111] In the face of criticism, M.A.C. argued that its agricultural program had always received much larger appropriations and employed more people to teach and research agricultural subjects than had the mechanical. For example in 1912–13 the agricultural division received an appropriation of $169,410.70 compared to $52,347.44 for the engineering division.[112]

As M.A.C. taught students to design, build, manage, and operate the machines, structures, and technical systems that drove American industry, agriculturists rightly grumbled that it failed to train students to meet engineering needs on the farm. Agricultural engineering required a knowledge and understanding of farm life and its rural surroundings that extended beyond technical skills.[113] Most land-grant colleges, including M.A.C., were not producing engineers interested in agriculture, with the result that little attention was being paid to such concerns as rural water systems, sewage disposal, drainage, farm machinery, and the heating, lighting, and ventilation of farm buildings.[114] Questions arose as to whether agricultural engineering courses should be offered by the engineering or the agricultural division. Since the passage of the Smith-Lever Act in 1914 expanded the role of extension work, the need for agents trained to address rural needs had become acute.[115] As a result, engineering and mechanical classes began to be offered for students enrolled in the agricultural course. The 1918–19 catalog listed a number of electives open to agricultural and forestry students. Under the rubric of farm mechanics appeared farm structures, farm conveniences and construction, power machinery, farm machinery, and farm drainage. In addition, the civil engineering faculty made available courses in agricultural engineering and road construction.[116]

As the mechanical arts course evolved into the engineering course, it grew along the major divisions that were occurring in the discipline across the country. Civil, mining, and mechanical engineering had dominated the nineteenth century, but at the beginning of the twentieth century the nascent disciplines of electrical and chemical engineering became a vital part of the engineering world. In the nineteenth century, Oliver Evans, Charles Talbot Porter, John Edson Sweet, Francis Asbury Pratt, Amos Whitney, and others designed and built high-speed engines, high–steam pressure engines, and machines that manufactured guns, sewing machines, and bicycles, among other things. Electrical engineering came about after William Sturgeon and Joseph Henry made practical electromagnets and Michael Faraday demonstrated magnetic induction that produced an electric current. These discoveries gave rise to the telegraph, cables, arc lighting, telephones, electric street cars, and the design and construction of hydro-electric plants. Scientists used electricity and principles from natural science to invent new devices and adapt them into practical, useful products that were manufactured and marketed. But the process did not stop there, for scientists kept making improvements in their inventions that in turn made their products more useful and often less expensive, and fueled the growth of new industries. Chemical engineers spurred on the growth of a number of industries including paper, rubber, dyestuffs, fertilizers, explosives, drugs, and plastics.[117]

The engineering course had to train people to make their way in the rapidly growing industrial complex in the United States. Specialization soon characterized the engineering courses of study. In 1890 two new departments—civil engineering and mechanical engineering—appeared in the catalog. These were later followed by electrical engineering and chemical engineering. The 1925 catalog opened its description with a long-winded sentence that began by establishing "A knowledge of the fundamental principles of the sciences" as "the basis of engineering practice." Science, not practicality, formed the basis of engineering studies at Michigan State College of Agriculture and Applied Science. Freshmen all followed a prescribed course, and all students in their first two years took "work in drawing, machine design, shop-work, surveying, mathematics, chemistry, physics, and English." During their junior and senior years, students specialized in one of four groups: chemical, civil, electrical, or mechanical engineering. A fifth option was the engineering administration course, where students could substitute classes in economics, banking, industrial relations, industrial management, corporation finance, etc. in place of advanced technical subjects. These students would manage the businesses spawned by scientific engineering.[118]

Although the engineering course fulfilled one of the Morrill Act's requirements for the land-grant college, determined opposition to the program's existence persisted. In 1912 some alumni proposed to change the name of Michigan Agricultural College to Michigan State College. Their effort motivated legislators, who were hostile to the engineering program or disillusioned with President Snyder's management of the school, to try to strangle engineering at M.A.C.

through legislation.[119] Subsequently, on 13 May 1913, the legislature passed a state budget that called for an increase in the mill tax from one-tenth of a mill to one-sixth of a mill. (In 1901 the legislature had levied a tax of one-tenth of a mill "on all taxable property" in Michigan to support M.A.C.) The new law put a limit of $35,000 for expenditures, from all sources—including money from the Morrill Act land grant-fund—that could be used to pay for the costs of the engineering program.[120] College administrators contended that the legislature could not regulate the use of money received from the federal government, and they challenged the constitutionality of the law in the Michigan Supreme Court. In March 1914 M.A.C. exceeded the $35,000 limit, a totally inadequate sum to finance engineering. O. B. Fuller, the auditor general of Michigan, then prevented M.A.C. from drawing any money from the state treasury until the Supreme Court acted. On 29 May 1914, the court ruled in favor of M.A.C. when it struck down the 1913 law.[121]

The power of the belief that Michigan Agricultural College existed primarily, even exclusively, to train practical farmers still prevented some people from accepting the necessity that the college teach engineering. Snyder thought that most farmers were not against the engineering course, but that some individuals who expressed strong negative feelings towards teaching engineering at M.A.C. were particularly influential. He argued that the friends of M.A.C. needed to point out to the public that the institution was "much more than a college for one class of people; that it truly represents the great industrial classes." He also believed that the opposition to the engineering program in the legislature was limited to "a few men [especially Senator Fred Woodworth, '98] who were governed quite largely by personal motives," who did not like it when members of State Board of Agriculture took issue with their opinions relative to the institution.[122]

The editorial that appeared in the *Adrian Daily Telegram* on Monday, 1 June 1914, summed up, quite passionately, the belief held by many farmers (contrary to Snyder's belief) that engineering at M.A.C. was subversive. The editorialist lambasted the engineering department as being worse than a "fifth wheel on a wagon," accusing it of being "more like a third horse pulling in a different direction from the regular team in front." The college existed for "the farmers, farms and farming" and should "draw students from the farms, and . . . give them such an education that they will want to go back to the farms, and make the business of farming more profitable and more popular." The engineering department not only attracted students from cities, but it also encouraged students from farms not to return to the business of their fathers. If a young man wanted to study engineering, he should attend the University of Michigan, which had a "magnificent" engineering school "where every branch of engineering is taught to an army of nearly 1,500 students."[123]

As the *Telegram* stated, some people thought that the solution to the controversy over engineering at M.A.C. would be to simply consolidate all instruction in engineering at the University of Michigan. Snyder answered that proposal in the 1914 *Annual Report* of the State Board of Agriculture with a strong defense of the engineering program that addressed the question "Why is engineering taught in an agricultural college?" As he had said on hundreds of other occasions, when the legislature accepted the land-grants made available by the Morrill Act, it mandated the State Agricultural College to teach the "mechanic arts." He pointed out that at M.A.C. students took what "might be termed industrial engineering courses," as opposed to the "professional engineering courses at the University." Once again Snyder contrasted the practical with the theoretical. He also noted that since both institutions were accommodating a maximum number of engineering students (1,400 in Ann Arbor and 600 in East Lansing), new facilities, at considerable costs, would need to be built at the University of Michigan to accommodate the college's engineering students.[124] It appeared that consolidation would not save money—until fire consumed the Engineering Building (built in 1907) and the adjacent mechanical shops (built in 1885) early on Sunday morning, 5 March 1916.

Quick action by Acting President Frank S. Kedzie and Dean of Engineering George W. Bissell not only rescued engineering from what appeared almost certain demise, but also laid to rest the controversy over the program's existence at M.A.C. In its Tuesday edition, the *Detroit Free Press* reported: "On all sides the prediction was freely made that Sunday's fire had wound up the career of the engineering department."[125] The *Free Press* reporter obviously had not consulted with Kedzie, Bissell, or others at the college. Before the sun had set on

Ransom E. Olds. Courtesy of Michigan State University Archives and Historical Collections (*M.A.C. Record* 21, no. 28 [25 April 1916]: 5).

Ransom E. Olds Hall of Engineering was built in 1916 after a fire destroyed the Engineering Building that had been constructed in 1907. The engineering program at M.A.C. had been quite successful from the time of its inception in 1885. Dean George W. Bissell wrote in the *Holcad* on 20 March 1911 that 87 percent of the graduates from the engineering course through 1907 were "in engineering work." Ransom Olds knew this when he donated $100,000 to replace the ruins of the burnt structure. Olds's gift enabled many more young people to study engineering at the college and sustained Bissell's observation that most M.A.C.–trained engineers looked to pioneering enterprises in civil, mechanical, and electrical engineering. *Courtesy of Michigan State University Archives and Historical Collections (Michigan State University, Buildings, Olds Hall).*

5 March, the faculty had made arrangements for all engineering classes to meet on Monday in makeshift classrooms. Several other institutions, including the Universities of Michigan and Illinois, sent "apparatus" to be used until M.A.C. could replace the equipment lost in the fire—which was nearly everything.[126] Within several days of the fire, Secretary to the Board Addison M. Brown, in conjunction with state officials at the Capitol, worked out a strategy to pay for the reconstruction of the building. While the ruins of the engineering building still smoldered, Kedzie played his ace. He told board member John W. Beaumont:

> Confidentially I want you to know that I have called on Ransom E. Olds for help; whether I succeed or not will depend upon his judgement in the matter of course but I have known for some time that he desired to do something worthy for M.A.C. and it seems that now is his best opportunity. I have communicated with him on his yacht

cruising off the coast of Florida and it is possible that something may come of it. I have asked him outright for one hundred thousand dollars for the replacing of the Engineering building. This I did Sunday and so far I have not received his refusal. Some times no news is good news and sometimes it is not.[127]

Olds responded affirmatively to his old friend's request, thereby removing many financial uncertainties for the college. Kedzie and Brown did not have to reallocate existing funds to rebuild, nor did they have to ask the legislature for additional money. The argument that moving engineering studies to the University of Michigan would save the state money lost its appeal and any validity it might have had. Even more significant, the entire M.A.C. community—administration, faculty, students, and board members—had acted in a manner that was consistent with the conviction that Michigan Agricultural College had to be more than an agricultural school if it intended to meet the needs of all of the people in Michigan.

For a final word relative to tension between agriculture and engineering, we turn to a student analysis to give clarity to what appeared to many to be contradictions between teaching agriculture and engineering, and the practical and the theoretical at Michigan Agricultural College. An editorialist writing on 9 June 1913 for the student newspaper, *Holcad,* concluded that "working side by side, the engineer and the farmer have brought agriculture up to its present high plane."[128] Examples of this partnership abounded. Electricity ran small machines and lit the farmers' houses. Gasoline engines powered tractors that allowed farmers to increase their productivity and "automobile trucks" to move their farm goods quickly over hard-surface roads and concrete bridges. Other gasoline engines ran pumps that sent water to farm buildings or irrigation ditches fashioned by engineers. The designers of all of these technological wonders had to have a theoretical knowledge of physical science and mathematics before they could create devices that had enormous practical applications on the farm or in town. Practicality grew out of the theoretical. Justin Morrill knew what he was doing, after all, when he wrote into law the requirement that the land-grant college teach both agriculture and the mechanic arts.

THE WOMEN'S COURSE

The establishment of the women's course came about as changes in American society reinforced the demand of women that they be given the same opportunity as men to receive an education at Michigan Agricultural College. Mary Mayo's long years of hard work calling for a course of study designed specifically for women finally bore fruit in 1896. Young women could now learn how to perfect the art and science of homemaking or pursue careers outside of the home. Social change in the late nineteenth century had brought about the necessity that young women living in Michigan be empowered to transform knowledge taught at M.A.C. into ways that let them take their places in the emerging middle class in rural and urban Michigan and throughout the United States. The new course of study drew upon many of the same sciences that gave life to the agricultural course.

The women's course fit into movements that saw the rise of nutritional science, the desire to improve home life in America, and the formation of domestic science or domestic economy programs at other land-grant colleges, notably in Iowa, Kansas, and Illinois. Whereas M.A.C. broke much new ground as it developed its agricultural course, these other institutions had opened the way for women at land-grant schools. But once President Snyder inaugurated the course, he devoted significant resources to it, and women finally became inte-

Female students at Michigan Agricultural College gathered for this photograph in 1909. Not all of the approximately 250 women (out of the total enrollment of about 1,500) appeared in the picture. Courtesy of Michigan State University Museum (Framed Collection).

gral to the M.A.C. community. The women's course (which became home economics in 1909) not only provided opportunities for young women to get a college education, but it also served to slowly integrate them into the institution as a whole. Like the agricultural course, the practicality taught in the new course of study obscured its solid academic base, as it, too, was rooted in the sciences that went beyond practice.[129] By adopting the women's course, M.A.C. finally joined in the revolutionary cause that was empowering women to play greater roles in American life.

Interest in nutritional science grew throughout the nineteenth century. Technological advances enabled people to preserve food in glass jars and tin-plated cans and shippers to move fresh meat in refrigerated rail cars.[130] Justus Liebig's work in agricultural chemistry stimulated research into the chemical composition of foods.[131] In the United States, Wilbur O. Atwater's career at Wesleyan University, the experiment station at Storrs, Connecticut, and the USDA's Office of Experiment Stations did much to establish the science of human nutrition that was rooted in chemistry. Atwater and his colleagues held that "pure science or research" was essential to discovering the secrets of nutrition.[132] By the end of the century, scientists knew that carbohydrates, fats, and proteins make up the three major organic substances in food. During the first quarter of the twentieth century, researchers learned that vitamins, even though they occurred in very

Women planting vegetable garden, circa 1910. The small building to the left was a tool shed for the horticulture department. Courtesy of Michigan State University Archives and Historical Collections (Horticulture, Field Scenes, Folder 4).

small quantities, formed a vital element in nutrition, and that deficiencies in them caused disease.[133]

The study of nutrition stimulated the creation of national organizations that encouraged the development of the discipline. Ellen Swallow's work in consumer education and scientific research helped bring concerns related to nutrition before the public in the 1880s.[134] When the federal government gave its support to the scientific study of nutrition, it erected the framework for research to become a nationwide effort. Congress appropriated $10,000 in 1895 for nutritional studies. This money and subsequent funds were administered by the Office of Experiment Stations in the USDA. The USDA's Office of Home Economics carried on this work between 1915 and 1923, until the Bureau of Home Economics replaced it.[135] Swallow provided important leadership to the home economics movement through conferences held at Lake Placid, New York, each year from 1899 until 1908, when the American Home Economics Association came into being. The Association's purpose was "the improvement of living conditions in the home, the institutional household and the community."[136] The organization brought together teachers, institutional managers, social workers, and others committed to its mission. Within a year it published the *Home Economics Journal.* During World War I, dietitians faced many challenges to meet the food needs of hospitals in the United States and overseas, which led them to form the American Dietetic Association in 1917. This organization made it easier for researchers to share their findings with others and for dietitians to influence society as a group rather than as isolated individuals.[137]

Post–Civil War America witnessed a growing poverty in both rural and urban communities that weakened family life and contributed to a host of social problems that were relevant to the establishment of the women's course. Such writers as Jacob Riis, Jane Addams, and John Spargo dramatically exposed the horrors of big city life to a national audience.[138] Crowded living quarters and poor sanitary conditions gave rise to tuberculosis, diphtheria, scarlet fever, measles, and other communicable diseases that sapped the health of adults and children who often were suffering from malnutrition or hunger. Alcohol abuse, unsafe working conditions in factories, and lack of good public health services also contributed to the pains of America's poor. Economic hardship coupled with ignorance of proper diet and causes of disease undermined family life on farms and in towns and cities.

The deterioration of the health and well-being of families fostered the demand that land-grant colleges train women to manage their own households more effectively and to teach others how to improve the health of their families. President Theodore Roosevelt was acutely aware of this concern when he appointed the Commission on Country Life in 1908. Two of its seven members were M.A.C. graduates— Liberty Hyde Bailey and Kenyon L. Butterfield. In his invitation to Bailey, the president lamented that most efforts to improve life in rural America were directed to improving crop production rather than looking after human interests. Roosevelt then made this point: "There is no more important person, measured in influence upon the life of the nation, than the farmer's wife." He proposed that women educated at land-grant colleges could implement recommendations made by the commission to learn more about country life, establish extension work, and hold conferences promoting "rural progress."[139]

Much of the nineteenth century saw women actively working to improve their station in life by offering girls and women education that was equivalent to that available to boys and men. It must be noted that reforms in education were just one of many movements, including female suffrage, temperance and prohibition, and the right to enter professions reserved for men, that were designed to extend more democracy to women. Ellen Swallow gave definition to an important underlying assumption to efforts to improve the lives of women when she expressed the belief that all people, including the homemaker, need an education to be able to influence their environment.[140] In other ways, the thinking and efforts of Emma Willard and Catharine Beecher laid the groundwork for the establishment of domestic economy programs at land-grant colleges. Their strong articulation that women were destined to operate in the domestic sphere of life made a strong a case for teaching girls how to be good homemakers, and their commitment to female education gave them considerable influence. Willard, in 1819, stated that women have a duty "to regulate the internal concerns of every family," and that if they fail in that, they can not be good wives or mothers.[141] Beginning in 1821 and for the next seventeen years, she put into practice many of her ideas at the Troy (New York) Female Seminary, where she trained many female teachers.

Catharine Beecher believed that God had created men and women to operate in distinctive spheres, and that males and females should be educated in separate institutions to fulfill their respective roles. In 1870 she told an audience in Boston: "For young men we find endowed scientific schools to teach them agricultural chemistry, that they may learn wisely to conduct a farm; why should not women be taught domestic chemistry and domestic philosophy?"[142] Beecher's devotion to female education grew out of her belief that many young wives and mothers suffered greatly because "of *poor health, poor domestics, and a defective domestic education.*" The way to alleviate this situation was to train them properly "for their profession." Beecher wanted "to place *domestic economy* on an equality with the other sciences in female schools."[143] While the adoption of the women's course at M.A.C. was not in line with Beecher's wish for separate institutions for men and women, it is significant that the domestic economy program was founded on the scientific basis for which she had been a strong advocate.

The emergence of domestic science and domestic arts in the late nineteenth century reinforced the belief that women played the central role in sustaining healthy families and making safe and stable homes. Some proponents of these disciplines saw domestic science as a way to reform Americans by improving their diets. Others believed that the role of women in society could be strengthened by teaching girls about textiles and how to sew their families' clothes at home. Once a woman had acquired a working knowledge of the scientific principles necessary to practice domestic science and domestic arts, she could apply her skills in her own home, and she could teach others.[144]

A look into the establishment of domestic science or home economics programs at other land-grant institutions puts M.A.C.'s program into historical perspective. Iowa Agricultural College gave instruction in "domestic chemistry" as early as 1871, and Kansas State Agricultural College opened a "School of Home Economics" in 1873. Neither of these programs possessed much sophistication at first.[145] But the Illinois Industrial University began a "School of Domestic Science" in 1874, under the direction of Louisa C. Allen, that may have been "the first college course of high grade in domestic science organized in the United States." At Illinois, female students combined scientific courses in chemistry, anatomy, and physiology with liberal arts subjects. Science created the basis for technical studies.[146] Other colleges that started home economics programs later included the University of Wisconsin in 1895, Ohio State University in 1897, and Purdue University in 1905.[147] The University of Michigan left domestic science and home economics to M.A.C.[148] At Cornell University, home economics commenced in 1900, as a result of Liberty Hyde Bailey's leadership.[149] Cornell's home economics course served as "the appropriate place for women" in the curriculum and as "a means of removing women from the academic and professional mainstream." This, in turn, led to separation by gender at Cornell, which emphasized the woman's separate sphere in the home.[150]

The women's course at M.A.C. amplified the institution's purpose both as an instrument of change and as a training ground for agents of change. The program was born in an atmosphere where people—both men and women—assumed and advocated that the proper sphere—even the only one—for women was in the home. A woman's highest calling was to serve her husband and children, and her duties entailed caring for them and managing her household in a manner so that proper diet, good health, and virtuous habits prevailed. Although President Snyder and the faculty did not challenge this view, they consciously intended to educate female students in such a way that they could also seek and have a professional life outside of the home. Many young women welcomed opportunities to transform their newly acquired knowledge into betterments in their homes and in society as a whole. When they did either of these things, they became agents of change and participated more fully in American democracy.

For years prior to 1896, male students at M.A.C. had voiced a wide range of opinions about the presence of female students on their campus.[151] The views of three young men as they appeared in the *College Speculum* in the 1880s and 1890s reveal arguments for and against co-education and the obstacles that stood in the way of women being accorded their rightful place in Michigan Agricultural College. Women encountered these attitudes as they pursued their studies. All three men assumed that a woman's proper place was in the home. In 1881, John Evert, '82, put his thoughts forth quite forcefully:

> To refuse women equal educational advantages with men is contrary to the spirit of liberty, and implies minds devoid of the faculties of reason and judgment on the part of the "weaker sex." But women do possess as much common sense, at least, as men. Then why not acknowledge the fact by giving them full freedom in acquiring an education?[152]

Evert stressed that it was important to educate women to have both strong bodies and strong minds so that they would make "good housewives." He felt it appropriate that M.A.C. extend to women, just as it did to men, the opportunity to participate in its manual labor system. Seven years later, R. C. Clute, '86–'89, argued that women did not need to learn the sciences that men required for their work, and besides, their natural preferences led them to study music, art, literature, and languages—all subjects that M.A.C. should not have to spend its scarce resources to support. Clute also expressed alarm that great cost would result from the construction of separate accommodations and the salaries needed to pay new instructors to teach the women.[153] Just a year before the start of the women's course, E. J. Heck, '95, minced no words when writing about the education of women. Heck lamented that women sought to find their vocations outside of the home "in the conflicts of the exterior world," which led to neglect of their children. He charged: "As a result, our homes are vomiting forth thousands of young men for the work house, thousands of young women for courses of debauchery and shame." After denouncing the suffragettes and reformers he concluded: "To make home attractive is the highest triumph of woman." It was in the home where "the true sphere of woman's influence" resided.[154]

The Edwards, Smith, and Kedzie report of 1896 underscored the importance of educating women to be good wives for their farmer husbands. It emphasized the centrality of the home in society and the responsibility that women had to make a healthy home environment that promoted their children's physical, mental, and moral well-being. Furthermore, it noted, it was time for M.A.C. to join her sister institutions that already had courses in domestic economy. Edwards, Smith, and Kedzie also believed that a larger presence of women on campus would be "extremely helpful in elevating the moral tone of the [male] students."[155] At Snyder's urging, the State Board of Agriculture approved a women's program to begin in 1896. Like in so many other instances, this course had been presented as an enhancement of agricultural education, which it was, but Snyder saw it as being far more than something that would simply improve the lot of farm wives.

President Snyder demanded that the program educate young women to meet the challenges of life inside and outside the home. Snyder's vision set the new program on a successful journey. At the outset, the "ladies' course" followed the agricultural course but substituted domestic science subjects for "the purely agricultural studies."[156] In 1899 Snyder crowed that the course was "especially strong in the natural sciences and their adaptation to the practical affairs of the home."[157] Once again, notice that when Snyder links science to practicality in his private correspondence, he puts science first. Two years later he commented to C. A. Jewell Jr. that the college always gave "something more than the practical, although almost every subject is practical in a sense." Snyder went on to say:

> You know something of the training young men receive in science. Our young women receive practically the same, but instead of working it out in a practical way with reference to agriculture, it is worked out with reference to the home. The botany is carried out in

floriculture and landscape gardening. The bacteriology is carried
into communicable diseases, house sanitation, etc. Chemistry and
physics are applied in the same way.[158]

Snyder advised Bess Howell of Ypsilanti that if she wished to teach she
should to take the full course, because she had to be able to cover her
subject from both scientific and practical standpoints.[159] In 1905, Snyder bluntly told Professor W. C. Latta, '77, at Purdue that "we did not
establish a Domestic Science Course; we established a Course for
Women to prepare them for life."[160] He wanted M.A.C.–educated women
to have the scientific training required to pursue research and service
positions in both the public and private sectors of American society and
faculty appointments in colleges, universities, and public schools.[161]

President Harry B. Hutchins of the University of Michigan asked
Snyder to describe M.A.C.'s home economics course to him in 1912,
because Michigan was considering starting one of its own. Snyder
pleaded with Hutchins not to, since he felt that if the university did
so, many students would go to Ann Arbor instead of come to East
Lansing. But this also gave Snyder an opportunity to wave the green
flag as he boasted that M.A.C.'s program was on a par with Illinois',
Ohio State's, Minnesota's, and those at "other good institutions." But
then Snyder chose his words carefully:

> We offer no other course for young women. It is a technical course
> and in a general way is of the same rank as our technical courses in
> Engineering and Agriculture. You will notice from our catalog that as
> a basis for technical work in this course we give very thorough
> instruction in the natural sciences. We offer particularly strong
> courses in Chemistry, Physiology and Bacteriology. . . . The work in
> Domestic Science or Home Economics has developed almost entirely
> in this country in connection with Agricultural education.[162]

Snyder staked out his territory with precision vis-à-vis a potential
competitor by highlighting the technical and aligning his home economics program with agriculture. All of this was the proper business
of the agricultural college—the land-grant college in Michigan. If the
University of Michigan were to start its own program, M.A.C. would
have to expand its scientific offerings to keep its students. The
taxpayers would not be happy to see duplication of courses at two
public institutions.

In 1896, Edith F. McDermott, professor of domestic economy and household science, welcomed thirty-one freshmen women into the new women's course. They came armed with the State Agricultural College's promise that science would vastly improve their lives. The college catalog for 1896–97 assured female students that applied science would emancipate them "from the tyranny of the kitchen and the nursery."[163] True to Snyder's commitment to science, the freshman and sophomore women took many of the same courses as men enrolled in the agricultural course, including algebra, geometry, botany, physics, chemistry, anatomy, English, and plant histology. But women studied cooking, sewing, and household economy instead of livestock, soils, dairying, and soil physics, and they substituted calisthenics for drill and military science. They also took more drawing. Freshman cooking classes provided instruction "in the preparation, cost, composition, and dietetic value of foods." Sophomore courses in domestic art taught women how to make different stitches, to operate sewing machines, to cut cloth and make garments, and to assess the quality of materials.[164] The junior and senior years saw women studying literature, German,

Edith F. McDermott, professor of domestic economy and household science, instructing members of the first class of women who enrolled in the new women's course in 1896. McDermott has the distinction of being the first woman ever to be appointed professor in M.A.C. This classroom was located in Abbot Hall, which had previously been a men's dormitory and had been reconfigured to house the women's program. McDermott had taught domestic science in Allegheny's Fifth Ward Manual Training and Domestic Science School, which was part of the system that President Snyder had overseen before coming to M.A.C. She had received her training at Allegheny College, Drexel Institute, and Cornell University. Courtesy of Michigan State University Archives and Historical Collections (Michigan State University, Human Ecology, Cooking).

A statue of "Diana and her fawn" stands in the entranceway of the Women's Building. The art was apparently procured in the 1910s after the State Board of Agriculture requested that some "good art" be put on display in the building. Courtesy of Michigan State University Archives and Historical Collections (Michigan State University, Buildings, Morrill Hall, Interior).

bacteriology, pomology, floriculture, kitchen gardening, and history of art, among a host of other subjects. Mrs. Maud Marshall taught piano to first and second year female students, which was the beginning of formal music education at M.A.C. The catalog for 1898–99 listed a four-year sequence for instrumental music and a weekly chorus class. At this time, M.A.C. had no intention of starting a conservatory, but perceived music education as another way to help young women consider "other lines of work."[165]

The lack of separate accommodations for female students had long prevented M.A.C. from offering a course of study designed for women. Although the use of a facility only for women may have impeded their integration into the college as a whole, it enabled the faculty to put together a thriving program. To make room for the "co-eds," Snyder converted Abbot Hall, a men's dormitory, into a women's dorm for forty, with a sewing room, a dining room, and cooking laboratory added on to the structure. When Abbot Hall proved to be too small for the growing program, M.A.C., with the help of the Women's Clubs of Michigan and Mary Mayo's prodding of the Michigan State Grange, convinced the legislature to appropriate $95,000 to build a four-story structure to

house 120 students and facilities to be used in the women's program.[166] Rooms for domestic arts, lectures, storage, offices, kitchen laboratory, laundry, dining, woodwork shop, piano practice, gymnasium, and literary society meetings met both academic and social needs. The second floor parlor, with its fireplaces and Steinway grand piano, hosted numerous gatherings over the years, including meetings of the East Lansing Women's Club, lectures, and recitals.[167]

The State Agricultural College emphatically and publicly stated in 1900 that the women's course was to educate women to be good housewives. Again, M.A.C. stressed the objectives that would arouse much support among the farming community, while at the same time, it proceeded quietly to train young people to assume roles outside of the home. The *M.A.C. Record,* in its 2 October 1900 issue, asked the question: "Education for our Women—Of What Shall it Consist?":

The question that we have had to deal with in formulating the course for our women is what constitutes complete living, not for

Students walking in front of the Women's Building (Morrill Hall) in the 1920s. The construction of this facility, in 1900, enabled M.A.C. to meet its commitment to offer a full course in domestic science and domestic arts and to provide adequate dormitory space for female students. The Women's Building also symbolized the college's resolve to offer an education, rooted in the sciences and the liberal arts, designed to prepare young women to be informed, productive citizens both at home and in their communities. In addition, the building became a campus and community center for concerts, lectures, and dinner parties. Courtesy of Michigan State University Archives and Historical Collections (Michigan State University, Buildings, Morrill Hall).

Built in 1888, the first Abbot Hall was a men's dormitory. After President Snyder initiated the women's course in 1896, he housed that new program in Abbot Hall, giving women a dormitory and the new domestic science and arts courses a facility. The women's dormitory and the women's course moved to the Women's Building after its construction was completed in 1900. Located to the left of Abbot Hall was the Bath House erected in 1902. W. J. Beal gave this description of its interior: "The plunge bath is 35 feet by 17 feet, having an average depth of 5 feet, 6 inches." A corridor connected the second Bath House to the Armory. Courtesy of Michigan State University Archives and Historical Collections (Michigan State University, Buildings, Albums, 1918 or 1919).

all women, but for the women whose homes are not on our farms, the women who are to be the homemakers in the houses of the great middle class in our cities, the women who expect neither to teach, nor to preach, nor to write, nor to agitate, but desire to prepare themselves with just as much system and definiteness of purpose to fulfill their duties within the household—the cultivation of economy, of sanitation, of wise dietetics, of a cheerful fireside, a broad-minded, helpful intellect and a cogent moral character—as the man uses in preparing himself to establish and maintain the household. What is it that constitutes preparation for complete living for this class?"[168]

Interestingly, the *Record* had answered its own question a week earlier in its 25 September issue:

We, here, are presumptuous enough to think we are helping the pendulum to come to the perpendicular. The old-fashioned idea of femininity, the new fashioned idea of a man's education for a woman, are giving way before the national ideal whose end we may call womanliness. Women are not men and it is foolish to act as if they were. Their work is as important if they will only take their place and do it. The term home-making may sound sentimental to some, but what home-making may mean in its highest sense we now see only in part. The women in our homes may have sweetness, mental discipline, strength of character, but they do not always know why or how to avoid the evils to be shunned in housekeeping.[169]

Snyder's emphasis on science is noticeably lacking here. Even if M.A.C. hoped to educate young women to be good housewives *and more,* it still was prudent to emphasize publicly the domestic over the science.

In 1913, Hazel Ramsay, '14, and Clara Jakway, '13, described for readers of the *Holcad* how women taking domestic science and domestic arts courses prepared themselves for one of two occupations: teacher or housewife. Domestic science saw them studying kitchens, cooking, baking, dietetics, house architecture, household management, home nursing, and institutional management. Subjects covered in domestic arts included sewing, textiles, color and design, woodworking, household art, millinery, and teaching domestic art.[170] Field trips reinforced classroom and laboratory work. For example, in June 1911, sixteen seniors spent a day in Battle Creek observing Alice Cimmer, '00, while she taught domestic science at the new high school. Later the group went to "Postville [C.W. Post Cereal Company] where the predigested foods are made." Here the students discovered that drinking "Postum" might "stimulate the brain to greater activity" during final examinations.[171] Students acquired practical teaching experience when they taught domestic arts to girls at East Lansing High School. The women put into practice skills learned in domestic science classes by preparing the meals that they served to their classmates in Club C, their boarding club in the Women's Building.

By the mid 1920s, the home economics course encompassed eight majors that were intended to equip graduates to apply scientific principles "fundamental to living and utilization of all modern resources in the improvement of the home."[172] The catalog for 1925–26 clearly reveals that home economics students were offered opportunities to train for a career that got them out of the house. The general course trained elementary and secondary school teachers; majors in foods

Domestic Science Laboratory, Women's Building, circa 1910. Courtesy of Michigan State University Archives and Historical Collections (Michigan State University, Human Ecology, Cooking).

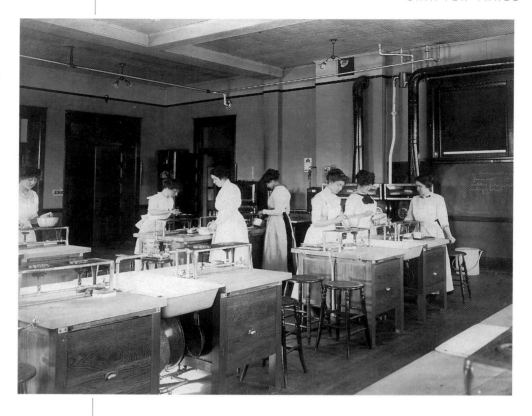

A sewing class in the Women's Building, circa 1908–9. Courtesy of Michigan State University Archives and Historical Collections (Michigan State University, Human Ecology, Sewing).

and nutrition anticipated working as dietitians for hospitals, private companies, or clinics, or even writing for women's magazines; and institutional management majors found jobs as managers of school lunch programs, cafeterias, and tea rooms. Clothing majors became designers of clothes, interior decorators, or editors for women's magazines; textile students took employment in advertising, teaching, and textile chemistry; and women in both groups found employment as buyers for stores. Women taking the major in related arts also minored in art to prepare for careers as advisors in house furnishing and clothing departments in stores, and with more training in art, they could design theatrical costumes. Students who completed the vocational education requirements in the homemaking course met the provisions of the Smith-Hughes Act to teach vocational home economics in the public schools. Finally, in conjunction with Edward W. Sparrow Hospital in Lansing, young women could earn a bachelor of science degree and receive certification as a general nurse through a five-year course.

Students posing in the woodworking shop in the basement of the Women's Building. Courtesy of Michigan State University Archives and Historical Collections (Michigan State University, Human Ecology, Woodworking).

Students kept notebooks in which they recorded notes from lectures and laboratory experiments and copied relevant information to the subject they were studying in each class. Women taking cooking classes acquired a collection of recipes, which they could use when teaching or making meals for their families. Pictured above is a recipe for chicken pie taken from Emily L. Kilbourne's "Recipes" notebook, 1900. Courtesy of Michigan State University Archives and Historical Collections (Emily L. Kilbourne Papers, 1899–1900, UA 10.3.12, "Recipes," 33, Folder 2).

The women's course required female instructors to teach the classes offered in the new program, and over time it slowly paved the way for women to join the faculty in other departments. In 1896 Edith McDermott became the first woman to hold the title of professor on the faculty of the State Agricultural College. She did not have an easy time during her two-year tour of duty. In addition to her teaching responsibilities she supervised the female students who resided in Abbot Hall. Snyder evaluated her teaching of domestic science as excellent, but he felt that she had failed in her duties as matron by being "too young for such a place" as M.A.C. Although she was "a conscientious young woman" and took part in religious activities, she danced in spite of Snyder's efforts to get her to stop. He believed that a woman in McDermott's position undermined her moral position with the students when she danced.[173]

Several outstanding administrators served the home economics program over its first thirty years. In 1901, Maude Gilchrist succeeded Maud Ryland Keller, who had become dean of women in 1898. Keller had outfitted the new Women's Building before it was occupied in 1900. Gilchrist held a bachelor of science degree from Iowa State Teachers College and a master of arts degree from the University of Michigan, and had studied for a year at the university in Göttingen, Germany. She remained at M.A.C. for twelve years, providing stability at the helm of the women's course. When Gilchrist became the dean of home economics, she took her place with the male deans of the divisions of agriculture, engineering, and veterinary science as one of the top administrators in the college.[174] Gilchrist tirelessly promoted the work of her division by speaking to women's clubs, high schools, and other public groups across the state.[175] Dr. Georgia Laura White, Ph.D., Cornell, provided inspired leadership as dean of home economics from 1914 until 1918. Other noteworthy administrators were Mary E. Sweeny, Louise Hathaway Campbell, and Jean Krueger.

Three instructors provide some insights into the pioneering efforts made by early women faculty. Louise Freyhofer, who had a good knowledge of music theory and history, became synonymous with music on the campus. She gave piano lessons and directed the college chorus as it performed such masterpieces as Haydn's *Creation* and Handel's *Messiah* before thrilled audiences. For seventeen years (1902–1919), Freyhofer touched the lives of thousands of students and concertgoers.[176] Jennie L. K. Haner put the domestic art program on a firm footing as she taught sewing, textiles, and woodworking from 1897 until 1908. Her high standards demanded that her students learn how to make things properly.[177] Agnes Hunt came to M.A.C. in 1910 as professor of domestic science. Her background included a bachelor of science degree from the University of Illinois, two years at the College of Hawaii, where she started the domestic science program, and a trip to China and Japan. Hunt was known for giving her students challenging problems to solve in the kitchen. She left M.A.C. in 1915.[178]

Although a few women entered the faculty outside of the women's course, they made slow progress and found themselves in the lower levels of the instructional hierarchy. The faculty was divided according to gendered spheres, and that division gave way very slowly. Georgiana Blunt, Ph.M., was appointed assistant professor of English and modern languages for the 1898–99 school year, thereby becoming the first woman to hold that rank outside of the women's program. Blunt stayed for three years. It was not until the 1914–15 academic year that Lydia Zae Northrup, '06, became the second woman to hold the rank of assistant professor outside of home economics when she received an appointment in bacteriology and hygiene. Northrup had graduated from M.A.C. in 1906, then worked for the college as an assistant in bacteriology and earned her master

Maude Gilchrist guided the women's course as dean of the Women's Department and later dean of home economics during her tour of duty at the college between 1901 and 1913. She introduced student teaching into the curriculum of the women's course which helped to prepare graduates to teach domestic arts and domestic science in the public schools. Gilchrist left M.A.C. in 1913 to accept the professorship of botany at Wellesley College. Courtesy of Michigan State University Archives and Historical Collections (People, Gilchrist, Maude, no. 1095).

Elida Yakeley was born in Trenton, Michigan, spent some of her childhood years in Montana, and received an education in business at Ferris Institute. Yakeley fielded many requests from students for information and advice. *The Holcad* in its 22 May 1911 issue articulated Yakeley's sensitivity to students and her relationship to them: "Being a business woman, she has perfected herself in deciding how much an inquisitive person should be allowed to know of her various 'experiences' and how much she had better keep to herself. Hence we could not learn her attitude toward the various perplexing problems which face the feminine mind today, such as the militant suffragettes, harem skirts and spring hats." Courtesy of Michigan State University Archives and Historical Collections (*Wolverine*, 1914, 17).

of arts degree from the University of Michigan before her promotion. She also published an article, "A Bacterial Disease of Larvae of the June Beetle," in the German journal *Centralblatt fur Bakteriologie.*[179] Six years later, three longtime instructors finally attained the rank of assistant professor. Norma L. Gilchrist, A.B., had taught English since 1905–06, Mary Amelia Hendrick, B.A., had taught history since 1907–08, and Eugenia Inez McDaniel, A.B., had taught entomology since 1910–11 before their advancements to assistant professor. In 1923 McDaniel had the distinction of being the first woman to attain an associate professorship outside of home economics. As of 1925, no woman had become a full professor except in the home economics division.

A look at the catalog for 1925–26 unveils the persistent dominance of males on the faculty. Even then, only 43 (16 percent) of the 269 teachers were women (see figure 5). Eugenia Inez McDaniel was the only associate professor outside the home economics program. Female instructors had appointments in English, Spanish, French, and history, but most still taught courses in music, drawing, and home economics. Every seat on faculty committees was held by a man except for one on the library committee, held by Linda Landon, librarian, and seats on the scholarship, substitutions, social, advanced credits, and home economics course committees held by Jean Krueger, dean of home economics. Clearly, female faculty had very little influence compared to their male colleagues.

Elida Yakeley held one key administrative office, and other women occupied clerical positions throughout the college. Yakeley became President Snyder's secretary in 1903 and held that job until she was appointed registrar in 1908. She kept student academic records, updated and published M.A.C.'s annual catalog, and represented the institution at national meetings of college registrars. Yakeley's interest in the students' academic progress led her to follow many of their careers after they left campus. Students described her as "a thoroughly just, broad-hearted friend."[180] Throughout the campus, women performed other essential tasks that kept the college functioning. For example, Agnes Jones did stenographic work for Richard Lyman and other professors.[181] Miss Goodhue was a stenographer for the chemistry department in 1910, and in 1912 Carolyn Goritz was a stenographer and bookkeeper for the bacteriology and

Figure 5. Faculty by Rank and Gender, 1925

RANK	MEN	WOMEN
Professor	40	1
Associate Professor	40	7
Assistant Professor	47	9
Instructor	60	24
Teaching Assistants	39	2
Total	*226*	*43*

Source: *Catalog of the Michigan State College of Agriculture and Applied Science* for the year 1925–1926 (East Lansing: The College, 1926), 7–14.

hygiene laboratory[182] While most of these women and their work remain invisible, they contributed greatly to the success of M.A.C.

Low salaries plagued M.A.C.'s faculty during the college's first seventy years, with women receiving less compensation than their male counterparts. Ruth F. Allen's experience illuminated the cruelty of gender inequity in pay and academic rank, which helped to determine salary. Allen had earned a Ph.D. from the University of Wisconsin in botany, and she came to M.A.C. as an instructor. Her salary in 1912 stood at $850. Her colleague Dr. Richard de Zeeuw received $1,400 per year as an assistant professor. Ernst A. Bessey, chairman of the botany department made an impassioned plea that Allen's salary be raised to at least $1,000. He summed up the situation: "Were it not for the unfair discrimination against women teachers so far as salaries are concerned she would be receiving $1200 to $1500 elsewhere in view of her training and ability as exhibited by her exceptionally fine doctor's thesis."[183] Even if Allen had been an assistant professor, her salary almost certainly would have been significantly less than de Zeeuw. Helen Isabel Michaelides, instructor in French and English, was denied a raise because she taught fewer students in her French classes than other teachers had in some of their courses.[184] Lillian Loser Peppard had risen to the rank of associate professor of domestic art in 1917, but she left for a new job that paid her $2,500— an $1,100 increase from M.A.C.'s meager wage.[185] Even the exhortations of board member John W. Beaumont that the institution spend

Marie Dye joined the faculty as assistant professor of nutrition in 1922 after she had completed her doctorate at the University of Chicago. In 1929 the college elevated her to professor of nutrition and dean of home economics. Dye introduced research into the home economics program, thereby strengthening both the curriculum and outreach. She served the college with distinction until her retirement in 1956. Courtesy of Michigan State University Archives and Historical Collections (College of Human Ecology, People, Dye, Marie).

more money on salaries than buildings did not bring about any significant increases in faculty pay.[186]

The appointment of Dr. Marie Dye, who held a doctorate from the University of Chicago, as assistant professor of nutrition in 1922 foreshadowed a bright future for the home economics division. After the home economics program moved into its new building in 1924, Dye had the facilities at her disposal to initiate scientific research and a graduate program in nutrition. She pursued her own interests in the nutritional value of canned peas and cod liver oil. In the years that followed, Dye stimulated research into nutrition for infants and young children, the effects of freezing on the color of fruits and vegetables, and how homogenized milk could be utilized in cooking.[187] She was appointed dean in 1929, and provided leadership in that capacity until 1956.

THE FORESTRY COURSE

A course in forestry evolved out of the agricultural course and a growing interest in using Michigan's forest lands more wisely. Professor W. J. Beal's experiments with the growth of different species of trees motivated him to teach a lecture class in forestry in 1881.[188] In 1902 M.A.C. inaugurated a four-year forestry course, under the direction of Professor Ernest E. Bogue. It required students to follow the regular agricultural course for two years before they took a specialized curriculum that included "considerable technical work in forestry as well as practical work in the woods" in M.A.C.'s forest preserve in Oscoda County.[189] Students acquired much practical experience on campus as they studied silviculture, forest valuation, mensuration, and the identity of trees in M.A.C.'s three wood lots that covered 175 acres. During the last two years of the course students took a range of subjects in addition to forestry, including botany, English, German, military science, bacteriology and hygiene, horticulture, civil engineering, zoology, geology, and meteorology. The scientific classes gave students an understanding of soils, insects, and wildlife, among other things they needed to function effectively as foresters.[190]

Forestry at Michigan Agricultural College taught the practical side of the subject to undergraduates, which was still another expression

of its land-grant philosophy. This stood in stark contrast to the University of Michigan, where Professor Filibert Roth and his colleagues taught the theoretical side of forestry to graduate students. Despite this difference, President Snyder got a ringing endorsement for his college's approach to forestry from, of all places, Ann Arbor. In June 1902, Charles A. Davis, an instructor in forestry at the university, very succinctly placed M.A.C.'s method of teaching forestry into its land-grant philosophy in a letter he wrote to Snyder on the subject. Davis encouraged Snyder to train men "who shall act as foremen and superintendents of lumber camps, who shall know the practical side of lumbering thoroughly, and who shall have enough training in the theoretical side of Forestry to handle their lumbering in such a way that the future regeneration of tracts cut over will be provided for and as little damage done to the young growth as possible while lumbering

Field work comprised an important part of the work done by forestry students. Summer camps provided opportunities for them to get practical experience that they needed to manage forests for their future employers. An M.A.C. student is pictured sitting in his tent during summer term, 1910, at the headquarters of the Public Domain Commission, Cold Springs, Higgins Lake, Roscommon County, Michigan. "These tents were 10 x 12 of 8 oz. duck. Each tent was furnished with two iron bed steads, a table, two chairs, board floor, broom, lantern and matches." Courtesy of Michigan State University Archives and Historical Collections (Michigan State University, Department of Forestry, Collection, Forestry Camps, Summer 1910).

is going on." Davis went on to point out that Cornell, Michigan, Yale, and other colleges prepared a "crop of Forest Managers and scientific investigators" who would be neither willing nor prepared to do the practical work needed in the woods. Davis acknowledged the leadership that M.A.C. had demonstrated throughout its history and noted that "in developing the forestry work in the Agricultural College, you would be but following the traditions of the Institution and carrying out its well defined policy if you established the work on these lines of practical work rather than upon the basis attempted by the schools already at work."[191] M.A.C. departed from the established ways of higher education as it had before as it fashioned its forestry program to emphasize the transformation of theoretical knowledge into practical applications, while at the same time committing itself to a scientific approach to forestry.

Gifford Pinchot argued against teaching forestry to undergraduates and for combining the two forestry programs in Michigan. He suggested that if M.A.C. dropped its undergraduate program and adopted a graduate course united with the University of Michigan's program an "immense advantage to the progress of forestry in Michigan" would result, and that if the two programs became one and dropped the graduate courses in favor of the undergraduate "the result would be destructive."[192] Snyder believed that the university's forestry professors might be willing to move to M.A.C., but the idea never received support from the university's Board of Regents, and both institutions built viable forestry programs, with the University of Michigan emphasizing the theoretical and M.A.C. the practical. As M.A.C.–trained foresters went to work for the U.S. Forest Service, lumber companies, nurseries, and state forest services, they laid to rest Pinchot's fears that teaching undergraduates was a mistake.[193] The 1925–26 catalog still proudly proclaimed that "Special stress is laid on the practical side in every phase of Forestry."[194]

VETERINARY SCIENCE

As Michigan slipped into the twentieth century, the needs of agriculture cried out for of a complete course of study to train veterinarians at Michigan Agricultural College. The college had offered classes in

animal physiology and zoology since the 1860s. In 1883 Dr. Edward
A. A. Grange came from the Ontario Agricultural College to teach veterinary science, and M.A.C. built a veterinary laboratory in 1885. Still,
the first veterinary medicine colleges in Michigan were private institutions. The Detroit Veterinary College offered a two-year course
between 1891 and 1899, and the Grand Rapids Veterinary College
trained veterinarians from 1897 until 1918.[195] But the public wanted
a veterinary school at M.A.C., and in 1907 the Michigan legislature
authorized the State Board of Agriculture to offer a four-year
course.[196]

In 1910 M.A.C. established a veterinary science division designed
specifically to train veterinarians and convert discoveries in the laboratory into effective treatments for animal diseases.[197] Initially, students took the regular agricultural course through the first term of
their sophomore year, then followed a distinct curriculum until graduation. Students enrolled in classes offered by the departments of
anatomy, medicine, pharmacology, and surgery in the veterinary division, and received instruction from professors in other departments
including botany, bacteriology, zoology, physiology, chemistry, pathology, and entomology. Richard P. Lyman, dean of the Division of Veterinary Science, worked to link his students and his program to the
agricultural community in Michigan. He set forth his goal that "we
shall aim not alone to train men to deal with the problems in animal
husbandry and sanitation, but likewise to cooperate with stock owners and veterinary surgeons in the investigation and prevention of
animal ailments as far as our resources and facilities will permit."[198]
Lyman pushed hard to expand the resources available to him.

Tension and conflict played significant roles in the evolution of
the veterinary science course and improvements in its physical plant.
Lack of good facilities and Lyman's abrasive manner motivated some
students in 1912–13 to transfer to Ohio State to complete their studies. When President Snyder got wind of student dissatisfaction in July
1912, he made a forceful argument that M.A.C.'s four-year program
had higher standards than any other institution's three-year course,
and that any student who left for another school would be making an
"awful mistake." Furthermore, he noted that Lyman's stature in the
profession would be a great asset to M.A.C. students as they sought
employment after graduation.[199] Nevertheless, some left, including

The first Veterinary Laboratory was built in 1885. The Veterinary Clinic pictured above was occupied in January 1915. The structure included a hospital for large and small animals; a large consultation room; operating rooms for large animals and dogs; rooms for contagious diseases, skin diseases, drugs, and sterilization of instruments; a lecture and medical clinic room; a diagnostic laboratory; a cat ward; a general dog ward; and offices. This was a busy facility. The 1915 *Annual Report* stated that between 1 September 1914 and 1 July 1915, 416 surgeries were performed and 105 medical cases were treated by college veterinarians. The Veterinary Clinic is now part of Giltner Hall. Courtesy of Michigan State University Museum (Volume of Buildings, 1 November 1934, by Herman H. Halladay).

Max Wershow, who charged that Lyman was "able professionally," but "proved an absolute failure as an executive."[200] Lyman argued that the cause of student unrest was their frustration with M.A.C.'s inadequate facilities for "teaching veterinary clinical work," a position that had considerable merit.[201] In January 1915 the veterinary division occupied a new building.[202]

A nasty dispute in 1914 and 1915 between Dean Lyman and Dr. Ward Giltner, professor of bacteriology and hygiene, over the work of the bacteriology department exposed tensions that the college went through when it instituted new programs. Until the veterinary course came into being, the bacteriology department had handled requests to investigate all matters pertaining to communicable diseases suffered by domesticated animals.[203] Lyman, however, wanted to create a department of pathology and veterinary bacteriology in the veterinary division by transferring Assistant Professor Elam T. Hallman from the Department of Bacteriology to the veterinary division, thereby placing control of teaching and researching animal diseases under Lyman's direction.[204] This arrangement, Lyman argued, would enable "all veterinary work" to pass through his division.[205] His move to enlarge the veterinary division by transferring responsibilities from another department to his own went against the policy prescribing that the existing departments were to provide as much instruction to veterinary students as possible.[206]

After a contentious year, President Snyder crafted a solution that addressed the teaching, research, and extension functions of M.A.C. He recommended that the teaching of animal disease be moved to the veterinary division, and that it be handled by a new department called Animal Pathology that would initially be under the charge of Hallman, who would be promoted to associate professor. All extension work in animal pathology was passed to the new department, including "the examination of all diseased animal tissues." In addition, the veterinary division assumed the responsibility of producing and distributing hog cholera serum.[207] But "the research work in tuberculosis, contagious abortion, hog cholera and other infectious animal disease also the instructing of Veterinary students in Bacteriology" were to remain in the bacteriology department. At the same time the Department of Veterinary Anatomy was formed under the care of Associate Professor Frank W. Chamberlain. Finally, after a blizzard of pointed correspondence and a number of testy meetings, a new equilibrium was found.

Enrollment in the veterinary science course grew for the first few years but declined after World War I. Seventy-seven students studied in the program in 1916, twenty-five in 1920, and thirty-four in 1925. These figures reflected the national trend of shrinking numbers of students in veterinary medicine colleges. The demand for veterinarians appeared bleak, as gasoline engines displaced horses, the number of farmers decreased, and hard economic times overwhelmed segments of American agriculture. But Ward Giltner, who succeeded Lyman as dean, expressed optimism that veterinarians would find careers in public health, dairy hygiene, household sanitation, and industrial hygiene.[208]

A Changing Curriculum

Changing times in the United States brought on additions and adjustments in M.A.C.'s curriculum as the institution evolved from an agricultural college into a state college in 1925. New subjects and courses of study grew out of M.A.C.'s constant commitment to transform knowledge into practical applications in the world. The exploding demands of American business, industry, schools, commercial

Dora Hall Stockman became the first woman to sit on the State Board of Agriculture upon her appointment in 1919. She and her family knew rural northern Michigan quite well, since they engaged in farming, storekeeping, and logging over the years. A writer, Stockman published a couple of books and contributed to the *Michigan Farmer* and other serials in addition to editing the *Patron* for the Grange. She held the position of State Lecturer for the Grange in Michigan for sixteen years (1914–30). After her election in 1939, Stockman served in the Michigan House of Representatives. She lived from 1872 until 1948. Courtesy of Michigan State University Archives and Historical Collections (People, Stockman, Dora, no. 2620).

agriculture, cities, government, and public health forced the land-grant institution to allow students to take different combinations of subjects to learn how to direct the benefits of knowledge toward new problems and challenges. As a result, the courses of study in applied science, liberal arts, business administration, and a graduate program that granted doctoral degrees found their way into the college catalog by 1925. In addition, the faculty built collections in the library, museum, and botanical garden that were essential to M.A.C.'s academic program and its service to the public.

THE APPLIED SCIENCE COURSE

In 1921 Michigan Agricultural College inaugurated its applied science course, which had been made possible by classes already being offered in a large number of subjects. Because the requirements for jobs were so varied, the applied science course was predicated upon giving students "a large measure of liberty in the selection of the courses to be pursued." Furthermore, the new course was made up of subjects "fundamental to M.A.C.": agriculture, engineering, home economics, and veterinary medicine.[209] The course required freshmen to take a year of chemistry and English, two terms of mathematics and another term of mathematics or economic geography, a year of botany, zoology, French, German, or Spanish, and one term each in drawing and current history. The next year they began to specialize. In 1925 students chose majors in bacteriology, botany, chemistry, economics, entomology, geology, history and political science, mathematics, physics, or zoology.[210] The applied science course proved popular, as enrollment jumped from 122 in 1921 to 284 in 1925.

M.A.C.'s applied science course equipped men and women to make practical application of science in careers in "scientific agriculture, engineering, home economics and other lines of work." True to the ideals of Joseph R. Williams, T. C. Abbot, and others, this course gave students opportunity to take subjects that enabled them to become "efficient" citizens.[211] Administrators expressed their belief that the applied science course would produce specialists to fill positions in all levels of government, particularly the U.S. Department of Agriculture, and private industry, as well as educate science teachers for high schools and colleges. Bacteriologists played vital roles in the protection of public health by overseeing water supplies and identifying causes of disease in humans and animals. The chemistry department taught students who wished to pursue jobs in such fields as electrochemistry, industrial chemistry, textiles, and physiological chemistry. Plant pathologists studied seeds, directed botanical gardens, and researched plant diseases. A major in history and political science seemed to be a good grooming for a career in public service. Entomologists applied their knowledge of insects to the health of the fruit and vegetable industries, while economists worked for corporations in marketing, finance, accounting, and tax matters.[212] Students could also enroll in a six-year course in applied science and veterinary medicine that yielded a bachelor of science degree in four years and a doctor of veterinary medicine degree two years later.

LIBERAL ARTS

In response to growing student demand for a liberal arts course of study, the State Board of Agriculture established one in 1924 that offered both bachelor of arts and master of arts degrees. Once again, the land-grant philosophy of education expanded in a way that allowed young men and women to acquire more intellectual skills to better serve the public. This course appealed to many who were undecided regarding their future goals and to those wishing to teach subjects in the liberal arts. The course offered eights majors: English, economics, drawing and design, music, history and political science, mathematics, modern languages, and sociology. Students chose a minor field from one of the major fields of study or education, physical education, military science, or one of the natural sciences. All students had to complete two years of a modern foreign language.[213] Liberal arts proved to be an instant success with 335 students enrolled in the first year and 716 (out of a total of 2,352 undergraduates) in the second.

The establishment of the new course did not require the formation of any new departments. Instruction in liberal arts subjects throughout the college over the years had led to the development of strong offerings in the humanities and social sciences. English courses included writing, public speaking, literature, and drama. Modern languages were French, German, and Spanish. The economics department taught accounting, marketing, finance, management, and agricultural economics. Drawing and design offered a range of subjects, including history of art, freehand drawing and painting, decorative design and composition, and advertising arrangement. Electives in history and political science were fewer than most of the other departments. Students were only offered American, English, and European history, the three levels of government in the United States, and one course in international organization. The sociology department gave instruction in rural and urban sociology, community development, family and child welfare, social problems, and social work. In addition to such traditional subjects as algebra, geometry, trigonometry, statistics, and calculus, the mathematics department also listed courses for business administration and finance and investment.[214]

BUSINESS ADMINISTRATION

Growing demand for specially trained managers to oversee the increasingly complex operations of American agriculture, business, industry, and commerce led to a course offering a bachelor of arts degree in business administration. Students learned about "the five common phases of every business—production, accounting, finance, insurance and personnel" as they specialized in one of three groupings: personnel, secretarial, and welfare management; accountancy; or production or general business management. During their freshman and sophomore years, students took English composition; a year of German, French, or Spanish; agricultural and industrial history; sociology; physical education; military science; and a year of chemistry, botany, or zoology, in addition to an economics class each term. Graduates of the business administration course found jobs as personnel managers, executive secretaries, accountants, and proprietors or managers in privately owned industries and businesses or in publicly owned utilities.[215]

GRADUATE STUDIES

Graduate studies occupied a secondary status in the curriculum at Michigan Agricultural College until the mid 1920s, when a doctoral program, which led to the doctor of philosophy degree in seven scientific disciplines, was instituted. From 1861 until 1878, the college granted master of science degrees to graduates "who made a proper proficiency in scientific studies." From 1878 until 1881 students were required to present a thesis in order to receive a master's degree. Beginning in 1882 the faculty required a one-year residency in addition to a thesis from recipients of the advanced degree, though two years of work "in scientific study" and a thesis was still sometimes considered sufficient to fulfill the requirement up until 1898, when the residency provision came to be strictly enforced.[216]

Ever mindful of its agricultural constituency, M.A.C., in 1894, engaged in a debate over the thinking behind its awarding of the master of agriculture (M.Agr.) degree. The M.S. degree committee told the State Board of Agriculture that it would be wise for the college to honor graduates who had taken up careers in practical agriculture such as farming, fruit-growing, and stock-breeding. Until then the only college honor available to graduates of the agricultural course had been the master of science degree. These honors went to students of chemistry, botany, entomology, and the other sciences, but excluded practicing farmers. The committee went on to say: "The College needs to come into touch with the soil in conferring post graduate degrees. In bestowing this degree upon graduates who have distinguished themselves in the special lines of work for which the Agricultural College was created. This will awaken the enthusiasm of the Alumni and especially the reason for her existence."[217] M.A.C.'s faculty, administration, and the board, found it to be good politics, then, to confer the M.Agr. degree (in some cases master of horticulture) upon deserving alumni. In 1907 M.A.C. began to grant other professional degrees: mechanical engineer (M.E.), electrical engineer (E.E.), and civil engineer (C.E.) to alumni who had at least five years experience, had exhibited "some degree of distinction," and had written a thesis.[218]

Although M.A.C.'s advanced degrees recognized graduates who either had undertaken successfully additional study or had made worthy contributions to society, they did not require the rigor needed for viable post-graduate training. Professor of Bacteriology and Hygiene Charles E. Marshall argued in 1909 that departments engaged in agricultural scientific investigation should offer work that led to a "doctor's" degree. Persons receiving this advanced training could contribute more to the work at the experiment station, where master's students currently followed a largely "systematic" course of study and their theses were "usually very immature." Doctoral students who had received their undergraduate degrees at agricultural colleges were preferable to people who had been trained at universities who knew little about agriculture and needed a few years to learn how to apply their knowledge to agricultural research. A doctoral program, Marshall suggested, would upgrade and improve the efficiency of the scientific work performed at the experiment station.[219] At the urging of Marshall and some of his colleagues, President Snyder asked Robert S. Shaw, dean of agriculture, to study the matter with the recommendation that M.A.C. might offer a doctor's degree in one or two departments that carried on an "advanced line of research work."[220]

Although it took another fourteen years before a doctoral program was established, by 1915 the master's program had been strengthened. A few qualified students received graduate assistantships, but the program did not run smoothly. Ward Giltner complained that M.A.C. made offers too late in the year to appoint the best candidates, even though it cost no more to issue contracts in a more timely fashion.[221] Work done by graduate assistants did not count towards their degree requirements.[222] In 1915 twenty students were enrolled in the graduate programs in the agricultural course and two in home economics. The number of professional degrees expanded by 1918 to include master of chemical engineering (Chem.E.), master of forestry (M.For.), master of home economics (M.H.E.), and master of veterinary science (M.V.S.). The 1922–23 catalog announced that a Ph.D. could be earned at the Michigan Agricultural College in seven fields: bacteriology, botany, chemistry, farm crops, soils, entomology, and horticulture.

THE LIBRARY

In 1857 the faculty began collecting a body of scholarship that enabled professors, students, and the public to enhance their studies and research. By 1900 "books containing the best thought of every age" filled the shelves with accounts "of man's triumphs in art, literature, science and invention."[223] Librarians and faculty members worked together to add new scholarship by purchasing recently published books and periodicals that were relevant to the courses being taught. As the collection expanded, the need for more space to house it and for patrons to use materials also grew. While M.A.C. struggled to meet the demand for larger facilities, librarians continuously worked to make the collection more accessible to the college community.

Lack of funds contributed to the library's humble beginnings on the third floor of College Hall in 1857, where it stayed until it moved to rooms on the first floor vacated by the chemistry department in 1871. In 1881 the collection took up residence in the new Library-Museum (now Linton Hall). The collection in 1861 included about 1,200 volumes and newspapers that cost the institution nothing beyond postage. President pro tem Fisk judged its holdings of government documents, Patent Office reports, and agricultural society transactions as being "of comparatively little use to the college." Fisk lamented the absence of works of science, literature, history, and books for general reading as being "severely felt by the students."[224] The next year the State Board of Agriculture directed that money generated by student matriculation fees ($570) be used to buy materials for the library, including a set of the *Encyclopedia Britannica*.[225]

The library provided the first opportunity for women to assume administrative roles at Michigan Agricultural College, with Mrs. Linda E. Landon serving with distinction for forty-one years after her appointment as librarian in 1891. The first two librarians, George T. Fairchild and Elias John MacEwan, added the responsibilities of the position to their primary appointments as professor of English literature. Mrs. Mary Jane Cliff Merrell, the second woman to graduate from M.A.C. (in 1881), became the first full-time librarian in 1883, and she held the position until 1888. Mary Mouat Abbot, '78–'81, and Jane Skellie Sinclair followed Merrell, occupying the position for one and

George T. Fairchild joined the faculty in 1865 as instructor in English literature and was promoted to professor in 1866, a position he held until 1879. When President Abbot took a year off, due to ill health, to travel in Europe, Fairchild filled in as acting president from May 1873 until May 1874. He also had the distinction to be the college's first librarian, starting in 1872. In 1879 he moved to Kansas to become president of Kansas State Agricultural College for the next eighteen years. W. J. Beal credits him with playing a key role in the passage of the second Morrill Act in 1890. *Courtesy of Michigan State University Archives and Historical Collections (People, Fairchild, George T., no. 938).*

Mrs. Linda Eoline Landon standing in the library located in the Library-Museum building, circa 1907. During her forty-one years (1891–1932) as college librarian, Landon provided several generations of faculty and students with the latest books and periodicals needed for their research projects and reading interests. She also built a library staff capable of serving a growing student body and an expanding curriculum. Landon's leadership and work prepared the way for the creation of a library that would meet the needs of Michigan State College and Michigan State University as it evolved into a major research university in the second half of the twentieth century and into the twenty-first century. Note the Experiment Station "branch library" to the left of Landon. Courtesy of Michigan State University Archives and Historical Collections (Michigan State University, Buildings, Linton Hall, Interior).

two years respectively. But Landon elevated the library's status as she built a specialized staff of nine by 1925 to care for the materials and to respond to requests from patrons. The staff included a reference librarian; a cataloger and an assistant; assistants in charge of periodicals, circulation, and assigned reading; and two library assistants.

As the collection and the student body grew, space never seemed adequate. The *M.A.C. Record* described the chaotic situation in the library in 1915:

A look in the library during the noon hour . . . will convince even one who knew nothing of the situation that one of the most urgent needs of this institution, in the way of buildings, is added library facilities. Students will be found sitting on the tables, on the window sills, on piles of books and papers for which there is no room, and leaning against the walls, all earnestly endeavoring to "make use of the library." With shelf room and seating facilities intended for a college of 300 students, it is no wonder that they find it crowded when the enrollment is around the 2,000 mark.

The need for enlargement was first felt 22 years ago, and at that time the galleries were built and further facilities talked of. Since that time rooms in four other buildings have been added as store rooms, and these books and bulletins are unavailable for use. Under the present conditions 38,000 bound volumes are crowded on the shelves of the library. The building is no wise fireproof and no one can estimate the loss which would be entailed if fire should start. It is hoped that among the many things which should be considered in the appropriation for M.A.C. the next two years, this subject will receive its proper attention.[226]

By 1921 the collection had grown to 41,524 volumes, with another 4,933 books in the experiment station's library. Finally, in 1924, Landon's efforts to build a new library were rewarded with the construction of an edifice (now Michigan State University Museum) that stood on part of the site of (the first) Williams Hall, which had burned down on 1 January 1919. She supervised the move into the new library during the summer of 1924.

Librarianship, like other professions, never stands still, and new, improved methods for organizing collections needed to be implemented from time to time. Linda Landon and her assistants began a three-year project in 1894 to catalog the library's holdings according to the Cutter system, which divided the collection by classes designated by a letter so that, for example, works in philosophy were assigned "B" and history "F." The librarians then prepared cards for the author and title or subject for each book, which were then placed in the "Catalogue of Works."[227] Thirty years later Charlotte Jackson and Esther Betz recataloged the collection according to the Dewey Decimal system.[228]

The faculty and the librarians built a credible collection over the years that supplied knowledge for broadening and deepening courses of study. The book selection policy required each department to buy books in its field, thereby ensuring that recent scholarly works found their way into M.A.C.'s library. Over time professors Beal, Bailey, Clute, and Hedrick, among others, contributed to the makings of a fine collection.[229] With much satisfaction, Landon reported in 1897:

Language, oratory, and the history of literature are generously represented; and the section devoted to economics and social science is so filled with good things that one interested in these subjects is reluctant to turn from it even to greet their old friends, the poets, who are here in great numbers.

The philosophical, biographical and historical collections, also literary essays, criticisms, and works on Shakespeare and the drama, are particularly good and give evidence of more careful selection.... In the reading room are found the periodical literature which keeps us in touch with the most advanced thought of the day, and many agricultural and horticultural papers.[230]

The collections were also strong in civil, mechanical, and electrical engineering, agriculture, horticulture, veterinary science, zoology, botany, chemistry, mathematics, physics, fine arts, and history of arts. When researchers needed materials not in the M.A.C. library, Landon borrowed books and periodicals from other institutions (in 1925 she filled 162 requests) including the universities of Michigan, Illinois, and Chicago, and Johns Hopkins University.[231]

THE MUSEUM AND BOTANICAL GARDEN

From its earliest days, the State Agricultural College built a collection of specimens for a "general" museum that was intended to provide examples to be used for research and teaching the natural sciences. Mounted birds and animals, fossils, and skeletal remains formed most of the collection. Initially housed in College Hall, the museum room moved to the new Library-Museum in 1881. Many of the items were exhibited in glass cabinets and display cases, but not enough resources were devoted to the care of the collection. In 1916 the curator, Walter B. Barrows, professor of zoology and physiology, lamented that lack of space prevented the "re-arrangement of the collection even had the time for such work been available." Consequently, Barrows struggled to keep the exhibits in order and pests away from the objects.

Cramped quarters did not prevent teachers and students from using the collections. M.A.C. professors required students to study specimens as part of their work in zoology, physiology, biology, and geology. High school students came to the museum on field trips to

see mounted animals and fossils that were not available in their own communities. True to the land-grant philosophy, the zoology department prepared exhibits to send to meetings and interested organizations throughout the state. For example, in February 1916 an exhibit of 150 birds of both beneficial and harmful species found in Michigan went to the Wild Life Conservation Association meeting in Saginaw. The zoology department sent one of its members to speak to the conferees regarding the state's natural history "and the work of the college."[232]

W. J. Beal built botanical collections that served the campus community, Michigan, and the nation. Dennis Cooley, a physician, had started the herbarium in 1863 when he donated his collection of 20,000 specimens to M.A.C., and Beal added to the collection until he

The "General Museum" was located on the second floor of the Library-Museum (Linton Hall). It consisted of mounted mammals, birds, reptiles, fish, and other animals, fossils, skeletons, and minerals. W. J. Beal in his history of M.A.C. refers to the largest mammal as "a fine bull moose from northern Minnesota," which can be seen in this photograph along with many other specimens that are also on display. Students and the public made frequent use of these collections.
Courtesy of Michigan State University Archives and Historical Collections (Michigan State University, Buildings, Linton Hall, Interior).

Professor W. J. Beal in his garden, circa 1910.
Courtesy of Michigan State University Archives and Historical Collections (People, Beal, William J., Folder 4, no. 256).

retired in 1910. By then he had assembled a total of 105,688 specimens of various types of plants for the college's herbarium:

Seed plants, ferns, and their allies 82,069
Mosses and liverworts 2,010
Lichens 1,183
Fungi 17,953
Algae 2,470
TOTAL . . . . .105,688[233]

Fortunately, the collection survived the fire that destroyed the Botanical Laboratory in 1890. Beal, his colleagues, and their students drew upon examples in the collection to identify plants and seeds sent to them by farmers, nursery operators, entomologists, foresters, and gardeners who wished to know if their plant was poisonous, a weed, or noxious.

In 1873 Beal started a botanical garden, with grasses and clovers, which had expanded to 2.1 acres by 1910. By that time Beal had planted over 2,100 species, many native to Michigan. Arranged by families, each plant was identified by two labels—one above ground and the other a zinc tag in the ground containing its number, which had been written in the garden's record book. People studying plants visited the gardens from M.A.C., other Michigan colleges, and out-of-state. New species came by donation or purchase, and Beal shared M.A.C.'s collections to help build gardens in other states.[234] The W. J. Beal Botanical Garden, which has more than 5,000 different kinds of plants, still functions in the manner established by Beal nearly a century and a half ago.

The faculty had transformed units of knowledge into courses of study used to teach their students the content of knowledge and its application in the world. The land-grant way demanded that knowledge discovered through scientific research and course work be used by professors, students, and alumni to serve men, women, and children living throughout the state, nation, and world. M.A.C.'s alumni, faculty, and administrators worked tirelessly to develop an amazing outreach program that touched the lives of thousands upon thousands of Michigan citizens. And, in the process, the people of Michigan, including the practical farmers, came to accept the college as an institution that served them well.

4

Reaching the People

FROM ITS BEGINNING, MICHIGAN AGRICULTURAL COLLEGE WORKED hard to find ways to transform knowledge gained at the college into practices that would improve ordinary or common things in the everyday lives of people residing in Michigan and elsewhere. Over the course of sixty years, M.A.C.'s outreach to the public evolved from a campus-centered approach to one that also depended upon associations located in counties and high schools throughout the state. Initially, professors got in touch with the public by answering letters and sending exhibits to fairs, but these efforts did not get M.A.C.'s message into the hands of very many people. In the mid 1870s, the faculty in conjunction with local agricultural leaders organized farmers' institutes, where professors gave public lectures to people in their own communities and discussed their concerns with them. M.A.C. researchers interpreted their findings for the public through articles in newspapers and agricultural journals. The establishment of the experiment station in 1887 stimulated more scientific research on campus and the publication of bulletins that reported the results of experiments carried out by M.A.C. professors and scientists. Soon after the turn of the century, alumnae began to teach home economics in high schools, and they were soon joined by male classmates who introduced courses in agriculture in rural high schools. The advent of extension in 1908, which became the Cooperative Extension Service in 1914, marked the start of an orchestrated effort to transform rural Michigan. Agricultural and home demonstration agents and extension specialists helped people where they lived, to improve their lives and to thirst for even more information from their land-grant college. The Cooperative Extension Service gave M.A.C. the service component that it needed to accomplish its mission.

LETTERS AND EXHIBITS

The State Agricultural College had to communicate its purpose, the substance of its academic programs, and the findings of its research to the public in order to fulfill its mission. From the institution's beginning, administrators and faculty members faithfully answered a myriad of questions by mail. Queries from prospective students or their parents always got a quick response from the president. Editors of newspapers and agricultural journals demanded information about M.A.C., asked faculty to contribute articles, or asked that bulletins issued by the experiment station be sent to them. Alumni frequently corresponded with the presidents sharing any concerns that they had for the well-being of their alma mater. On numerous occasions, politicians unblushingly expressed their opinions and ideas for improving the college to the presidents. The exchange of letters between M.A.C. officials and individual citizens enabled the institution to begin meeting the personal and commercial demands of the world of agriculture.

A look at a few letters written to the presidents of M.A.C. in the 1890s unlocks one of the secrets to the ultimate success of the institution's land-grant philosophy and its eventual acceptance by the people of Michigan. Inquiries, no matter how small or seemingly insignificant, merited replies. The concerns expressed by common men and women usually found a respondent at M.A.C. who showed a genuine interest in providing helpful information. In 1912, Dean Robert S. Shaw reminded the heads of the departments in the agricultural division and the experiment station "to answer correspondence as promptly as possible" and "to give as full information as possible, thus satisfying the questioner rather than dealing with the matter in the easiest way."[1] (W. J. Beal reported in his *History of the Michigan Agricultural College* that 5,000 or more questions might come to the college in one year.[2]) The benefits of knowledge gained on the campus were not reserved for the elite but could be put to use by common people in their barns, fields, gardens, kitchens, or shops.

On 27 October 1894, for instance, W. L. Webber sent, by American Express, a package to President Lewis G. Gorton containing a plant found on one of his farms in Huron County. Webber wanted to know if it was a Russian thistle or prickly lettuce. Clearly troubled by the presence of this weed in his field, Webber looked to M.A.C. to receive "definite advice" about its identification. Three days later Gorton assured him "it is nothing worse than a prickly lettuce." For good measure Gorton also sent Webber two copies of a bulletin on Russian thistle.[3] Earlier in the year M. Elva Worden of Oden pleaded: "I write to ask you about a stump puller we need one (or many) in this community very much but do not know what kind to buy." Two days later Gorton advised Worden that there was "no better stump puller than the screw" and he could purchase one from Wm. McDonald of Lapeer.[4]

Others calling upon Gorton for help included Mrs. C. H. Strand of Jackson, who wanted to know how many times she needed to spray her apple orchard for army worms and "what preparation do you use to spray with."[5] On 22 June 1894, insects feasting on rye heads climbed out of Gorton's mail. W. L. Steele, of Summerton in Gratiot County, had sent the pests to the college wanting to know: "What are they & what damage (if any) do they do? What is the remedy (if any)?"[6] State Senator F. M. Briggs of Plymouth touched upon the subject of ornamental flowers when he relayed his daughter's need to know how to care for her petunias during winter. Brigg's words brimmed with fatherly pride as he reported that her flowers were "wonderfully fine" and that "Every one of them lived."[7]

In April 1893, J. W. Sharp of Kenton, Houghton County, articulated a wide range of issues being addressed by M.A.C. that were essential to Michigan agriculturalists. His letter to President Oscar Clute reveals questions about planting relative to climate, economics, publications, and myth. Skeptical of his neighbor's belief in the effects of the moon, Sharp called upon Clute to relay to him the practical application of scientific knowledge known at M.A.C. to his farm:

> Will you kindly give me some information regarding the time to plant seeds. I have only one neighbor and he says that seeds planted in the fading of the moon will not bear but have all false blossoms. I am just starting a farm and I would like to run it as profitably as possible also is there anything in the story that potatoes and all roots does better if planted in the Dark of the Moon. I am going to set some Strawberries plants and some fruit trees this spring if you have a circular or some pamphlets bearing on farm notes you would do me a great favor by sending me one I am so far north that I will have to give Tomatoes a good start to get them ripe before the frost

sets in the fall. I planted my Tomatoes yesterday and my Neighbor told me I would have nothing but false bloom hence my writing to you on the subject.[8]

It is noteworthy that Sharp called upon the agricultural college for what he apparently believed would be more reliable information than that provided by his neighbor. As more farmers looked to M.A.C., they exhibited a willingness to let go of old ways in hopes of improving their businesses.

Other citizens, however, were not pleased with M.A.C.'s response to their concerns, and they did not hesitate to point out where they felt that the college was not doing its job. In 1894, J. L. Buell wanted to have soil from his Dickinson County farm tested at M.A.C. He believed that these analyses "would be on a tract of all the soil in this county." When his request was denied on the grounds that he intended to use the results for "mercenary" purposes, he lashed out at what he saw as a slight to the Upper Peninsula. He noted that M.A.C. had published "chemical tracts for the soils from some 25 of the Lower counties," but "there is but a single analysis from the U.P.—that for muck from Luce County." He thought that rather than reject his application for soil analysis, M.A.C. should be eager to extend its experimental work to more regions of the state. Upper Peninsula tax-payers were not getting a good return on their investment in the college. The public cared about what went on at the agricultural college.[9]

Some citizens wished to purchase things offered by M.A.C. for their personal or commercial use, and the nature of their requests shows that the public was coming to expect more from the college than training practical farmers. Mrs. Belle Tanner, of Charlotte, inquired of the "Supt. of Agricultural College": "I have been informed that you have the Silk worm at the Collage and if So could i get a cuple of Coocoons or Some Eggs and if So how much would it cost me."[10] Charles Ziem, who operated the "Sallon and Restaurant" in Saginaw, spelled out the history of his problem and made his need known in a few words: "please Send me one Setting of Houdan Chicken Eggs, Had no luck with last years Setting, only two Hens left rest all Stolen: Enclosed please find Money order for $1.50 Yours Respectfully."[11] Two cocoons and a dozen eggs made no difference in the state's economy, but they enriched the lives of two of its citizens.

Public interest in petunias, silk worms, and Houdan eggs function as examples of the changing nature of agriculture and life in America and the land-grant response to it. The growth, maintenance, and use of ornamental flowers and plants evolved into an important enterprise that depended upon people trained and knowledge generated at Michigan Agricultural College and its sister institutions across the country. People grew flowers not only to enjoy their beauty, but also to help commemorate important times in the lives of their families such as weddings, funerals, and anniversaries, in addition to decorating the graves of their loved ones. Women and men who grew their own petunias, such as Miss Briggs, increasingly called upon the college for more information relative to their successful care. Many others purchased needed floral arrangements at their local nurseries and greenhouses, which stimulated the creation of more of these businesses. Men and women trained in horticulture at M.A.C. played key roles in the design and operation of nurseries and greenhouses.

The agricultural community's demand that M.A.C. create exhibits for fairs held in Michigan offered another avenue for it to demonstrate its worth to the public. Since early in the nineteenth century when Elkanah Watson introduced the idea of agricultural exhibits to the United States, farmers eagerly showed off their livestock and produce at local and state fairs.[12] M.A.C. regularly sent exhibits of livestock, horticulture, floriculture, soil science, and other relevant subjects to state and local fairs. When the college showed reluctance (often due to costs) to participate in fairs, fair officials did not hesitate to voice their displeasure. In August 1871, P. R. L. Pierce complained to Governor Henry P. Baldwin that the State Board of Agriculture acted "very unwisely" when it refused to send farm produce from M.A.C. to Grand Rapids for the Northern Agricultural Society's Exposition. Pierce impressed upon the governor that "great harm" would accrue to the institution if it did not change its mind.[13] Twenty-two years later, Huron County officials enthusiastically invited M.A.C. to take part in the county fair that they were planning for their new fairgrounds, pointing out the benefits it would gain by sending exhibits:

> If you would consent to exhibit here it would accomplish two grand
> objects. It would practically insure the success of our fair. And it
> would show the people of Huron County as nothing else would the

good that the college is doing the farmers of this State. Nine-tenths of the farmers of Huron County have but a very meager knowledge of what the college is doing, or of the advantages that it presents to the farmer boys of this State.[14]

As this passage shows, supporters of M.A.C. relentlessly called upon its officials to take every opportunity to let the farmers of Michigan know what their agricultural college had done for them lately.

Meeting demands to show a presence at scores of fairs held each year throughout Michigan could have overwhelmed M.A.C., but President Snyder spared no effort for the Michigan Agricultural Society's annual fair in 1903. Each department exhibited its work: engineering showed "work in wood, pattern making, forge work, machine work and draughting"; the women's department displayed textile and food products; and the agricultural department brought fruits, grains, grasses, and other things grown in its fields and gardens. The livestock exhibit included "one typical animal from each breed of cattle, sheep and swine" accompanied by a knowledgeable person who would explain each breed to the public.[15] Valuable as fairs were, they formed only a small part of the effort to reach the public. Farmers' institutes connected with far more people beginning in 1876.

FARMERS' INSTITUTES AND FARMERS' WEEK

The origins of farmers' institutes in Michigan brought to light the State Board of Agriculture's recognition that the agricultural college had formed ties with very few Michigan farmers during its first sixteen years. As a result, the board and the faculty proposed to undertake measures to remedy this sad state of affairs. In 1875, M.A.C. decided to hold farmers' institutes in six communities across the Lower Peninsula. Michigan followed the examples of Kansas and Illinois, where successful institutes had already been held.[16] In the *Annual Report* for 1875, professors R. C. Kedzie, W. J. Beal, and R. C. Carpenter concisely summed up for the board the fundamental challenge facing the State Agricultural College:

Whatever may be the cause, we think the *fact* is sufficiently evident that there is a want of sympathy between the farmers and the

Agricultural College. By reason of this want of sympathy the farm- ers are deprived of much of the good which they may secure from the Agricultural College, and which they have a right to demand; and the College is crippled in its work for the same reason. We believe that this want of sympathy and lack of interest are because the farmers, as a class, know but little of the real working of the Col- lege, and that if the Board and Faculty could be brought into more intimate association with farmers in all parts of the State, these evils might be removed. If the College is not doing such work as ought to command the confidence of intelligent farmers in all parts of our State, then our system should be altered so as to meet the just demands of the farmers; if we are doing such work, we may still fail of our duty if we fail to make this fact known. There is something wrong when the College, after 16 years of continuous work, is still denounced and decried in some of the most flourishing sections of our State.[17]

Kedzie, Beal, and Carpenter blamed the college for its shortcomings, not the farmers. If farmers only knew of M.A.C.'s good works on their behalf, they could benefit greatly. M.A.C. had to find ways to engage farmers in their own education if the institution hoped to achieve its mission.

Local farmers played an active role in the institutes, for they were responsible for choosing topics that would interest their neighbors. Professors went to Armada, Rochester, Allegan, Decatur, Adrian, and Coldwater in January 1876 "to meet and talk over, in a common-sense way, matters of vital interest to the farmer." Farmers were encouraged to share their knowledge and concerns with the visiting lecturers but also "to put in practice on the farm" some things they learned at the meetings. Furthermore, leading local agriculturalists gave talks, pre- pared essays, and participated in discussions. While the community paid for a heated and lighted hall, M.A.C. picked up the other expenses. The board encouraged extensive coverage by the press.[18]

The State Board of Agriculture published the lectures, essays, and points of discussion at institutes in its annual reports. This met the col- lege's commitment to make available to the public, in a widely circu- lated format, M.A.C.'s informed responses to problems and concerns of Michigan farmers. A sampling of topics discussed at meetings, which ran for two nights and one day, suggests the wide range of agri- cultural issues that demanded attention. President Abbot presented

"The Prejudice Against Industrial Schools" at Armada, Rochester, and Coldwater. R. C. Kedzie's talk on lightning rods elicited considerable discussion in Allegan, Decatur, and Coldwater. Professor George T. Fairchild challenged audiences in Allegan, Decatur, and Adrian with his lecture "Education—Who Needs It? Who Can Afford It?" Entomology professor A. J. Cook told farmers in Decatur, Adrian, and Coldwater that "cut-worms, May-beetles, and army-worms" were the most serious pests threatening crops in Michigan. Professor R. C. Carpenter lectured on "Road Making" in Allegan and on "Farm Machinery and Implements" in Decatur.

The contributions of local presenters demonstrated that not all agricultural knowledge in Michigan resided along the banks of the Red Cedar River, and their interests foretold both future experimentation at M.A.C. and subjects for outreach. Across southern Lower Michigan, farmers exhibited a thirst to know more about growing fruit, pest control, breeding livestock, producing hay, growing corn, draining wetlands, increasing the fertility of soil, and ways to improve the efficiency and profitability of their operations. For example, in Allegan farmers and professors heard E. C. Reid read Mr. Clubb's essay "Fruit-Growing in Michigan," as well as speeches given by other local men titled "Shall Farmers Keep Bees," "Long-Wooled Sheep," "Breeding and Feeding Swine," "Raising Hay and Fattening Cattle," "Cutting and Curing Hay," and "Thinning Fruit." When farmers returned home, they took with them new ideas that might help them become more efficient and productive. President Abbot and professors Kedzie, Carpenter, Beal, Cook, Gulley, and Fairchild, brought back to M.A.C. a newly informed perspective of the practical needs of the agricultural community where many viewed them as being irrelevant. The State Agricultural College had taken an important step toward connecting with the practical farmers of Michigan.[19]

The Rochester Farmers' Institute is of particular significance, because two women made presentations that foreshadowed the inclusion of issues relevant to the home as well as the field and the barn, and the future role of women in M.A.C.'s outreach programs. On Friday morning, 14 January 1876, Annie Hall told her audience: "No plant, tree, or flower is more sensitive to sun, shower, or soil than is every human life to the thousand trifling things which make up its daily existence. Yet how much more attention is paid to the first than the last. . . . Home is the place where a life is rooted and from whence it draws its greatest good or deepest ill." She went on to challenge parents to make their homes "attractive," just as the proprietors of saloons and theaters had done, in order to instill good moral values in their children. She also urged that parents encourage their daughters and sons to read and to subscribe to periodicals and buy books for their families.[20] In the afternoon, Mrs. Keeler, of Disco, encouraged farm men and women to participate in the Grange because its activities enabled them to develop social skills needed to engage the world away from their farms. She advised each family "to take at least one good newspaper, with one or more standard magazines, and season the mixture with a liberal sprinkling of the latest publications of the most popular authors and poets" to prevent "the mind from retrograding."[21]

Hall and Keeler opened the way for women to introduce more of their concerns into public agricultural education offered at the institutes, which helped to prepare the way for the establishment of the women's course at M.A.C. in 1896. Mary Mayo, Margaret Sill, and others used the institutes as opportunities to elevate the attention paid to life in the farm house to match the interest given to life in the barn. Human nutrition, the well-being of children, and the quality of life for women who operated households became as highly regarded as animal nutrition, the health of calves, and the quality of life for men who tilled the soil. At Battle Creek in January 1881 Mary Mayo argued forcefully that education of both sexes was "at the bottom of all society."[22] During the 1895–96 year, Mayo led separate sessions for women at twenty institutes with a total attendance of 5,309.[23]

Other women also came to the podium to discuss publicly ways to improve the ordinary and regular routines of daily life and, at the same time, laid the groundwork for monumental changes in the education of women in Michigan. In the process, they made sure that the land-grant way in Michigan included women. On 10 January 1883, a nervous Mrs. Albert Granger shocked no one when she read her essay "Management of the Dairy" at the Hastings Farmers' Institute. Everyone in the hall knew that farm women made butter from cream and that cream came from milk given by cows. She noted that the production of good butter required cleanliness, cool temperatures, and a proper room for the keeping of cream and milk. No doubt, some butter makers learned some ways to improve their methods, but no one could have missed Granger's most poignant point as she concluded her essay:

> I hope the day is not far distant when creameries and butter factories will be established at convenient distances through the country to take the milk from the farms, thus dispensing with the drudgery incident to the care of milk, and leaving more time for the wife and daughters to read and study and cultivate their minds, that if called upon to write an essay for an occasion like this, they would not so keenly feel their inability to respond to the call.[24]

Every farm wife could identify with "drudgery." Granger looked to improved technology not only as a way to escape drudgery, but also to open the door to a life enhanced by an educated mind. She linked women's lives on farms with the changing ways of the emerging commercial agriculture that saw more farmers producing more items for sale on the market. Granger insisted that women benefit from new technology.

Margaret Sill, of Detroit, advanced the practical instruction of the institutes when she introduced cooking classes to the Battle Creek Institute in the mid 1890s, where she struck a responsive chord among local women. She gave a series of lectures (one titled "The Elevation of Household Labor") and cooking demonstrations over a three-day period in January 1894.[25] Sill then put on a cooking school in April to expand the instruction that she had given at the institute. Pleased with the opportunity to improve her kitchen skills, Nellie H. Mayo told President Gorton "I dont think you can realize the good you have done in getting her here at the Farmers Institute. . . . and I am very glad I have the oppertunity of a course of her lectures."[26]

Farmers' institutes grew in number after 1876 and reached tens of thousands of people who enjoyed the lively gatherings. Men and women often fought their way through snowstorms and frigid temperatures to meet with their friends and neighbors. Local musical groups entertained audiences with songs and instrumental music before each session. As a result of legislation passed in 1895, the State Board of Agriculture appointed Kenyon L. Butterfield, '91, superintendent of farmers' institutes, and he provided leadership that stimulated new growth.[27] In 1896, sixty-eight counties hosted meetings with an estimated attendance of approximately 91,000.[28] The institutes' success prompted Butterfield to say that they had brought about a "marked change of attitude toward the College" by the farmers. He emphasized that the institutes "brought the college down to the people, where they could see it work; they have enabled the college professor and practical farmer to face each other on a common platform."[29] Growth continued under Butterfield's successors, Clinton D. Smith and Levi R. Taft. In 1906 there were 74 two-day institutes and 257 one-day institutes with a combined attendance of 122,433.[30] Ten years later, 179,210 people took part in 357 sessions at 78 county institutes and 1,152 sessions at 470 one-day institutes held across the state.[31]

Even though the institutes did much to bring together the State Agricultural College and the people of Michigan, the college could never rest from its efforts to reach the people. In 1896 a Board of Visitors that had been appointed to evaluate M.A.C. lamented that it was not "fully appreciated by the people of our State."[32] Four years later the Michigan State Grange acknowledged that the "Institutes are the means by which the College reports most direct to the people." At the same time, the Grange recognized that most farmers benefited from M.A.C.'s work, whether they were aware of it or not. Bulletins, press accounts, institutes, agricultural organizations, and just plain talk between farmers and college personnel all served "to mix experimental results with the home, thoughts and methods of people."[33]

The creation of the annual Round-Up Farmers' Institute in 1896 further strengthened the institutes' impact on Michigan farmers. Farmers came from across the state to the four-day meeting in Grand Rapids to hear lectures and participate in discussions of a wide range of topics including marketing peaches, growing vegetables in hothouses, stock breeding, and growing potatoes. Mary Mayo led three sessions of the women's section designated "The Kitchen," "The Rural Home," and "Mother and Daughter."[34] Nearly fifty counties were represented. In order to stimulate more public interest in the following year's meeting, lecturers gave different presentations than those delivered at local meetings.[35] Farmers and college officials had sown the seeds that saw the Round-Up grow into Farmers' Week, first held at M.A.C. in March 1914.

Farmers' Week represented the success of nearly forty years of institute work. Now farmers, by the hundreds, came to *their* agricultural college for the major outreach event of the year. For four days in

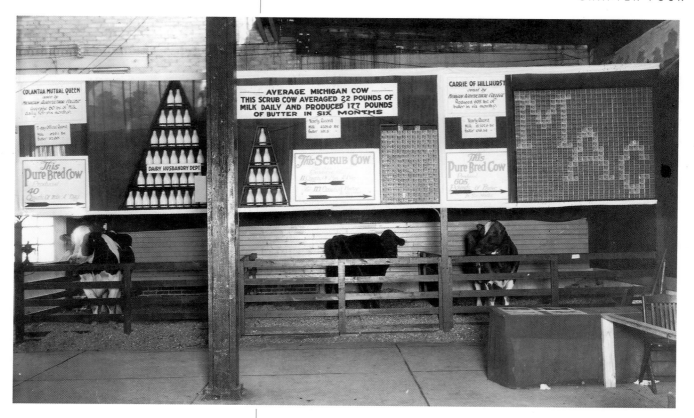

Agents of the Cooperative Extension Service encouraged dairy farmers to improve their herds by getting rid of their scrub cows and replacing them with pure bred animals. Displays like this one at Farmers' Week, circa 1921, graphically illustrated that pure bred cows gave much more milk and butter fat, which meant larger milk checks for the producer. Courtesy of Michigan State University Archives and Historical Collections (Michigan State University, Cooperative Extension Service, Farmers Week, Folder 6).

early March farmers took in exhibits, attended lectures, watched demonstrations, and discussed a wide range of agricultural matters with M.A.C. faculty and non-college presenters, and among themselves. The farmers found the new format superior to that of the old Round-Up.[36] Jennie Buell, the lecturer for the Michigan State Grange, called the first Farmers' Week "the finest demonstration of 'reaching the people' I have ever seen in our state."[37] The second Farmers' Week, held in 1915, drew about 1,000 people. In 1922 nearly 5,000 attendees witnessed "M.A.C. on Parade," a mile-long procession led by the varsity band. Representatives from athletics, student organizations, and academic departments marched with livestock and farm machinery to the delight of the crowd.[38] The event was so impressive that the *Detroit Free Press* made a film of it and showed it in theaters around the state.[39]

Buell recognized one of the key elements to the success of outreach—discussion of "the relation of costs and incomes." Reflecting upon the poultry sessions, she surmised that most farmers who kept a flock of chickens did so at a "positive loss if an account were kept of feed, etc." The presenters' emphasis on the financial aspects of raising poultry helped farmers to take a critical look at equipment and feed needed to keep a chicken coop healthy and profitable. Poultry growers needed to know how to sort their way through the advertisements of "all manner of patented contrivances and ready-mixed foods" that appeared in journals and catalogs. Buell hoped that future meetings would provide poultrymen with more information by addressing the topics of "Comparative prices of poultry supplies and equipment" and "Home made devices and contrivances."[40] Since M.A.C. understood that agriculture was a business, it structured some of its outreach programs to teach farmers how to keep their costs in line with the earnings received for their products.

READING CIRCLES, DEMONSTRATION TRAINS, ON-CAMPUS MEETINGS, AND WEATHER

As Michigan Agricultural College sought new methods to reach the public, it organized reading circles, outfitted demonstration trains, and hosted meetings and gatherings on campus to touch people where they lived or to encourage them to visit M.A.C. These programs expanded the scope of M.A.C.'s work to enlarge the number of men, women, and children directly involved in activities designed to introduce them to new ideas, new technology, and new ways. As outreach grew, it sought to meet more of the physical, intellectual, emotional, economic, and social needs of individuals, families, and the communities where they lived.

Putting books and other printed materials about homemaking and agriculture in the hands of rural Michigan men and women became one important means of outreach. Hoping to achieve that objective, the Michigan State Grange and Michigan Agricultural College created the Farm-Home Reading Circle on 13 December 1892. Men and women could purchase books by mail from the college at cost. These works could be used individually or by members of small reading groups. Readers could choose one of five courses: soils and crops, livestock husbandry, garden and orchard, homemaking, or political science. Each course had a reading list of five books, and people could choose, but were not required, to take an examination on each book that they had read. After passing the exams, they received a certificate signifying that they had completed the course.[41] M.A.C. and the Grange actively promoted the program, distributing 5,000 advertising packets and 1,500 circulars at institutes during 1896. By the end of the year, 221 people had enrolled, but many more had access to the materials purchased by members.[42] In 1908, Professor of Agricultural Education Walter H. French changed the program's name to Farm Home Reading Courses and offered separate four-year courses for men and women. Although a lack of funds stymied the program's growth, 193 people enrolled the first year and fifteen granges and six farmers' clubs did the reading as a group.[43] Four years later, the time needed to complete the course had shrunk to three years, and the program was called the College Extension Reading Course. M.A.C. encouraged people to meet as reading clubs with members of the granges, farmers' clubs, and school districts "for discussions, debates, readings, etc.," which enriched the social life of the community.[44]

Reading programs evolved to include subjects that would challenge the intellectual interests of growing numbers of people. Jennie Buell, when she served as a specialist in continuing education, linked critical reading to the growth of democracy. She stressed the goal of the Home Reading Courses in 1926: "The aim of this effort is to encourage a definite educational use of daily margins of time; to set people consciously to think of what they read, and then to discuss it with others. Only so shall we attain in the course of time to an intelligent democracy."[45] Nineteen courses explored topics such as "Child Study," "Practical Helps for Parents," "Other Folks," and "Wild Life." A few examples of books on the lists reveal that readers in 1926 could expect to learn about topics that were still popular eighty years later. Elizabeth Harrison's *A Study of Child Nature* gave "many incidents to illustrate how to help a child develop," and Herbert A. Gray dealt with "problems of sex in a positive, wholesome manner" in *Men, Women, and God.* The "Joy-of-Country-Living Literature" course included two books by M.A.C. alumni. *Adventures in Friendship* by Ray Stannard Baker, who wrote under the pseudonym David Grayson, told that "When a man's heart really opens to a friend, he finds there room for two." Liberty Hyde Bailey's *The Holy Earth* argued that "One does not act rightly to one's fellows if one does not know how to act rightly toward the earth." Men and women enrolled in "Other Folks" read books that offered critical insights into American social history, including Jacob Riis's *The Making of an American* and Booker T. Washington's *Up from Slavery*.[46]

Perry G. Holden, '89, professor of agronomy at Iowa Agricultural College, introduced the agricultural demonstration train to Iowa farmers in 1904. Two years later, M.A.C., in conjunction with the New York Central Railroad, put its first "rolling classroom" into motion. During the next thirty-one years, the college, the railroad, and local communities pooled their resources to outfit sixty-three trains to travel across both the upper and lower peninsulas. The railroad supplied the trains and M.A.C. constructed exhibits and provided staff to give lectures and demonstrations. Local officials at each stop made the arrangements necessary to enable as many people as possible to benefit from the instruction offered when the trains stopped in their towns. Each train focused on one subject. The first train in 1906 dealt with corn, and for the next three years the theme shifted to general farming. In following years other topics included health, land clearing, dairy, sheep, forest fires, weed control, soils, poultry, and better dairy sires.[47] College officials expressed great pleasure when the local press advertised the coming of a train and published favorable reports about the accomplishments that resulted from visits.[48]

M.A.C. staged numerous functions on campus, where people could benefit from campus facilities and faculty and staff expertise. In the early years, farmers came to participate in events such as the sheep shearing exhibition held on 23 and 24 May 1866 that featured a speech by Governor Henry H. Crapo, and the public sale on 18 April 1888 of shorthorn and Hereford cattle bred at the State Agricultural College.[49] In the early twentieth century, M.A.C. sponsored conferences, like the one-week poultry institute in 1911, that covered virtually every aspect of the poultry business from hatching to marketing.[50] Operating on a larger scale, the Michigan Improved Stock and Breeders' and Feeders' Association brought together a number of different groups for their annual meeting on campus. In 1916 members from the following state associations of breeders of horses, cattle, sheep, and swine came to East Lansing to learn ways to improve their businesses: horses, Holstein-Friesian, shorthorn, Aberdeen Angus, Guernsey, red poll, Jersey, Hereford, merino sheep, Shropshire sheep,

A large crowd flocked to a stock auction held in the agricultural pavilion, located at the back of Agriculture Hall, in 1915. Many farmers and cattle buyers came frequently to M.A.C. to attend stock judging events and auctions. Courtesy of Michigan State University Archives and Historical Collections (Michigan State University, Agriculture, Livestock auctions).

Hampshire sheep, Oxford sheep, Poland China, Duroc-Jersey, and Berkshire.[51] Meetings like these provided venues for farmers to be introduced to new techniques, to acquire improved breeds, or to learn to manage their operations more profitably. These get-togethers gave farmers a chance to see for themselves M.A.C.'s facilities and to observe its faculty and students at work. But specialized gatherings reached narrow audiences and relatively few people.

Since Michigan Agricultural College saw the world of agriculture as encompassing more than the farmstead, it hosted meetings designed to uplift the social as well as the material side of rural life by focusing on the community and the farm family's place in it. Conferences for rural ministers, first held in 1910, serve as examples to illustrate this function.[52] The one-week program for the 1919 annual conference for ministers and laymen and the two-week short course that followed for ministers stressed that they needed to do more than preach sermons and officiate at weddings and funerals. Pastors

THE FARM CONVENIENCE TRAIN

OPERATING OVER THE NEW YORK CENTRAL LINES IN SOUTHERN MICHIGAN WILL SOON BE IN YOUR COMMUNITY.

It will contain the most complete exhibit and information on Farm Water Systems — Septic Tanks — Farm Construction — Concrete — Farm Electrical Equipment — Lightning Protection and Handy Devices for the Farm, ever to be taken directly to your door.

The Farm Conveniences listed above are becoming more of a necessity every day. They save time and labor and make farming more enjoyable. Most of the equipment will be inexpensive, and easily installed. Much of it may be made on the farm.

SELECT THE PLACE MOST CONVENIENT TO YOU AND VISIT THE TRAIN.

SCHEDULE OF MEETINGS
(All Central Time.)

Eaton Rapids, Monday, October 1......12:00 Noon to 4:30 P. M.	Allegan, Monday, October 8............9:30 A. M. to 12:00 Noon
Albion, Monday, October 1 ....6:30 P. M. to 9:30 P. M.	Byron Center, Monday, October 8.... 1:30 P. M. to 4:30 P. M.
Hillsdale, Tuesday, October 2............9:00 A. M. to 12:00 Noon	Caledonia, Tuesday, October 9.......... 9:00 A. M. to 12:00 Noon
Adrian, Tuesday, October 2............1:30 P. M. to 5:00 P. M.	Hastings, Tuesday, October 9........ 2:00 P. M. to 9:00 P. M.
Hudson, Tuesday, October 2............7:15 P. M. to 10:00 P. M.	Charlotte, Wednesday, October 10.... 9:00 A. M. to 12:00 Noon
Deerfield, Wednesday, October 3....8:30 A. M. to 11:00 A. M.	Nashville, Wednesday, October 10.... 1:30 P. M. to 4:20 P. M.
Monroe, Wednesday, October 3........ 1:00 P. M. to 4:00 P. M.	Dowagiac, Thursday, October 11...... 9:00 A. M. to 2:00 P. M.
Manchester, Wednesday, October 3.... 6:00 P. M. to 9:00 P. M.	Niles, Thursday, October 11............ 3:30 P. M. to 9:00 P. M.
Blissfield, Thursday, October 4.... 9:00 A. M. to 12:00 Noon	Cassopolis, Friday, October 12......10:00 A. M. to 3:00 P. M.
Tecumseh, Thursday, October 4.... 1:00 P. M. to 3:00 P. M.	Lawton, Saturday, October 13........ 8:00 A. M. to 11:00 A. M.
Sturgis, Friday, October 5 .... 8:00 A. M. to 10:45 A. M.	Marshall, Saturday, October 13.... 1:30 P. M. to 4:30 P. M.
Coldwater, Friday, October 5 .... 1:00 P. M. to 5:00 P. M.	Battle Creek, Monday, October 15...... 9:00 A. M. to 12:00 Noon
Three Rivers, Saturday, October 6.... 9:00 A. M. to 12:00 Noon	Jackson, Monday, October 15...... 2:05 P. M. to 6:00 P. M.
Kalamazoo, Saturday, October 6........ 1:00 P. M. to 5:00 P. M.	Mason, Tuesday, October 16 ....10:00 A. M. to 3:30 P. M.

Look for the Cars Near the Depot or Freight House.

THE MICHIGAN STATE COLLEGE
In Cooperation with
THE NEW YORK CENTRAL LINES

Six thousand copies of this advertisement for an agricultural demonstration train in 1928 were distributed by county agents, banks, and local businesses. Courtesy of Michigan State University Archives and Historical Collections (Madison Kuhn Collections, UA 17.107, Box 1136, Folder 58).

attended sessions to study their relationships with farmers, community organizations, and schools. They heard lectures on technological advances in gasoline engines, electrical systems, and batteries. Wives were given the option to go to the regular sessions or to participate in programs offered by the home economics faculty on "food, health, home decoration, etc." Ministers who attended the short course heard a series of lectures on church history, "The modern preacher and his job," Paul and Corinth, and "present tendencies in literature," all of which enhanced their scholarly and professional training. But they devoted equal time to studying subjects that were important to their parishioners: gardening, animal husbandry, poultry, and farm crops.[53]

In Augusts during the late 1890s and early 1900s farm families living throughout the Lower Peninsula hitched up their buggies or boarded special trains to travel to an outing at their agricultural college. For instance, on 15 August 1899 an excursion train pulled out of

Big Rapids at 6:50 A.M., stopping at the farming communities of Rod-ney, Mecosta, Remus, Millbrook, Blanchard, Wyman, Edmore, McBride, Stanton, Sheridan, Fenwick, Shiloh, Stanton Junction, Ionia, Lyons, Collins, Portland, Eagle, and Grand Ledge before arriving in Lansing at 10:45. Guides greeted men, women, and children, who came armed with picnic baskets brimming with tasty food.[54] The vis-itors enjoyed the gardens, inspected the barns and fields, met profes-sors, and toured buildings as they acquainted themselves with the grounds and people of Michigan Agricultural College. People came in great numbers, as attested to by the *M.A.C. Record* in 1897:

> To record the names of visitors to M.A.C. during August would sim-ply be impossible. They came in two, three, fives, dozens, wagon loads, car loads, and they kept coming. They tramped about, seemed interested and got tired, but, so far as learned, we heard of no growlers. It is estimated that at the Sunday-school rally alone between 6,000 and 8,000 visitors were on the grounds. Teams were hitched along the river below the botanical garden to every avail-able yard of fence south of the barns and down the lane beyond the bridge, and to every hitching post on the campus.[55]

Since weather played a vital role in the business of farming, Pro-fessor Robert C. Kedzie made meteorology an important part of M.A.C.'s agricultural research by systematically keeping records of daily weather conditions. Starting in 1863, Kedzie noted the tempera-ture, relative humidity, barometric pressure, clouds, winds, and pre-cipitation over the course of every day of the year. Chemists at the experiment station carried on this work after Kedzie's death in 1902, and in 1910 the United States Weather Bureau opened a station on campus directed by one of Kedzie's students, Dewey A. Seeley, '98. Kedzie first taught an elective course in meteorology in 1873 that opened the way for a number of students to follow careers in meteor-ology. Kedzie and his successors made their records readily available to the public by publishing them in the State Board of Agriculture's annual reports. Agriculturalists could ascertain information that was important to their work such as the last and first days of frost, levels of precipitation, and normal amounts of sunshine in the south central part of Lower Michigan.[56] Kedzie's work also served as an example for farmers to keep similar records in their own localities.

Teaching Home Economics and Agriculture

In the early twentieth century M.A.C. began to train women and men to teach, respectively, home economics and agriculture in the public schools. The college's commitment to meet society's need to educate high school girls and boys in the basics of homemaking and farming led to collaboration between M.A.C. faculty and local school officials. M.A.C. hired professors to teach the history and philosophy of education, to design courses of study for both college and high school students, and to observe the work of M.A.C. students in high school classrooms. As a result, thousands of Michigan youth learned how to live more productive and healthy lives and to adapt to changes that they encountered throughout their lives.

Even before women educated at M.A.C. received formal training in the craft of teaching, they gave instruction in domestic science and domestic arts in the public schools. They stressed the centrality of ordinary things in the daily life of children. Hariette Robson, '00, worked in the manual training department of the public schools in Detroit in 1901. She and her nine colleagues gave instruction in sewing and cooking to girls and in the use of hand tools to boys in grades five through eight. Robson divided her time between nine different schools, where she found that the girls looked forward to their sewing classes. Consequently, Robson met children from "all classes and conditions of society." As she taught the daughters of the rich as well as poor German, Polish, and Jewish girls, Robson could not help "growing very cosmopolitan in sympathy." She shared with the readers of the *M.A.C. Record* in November 1901 the importance of the small things in life as she explained why she preferred working with the poorer students: "Their cramped, soul-starved lives explain so much, and they are so appreciative of every little kindness." She noted that it was important to compliment a little girl who was wearing a clean apron or possessed "a pair of clean hands when on a tour of inspection."[57] Clean hands in the kitchen, of course, led to a healthier family, and even that small detail needed to be recognized, valued, and taught.

School administrators sought out graduates of the women's course, and M.A.C. women filled positions across the state and country. Beginning in 1902 seniors received instruction in educational philosophy and history from Professor Maude Gilchrist that helped to prepare them for the classroom.[58] In July 1903, President Snyder recommended Hettie Wright, '03, to Superintendent E. E. Ferguson to teach sewing and cooking in the schools of Sault Ste. Marie.[59] Snyder suggested several candidates to E. W. Blackhurst, principal of the Racine High School, Racine, Wisconsin, to instruct domestic science classes in his school. Marguerite A. Nolan, '02, had a year of teaching experience, but not in domestic science. Snyder saw fit to tell Blackhurst that "She is Catholic in religion." Others who could fill the need in Racine were recent graduates Bessie Buskirk, '03, and Edna Smith, '03, or the experienced Alice Cimmer, '00, and Fleta Paddock, '01, both of whom would command a higher salary.[60] Alice Gunn, '01, lived in Iron Mountain, where she taught sewing, household economy, and cooking classes in 1903.[61] Myrtle B. Craig, '07, the first African-American woman graduate of M.A.C., joined the faculty of Western University in Quindaro, Kansas, in 1907, and in 1911 she taught domestic art at Lincoln Institute in Jefferson City, Missouri.[62] Nina Andrews, '08, outfitted the classrooms where she offered instruction in cooking and sewing to girls in both grammar and high schools in Mesa, Arizona in 1911.[63] Alumnae filled teaching positions in other states including Texas, Illinois, and Pennsylvania, as well as in many communities in Michigan.

Ever the leader, Jonathan L. Snyder believed it was imperative that M.A.C. prepare young men to teach agriculture in Michigan's high schools. But before he could turn his vision into reality, he had to convince the public to follow his path to achieve this objective. At the turn of the century, any approach to improving agricultural education depended upon the public being receptive to committing sufficient resources to create successful programs. From M.A.C.'s perspective, the central issue was the role that it could play in initiating and supporting efforts to introduce the state's youth to better farming practices. M.A.C. weighed different options that called for close affiliation with grade schools, or offering an agricultural course at the college that was not rooted in science, or introducing the teaching of agriculture in Michigan's high schools.[64]

M.A.C. decided to persuade communities to establish agricultural courses in their high schools, and to train young men to teach the appropriate subjects. Once again, President Snyder led the charge with a forceful proposal. In June 1907, Snyder shared his plans with

I. Roy Waterbury, of the *Michigan Farmer*, shortly after he took a seat on the State Board of Agriculture. The president acknowledged that board members had provided great assistance to the college "from the business standpoint, but their life-work has not led them to investigate very thoroughly the educational phases of the great problems we have to meet." Now was the time "to push the introduction of agricultural education in our public schools." Some people advocated establishing rural high schools devoted to the study of agriculture. Snyder, however, argued that since most youth in Michigan lived near one of the state's several hundred high schools, they could take instruction in agriculture in these existing institutions. In one sentence Snyder articulated what became the essence of the land-grant philosophy of agricultural education for public schools in Michigan in the early twentieth century: "If we could work out a good course in elementary agriculture for a high school and then were in a position to prepare teachers for the same, it seems to me that in a few years we ought to have such courses in operation in perhaps a hundred high schools in the State." Snyder believed that opposition to this approach no longer existed, and that educators and the public would be receptive to expanding high school curricula to include agriculture.[65]

Snyder's vision saw the preparation of agricultural teachers move from county normal training schools to Michigan Agricultural College. In 1907 there were about forty local normal schools, each connected to a high school, that tried to include in their curricula courses to prepare instructors to teach agriculture in country schools.[66] In addition, there was an agricultural high school in Menominee.[67] Several years later, the Dunbar School of Agriculture and Domestic Science opened in Chippewa County.[68] Lack of qualified personnel prevented the normal schools from training enough teachers of agriculture. M.A.C. offered summer courses on campus to female teachers from normal schools, but the college could not meet the need.[69] Although M.A.C. provided some assistance to the Menominee County School of Agriculture and the Dunbar School, Snyder channeled his institution's resources toward the public high schools. The centerpiece of his approach was the formation of a department devoted to agricultural education at M.A.C.

Guided by Snyder's vision, Walter H. French energetically launched the Department of Agricultural Education at M.A.C. in 1908.

Prior to joining the faculty he had been the deputy superintendent of public instruction for the state of Michigan. While in that position, French had come to see the purpose of agricultural education as being "to improve the condition of rural life; to increase the productive power of the soil."[70] Before 1908 ended French published his new department's first mimeographed bulletin, "A Course in Agriculture for the Public Schools of Michigan."[71] Not surprisingly, this document mirrored the college's agricultural course, requiring a sound scientific foundation for practical agriculture. The syllabus for a high school course in farm crops called for students to interview farmers and observe their practices in six realms: planting, cultivation, harvesting, crop rotation, adaptation, and marketing. Students were to be shown how to apply their study of chemistry, physics, and biology to agricultural concerns. The bulletin suggested that a boy take one agricultural course each semester, beginning with botany, followed by agricultural botany, horticulture, crops, animals and poultry, soils, dairying and farm mechanics, and ending with farm management. Students gave recitations four times a week and spent time in the laboratory, where, among other activities, they experimented with the effects of different temperature and moisture levels on the germination of seeds.

As French trained teachers and convinced high schools to offer instruction in agriculture, M.A.C. expanded and strengthened its outreach to Michigan youth. He offered courses in the science and history of education to men and women interested in teaching agriculture and home economics.[72] In April 1908, F. P. Knapp, superintendent of North Adams schools, asked French to recommend a person to teach agriculture in the district's high school.[73] Roswell G. Carr, '08, went to North Adams High School to inaugurate an agricultural course, thereby starting a new connection between M.A.C. and the farming community. Snyder and French had read correctly the mood of the schools and the farmers, as districts requested more teachers than M.A.C. could produce. During the 1912–13 academic year M.A.C. alumni taught agriculture in twenty-one Michigan high schools and another eleven did likewise in six states spanning from New York to California. French and his assistants oversaw the work of their former students by observing them and visiting with them in their schools.[74] By September 1917 sixty-seven schools had an agricultural course, but that number shrank to forty-seven as teachers joined the military to participate in World War I. Approximately 3,300 youth took agricultural classes in 1917. French was especially proud of the boys' efforts to promote the production of food needed for the war.[75]

The Smith-Hughes Act of 1917 gave a big boost to vocational education at M.A.C. and across the country at the same time that World War I pulled many male teachers out of the classroom. Federal funds provided by the legislation, along with matching state monies, helped to pay for training secondary teachers of home economics and agriculture, a task that fell to M.A.C.[76] French worked out an arrangement with East Lansing High School whereby students could practice teaching under the supervision of Elizabeth Frazer (home economics) and Professor E. Lynn Grover (agriculture). At the beginning of the 1917–18 school year, sixty-four men and seventy-two women were enrolled in the education program; forty men and sixty-one women were graduated at year's end. Thirty-two of the males entered the military rather than take teaching positions.[77] In 1924 the Department of Agricultural Education was reconfigured as the Department of Education and began to offer professional training to students wishing to teach arts and sciences in secondary schools in addition to teacher training in home economics and agriculture.[78] Thirty-seven men received teaching certificates in agriculture and forty-four women earned certificates in home economics in June 1925.[79]

Robert L. Clute's, '96, experience on the Philippine Islands in 1905 foreshadowed M.A.C.'s undertaking to introduce a formal program of agricultural education to communities in that he collaborated with local authorities and worked with people where they lived to improve the practice of agriculture. Male and female students between the ages of fifteen and twenty-two planted, cultivated, and irrigated vegetables on their five-by-thirty-foot plots of ground. They grew yellow dent corn, eggplant, onions, lettuce, beets, radishes, cabbage, okra, carrots, string beans, and tomatoes, all from American seeds. To their dismay, the cotton boll weevil chewed up "all buds of cotton." In the laboratory, students studied "seed germination, plant food, fruits, soils, insects, drainage and irrigation." Each month they attended a meeting to receive instruction from specialists and employees of the colony's agricultural bureau, who discussed "the culture of sugar, hemp, tobacco, and rice." This activity led the chief of the bureau to advocate the establishment of an agricultural school in the Philippines.[80] An alumnus of M.A.C. introduced the land-grant emphasis on agricultural education halfway around the globe even before an agricultural education department or an extension service was established in Michigan.

THE EXPERIMENT STATION

The organization of the experiment station at Michigan Agricultural College in January 1888 reaffirmed commitments by presidents T. C. Abbot and Edwin Willits to science, and led to a growing program of experimentation. For several years, Abbot and Willits worked hard to bring about the Hatch Act, which Congress finally passed in 1887. This measure called for the formation of agricultural experiment stations in conjunction with land-grant colleges in each state.[81] Section 2 laid out in detail the extent to which Congress intended M.A.C. and other land-grant institutions to engage in agricultural research:

> That it shall be the object and duty of said experiment station to conduct original researches or verify the experiments on the physiology of plants and animals; the diseases to which they are severally subject, with the remedies for the same; the chemical composition of useful plants at their different stages of growth; the comparative advantages of rotative cropping as pursued under a varying series of crops; the capacity of new plants or trees for acclimation; the analysis of soils and water; the chemical composition of manures, natural or artificial, with experiments designed to test their comparative effects on crops of different kinds; the adaptation and value of grasses and forage plants; the composition and digestibility of the different kinds of food for domestic animals; the scientific and economic questions involved in the production of butter and cheese; and such other researches or experiments bearing directly on the agricultural industry of the United States as may in each case be deemed advisable, having due regard to the varying conditions and needs of the respective States or territories.[82]

This legislation called for Congress to appropriate $15,000 annually to each state for its land-grant college, which in turn, generated a national system of agricultural research.

The Hatch Act pushed scientific investigation to the forefront at M.A.C., and it necessitated the development of a sizable research staff

and a statewide presence as the college built upon its earlier work. Faculty had conducted experiments from the institution's first days, and in 1885 the Michigan legislature made money available for the publication annually of bulletins pertaining to agriculture written by professors.[83] M.A.C. chose to integrate its experiment station into its existing structure. President Willits became director, and professors Robert C. Kedzie, Albert J. Cook, William J. Beal, Samuel Johnson, Edward A. A. Grange, and Liberty Hyde Bailey contributed one-third of their work to the station's research program in chemistry, zoology, botany, practical agriculture, veterinary science, and horticulture.[84] In 1895 Clinton D. Smith, professor of agriculture, became director, and the staff included three assistants in agriculture, three in horticulture, two in chemistry, and one in botany. Further growth occurred as a result of the Adams Act, enacted in 1906, which after six years raised federal funding for agricultural research by another $15,000 per year. The station's staff kept expanding, and in 1925 fifteen scientists served on the council with its director, Dean Robert Shaw. In addition,

In June 1888 Professors W. J. Beal and Liberty Hyde Bailey trekked across the northern Lower Peninsula in order to study cutover forest lands and to determine their potential for agriculture. They left Harrisville on 12 June, battled mosquitoes, rain, and poor roads, before arriving in Frankfort on 22 June. Along the way they collected botanical samples, observed where crops did well, and concluded that the region likely faced a bright future. M.A.C. students Lyster H. Dewey, '88, Daniel A. Pelton, '88, and Charles F. Wheeler, '91, along with W. W. Metcalf, a deputy sheriff in Crawford County, who served as driver, and two reporters from Detroit made up the rest of the party. In this photograph, taken at Metcalf's house in Grayling on 17 June, Bailey and Metcalf are standing next to their "prairie schooner," with Beal, Dewey, and Wheeler seated on the rig. A few weeks later Bailey left M.A.C. to assume his new position as professor of horticulture at Cornell University. Courtesy of Michigan State University Archives and Historical Collections (Michigan State University, Department of Forestry Collection, Beal Botanical Expedition, 1888, Folder 7).

Michigan Agricultural College hoped to find ways to turn the infertile "Pine Barrons," situated in fifteen counties in the northern Lower Peninsula, into soil that was agriculturally productive. After the college established the experiment station in 1888, it created a sub-station in Grayling on eighty acres of land donated by the Michigan Central Railroad. Professor Robert C. Kedzie supervised the work at Grayling, where he and Professor W. J. Beal planted a large number of forage plants and grasses hoping to find species that would grow in this hostile environment. One of the challenges they faced was to find or create an adequate source of fertilizer. In this picture three men are removing marl from a lake near the station. I. H. Butterfield, secretary to the State Board of Agriculture, observed in 1893 or 1894 that the marl "seems to have had the best effect, of any fertilizer yet applied, but so far is too expensive to obtain" (I. H. Butterfield, n.d., Oscar Clute Papers, UA 2.1.5, Box 863, Folder 81). Unfortunately, many of Kedzie's and Beal's experiments proved unsuccessful. Courtesy of Michigan State University Museum (Framed Collection).

sixty assistants and research assistants carried on work in chemistry, plant physiology, soils, horticulture, bacteriology, dairying, farm crops, parasitology, entomology, economics, botany, agricultural engineering, and animal pathology. T. T. Lyon directed research into fruits at the substation in South Haven from 1889 until his death in 1900. He willed his land, which had been used for the station, to M.A.C.[85] Two more substations were established, one near Grayling and the other in the Upper Peninsula at Chatham.[86]

Tensions existing at the experiment station threatened to hinder its research functions. Professors with teaching responsibilities often found it difficult to spend enough time doing the research expected by the station. In 1901, for instance, Director Smith complained that Rufus H. Pettit's and Charles F. Wheeler's teaching loads prevented them from giving the station "its due proportion."[87] Three years later, A. C. True, director of the USDA Office of Experiment Stations, reminded Snyder that under the Hatch Act, experiment stations were to be "research departments of the colleges," and that "other departments of the college [were] to do the educational work in an ordinary sense." M.A.C. could not use Hatch Act funds to pay for a disproportionate share of college officers' salaries, for upkeep of the college farm, or to carry on "general farming or horticultural operations."[88]

The anxious relationship between the scientific and the practical produced conflict at the station. This was particularly evident in 1908, when Snyder had to replace Smith as director. Since the Adams Act sent additional federal money to East Lansing for scientific research, it became more important than ever that a "scientific man" oversee scientific experimentation. The director had to be someone who understood that pure science does not produce "immediate results," and he had to be able to withstand public demands for premature answers to pressing agricultural problems. On the other hand, experimental work of a more practical nature also occupied an important place at the station. Snyder put it succinctly:

> A large amount of experimental work must be carried on which does not come under the head of research work. The livestock demonstration in up-breeding and cheap beef production as carried on recently by Prof. Shaw are good examples of this class of work. This type of work is of great practical value and should be undertaken with as much care as the scientific work.

Furthermore, experimenters such as Shaw needed to understand farming and farmers if they hoped to "inspire their confidence" when working with them. To meet this dual nature of the station's work, Snyder appointed Shaw as director and Charles E. Marshall, professor of bacteriology and hygiene, as vice director in charge of scientific research.[89] By taking this approach, Snyder held true to his belief that the experiment station should be for research, and research "should be the main feature of the farm."[90]

M.A.C. scientists created research agendas that led them to seek answers to thousands of questions pertaining to Michigan agriculture. The list of projects for just one department for one year reveals the extent of research conducted at the station and the researchers' concern that the results have practical applications in the state. In 1893 zoologist G. C. Davis laid out the following plan for scientific experimentation in entomology:

1. Study of parasites and their effect upon injurious species.
2. A study of the pests of the greenhouse from an economic point of view—that is, to learn about habits, life histories, remedies, parasites, etc.
3. Study of the habits of the ground mole with view of ascertaining the most desirable remedy in exterminating them.
4. To try carbolyid lime with other remedies for the striped cucumber beetle.
5. To test kerosene emulsion and other remedies for the cabbage, onion and radish Anthomyia.
6. To experiment with any insecticides that may appear desirable.
7. To attend to any insect outbreak in the state that seems to need special attention.[91]

It was not enough simply to identify pests and the plants they attacked or to find ways to kill them—the land-grant way demanded that M.A.C. researchers make their remedies known to the public so that growers could use them to eradicate harmful insects. As the scope of research broadened and deepened, scientists at the station provided much of the "stuff" needed by cooperative extension.

Experimentation at the station generated an incredible body of research that the college shared with the public. This work shaped many of the blocks that formed the foundation of Michigan Agricultural College's outreach. Each year six, eight, ten, or more bulletins rolled off the presses reporting the results of scientific analyses and offering advice on its practical benefits for farmers and gardeners across Michigan. The range of topics included analysis of fertilizers, identification of insect pests and methods of eradicating them, dairying, growing and caring for potatoes, growing corn, the chemistry of soils, and many more. In 1906 the station printed and sent out 45,000 copies of Andrew J. Patten's *Fertilizer Analyses,* thereby "preventing by this very publicity any successful attempts to defraud the people" on the part of producers of worthless fertilizers.[92] The *Fertilizer Analyses* written by Patten and three others ten years later (there were others in between) was issued as Regular Bulletin No. 275, with a press run of 50,000.[93]

Several bulletins issued over time provide insight into the range of content made available to agriculturalists in Michigan. Clinton D. Smith gave instructions for dairy farmers on keeping records and the use of the Babcock test for butterfat in *Dairy Records* (1896).[94] A highly technical bulletin by Charles E. Marshall and Bell Farrand, *Bacterial Association in the Souring of Milk* (1908), built upon their earlier published work that appealed to scientists rather than farmers.[95] In 1914 onion growers found a wealth of information in *Onion Culture on Muck Lands,* by C. P. Halligan, from which they learned how to prepare soil, select and plant seeds, cultivate and weed fields, and harvest and store produce.[96] Halligan reached a much larger rural and urban audience in 1918 through his bulletin *Trees, Shrubs and Plants for Farm and Home Planting.* He gave advice on choosing a good site on which to build a house, properly planning the grounds, and selecting shrubs, trees, vines, perennials, and annuals.[97]

The emergence of extension in the Upper Peninsula in 1912, out of the work done at the Chatham experiment substation in Alger County, illuminated the evolving relationship between the experiment station and extension. Leo M. Geismar, who had served as the station superintendent since its origin in 1899, put up a stiff, but unsuccessful, resistance to Snyder's efforts to transfer him to full-time extension in 1912. Through his correspondence with Snyder, Geismar revealed that since 1904 some of his activities had been, in fact, extension work. During the station's first years, Geismar focused his research on determining which crops would grow in the cold Upper Peninsula, but his emphasis had shifted toward working with farmers to help them improve the productivity of their fields.[98] In 1904 he distributed seed corn in Menominee County and induced farmers to increase their acreage from 152 to 1,007 acres in five years. As a result of Geismar's effort, yields in the pea crop in Chippewa, Mackinac, Luce, and Schoolcraft counties jumped from 70,000 to 310,000 bushels over five years. He had also encouraged farmers to grow alfalfa. As production increased, the trust between Upper Peninsula farmers and Geismar deepened.[99]

When word leaked out that Geismar's tenure at the experiment station was to end, people in the Upper Peninsula, who mistakenly believed that the station was to be closed, expressed outrage. L. C. Holden, chairman of the executive committee of the Michigan State Grange, touted Geismar's accomplishments and gave a concise statement of his relationship with U.P. farmers: "Our farmers, who of course at first, took him to be an educated fool—as they are inclined to do when any one seeks to impart scientific knowledge—now regard him as their greatest benefactor."[100] Thomas B. Wyman, secretary-manager for the Northern Forest Protective Association, voiced his belief that "the breeding, development and distribution of crop seeds not now believed to be adapted to our climatic conditions" was more important than extension. Besides, the Duluth, South Shore and Atlantic Railroad planned to hire an extension worker (Henry W. Geller, '04) to assist new settlers moving into the Upper Peninsula.[101] Joining the chorus, Thornton A. Green, president of the Upper Peninsula Development Bureau, put it bluntly: "Nobody wants the experiment work that Mr. Geismar has been carrying on, stopped, even for one year. . . . It is not as important however to go among the farmers, as it is to show the farmers what they can grow in this territory and to experiment with various fruits, grains, etc. to show which are hardy in this locality and which are not."[102]

While there may have been confusion in the mind of the public, President Snyder understood the distinction between the purpose of extension and the experiment station, as well as the close relationship between them. Clearly stung by accusations that M.A.C. intended to do less for the Upper Peninsula, he told Holden that it in fact planned "to do a great deal more." According to Snyder, Geismar failed to recognize the opportunity to encourage people he already knew to try new ways. The work at Chatham demanded a full-time researcher who could devote fourteen hours a day to his work.[103] Beneath the surface of this controversy laid an emotional, personal issue for Geismar. His wife had recently died, and he did not want to move his young daughters to a new town. But in order to give needed space to his replacement, the State Board of Agriculture required Geismar to relocate from Chatham.[104] Geismar soon put his energies into his new job organizing farmers' and boys' clubs, made plans to show farmers how to increase yields and profits, and intended to help staff an agricultural train that would run across the Upper Peninsula—all the stuff of extension.[105] In just a few weeks, Geismar told Snyder "that there are almost unlimited possibilities in extension work" in the Upper Peninsula.[106]

THE COOPERATIVE EXTENSION SERVICE

The establishment of cooperative extension finally enabled Michigan Agricultural College to transform its research and teaching into the practicality that had been demanded by the public since 1857. Farmers' institutes had brought professors and researchers to grange halls and community auditoriums to meet face to face with the men and women of rural Michigan. After the meetings, however, the specialists usually moved on to the next institute or returned to campus. In contrast, the county agricultural agents and home demonstration agents actually lived in the counties they served and worked with people on their farms, in their schools, and through their own local associations. One of the hallmarks of extension work became specialists in the field helping local people to organize for specific purposes and then letting them carry out the association's mission. The agents established personal contacts with thousands of men, women, and children as they communicated through the spoken word, demonstrations, and letters information and techniques that could improve the efficiency and profitability of farming and homemaking. The agents did not just tell women how to can fruit or sew dresses or tell men how to prune trees or apply insecticide; rather they showed them how to do it.

Four fundamental characteristics of extension worked together to give clearer definition to the land-grant philosophy in Michigan: it was cooperative, it was democratic, it was organized, and it dealt with ordinary things. Extension grew out of the cooperative efforts of M.A.C., the U. S. Department of Agriculture, local units of government, and volunteer organizations. It promoted democracy by equipping thousands of people to assume leadership roles in their communities and to acquire and transform knowledge needed to improve their personal well-being. Extension thrived because it was highly organized at the federal, state, and college levels, and its agents worked with and through local government and associations, which they often had helped to organize in their counties. Paying attention to the details of the activities of people's ordinary lives such as sewing clothes, preserving food, harvesting more potatoes per acre, and raising healthier chickens, made up the "stuff" of extension. People welcomed agricultural and home demonstration agents into their homes, fields, schools, churches, and most important into their lives.

Extension work at M.A.C. was part of a national movement to take the knowledge learned at land-grant colleges directly to the people. Some other states had acted more aggressively in extension before Michigan hired W. F. Raven to work as a field agent in livestock in 1908. Rutgers University started an extension program in New Jersey in 1891.[107] Lewis E. Reber joined the University of Wisconsin in 1907 to lead its newly created extension division. Reber articulated a land-grant philosophy for extension that encompassed key elements that informed extension programs nationwide. He believed that extension did not "disdain the simplest form of service," and that it literally carried "the University to the home of the people."[108] The Iowa Agricultural College formed an extension entity in 1906 that became a "separate service with a director responsible directly to the president" in 1912.[109] Cornell employed its first county agricultural agent in 1911.[110] C. R. Barns of the University of Minnesota responded to Snyder's interest in extension work in Minnesota in 1912 by telling him that if "Michigan was 'to keep up with the times,' her endeavors in that direction would have to be very considerably broadened."[111] Snyder intended to do just that.

The first six years of extension at M.A.C. laid the groundwork for the official establishment of the Cooperative Extension Service in 1914. Raven's efforts foretold the enormous amount of work that agents in the field would be required to do to reach Michigan families.

For example, during the 1912–13 year, he oversaw the clearing of sixty-five acres of land at the Chatham experiment substation, helped with fairs and exhibits throughout the state, and organized six cattle breeding associations. (By 1913 he had helped to set up sixty-six associations in twenty-nine counties.) The bulk of his job brought him into direct contact with thousands of people, as was reported in the 1912–13 *Annual Report:*

> During the year Mr Raven visited 408 farms owning 3220 cows, spent 57 days as local adviser, 55 days as demonstrator and 153 days in other extension work. He delivered 106 lectures, attended three one-week schools of agriculture, and directed the three weeks' trip of the demonstration train over the Duluth, South Shore and Atlantic Railroad in the Upper Peninsula.[112]

Much thought, planning, and organization took Oliver K. White, '07, agent for horticulture, into Michigan's orchards, where he made personal contact with many farmers. Between 1 July 1911 and 30 June 1912 he gave a series of lectures in twenty-five fruit-growing communities. He also carried out demonstrations in orchards located along the well-traveled roads, making it easier for distant neighbors to attend and to observe the results of his work. Starting in spring, growers learned, firsthand, when and how to prune and spray for scale. Throughout the growing season, other sessions focused on "early foliage spraying, cultivation, thinning and later sprayings, cover crops and packing."[113] White advertised his activities by poster and in local newspapers. The method of public demonstration motivated more farmers to take better care of their fruit trees than did visits to individual farms.[114] Other agents did extension work in forestry, farm crops, and soils.

Proponents of extension recognized that its success depended upon solid organization, and they worked tirelessly to build a structure that welded together local units of government and associations with M.A.C. and the USDA.

A major step along this path occurred in 1912 when the chamber of commerce of Alpena County; the boards of supervisors of Alpena, Presque Isle, and Montmorency counties; M.A.C.; and the USDA Bureau of Plant Industry signed an agreement calling for a cooperative work "to conduct farm management field studies and demonstrations to the end of upbuilding agriculture" in the three counties. The Bureau of Plant Industry pledged to pay half of the $2,200 required to employ an agent to "devote his entire time to these investigations and demonstrations." Each of the local units agreed to fund a portion of the remaining $1,100. M.A.C. said it would give "advice and encouragement to the agent."[115] The birthing of a system of county agricultural agents in Michigan began with the appointment of Harvey G. Smith as agent.

At the same time, M.A.C. and the Bureau of Plant Industry joined together to hire a state farm management leader to oversee "all farm management field studies and demonstration work" in Michigan. The bureau and college agreed to share the cost of salary, and M.A.C. provided office space and equipment. The contract also called for the employment of district supervisors to help the state leader to carry out studies and demonstrations. At the bottom of the pyramid stood county agents, who were to take the result of the work directly to the farmers through demonstrations and cooperation with local agricultural clubs and associations. Funds for the county agents' maintenance were to come from the bureau, the college, or local entities.[116] Dr. Eben Mumford received the appointment as the state leader.[117]

H. F. Williams, a district supervisor, inaugurated the field studies and demonstration work in south central Michigan in 1912 and 1913. His report unveils the democratic nature of the fast-evolving extension movement. At each turn, extension worked to empower ordinary people to shape their own destiny. Participation by farmers was always voluntary. Williams noted that no farmers had taken any action to form a farm bureau, although most seemed agreeable to the idea. For them to do so would have required a complete change in attitude that would come about only if they could be convinced of its value. He stressed the importance of agents surveying the types of agriculture, soil and geological conditions, and other situations present in the counties where they worked to identify existing problems so they could propose projects to deal with them. He believed that solutions should derive from local circumstances and not be unilaterally imposed by outside authorities. Williams, a native of Missouri, felt that he lacked sufficient knowledge of Michigan to do competent extension work in the state. In his opinion, only men thoroughly acquainted with the local physical and social landscape were qualified to become county agents. He helped to organize in Litchfield a cooperative association for shipping livestock. The local farmers' club brought the idea to fruition with the result that they sent off two railroad carloads of cattle in the association's first month. As a result, the farmers made $210 more than they would have had they had sold their animals on the local market.[118]

Extension work in Michigan and across the United States paved the way for passage of the Smith-Lever Act in the spring of 1914. Representative James C. McLaughlin of Michigan, who had first introduced the measure in the House of Representatives in 1909, summed up the thrust of the new law to President Snyder: "Every reasonable effort ought to be made to get in actual touch with farmers right on the farms." But for this to happen the "right kind of men," who could communicate with farmers on the problems they faced in their everyday work, had to be trained and employed.[119] The act required that cooperative extension work provide "instruction and practical demonstrations in agriculture and home economics to persons not attending or resident" in colleges.[120] M.A.C. alumnus Kenyon L. Butterfield, president of Massachusetts Agricultural College, played a key role in the drafting and passage of the Smith-Lever Act. Butterfield, along with Liberty Hyde Bailey, had served on President Theodore Roosevelt's Country Life Commission that recommended that a "nationalized extension work" be undertaken by agricultural colleges.[121] Fortunately, Butterfield's and Bailey's alma mater was philosophically intune with the measure and was already educating both men and women to carry out its provisions.

Michigan Agricultural College's insistence that students who studied agriculture and home economics be well grounded in science and the liberal arts prepared many to teach farmers and homemakers to apply the results of experimentation to their own farms and homes. The "right kind" of men and women, who knew and understood Michigan and its people, could be found in East Lansing. It was no accident that M.A.C. was in a position to initiate extension work in 1908 and to expand it dramatically after 1914 through its Cooperative Extension Service. It had taken over half a century for land-grant colleges, including M.A.C, to evolve to the point where they could take their message to thousands of people where they lived. In the *Annual Report* for 1913–14 the college made clear its commitment to extension and why and where it fit into its program:

Extension work is the natural outgrowth of the college and experiment station. The first and the highest work of the college is to develop men. These men in turn through the experiment station develop new information. If the work stopped here it would be incomplete. Now agricultural knowledge is worth nothing unless applied in everyday practice. Men trained in an agricultural college must not only discover new truths and new methods but they must teach the farmer how to use the new discoveries to best advantage. If the farmer cannot come to the college for this new information—and but few of them can—the college must take it to them. This the College is endeavoring to do in the development of extension work.[122]

Teaching, research, and outreach formed the equilateral triangle of land-grant education. For the land-grant to be effective, each side of the triangle had to be kept strong.

Without solid organization even the most talented men and women could not carry out extension's objectives. Robert J. Baldwin was appointed director of extension work in 1914—a post he filled ably for many years—and Eben Mumford continued as state leader for county agricultural agents. Since the federal government had determined that the county should be the basic unit of organization for extension, adequate supervision and support for each county agent was essential. Concerned about the health of extension in Michigan in October 1915, C. B. Smith, '94, of the USDA State Relations Service admonished M.A.C. to hire assistant state leaders. Smith went right to the point in a letter to acting president Frank S. Kedzie:

The main weakness of the Michigan work would seem to be the lack of well organized counties with a strong cooperating association in each county standing back of the movement financially, morally and constructively, and the lack of adequate over-head administration which gives the agents help along the lines they need help at the time they need it.[123]

The assistant leaders could help county agents build support organizations, make sure the agents' plans fit local needs, pass new information along to agents, and address complaints made by local residents. Within three years M.A.C. appointed three assistant state leaders. The home demonstration agents also had a state leader and three assistants.

Extension agents relied upon the farm bureau to introduce them to county residents and to help them organize local associations. Once a farm bureau was set up, agents worked through it to reach as many people as possible. Local organizations selected representatives to form a cabinet that gave information about the county's needs and advice to the agent, who then worked through the bureau to arrange meetings, conferences, and demonstrations.[124] For example, Clarence B. Cook, '88, found the farm bureau in Allegan County to be off to a good start in 1914. Farmers, lawyers, doctors, businessmen, and teachers joined together "for the betterment of agr'l conditions." Members paid annual dues of two dollars, with local banks promising to cover cost overruns during the bureau's first two years. Members came from every township, and they elected a vice president to represent them at the county level. The bureau's constitution made "provision for soil, seed & stock improvement." Cook complained that not all subscribers paid their dues and only "a few willing hands do most of the hustling." He expressed hope that a growing membership would lead to a smoother-running system. Whenever possible, Cook tried "to supplement & work in cooperation with all organized farm effort." Although some granges (those with older members) and the Allegan Board of Trade resisted the agent's assistance, other associations and boards cooperated with extension, especially the YMCA county secretary, the county school commissioner, and the farmers' club.[125]

The national needs generated by World War I spurred rapid growth of farm bureaus and extension in Michigan by mid 1918.

County agents taught farmers techniques and practices intended to improve the productivity and profitability of their enterprises by putting on demonstrations on local farms. Demonstrations covered a wide spectrum of topics ranging from breeding livestock to spraying fruit trees. This photograph shows an agent instructing a group of women and men on how to breed poultry selectively and how to cull their flocks of "slacker hens." Courtesy of Michigan State University Archives and Historical Collections (Michigan State University, Cooperative Extension Service, Demonstrations).

Table 1. Selected Activities of County Agents, 1914 and 1918

	1914	1918
Business calls on agents in office	4,326	61,969
Telephone calls to and from agents' offices	8,334	53,546
Letters written	12,143	50,595
Copies of circular letters mailed or distributed	63,460	226,758
State and U. S. bulletins distributed	28,910	87,026
Articles published in local press	907	4,388
Farm visits	6,192	25,632
Meetings where agent took part	762	12,559
Total attendance at meetings	67,115	207,140
Associations organized for adults	28	187
Total membership for newly organized adult associations	1,256	7,003
Boys' and girls' clubs organized	31	759
Total membership in newly organized clubs	1,086	11,588
Total attendance at local short courses or extension schools	1,765	41,502

Sources: "Report of Accomplishments of the County Agents," "Summary for 1914," Michigan, Cooperative Extension Service Records, UA 16.34, Box 1, Folder 11; "Annual Report of County Agent," 1 December 1917 to 1 December 1918, Michigan Cooperative Extension Service Records, UA 16.34, Box 2, Folder 56; and "Report of County Agricultural Agent Work in Michigan for the Fiscal Year Ending June 30, 1918," Cooperative Extension Service Records, UA 16.34, Box 2, Folder 56.

From 1 July 1917 to 30 June 1918 the number of counties with a farm bureau rose from thirty-one to seventy-one; sixty-eight county agents served them. (Three agents worked in two counties.) A survey of bureau and agent activity in 1918 brings to light the breadth and depth of extension work in Michigan. (See table 1 for a comparison of county agent activity in 1914 and in 1918.) Agents in conjunction with the farm bureaus helped to organize associations for the following interests (the number of new entities formed in the 1917–18 fiscal year appears in parenthesis): farm loans (35), pure seeds (19), potato growers (22), threshers (10), cow testing (2), community breeders (15), milk producers (5), beekeepers (9), livestock shippers (8), and marketing (52). Agents carried out projects and conducted demonstrations that touched upon most aspects of agriculture in Michigan including soils, drainage, fruit, control of insects and plant diseases, livestock, poultry, farm management, farm labor, marketing, and farm crops.[126]

Before much land in Michigan could be transformed into productive farms, stumps had to be removed and water drained. From its first day, M.A.C. itself struggled against these same foes as the college created its campus from heavily wooded and swampy lands.

TOP: Agricultural engineers from M.A.C. operate a "Tile Drainage Demonstration Machine" to show farmers how tile could be laid to draw off water from lowlands. Courtesy of Michigan State University Museum (Acc. No. 152.7.19.1).

BOTTOM: Agricultural engineers from the college also taught students how to use dynamite to blast stumps. A man is drilling a hole in a pine stump, in preparation for blasting, on an unidentified farm in Michigan, circa 1920. Courtesy of Michigan State University Museum (Acc. No. 2003:152.7.26.16.1).

As county agents challenged farmers to increase profits, they encouraged them to plant certified seed, and they spent much time and energy trying to locate adequate quantities of seed for their constituents. During the war years of 1917 and 1918 extension procured and distributed to Michigan farmers seeds of barley, oats, corn, beans, alfalfa, soy beans, sweet clover, sugar beets, and potatoes, and urged them to use pedigreed seeds resistant to smut and rust. Agents organized nineteen pure seed associations around the state, hoping to increase production, to have farmers plant standard varieties, and to improve marketing of crops.[127] The farm crops department at M.A.C. and extension cooperated with the Michigan Crop Improvement Association to have its members plant samples of grain in their fields to test their suitability for local conditions.[128] Further aid came from Bertha A. Hollister, state seed analyst, who analyzed seeds for their purity at the seed laboratory at M.A.C.[129]

Graduates of Michigan Agricultural College filled the ranks of the extension service as it grew rapidly to meet demands generated by American involvement in World War I. An important part of extension's program was to aid the federal government's effort to conserve and to preserve food.[130] Bad harvests in 1916 and 1917 depleted wheat reserves, making it difficult to ship adequate supplies to America's European allies. As a result, increased production and especially conservation of wheat, pork products, and sugar formed an important segment of extension work. President Woodrow Wilson created the Food Administration in 1917, and, under the leadership of Herbert Hoover, the agency worked through agricultural and home demonstration agents to enlist the support and cooperation of the public to meet the allies' food needs. M.A.C. had a close relationship with the Food Administration through Harry J. Eustace, professor of horticulture, who went to Washington, D.C., to become associate in charge of the perishable food division within the Food Administration.

In the 1917–18 fiscal year, four components made up extension, and M.A.C. alumni occupied the majority of positions in each division: county agricultural agents (42 out of 68), boys' and girls' club work (8 out 9), home demonstration agents (15 out of 27), and extension specialists (9 out of 16). Agricultural agents worked in seventy counties, showing farmers how to increase production to aid the war effort and to improve generally their agricultural methods. The emergency

brought on by the war put home demonstration agents in twenty counties and four cities. They taught women to preserve food and to change diets to meet the requests of the federal Food Administration. Club leaders oversaw the activities of 30,000 boys and girls, who carried out projects according to guidelines drawn up at M.A.C. Specialists based at the college in home economics, dairying, farm crops, potatoes and vegetables, insect pests, household engineering, muck crops, poultry husbandry, markets, farm management demonstration, health, and publicity worked with agents to help men, women, and children transform the scientific knowledge discovered at M.A.C. into the practical application called for in the Smith-Lever Act.[131]

Home demonstration agents empowered individuals and local associations by teaching women and girls to better manage households, reaching out to different ethnic groups, and partnering with local institutions and organizations to achieve the objectives of extension. Insights into their operations can be gained from the activities of Frances Marguerite Erikson, '15, who focused on the city of Grand Rapids, and Clara G. Rogers, '14, who covered the rest of Kent County

Clubs for boys and girls paved the way for the 4-H movement. These youngsters from Wayne County made up the "French Landing Poultry Club," circa 1926. Courtesy of Michigan State University Archives and Historical Collections (Michigan State University, Cooperative Extension Service, 4-H Clubs, Folder 1).

Clara Grace Rogers. Courtesy of Michigan State University Archives and Historical Collections (*Nineteen Fourteen Wolverine*, 56).

Frances Marguerite Erickson [Erikson]. Courtesy of Michigan State University Archives and Historical Collections (*Wolverine*, 1915, 47).

in 1918. They gave lectures and demonstrations on such topics as baking different kinds of bread and canning before clubs and classes that either they or local leaders had organized. Erikson and Rogers visited homes, wrote articles for the local press, and got to know influential women and men in their communities. Even though war efforts determined much of the home demonstrators' program, they introduced the subjects and employed methods that formed the foundation of extension in home economics.

Erikson's influence reached deeply into Grand Rapids. She drew Polish women into the effort to conserve and preserve food by working with their leaders and priests. Father Petrasci and "influential Polish women" arranged for a hall and encouraged women to attend Erikson's meetings. Priests translated recipes into Polish and typed them for distribution at the gatherings, and a Polish newspaper encouraged participation by publishing information relative to Erikson's program. She connected with African-American women through Mrs. Theodore Burgess. In January 1918 a group of black women, at Erikson's instigation, began to meet every three weeks to study food. In May, Burgess presented a canning demonstration that was well received by her friends. Neighborhood groups of Italian women welcomed Erikson's demonstration on yeast bread and quick bread. Mrs. Philip Caruso provided valuable assistance by translating "war bread" recipes into Italian. Erikson also held meetings, arranged through the local social welfare agency, in the Assyrian district of Grand Rapids.[132]

Clara Rogers's routine resembled more closely that followed by most agents whose territories included entire counties. She got off to a good start because farmers' clubs, organized by the farm bureau, welcomed her to Kent County in December 1917. As Rogers organized her home demonstration program, she held a meeting of women representing farmers' clubs, the Women's Council of National Defense, the Grange, and churches, which provided the leadership needed to create projects in their communities. Rogers and Erikson, in conjunction with the Women's Council of National Defense, invited women from all over the county to Grand Rapids to see an exhibit that explained how foods in large supply could be substituted for commodities needed elsewhere to support the war. Showing women how to can fruits and vegetables, save wheat, and raise more poultry formed the heart of Rogers's summer agenda. An example of her success was shown by women in Chase Lake who conserved 374 pounds of flour and canned 533 quarts of vegetables and 516 quarts of fruit. Eight girls' clubs organized by Rogers gave canning demonstrations to the public, and the members of three poultry clubs improved their flocks by securing "eggs of a pure strain from the college." They soon demanded that culling demonstrations be held in their communities. Rogers also cooperated with the county nurse to promote health issues at meetings.[133]

Twenty-one other agents carried out the objectives and war support efforts of extension throughout the state during 1918. Hilda Pollari, an assistant at large, did "very splendid work," primarily among the Finns. In Kalamazoo County, Blanche Clark developed "sugar saving recipes," while Bessie Turner, '16, encouraged grocers in Benton Harbor to display sugar substitutes in their store windows. Extension cooperated with the Red Cross to hold classes that taught women about "elementary hygiene" and caring for the sick at home. Grace Hitchcock, '15, started "Red Cross Dietetic classes" in Grand Haven, and Clara Waldron, '13, reached new groups of women through meetings of the Red Cross in St. Clair County. In the Upper Peninsula, Helen Pratt, '16, put on canning and jelly making demonstrations in homes in Chippewa County. In September, Maude Bennett organized garment making clubs for women living in Iron County.[134] Home economics specialists Mary Madden Person and Coral Ruth Havens and health specialist Elizabeth Leonice Parker traveled across the state building up a network of women trained to instruct other women in better ways to feed, clothe, and care for their households.

After the war ended, the number of home demonstration agents declined significantly, and extension specialists took over much of the work in home economics. In 1925 agents served in only six counties: Allegan, Kalamazoo, Marquette, Oakland, Ottawa, and Wayne. Two specialists in clothing, one in nutrition, and one in home management carried much of the load in home economics extension. Agnes Sorenson, Carrie C. Williams, Martha M. Hunter, and Marion Rogers Smith trained local leaders to show women in their communities ways to improve the management of their households. Their number contrasted with fifty-five county agricultural agents and thirty-three specialists who traveled up and down Michigan providing extension services in agriculture.[135]

LEADER
AGNES MARTIN

−1ST YEAR
DOROTHY HALBEISEN
MARGARET HALBEISEN
MARGARET HUEY
DOROTA ZEIS
DORA ZEIS
KATHLYN FISHER
MILDRED MARSH
ETHEL FISHER

−2ND YEAR
MADALENE SMITH
VERDA RANSOM
HENRIETTA HUBER
BARBARA SMITH
KATHLEEN JORDAN
NORMA BUST

The Cooperative Extension Service organized many people, including children, in Michigan to contribute to the state's effort to support the U.S. government's role in World War I. These girls were members of one of the many 4-H canning clubs scattered across the state. Courtesy of Michigan State University Archives and Historical Collections (Michigan State University, Cooperative Extension Service, Demonstrations, Folder 6).

Extension reached young people between the ages of ten and twenty through boys' and girls' clubs. The club movement in Michigan began with corn growing associations in Mason and Muskegon counties in 1908.[136] Club work set out to "train the whole child: head, hand and heart."[137] By 1914 sixty clubs had memberships ranging from ten to twenty-five, with each boy or girl undertaking a project and participating in their club's exhibit at the county fair or in their school.[138] As he performed his job as state leader for club work, Eduard C. Lindemann, '11, formed close working relationships with county agents, who then encouraged thousands of girls and boys to join clubs.[139] While club leaders hoped to teach youngsters to plant and care for field crops, to raise animals, to garden, to sew, and to can fruits and vegetables, they also intended to prepare young women and men to assume positions of leadership within their communities as adults.[140]

As in all other phases of extension, effective club work depended upon good organization at the local level that drew upon resources available from the USDA and M.A.C. Schoolteachers served as leaders of most clubs, with county school commissioners coordinating the work of all clubs into countywide organizations. Club leaders relied upon the services and advice of YMCA secretaries, ministers, Sunday school workers, the high school agricultural teachers, businessmen, and local newspapers. Granges, gleaners, and farmers' clubs offered prizes and scholarships and arranged exhibits and joint meetings. Since many of the members' projects centered on raising a farm animal or a field crop, the local livestock and pure seed associations assisted boys and girls in selecting the right breed or seed for their projects. Bulletins issued by the Bureau of Plant Industry were a valuable tool in the leaders' hands. Lindemann and Anna Bryant Cowles, '15, state leader for girls' clubs, furnished bulletins, demonstrations, instruction, and lecturers for clubs throughout Michigan.[141] County agents urged that clubs ally themselves with the farm bureau so that young people trained in the clubs would become farm bureau members as adults.[142]

Tightly monitored projects functioned as the primary vehicle for teaching girls and boys how to do things and to do them profitably. The extension division issued bulletins giving detailed instructions for each project that guided both leaders and club members. Garment making and canning were two popular projects for girls' clubs. Leaders taught girls "to dress simply and attractively," and to work through the complete process of making a garment. Girls drafted and altered patterns, sewed, became knowledgeable about fabrics and principles of design, and learned to choose colors. As they engaged projects, they recorded the cost of materials and time invested in each phase of their work, thereby gaining an understanding of the monetary value of the garments they made.[143] Leaders instructed girls in how to can fruits and vegetables by the "cold pack method," which did not require cooking the produce before putting it into the jars. When club members exhibited their completed projects, they were judged on the quality of their work, the "character" of their written reports, and the monetary value of their finished products.[144]

Boys either grew a crop or raised an animal, and sometimes girls took on projects that usually fell in the boys' domain. The bean project required a boy or girl to select and prepare the ground, to choose and plant the seeds, to care for the plants through harvest, and to sell the yield. Club members kept thorough records of the costs of materials and the time spent in order to learn how to calculate the monetary profit or loss of their endeavors. Participants worked hard; children between the ages of ten and fourteen had to plant at least half an acre, while their older siblings tended a full acre.[145] Each club member was accountable to an impartial authority who judged his or her crop or animal at a fair or an exhibit. For example, the criteria for "pork production" included:

Best hog with respect to purpose it is to serve	30%
The greatest daily gain in weight	20
The lowest cost of production	30
Best kept records, story and report	20
	100%[146]

Club work proved extremely successful and evolved into the 4-H movement.

Extension schools proved to be another vehicle for M.A.C. to connect with people. Initially each school ran for four days, but the war limited the running time to two days in 1917–18. That year nearly 7,200 people listened to 618 lectures pertaining to agriculture in 94 communities, while another 1,843 women heard 226 home economics lectures in 38 towns across Michigan. The choice of offerings in agriculture included farm crops, soils, horticulture, farm management, insect pests, dairy husbandry, animal husbandry, farm engineering, poultry, muck crops, and plant diseases. A mix of talks on food, textiles, health, and the war formed the core of the home economics curriculum.[147] Presenters gave a series of related lectures or demonstrations within each course. For instance, a livestock sequence in 1914 included: "A Study of Breeds," "Stock Breeding," "Starting and Selection of a Dairy Herd," "Feeds and Feeding," "Feeds and their Composition," "Management and Care of Dairy Cattle," and a "Summary."[148]

It is difficult to assess the effectiveness of methods, but a 1928 study of the influence of "various factors" on the results of extension in Jackson and Menominee counties showed that cooperative extension was working well. A brief profile of the 451 farm operators or farm women who took part in the survey tells us much about the people living in rural Michigan. Seventy-four percent of the families surveyed had some contact with a staff member of the extension service, and 47 percent had been in touch with a specialist from the college. Local volunteer leaders had worked with 34 percent of the sample. Fifty-eight percent had telephones in their homes, but only 22 percent owned radios. Twenty-seven percent belonged to the farm bureau, 20 percent held membership in a local cooperative association, 16 percent had joined a home economics club, and children from 23 percent of the families participated in 4-H. Farm families were readers: 62 percent took a daily newspaper, 42 percent a county weekly paper, 88 percent a farm paper, 48 percent a home magazine, and 20 percent another kind of magazine. The formal education of the vast majority of the men (87 percent) and the women (80 percent) ended in elementary or secondary schools, and only 1 percent of the men and 3 percent of the women had "some college."

Of the people studied, 83 percent of farm operators or farm women said that they had incorporated, on average, 4.1 new practices learned from extension into their regular routine. Agricultural ways had changed on 80 percent of the farms, and home economics ways had changed in 24 percent of the homes. This great difference probably came about because many more resources were directed towards agricultural practices. On 69 percent of the farms where changes occurred, the county agent effectively reached the farmer or his wife, and subject-matter specialists received credit for initiating new behavior from residents on 51 percent of the farms. Direct contact with an extension employee resulted in the farm family making a change 95 percent of the time. When a farm hosted "some formal extension activity, such as a boy or girl in club work, a field demonstration, or a community meeting, the adoption of improved practices followed in 100 percent of the cases."[149]

Radio

Students at M.A.C. introduced the new medium of radio during the First World War, and in the early 1920s they built a radio station—WKAR. In the winter of 1917–18, Paul G. Andres, '18, put together

David L. Friday became president of M.A.C. in March 1922 and resigned in May 1923. After he graduated from the University of Michigan in 1908 Friday received a faculty appointment to the university's Department of Economics, where he achieved a national reputation. He remained at Michigan until his rise to the presidency at M.A.C. Under Friday's leadership the curriculum expanded in many disciplines, especially liberal arts, engineering, and economics. He also strengthened the horticulture program and successfully lobbied the legislature to fund a new horticulture building that opened in 1925. Friday revived the graduate program and set in motion doctoral programs in bacteriology, botany, chemistry, entomology, horticulture, soil, and farm crops. The farm community rejected Friday's view that cooperatives could do little to increase the profitability of Michigan farms, except to teach producers how to be more efficient. Friday resigned his post after unfounded allegations were published that he had traveled with his female research assistant.

Courtesy of Michigan State University Archives and Historical Collections (People, Friday, David, no. 1037).

and operated the first broadcast apparatus, which enabled him to communicate by telegraph code (using the call "8YG") with places in the eastern United States. Although Ralph Wyckoff, '20, carried on for a time after Andres left the college, Forrest I. Phippeny, '26, can rightfully be called the builder of radio at Michigan Agricultural College. He had served in the U.S. Navy as a wireless operator before enrolling at M.A.C. Skimpy resources hampered the construction of a viable station, but Phippeny, with the assistance of Arthur H. Sawyer, professor of electrical engineering, put together and operated a station in Olds Hall of Engineering for slightly over $3,000.[150] In October 1923 Phippeny installed a "sending set" of 500 watts, an upgrade from the 250 watt set then in use, and he oversaw the construction of a 145-foot "aerial mast." This new equipment enabled WKAR, which introduced M.A.C. as "The Oldest Agricultural College in the World," to beam its signal throughout much of the United States east of the Rocky Mountains.[151] In early 1924 the station was put in a room on the fourth floor of the new Home Economics Building even before construction on the building was completed.

Michigan Agricultural College extended its outreach to the living rooms of many people by presenting news, information, and entertainment to the public even before the 500-watt transmitter was installed. On Saturday night, 13 May 1922, President David Friday broadcast his Founders' Day speech, telling alumni and others across the Midwest that he intended to place the college "in a position of leadership in the field of agriculture, home economics, and engineering education." Listeners first heard daily crop reports and weather forecasts at 10 A.M. in December 1922.[152] Hook-ups to other campus and Lansing locations allowed WKAR to broadcast concerts, political speeches, and athletic events. On 7 March 1923 listeners were treated to a selection of opera songs, the first musical program aired by WKAR.[153] Five weeks later the M.A.C. band entertained an audience that reached as far as Beloit, Wisconsin, and Cohoes, New York.[154] On 6 February 1924, J. B. Hasselman, the college's director of publications, thrilled alumni across the country as he gave a play-by-play account of the Michigan-M.A.C. basketball game won by Michigan 31-20.[155] (The M.A.C.–Central Michigan game on 29 January 1924 had been the first game to be broadcast on WKAR.) The next month, R. J. Baldwin, director of extension, initiated a series of lectures on agricultural topics in an effort to determine if radio could be used effectively to teach short courses.[156]

The college contributed to a growing movement within land-grant colleges in the 1920s to use radio to communicate with rural America. WKAR soon assumed a position of leadership among college radio stations in the Midwest. In 1926 Kansas State Agricultural College was acknowledged by Hasselman to be the leader, with Michigan State College close behind. The University of Wisconsin (which started WHA in 1919), Ohio State University, and Iowa State College were other pioneers in educational broadcasting.[157]

1924–THE PRESENT; 1925–THE NEW PRESENT

The land-grant mission to reach the people always operates in the present. It can not be an end in itself, and its current efforts perpetually reveal the needs for the future—the next present. For outreach to be effective, it has to be vital, expansive, and personal. The cornerstone of extension rests on the principle that informed men, women, and children can initiate changes in their lives, and then show others that they could do likewise. As 1924 drew to a close, extension proudly looked back over its accomplishments for the year, but more important, it anticipated 1925—its new present. In the year just past, county agricultural agents had held 503 training sessions for local leaders that were attended by 5,610 people who could now show others how to improve their farming techniques. Specialists in home economics extension had equipped scores of women to teach classes in clothing, nutrition, and home management to their friends and neighbors. Although nearly 16,000 boys and girls had participated in clubs, leaders still hoped to involve in club work some of the 62,000 youth who were not enrolled in school. Even though much progress had been achieved in programs designed to upgrade seeds, breeds, farm buildings, human nutrition, methods of pest control, and marketing, extension saw the new year as a time to extend its work to transform the knowledge gained by college researchers into practical application on Michigan farms and in Michigan homes. Extension kept reaching further and deeper into life in Michigan.

Michigan Agricultural College Fulfills its Mandate

Michigan's ongoing experiment in agricultural education in the last half of the nineteenth and early twentieth centuries showed American democracy at work. The establishment of the Agricultural College of the State of Michigan in 1855 and the enactment of the Morrill Act in 1862 resulted from strenuous efforts by citizens who demanded that colleges be created to study, research, and teach agriculture. On the other hand, strident opposition to the way M.A.C. did its business came from farmers—the group of people the college hoped to help the most. It was a magnificent, democratic moment when visionary men and women stayed the course to fulfill the mandate, given to them by the people through their elected representatives, to make M.A.C. relevant—even necessary—to the same people who often saw the land-grant institution as irrelevant or even detrimental to their interests. In the mid 1870s and mid 1890s the college responded to crises brought on by its inability to meet the needs and the demands of the agricultural community by changing its ways, although not nec-

Through the work of extension, the college had a commitment "to be of service to the men, women and children of the rural communities in developing both the economic and social phases of their lives." This series of three photographs illustrates one of the means used by extension specialists and agents to reach rural families throughout Michigan.

OPPOSITE, TOP: The "Better Homes Truck" was used in part of the Lower Peninsula in 1924. Courtesy of Michigan State University Museum (Acc. No. 2003:152.7.24.1).

OPPOSITE, BOTTOM: An extension specialist speaking to a crowd from what appears to be same truck in 1927. Courtesy of Michigan State University Museum (Acc. No. 2003:152.7.28.6).

ABOVE: Extension specialists George Amundson and Margaret Harris stand in front of the "Upper Peninsula Home Convenience Truck" before making their presentation in 1927. The open right side of the truck reveals a "model kitchen" and "simple hot and cold water systems." Courtesy of Michigan State University Museum (Acc. No. 2003:152.7.28.7).

essarily in the manner that farmers preferred. During the second decade of the twentieth century, it finally became evident to farmers throughout Michigan that their agricultural college really was working in their best interests.

By 1925 the implementation of the land-grant philosophy that characterized Michigan State College of Agriculture and Applied Science confirmed the wisdom of its founders and fulfilled their vision in an amazing fashion. The bases of the land-grant philosophy as mandated by state and federal law became the non-negotiable purpose of the institution. Buried deep in the 1861 act to reorganize the agricultural college appear the following words, which guided M.A.C.'s administrators, faculty, students, and alumni through its first seventy years:

> The design of the institution, in fulfillment of the injunction of the Constitution, is to afford through instruction in agriculture, and the natural sciences connected therewith. To effect that object most completely, the institution shall combine physical with intellectual education, and shall be a high seminary of learning, in which the graduate of the common school can commence, pursue, and finish a course of study *terminating in thorough theoretical and practical instruction in those sciences and arts which bear directly upon agriculture and kindred industrial pursuits* [emphasis added].[158]

The Morrill Act of 1862 reaffirmed the design of Michigan's agricultural college and extended its spirit through federal law to the rest of the United States. The act's provision that land-grant colleges were "to promote the liberal and practical education of the industrial classes" meant that this form of higher education would be available to the youth of rural and urban working classes. But this was not to be an education just for the sake of an education; rather men and women were to use the knowledge they gained at the land-grant institution "in the pursuits and professions of life." This implies a life of working for the public good.

Due to the attention given to M.A.C.'s work relative to agriculture, its conjunction with "kindred industrial pursuits" can easily be overlooked. But it was these words that not only allowed but required M.A.C. to see more than farms in Michigan and the world. In reality, most things that went on in the state and the nation shared something in common with agriculture. Men and women trained to follow kindred industrial pursuits designed automobiles, tractors, highways, and sewage systems, they conducted experiments in scientific laboratories, marketed fruit and flowers, and taught home economics and agriculture. All of these vocations and many more broke down barriers between farm and city in ways that were imperceptible to many, ways that created vital interdependencies between rural and urban America. Even as Michigan and the United States became increasingly more urbanized and industrialized, it was the farmers who grew food for a country whose population grew from 23 million in 1850 to 123 million in 1930. By 1930 approximately 56 percent of the people in the United States lived in cities, and 68 percent of the population in Michigan resided in cities. Farms and cities, fields and factories, and farm houses, tenements, and bungalows all fit together into a complicated whole that no one fully understood.

Michigan Agricultural College's ironclad commitment to educate informed citizens in a way that made "theoretical and practical instruction" relevant to ordinary things and occurrences in their lives constituted the bottom line of its land-grant educational philosophy. Students read great works of literature not only to become acquainted with new ideas, but also to help them learn how to write proper English. Laboratory experiments enabled them to understand how to discover new knowledge and apply it to practical concerns. The land-grant philosophy's insistence that knowledge not stop with the theoretical, but that it be transformed into practical applications, prepared students to communicate with people who neither understood nor cared how science worked. The progressive efforts made by M.A.C. in its outreach to the public clearly showed that the college could not fulfill its mission until a good-sized population of land-grant educated students worked in the field.

By 1925 Michigan State College was a major force in helping the people of Michigan adapt to changes and challenges confronting them. Its outreach programs did far more than transform knowledge into practical applications, they also transformed the lives of thousands of people. Men, women, and children were reading such works as *Up from Slavery* in their homes and discussing Booker T. Washington's experiences with their friends. Thousands of boys and girls designed and completed projects and learned how to work effectively with their neighbors in 4-H clubs. Men and women—from Ironwood to Monroe—made tens of thousands of inquiries to their county and M.S.C. extension agents for advice and information needed to improve crop yields, to prepare more nutritional meals, and to make their dwellings safer and more efficient, among hundreds of other concerns. Women and men, trained at the college to teach home economics and agriculture in high schools, taught thousands of youths to make their homes healthier and to run more productive farms. As more young adults moved from their family farms to cities, they took the knowledge and values learned from their land-grant mentors with them. Women teaching home economics in urban schools instructed girls, many of whom came from immigrant families, in hygiene, cooking, sewing, and household management. The land-grant philosophy reached out to all people—not just those who lived in rural Michigan.

5

The Alumni at the Turn of the Twentieth Century

B Y 1900 IT WAS CLEAR THAT PRACTICAL FARMERS' COMPLAINTS THAT the majority of men and women educated at Michigan Agricultural College did not take up farming were warranted. While farmers, politicians, and college officials argued over M.A.C.'s purpose, students prepared themselves to pursue careers that enabled them to realize their personal aspirations, and at the same time reshape their world. Land-grant–trained men and women apparently had little intention of letting the circumstances of their birth and their family history dictate their future, even if this meant not returning to the farm. By the time they left M.A.C. most students had abandoned the romanticization of the yeomen farmer, and many introduced new scientific and commercial practices to agriculture or took up professions that bore little resemblance to farming. M.A.C. alumni were fulfilling Justin Morrill's hopes for the children of the industrial classes.

During the late nineteenth century, M.A.C. students did not see the land-grant way as either static or beholden to the past. An editorial in the *College Speculum* in June 1882 expressed an understanding of the land-grant philosophy that was sensitive to farmers' complaints that M.A.C. did not train enough practical farmers, but that also endorsed the idea of educating people to enter other professions.[1] The *Speculum* recognized that agricultural colleges existed to impart "a knowledge of those sciences which pertain especially to agriculture and the industrial arts related to it." The paper noted that 40 percent of the institution's graduates were farmers and claimed: "No other school of its kind can show so favorable a result." The majority of graduates, however, became merchants, lawyers, physicians, machinists, engineers, teachers, and editors. The *Speculum* lauded the work done by alumni in careers outside of agriculture, declaring that "the

The Class of 1861 was the first class to graduate from the State Agricultural College. No one in that first class received his bachelor of science degree in person, because the Civil War drew all of them away from campus before commencement. (*Left to right*) Henry D. Benham, Larned Vernal Beebe, Albert Nelson Prentiss, Gilbert A. Dickey, Albert Fuller Allen, Adams Bayley, and Charles Edward Hollister. Courtesy of Michigan State University Archives and Historical Collections (Michigan State University, Students, Class Pictures, Class of 1861).

College may well be proud of their occupation and the degree of success they are attaining." With each passing year, more and more alumni found their life's work away from the family farm. If the farmers expected the State Agricultural College to educate a significant portion of their children to be practical farmers, two numbers reveal that such a hope was beyond the college's capacity to do so. Approximately 200,000 people operated farms in Michigan in 1900; and as of that year only 4,612 men and women had attended M.A.C. during its entire 45-year history. If all of the alumni had become enlightened practical farmers in Michigan and were still alive, they would have constituted less than 2.5 percent of the farmers in the state.

In 1900 M.A.C. published *Michigan State Agricultural College: General Catalogue of Officers and Students, 1857–1900*.[2] This document lists not only the names of 789 graduates through 1899, but it also records the names of nearly 3,800 men and women who attended but never graduated from M.A.C. Through 1899, 712 students had completed the agricultural course; 69 had finished the mechanical course; and 8 women had earned degrees in the new women's course. A brief occupational history follows the names of each of the graduates and of more than 2,000 of the nongraduates. This list of alumni vocations reveals a group of energetic, vigorous men and women contributing in almost every conceivable way to the development and growth of Michigan and the United States from the Civil War until the turn of the twentieth century.

The *Catalogue* shows a group of people who engaged their world in ways that exposed the revolutionary nature of the land-grant philosophy and confirmed the practical farmers' fears that M.A.C. was educating their children in a manner that led many away from the farm. The *Catalogue* left no doubt that land-grant education transformed young men and women who had grown up on farms, in small towns, and in cities into citizens who had learned to live successfully and to contribute to a world that was continuing to become vastly different from the world of the previous generation. This was democracy at its best, as people educated at M.A.C. helped to bring about significant changes in American society in a peaceful, constructive manner. M.A.C. alumni applied the knowledge and skills that they learned at their agricultural college to make a living while serving other people.

MICHIGAN AT THE TURN OF THE TWENTIETH CENTURY

Michigan, the home state for most M.A.C. students, had seen many changes between 1857, when the college welcomed its first students, and 1900. Although the state was still predominately agricultural, urbanization and industrialization had already recast the state demographically. The population had grown from 749,000 in 1860 to 2,421,000 in 1900, but the proportion of rural dwellers to urban inhabitants had shrunk from 86.6 percent to 60.7 percent. In actual numbers, the rural population had grown from 649,000 to 1,469,000 and the urban population from 100,000 to 952,000. This trend accelerated in the twentieth century, and in 1930 only 31.8 percent of Michigan's 4.8 million people lived in rural areas (see table 2). Michigan's farm population stood at 843,000 in 1890, reached its peak in 1900 at 982,000, and declined to 785,000 in 1930.[3] Even though agriculture's relative place in the state's economy shrank, acreage under cultivation continued to grow until 1920 (see table 3). Michigan's largest cities included Detroit with a population of 285,704; Grand Rapids, 87,565; Saginaw, 42,345; Bay City and West Bay City combined, 40,747; Jackson, 25,180; and Kalamazoo, 24,404.

By 1880 Michigan farmers had cleared enough land to cultivate crops intensively rather than extensively, and by 1898 improved

Table 2. Michigan Population, 1850–1930 (in thousands)

YEAR	URBAN	RURAL	PERCENT OF POPULATION RURAL
1850	29	369	92.7
1860	100	649	86.6
1870	238	946	79.9
1880	405	1,232	75.3
1890	730	1,364	65.1
1900	952	1,469	60.7
1910	1,327	1,483	52.8
1920	2,242	1,427	38.9
1930	3,302	1,540	31.8

Source: "Population of States, by Sex, Race, Urban-Rural Residence, and Age: 1790 to 1970," *Historical Statistics of the United States: Colonial Times to 1970* (Washington, D.C.: U.S. Department of Commerce, Bureau of the Census, 1975), part 1, 29.

Table 3. Number of Farms and Acreage in United States and Michigan, 1850–1930 (in thousands)

YEAR	NUMBER OF FARMS		TOTAL ACREAGE	
	UNITED STATES	MICHIGAN	UNITED STATES	MICHIGAN
1850	1,449	34	293,561	4,384
1860	2,044	62	407,213	7,031
1870	2,660	99	407,735	10,019
1880	4,009	154	536,082	13,807
1890	4,565	172	623,219	14,786
1900	5,740	203	841,202	17,562
1910	6,366	207	881,431	18,941
1920	6,454	196	958,677	19,033
1930	6,295	169	990,112	17,119

Source: "Farm Population, Farms, Land in Farms, and Value of Farm Property and Farm Products Sold, by State: 1850 to 1969," *Historical Statistics of the United States: Colonial Times to 1970* (Washington, D.C.: U.S. Department of Commerce, Bureau of the Census, 1975), part 1, 459–60.

equipment enabled two workers to do what ten had done in 1860. As land became scarcer, producers invested more resources into smaller plots of ground to increase yields and to restore or conserve the soil's fertility. After 1880 a marked change in Michigan agriculture occurred. Farmers raised more dairy cows, horses, swine, and poultry, and fewer sheep. As a result, they grew oats and hay needed to feed their new stock, which could not sustain itself by grazing. Increased production in oats and hay led to the decline in the quantity of wheat harvested in Michigan. Orchards and gardens produced fruits and vegetables for the farm family, but more important, growers sold their surpluses in urban markets, especially in Chicago. Farmers used fertilizers and new machinery to increase the productivity of both soil and labor.[4]

The 1904 census of Michigan shows that at the outset of the twentieth century Michigan's farmers engaged in a diversified agriculture that included dairying, production of grains and hay, poultry production, an extensive fruit industry, vegetable production, and nurseries. Between 1894 and 1904 there were shifts in production from wheat to corn, oats, and beans and from sheep to cows, swine, and chickens, and there was a dramatic increase in yields from apple trees (see table 4). Dairying and the cultivation of corn, oats, and clover occupied a prominent place in the economy of southern Lower Michigan, where apple orchards stood next to many farm houses. As dairy farmers increased the quantity of milk that they sold, creameries and cheese factories manufactured more butter, cream, ice cream, and cheese for sale to the growing urban population. Many farmers kept swarms of bees. The production of peaches, plums, cherries, and pears in the southwestern counties of Allegan, Berrien, Van Buren, and Kent gave Michigan a thriving fruit industry. Peaches also grew in counties farther north along the shore of Lake Michigan and farther inland in Newaygo and Oakland counties. Growers in Grand Traverse County brought an increasing amount of cherries to the market. Sugar beets played a key role in the economies of Gratiot, Bay, Saginaw, Huron, and Tuscola counties. Counties bordering on a line from Ionia County east to Tuscola County formed the bean country in 1903, and farmers grew potatoes from Allegan to Grand Traverse to Tuscola to Wayne County. The early 1900s saw a rise in the production of such garden vegetables as cabbage, tomatoes, sweet corn, onions, cucumbers, and

**Table 4. Comparison of Farm Commodities in Michigan
Between 1894 and 1904**

COMMODITY	1894		1904	
	ACREAGE	YIELD (BUSHELS)	ACREAGE	YIELD (BUSHELS)
Corn	953,763	40,556,871	1,516,417	48,249,885
Wheat	1,646,246	26,805,600	847,552	14,227,220
Oats	976,727	25,506,662	1,330,746	43,004,248
Barley	61,202	962,515	77,315	1,818,493
Beans	143,265	1,128,078	350,895	5,049,990
		(TONS)		(TONS)
Clover hay	911,699	1,239,185	1,239,548	1,737,285
Other hay	1,351,277	1,717,672	1,370,349	1,823,832
	TREES	(BUSHELS)	TREES	(BUSHELS)
Apples	9,491,066	1,855,887	8,892,889	15,927,580
Peaches	3,449,171	1,263,167	7,314,035	1,843,513
	NUMBER		NUMBER	
Calves	242,220		425,369	
Steers (all ages)	312,352		283,666	
Cows	506,390		746,685	
Horses (all ages)	663,447		697,364	
Sheep (all ages)	3,443,975		2,340,804	
Swine	1,045,151		1,574,461	

Source: *Census of the State of Michigan, 1904: Agriculture, Manufactures and Mines* (Lansing: Wynkoop Hallenbeck Crawford Co., 1905), 2:xvi–xxxv.

celery and of melons for the city markets. Lenawee, Wayne, Kent, Allegan, and Oakland counties were among the leaders in vegetable growing. Significant greenhouse and nursery businesses had also arisen. The 1904 census reports the presence of other commodities in Michigan agriculture, including strawberries, raspberries, blackberries, gooseberries, alfalfa, geese, and ducks. The census shows that most of

Michigan's agricultural activity took place in the Lower Peninsula, although significant numbers of new settlers in the Upper Peninsula would soon join the one farm duck residing in Keweenaw County.[5]

Several industries transformed much of Michigan during the second half of the nineteenth century. Lumbering in the Lower Peninsula brought great wealth and many jobs to Muskegon and Manistee on the Lake Michigan shore, stimulated furniture manufacturing in Grand Rapids, and led to economic growth along the Saginaw River and its tributaries. In the Upper Peninsula, Escanaba and Menominee, in particular, benefited from lumber production.[6] Railroad development paralleled the explosion in the lumber business, connecting agricultural and industrial centers to one another and to the national market. In 1860 Michigan had less than 800 miles of track; in 1900 trains rumbled over a network of nearly 11,000 miles that covered the state like a giant spider web.[7] Copper and iron mining stimulated the economies in the counties bordering on Lake Superior west of Marquette.[8] Detroit became a center of manufacturing for heavy equipment, including locomotives, steam engines, and freight cars. Among other enterprises in Detroit were Parke, Davis and Company, which started producing pharmaceutical products in 1867; Dexter M. Ferry ran a farm and garden seed business, and the firm of Pingree and Smith made shoes.[9]

The Michigan 1904 census shows that industrial growth exploded in the state at the beginning of the twentieth century, creating scores of new vocational opportunities for college-educated men and women. Capital invested in Michigan's manufacturing enterprises jumped by over 25 percent between 1900 and 1904, reflecting a 28.4 percent increase in cities of at least 4,000 people but only an 18.1 percent increase in rural areas. Further evidence of the growing concentration of industry in urban areas is shown by a 9.5 percent growth in the number of establishments in the cities, as opposed to a 9.2 percent decrease in the country. In four years the number of salaried workers in the manufacturing sector shot up 31.9 percent (from 13,062 in 1900 to 17,235 in 1904), and their salaries surged ahead by an even more impressive 44.3 percent. All of this growth drove up the value of manufactured products by 34.8 percent in just four years.[10] (See appendix 2 for identification of the industries in Michigan that were reshaping the state's economy as they provided thousands of jobs to people migrating to urban areas from within the state and overseas.) Farm machinery, railroad locomotives, cheese and butter, furniture, and drugs, among other products, altered the material standard of living for thousands of people as new technology made travel easier, life more comfortable, and dairy products more pure. Men and women trained at M.A.C. found careers in these burgeoning industries. Of particular note is the existence of twenty-two automobile manufacturers in Michigan in 1904.

Immigration had created a complex mosaic of people living in Michigan at the turn of the twentieth century. The several thousand Native people who called Michigan home comprised a very small minority, with approximately 3,000 Chippewa, 500 Potawatomi, and 2,500 Ottawa or Odawa residing in Michigan in 1910.[11] Between 1882 and 1892 immigrants to Detroit pushed the Polish population from 1,200 to 35,000, and nearly 3,000 Poles made railroad cars for the Peninsular Car Company.[12] African-American communities with about 500 residents each existed in Kalamazoo and Battle Creek, and their members found employment in skilled trades and service work. Significant numbers of African Americans also lived in Detroit, Grand Rapids, Jackson, Lansing, Ann Arbor, Saginaw, and Flint. Other African Americans farmed in Isabella, Van Buren, Allegan, Cass, and Mecosta counties.[13] Germans made up the largest foreign-born group in Detroit and Saginaw in 1890. The majority of them found laboring jobs in Detroit's industries, while others worked in managerial and professional occupations.[14] In 1890 over three-fourths of Michigan's 3,100 Italians had settled in the copper and iron mining areas of the Upper Peninsula.[15] The Dutch thrived in western Michigan especially in Holland, Grand Rapids, and Kalamazoo, where they engaged in manufacturing and commerce.[16] The Jewish community in Detroit grew throughout the last half of the nineteenth century, especially after 1880, when large-scale immigration saw Detroit's Jewish population jump from 1,000 to 10,000 by 1910. Many in this community distinguished themselves in law, architecture, commerce, and medicine.[17] The Irish found their way to Wayne and Kent counties in the Lower Peninsula and Houghton and Marquette counties in the Upper Peninsula. They worked as laborers, merchants, and miners, and on the railroads.[18] The decade between 1880 and 1890 saw large numbers of French Canadians arrive in Michigan, where many worked in the

lumbering industry in the Saginaw Valley and Lake Michigan towns. Others mined in Marquette and Houghton counties or participated in Detroit's increasingly diverse industries, businesses, and professions.[19] Norwegian and Swedish communities had evolved in Manistee, Muskegon, Marquette, and Houghton counties, and Swedes had also settled in Kent, Gogebic, and Menominee counties. Finns came to the Copper Country to work the mines, as did the Cornish.[20] Greeks, Russians, Hungarians, and Danes had established small communities in Michigan on the eve of the new century.[21]

Michigan's population in 1900 was literate, and most M.A.C. students received their basic education in the state's common schools. A large percentage of males living in Michigan could read: 97.6 percent of native white males; 85.9 percent of native black males; and 91.9 percent of naturalized foreign-born males were literate. Foreign-born white youth between 10 and 21 years of age formed the only large group of illiterates. Presumably many of these young people would learn to read English in the state's growing and improving public school system. As Michigan entered the new century, approximately 70 percent of school-age children were enrolled in school, and the average number of days of school each year was over 160.[22]

Some members of the Class of 1892 posed for this picture. The presence of two men who came from very different backgrounds represented what the student body of a land-grant college might become. Leander Burnett, '92, (*top row, far left*) had a mixed Native American and Anglo-American heritage. He came from Little Traverse (Harbor Springs), Michigan, where he had been raised by his Native American mother. Leander apparently never knew his Anglo-American father. Burnett married Grace L. Fuller, '91, in 1899, but she died the next year. Leander was working for a utilities company in Pittsburgh, Pennsylvania, when tuberculosis took his life in 1906 (David Thomas and Marc Thomas, "Leander Burnett: Saga of an Athlete," *Michigan History* 70, no. 1 [January/February 1986]: 12–15). Kolia San Thabue, '91, (*third row, fifth from left*) came from faraway Burma to take the agricultural course. The 1930 *Alumni Catalogue* reports that he was director of the government-aided school of Kanthagon and pastor of a Baptist church in Tahyagon Village, Wakema, Lower Burma. The land-grant college held the promise of being a home for a community of people from different ethnic groups, religions, and social classes from all over the world. Courtesy of Michigan State University Archives and Historical Collections (Michigan State University, Students, Class Pictures, Class of 1892).

Men and women who attended the State Agricultural College came from both rural and urban settlements, where people relied upon newspapers for much of their information about their community, state, nation, and the world. In 1900 Michigan publishers issued over 750 papers across the state. Nearly every county had a least one weekly (the frequency of most newspapers). Most had a strong Democratic or Republican editorial position, and some claimed to be independent. Many towns, large and small, had rival papers that helped to keep political issues alive in the minds of readers. For example, Menominee County boasted two Democratic papers, the tri-weekly *Evening Leader* and the weekly *Menominee Democrat;* two Republican papers, the tri-weekly *Menominee Herald* and *Menominee County Journal* (published in Stephenson); and an independent paper, the German-language *Pioneer and Volksfruend.* In Grand Rapids, three daily papers vied for readers of different political persuasions—the independent *Evening Press;* the *Grand Rapids Democrat;* and the Republican *Grand Rapids Herald.* Some papers, such as *The Michigan Poultry Breeder* (a monthly out of Battle Creek), *Alpena Farmer* (a weekly out of Alpena), *Michigan Catholic* (a weekly out of Detroit), and *Michigan Sugar Beet* (a weekly out of Bay City), directed their content towards specific audiences. Numerous foreign-language papers provided invaluable information and support to immigrants who did not know English. The Swedish *Superior Posten* (Marquette), the German *Volkszeitung* (Manistee), the Polish *Gwiazda Polska* (Detroit), and the Dutch *De Hope* (Holland) were just a few of the many newspapers that played important roles in the growing immigrant communities across Michigan.[23]

A Profile of the Alumni in 1900

The 1900 catalog reveals only a little information about each person listed, but these people were neither abstractions nor mere statistics–they helped to fill the ranks of the new professions that emerged during the late nineteenth century. Armed with their land-grant education, M.A.C. men and women engaged in business and commerce, all aspects of agriculture, government, and public service. In addition, they were husbands or wives, parents, neighbors, voters, churchgoers, readers of newspapers, and consumers. The roles assumed by M.A.C. alumni supported Justin Morrill's and Abraham Lincoln's belief that a college education for the sons and daughters of ordinary folks was essential for American democracy to thrive.

Alumni took their education at the State Agricultural College seriously. On Thursday, 30 May 1907, Russell Allen Clark, '76, told an audience gathered on the campus to celebrate the college's Semi-Centennial that being an alumnus carried with it awesome responsibilities. The educated man was to "be a leader of broader thought and higher morals," and he had the duty to bestow "upon his fellow-citizens and upon the state that educated him the greatest benefit of his intellectual training." Clark stressed that "boys" who remained at home "reached their limitations," while young men who came to M.A.C. received training "for the larger life beyond."[24] The day before, Lucius Whitney Watkins, '93, president of the Farmers' Clubs of Michigan and a former member of the State Board of Agriculture, had warmed the hearts of the college's friends when he said: "It is a great college that can turn out a first-class governor [Fred Warner] from a poor farmer boy in less than a year of its agricultural course! It is an institution that will in every case develop men and women, in the fullest sense of the term, out of all those who have capacity and desire to learn."[25]

The careers of men and women educated at M.A.C. during its first fifty years demonstrated that Clark and Watkins were not speaking in platitudes. Clark's language underscored the decidedly male dominance of M.A.C., although that was changing. From the institution's founding through 1899, only 33 women had received degrees and only another 325 had attended. The establishment of the women's program in 1896 made it possible for more women to study at M.A.C. and to assume faculty positions. Over time the growing female alumnae made increasingly greater contributions to life in Michigan, the United States, and the world. Nine men from other countries had graduated by 1899, and nearly fifty more had studied at the school. As of 1900 very few African Americans had enrolled in M.A.C., and would be another four years before the first degree was awarded to an African American.

While a majority of students attending Michigan Agricultural College came from Michigan, nearly one-half (374 out of 789) of the

Throughout the college's early history fire presented the greatest threat to the safety of students and faculty and to the existence of the college itself. On 8 October 1871 the State Agricultural College fared better than Chicago and Holland, Michigan, when great fires ravaged large parts of Michigan, Wisconsin, and Illinois. Professors Manly Miles and Robert C. Kedzie directed teams of students as they successfully prevented forest fires from reaching campus. Five years later students lost many possessions when Saints' Rest succumbed to fire in December 1876. During the night, after M.A.C.'s thrilling football victory over the University of Michigan on 18 October 1913, Secretary Addison M. Brown's barn burned down mysteriously. *The Holcad* attributed the blaze to "defective wiring." Fortunately, Brown's cow was rescued ("The Barn is Gone," *M.A.C. Record* 19, no. 5 [28 October 1913]: 2; *Holcad* 6, no. 4 [20 October 1913]: 5).

TOP: The Botanical Laboratory, which opened in 1880, as it appeared on 24 March 1890. Two students lived in the attic in exchange for looking after the building. The fire started when an overheated stove pipe from their stove set a partition in the attic on fire. Still in their nightshirts, the students escaped harm. Students living in Wells Hall carried microscopes, books, specimens in the botanical collection, and other items out of the building. Courtesy of Michigan State University Archives and Historical Collections (Michigan State University, Buildings, Botanical Laboratory).

BOTTOM: The first Wells Hall was built in 1877 after Saints' Rest burned down. On 11 February 1905 Wells Hall met the same fate as the college's first dormitory. Courtesy of Michigan State University Archives and Historical Collections (Michigan State University, Buildings, Wells Hall (1), Fire, 1905).

graduates had moved out of the state by 1900 (see appendix 3). Available information indicates that at the same time only 27 percent of people who had attended classes at M.A.C. but had not graduated had left Michigan. Graduates who stayed in Michigan lived throughout the state, giving the taxpayers an immediate return on their investment in M.A.C. (see appendix 4). The hometowns or post offices for 701 nongraduates (out of 3,800) are known at their time of entry,

TOP AND CENTER: A fire started in the basement of the Engineering Building (built in 1907) on 5 March 1916 and quickly consumed the entire structure and almost all of its contents. Gallant efforts by students and firemen put out fires on the roof of Wells Halls and kept the blaze from gaining a foothold in the building. Residents of the dormitory removed their personal belongings and furniture in case the fire reached their homes. The loss of the Engineering Building precipitated an effort to move engineering studies to the University of Michigan rather than rebuild and re-equip the facility at M.A.C. President Frank S. Kedzie called upon his old friend Ransom E. Olds for help. Olds, who believed that M.A.C.–trained engineers were among the best in the country, donated $100,000 towards the construction of a new building. He pointed out: "Now any institution that turns out successful men must have the right something in it or behind it. Any such institution should have the support of the public" ("R. E. Olds Gives M.A.C. $100,000.00 for Engineering Building," *M.A.C. Record* 21, no. 28 [25 April 1916]: 5). With Olds's gift the college built the Ransom E. Olds Hall of Engineering and saved its engineering program. Courtesy of Michigan State University Archives and Historical Collections (Michigan State University, Buildings, Engineering Building (2) Fire 1916, Michigan State University, Buildings, Wells Hall (2)).

BOTTOM: Venerable Williams Hall burned on 1 January 1919. Fortunately, the Student Army Training Corps residents had left campus and no new students had arrived yet, leaving the dormitory vacant. The postcard view pictured here shows only the outer shell surrounding the gutted structure. The loss of the first Williams Hall created a housing shortage for men and sent many male students looking for housing in the growing community of East Lansing. Courtesy of Michigan State University Archives and Historical Collections (HC 947 [Arno J. Erdman Collection]).

and they suggest that this population hailed from scores of small cities, villages, and unincorporated hamlets from all over the state. Only 119 of these students came from the state's six largest cities, including 57 from Detroit and 47 from Lansing. Young people from the farms and small towns of Michigan found that M.A.C. was their gateway into the world.

GRADUATE-FARMERS

Graduate-farmers helped other agriculturalists to transit from primarily producing for their own consumption and a local market to growing for and selling on a national market, and to adapt to the industrialization and commercialization raging throughout the United States. More graduates of the State Agricultural College engaged in farming than in any other occupation, but most of them contributed to their world in other productive ways as well. Slightly more than one-fourth of the 756 male graduates listed farmer as an occupation, but only 82 identified themselves as just a farmer after they left the college. If the college's claim that about 60 percent of the students came from farms was accurate, an M.A.C. education apparently made it possible for many farmers' sons and daughters to seek careers away from home, or, if they did take up farming, to opt not to confine their professional activities to their farms. A look at occupational histories brings to light graduate-farmers who worked for commercial businesses, filled leadership positions in agricultural organizations, held public offices, and taught in the public schools. They helped to build the educated human infrastructure that America needed to confront the spectrum of challenges presented by the early twentieth century.

Some men combined farming with a variety of business interests in Michigan and other states that allowed them to contribute to and benefit from the commercial world that reached beyond their own communities. In 1900 Bartlett Augustus Nevins, '75, resided in Otsego, Michigan, where he farmed and sold real estate and insurance. Nevins previously had manufactured farm machinery and had been a lumberman. Politically active, he had served his community in several capacities: as deputy sheriff, member of the local board of education, postmaster, member of the Michigan House of Representatives, and justice of the peace. Leslie Albert Buell, '83, lived in Minneapolis, Kansas, from 1883 until 1891, working in real estate, insurance, and lending. After he moved to Highland, Michigan, he took up farming and stock raising while managing a business in Detroit that grew out of his invention of "'Photo Titles' or Photographic Abstracts of Deeds and Mortgages." Buell's classmate, Milton St. John, '83, taught school for three years before spending nine years on the road selling seeds for D. M. Ferry & Company. He then farmed in Yates, New York, and served as president of a creamery.

M.A.C. graduate-farmers invested their talents in the wide range of new technologies and industries that sprang up after the Civil War. Some of these men moved easily from one job to another and from place to place—mobility characterized land-grant educated people. Warren W. Reynolds, '70, ran the Cassopolis (Michigan) Telephone Exchange when not looking after his farm. Charles Howard Parker, '75, was a coal merchant in Colorado City, Colorado, before he moved to Sturgis, Michigan, to farm. After farming for four years, Albert Avery Robinson, '78, managed the Detroit Motor Company, after which he sold electrical equipment and supplies in Detroit. Charles Homer Hoyt, '85, farmed near Irving, Michigan for nine years before he moved to Cleveland, Ohio, where he handled sales in Ohio for the Detroit Graphite Manufacturing Company.

Graduate-farmers provided leadership in agricultural organizations that formed to assist farmers influence government and business on their behalf. George Daniel Moore, '71, of Medina; Elmer Otis Ladd, '78, of Old Mission; Jason Woodman, '81, of Paw Paw; and Charles Henry Todd, '89, and Ellsworth Albert Holden, '89, of Lansing all played important roles in the Michigan State Grange. Other M.A.C. graduates held important posts in organizations that promoted more specific agricultural interests in Michigan. Erwin Noble Ball, '82, of Hamburg, was secretary of both the Michigan Merino Sheep Breeder's Association and the American Tamworth Swine Record Association. Arthur Cranson Bird, '83, who lived near the college, served the Michigan Farmers' Clubs as secretary (1894–95) and president (1895–96); he was also the secretary to the State Board of Agriculture from 1899 until 1902. Albert Baldwin Cook, '93, followed in Bird's footsteps when he became secretary of the Michigan Farmers' Clubs for the year 1898–99.

A native of Quincy, Illinois, Louis A. Bregger, '88, majored in horticulture. As a freshmen, Bregger recognized the value of an education at M.A.C. He told his brother: "One who receives a full course here leaves with a splendid education" (Louis A Bregger to brother, 14 September 1884, Bregger, Louis A. Papers, 1884–1909, UA 10.3.93, Folder 2). After his graduation, he worked for Graceland Cemetery in Chicago before moving to Bangor, Michigan, with his wife, Anna Henjes, in 1900. Bregger grew fruit on his farm. Courtesy of Michigan State University Archives and Historical Collections (Class '88 M.A.C., Class Album).

Graduate-farmers put their educations to work in their communities as they held local public offices across Michigan. Dairy farmer William Dallas Place, '68, played a prominent role in Ionia, where he served in succession as clerk of Ionia Township, 1874–86; township supervisor, 1886–88; Ionia county clerk, 1888–92; and member of the Michigan House of Representatives, 1892–98. In Allegan County, Lyman Augustus Lilly, '77, spent two years as supervisor of Hopkins Township and four years as the county's register of deeds. Orrin Preston Gulley, '80, had a four-year term as supervisor of Dearborn Township followed by a tour of duty as Wayne County's superintendent of the poor. In Montcalm County, Fred Charles Snyder, '82, filled the position of register of deeds when not tending to his farm; he also found time to be on the county board of school examiners for a couple of years before his death in 1893.

At least twenty of the graduate-farmers taught school at some time during their careers, and others contributed to Michigan's growing public education system as principals, superintendents, and commissioners of schools. William Asa Rowe, '73, a farmer in Mason, Ingham County, first taught in, then held the post of superintendent of township schools for six years, after which he became county commissioner of schools from 1882 until 1889. Rowe left public education to become the county surveyor. Although he apparently never taught, Alford Joshua Chappell, '82, fulfilled his public duty as principal of schools in Alba from 1888 until 1896 while also having a seat on the Antrim county board of school examiners. In 1896 Chappel moved to Manton, Wexford County, to assume the position as principal of schools. After John Walker Matthews, '85, farmed for several years, he moved around devoting his talents to high school education during its formative years. He taught science at Grand Rapids High School between 1890 and 1895. Several years later he spent a year as principal of South Division Street School in Grand Rapids before moving to Detroit to become the principal of Western High School.[26]

Graduate-farmers got into businesses that helped agriculturists produce and sell more on the constantly expanding national market. Israel Hibbard Harris, '73, bought grain from farmers in Bathgate, North Dakota, and sold them coal to heat their homes and feed for their cattle. Alva Sherwood, '81, D.V.S., cared for his neighbors' animals with a veterinary surgery practice to in addition to running his own farm and livestock dealership in Three Oaks, Michigan. Charles McDiarmid, '84, farmed at Bear Lake, Michigan, until 1893, when he migrated to Hemet, California, to operate a ranch. While still in Michigan, he was manager of the Farmers' Exchange in 1891, and in California, he became a director of the Hemet Deciduous Fruit Exchange. Willford Jerome McGee, '96, used his land-grant education in agriculture as a grain broker in Chicago. As the manufacture of butter moved from the farm house to the creamery, M.A.C.-trained men, such as Harry Lorin Mills, '98, of Nunica, Michigan, applied the latest scientific knowledge to butter making in creameries.

LAWYERS AND DOCTORS

Growing up on a farm or in a small town did not prevent young people from wanting to practice law or medicine, and a M.A.C. degree opened the door for more than 15 percent of the college's graduates to enter these two professions. At least seventy-five alumni practiced law throughout Michigan and the country. John Justice Kerr, '71, received his LL.B. from the University of Michigan in 1879. He did editorial work before he became the county attorney for Valley County, Montana, in 1893. At the end of his second term in 1898, Kerr worked as a lawyer in Glasgow, Montana. Three years after the University of Michigan awarded William Henry Burgess, '81, his LL.B. in 1884, he served as prosecuting attorney of Sanilac County, after which he sat as judge of probate from 1893 until 1897. William Richard Rummler, '86, applied his legal training as a patent attorney in Chicago. Andrew Brace Goodwin, '88, moved out of farming in 1892 into a legal career in Carson City, Michigan. In 1894 he assumed the duties of village attorney, added the responsibilities of village clerk two years later, and became a justice of the peace in 1899.

Another forty-three graduates turned to medicine to fulfill their career aspirations. John Knox Gailey, '74, received his doctor of medicine degree from the New York College of Physicians and Surgeons in 1877. Gailey cared for the sick in Detroit, where he was superintendent of Harper Hospital from 1882 until 1887, after which he took up the cause of the Children's Free Hospital and the Home of the Friendless. In 1900 Stephen Prince Tracy, '76, practiced medicine in Milford,

Charles W. Garfield, '70, sat on the State Board of Agriculture from 1887 until 1899. Born on a farm near Milwaukee, Wisconsin, in 1848, he moved with his family to a farm near Grand Rapids when he was about ten years old. Garfield played a leading role in the forestry movement in Michigan by helping to introduce forestry into the curriculum at both M.A.C. and the University of Michigan. In 1893 Garfield became president of the Grand Rapids Savings Bank, an institution that his father helped to organize. Courtesy of Michigan State University Archives and Historical Collections (People, Garfield, Charles W., no. 1082, Folder 2).

Nebraska. He had farmed for a number of years before he attended the Hahnemann Medical College of Chicago, which awarded him his medical degree in 1886. Ambrose Everett Smith, '81, studied medicine at the Rush Medical College in New York earning his medical degree in 1885. Smith, an eye, ear, nose, and throat specialist, treated patients in Olean, New York. Edward Sawyer Antisdale, '85, attained his medical degree from the University of Michigan in 1890. In 1900 he was professor of ophthalmology at the Chicago Eye, Ear, Nose, and Throat College.

WOMEN GRADUATES

In the late nineteenth century, women pushed out the perimeters of American democracy, demanding more civil rights and opportunities to participate in society, including the prerogative to attend colleges and universities.[27] At the dawn of the twentieth century, young women confronted seemingly conflicting philosophies that encouraged them to become homemakers, on one hand, and to break into both old and new professions, mostly dominated by males, on the other. The land-grant education available at M.A.C. prepared young women to follow either path or allowed them to acquire the skills needed to blend together home management with active involvement—often voluntary—in community life. Many of the early female graduates accompanied their husbands to wherever their employment took them or interrupted their careers, usually as teachers, upon marriage.

After the State Agricultural College graduated its first woman, Eva Diann Coryell, in 1879, it awarded the bachelor of science degree to thirty-two more women before 1900. For about one-half of them, only their address and married name, if applicable, appears in the catalog. Most of these women's primary occupation probably was homemaker, but few likely confined their activities to their families. Chippie [Mary L.] Harrison, of the class of 1888, who had been a hospital nurse, lived with her husband Thomas Flower Jr., in Detroit in 1900. Some women married classmates. Mary Lucy Carpenter, '88, moved with her husband Nelson Mayo, '88, a veterinarian, from Chicago to Kansas to Connecticut to Cuba to Virginia and elsewhere. Jessie Beal, '90, the daughter of Professor W. J. Beal, married the writer Ray Stannard Baker, '89. Katharine Cook, '93, the daughter of Professor A.

J. Cook and Mary Baldwin, lived in Washington, D.C. with her husband Lyman Briggs, '93, and became a writer and an advocate for strong parental involvement in the education of children. Mary Lilian Wheeler, '93, worked as a scientific aid in the United States Department of Agriculture's Division of Vegetable Physiology and Pathology. Later she wed her classmate, Dick Jay Crosby, '93.

Nine female graduates became teachers. These women brought a badly needed level of education to the public schools, since many other teachers were poorly trained. Although low pay, unruly students, and inadequate facilities often made teaching a difficult profession, school districts in Michigan made improvements through the consolidation of small districts into union schools, construction of better buildings, and the creation of high schools. Some teaching careers were influenced by marriage. Grace L. Fuller, '91, taught in Lansing for five years before she moved to Newark, New Jersey, in 1898. After one year in a Newark school, she moved back to Lansing with her husband, Leander Burnett, '92. Fuller died on 2 January 1900. Mary Elizabeth Belliss, '99, taught classes at the Industrial School for Boys in Lansing from 15 August 1899 until 17 March 1900. She then accompanied her husband, Charles Johnson, '99, to Shawnee, Pennsylvania, where he had a job as a gardener. Marian Weed, '91, instructed grade school children in Grand Rapids for seven years before she moved to Indianapolis, Indiana, to join the staff of the Girls Classical School located there. Amy Bell Vaughn, '97, the first graduate of the new women's course, stayed on at M.A.C. to earn her master of science degree. Vaughn then relocated to Chicago, where she taught cooking at Brown School. During the 1897–98 school year, students in the Holland, Michigan, high school learned mathematics from Bertha Marquerite Wellman, '96, who had taken a bachelor of pedagogy degree from the Michigan State Normal College in 1897. Daisy Edna Champion, '93, assumed a job in school administration as principal of Bingham Street School in Lansing.

Other female graduates took jobs that foreshadowed future career opportunities that would open up for women educated at Michigan Agricultural College. Mary Jane Cliff Merrell, '81, a widow, was the college librarian from 1883 until 1888. She then married Louis G. Carpenter, '79, who was first an assistant in mathematics and then an assistant professor at M.A.C. The couple went to Fort Collins, Colorado, where Louis accepted a professorship at Colorado Agricultural College in 1888. Following her graduation in 1883, Sarah Ellen Wood studied nursing at the Farrand Training School for Nurses in Detroit. Before settling at Essexville, Michigan, Wood served as the college nurse at Earlham College in Indiana.

CAREERS IN INDUSTRY AND COMMERCE

Approximately 15 percent of male graduates found careers in industry and commerce. M.A.C. alumni held a variety of positions that illustrate the growing importance of science, particularly chemistry, in American industry. William Robert Hubbert, '81, (M.D., Detroit Medical College, '85) worked as a bacteriologist for Parke, Davis and Company, and later for Frederick Stearns and Company in Detroit. Herbert Eugene Harrison, '88, lived in Milwaukee, Wisconsin, where the Diamond Soda Works and Liquid Carbolic Acid Manufacturing Company employed him as a chemist. Lemuel Churchill, '89, (M.D., University of Michigan, '92) managed the Newbro Drug Company in Butte, Montana, in 1899. Thorn Smith, '95, served as an assistant chemist in the college's experiment station before taking a job as chemist with the Grasselli Chemical Company in Cleveland, Ohio. By 1900 Smith had moved on to a chemist's job with the Ducktown Copper and Sulphur Company in Isabella, Tennessee.

Sixty graduates contributed to the advance of technology in American industry as draftsmen, surveyors, and engineers, some even before the mechanical course was started in 1885. Francis Hodgman, '62, distinguished himself as a civil engineer and surveyor. In 1886, he and Charles Fitzroy Bellows published a significant book, *Manual of Land Surveying,* which appeared in several later editions. Theodore Percy Caulkins, '78, designed machinery, inventing a machine that automatically boxed and bagged matches. Will S. Holdsworth, '78, worked as a designer and draughtsman in Chicago until he returned to M.A.C. as an instructor in drawing in 1881. In 1887 he was promoted to assistant professor of drawing.

Beginning in 1885 young men were able to take the mechanical course, and more of them became engineers and managers in industrial plants. Orlando John Root, '89, superintended the Lansing Iron and Engine Works for three years until he relocated to New York, where the Watertown Steam Engine Company employed him as assistant superintendent for three years. In 1899 Root went into partnership with William H. Van Dervoort, '89, his college roommate, to build gasoline engines in Champaign, Illinois.[28] Louis C. Brooks, '92, spent two years as a public school administrator in Michigan before launching his career as a draftsmen, engineer, and construction superintendent. From 1896 until 1900, Brooks held three positions, one on the Great Lakes, another with the General Electric Company in Schenectady, New York, and a third with the U.S. Navy in Newport News, Virginia. John Pierce Churchill, '95, started out as a draftsman for the Illinois Central Railway shops, but after two years he took a job as assistant engineer with the Northern Pacific Railroad in Duluth, Minnesota. In 1899, the Northern Pacific promoted Churchill to resident engineer for their Yellowstone Division at Miles City, Montana.

WRITERS AND EDITORS

The literate population in Michigan and other states looked to newspapers and magazines for information about politics, the economy, world affairs, entertainment, and other interests. Farmers learned about new techniques, methods, varieties of seed, breeds of livestock, and equipment from countless articles in agricultural periodicals. Since M.A.C. insisted that its students study English, history, and philosophy, they were prepared to write informed stories for the public. The combination of language skills, agricultural training, and experience growing up on farms made many M.A.C. alumni effective at communicating with agriculturalists.

Alumni wrote for and edited prominent national publications and newspapers in large cities and small towns. Peter H. Felker, '71, began his career in the newspaper business as the managing editor of the *Grand Rapids Saturday Evening Post.* He moved to St. Louis, Missouri, where he edited the *St. Louis Grocer* and the *St. Louis Dry Goods Reporter* before assuming the presidency of Schultz Publishing Company in 1892. The *Detroit Journal* employed Howard Morrill Holmes, '81, as a reporter. Holmes then took the position as assistant city editor at the *Detroit Evening News.* Ray Stannard Baker, '89, served as an assistant editor for *McClure's Magazine* and published *Boys' Book of Inventions* and *The New Prosperity.* He became one of America's most prominent writers in the late nineteenth and early twentieth centuries. Albert Carleton Sly, '91, edited the *Roscommon Democrat* in Roscommon, Michigan, for three years after his graduation.

M.A.C. graduates connected with people from their home communities when they wrote for and edited agricultural journals. They used this medium to disseminate information to farmers that resulted from scientific investigations at agricultural colleges, experiment stations, and the USDA. Arthur Cranson Bird kept the agricultural community in Michigan informed through his efforts as the associate editor for *Michigan Farmer* from 1896 until 1900. Colon Christopher Lillie, '84, regularly wrote articles on dairying for the *Michigan Farmer.* Herbert Winslow Collingwood, '83, edited the *Rural New Yorker* for agriculturalists in New York. George Watt Park, '86, edited and published *Park's Floral Magazine* in Libonia, Pennsylvania, where he worked as a florist. Kenyon L. Butterfield kept in touch with farm families in Michigan through his job as editor of the *Grange Visitor* for five years after his graduation.

TEACHING AND RESEARCHING AGRICULTURAL SCIENCE

In 1902 W. H. Jordan, director of the agricultural experiment station in Geneva, New York, told the attendees at the annual meeting of the Society for the Promotion of Agricultural Science (SPAS) that they had a responsibility to train teachers of agricultural science and scientists to undertake original investigations. SPAS, established in 1880, was an organization of men (limited to one hundred members) committed to experimental agricultural science. Jordan's definition of his discipline, shared by many practitioners of scientific agriculture, quivered under the tension over the distinction between pure and applied research:

> It is obvious that *agricultural* science is *applied* science, that is, it is the application of the laws and facts of the various sciences to the phenomena and methods of farm practice. The point to be noted here is that we must distinguish between the science and the art.

The science consists in the knowledge which explains a phenomenon or dictates a method; and the art is the doing of something, or the prosecution of a method by mechanical means.[29]

Jordan's position found favor with many M.A.C.–trained agricultural scientists. It was no accident that in 1902 twelve M.A.C. graduates and two of their mentors, W. J. Beal and R. C. Kedzie, were active members of the society. Beal had played a key role in founding the society, served as its first president, and later took another term as its leader. Kedzie, Charles E. Bessey, '69, and Byron D. Halsted, '71, each served one term as president during the society's first twenty years.[30]

As evidenced by its representation in the Society for the Promotion of Agricultural Science, Michigan Agricultural College provided leadership in the emerging field of agricultural science. It had already fulfilled Jordan's call for training teachers and researchers. In 1900 over one hundred alumni of the State Agricultural College taught at agricultural colleges, conducted research at experiment stations, or worked for the USDA. These people probably had a greater impact upon life in America than any other group of graduates. M.A.C.–trained agriculturalists formed a significant part of the vanguard in agricultural higher education before 1900. Scientists trained at M.A.C. were in demand across the country (see figure 6). Consequently, they influenced the evolution of land-grant institutions and agricultural science in their formative years.

Two M.A.C graduates who pioneered agricultural education and research at land-grant colleges were no longer active in higher education in 1900, but are nevertheless worthy of mention because their careers give further evidence of roles played by M.A.C. alumni in the formation of other land-grant institutions. Charles Lee Ingersoll, '74, who died in 1895, left the faculty at M.A.C. in 1879 to accept an appointment as professor of agriculture and horticulture at Purdue. Although he stayed for only three years, he made a significant contribution to the establishment of a strong agricultural program there.[31] In 1882 Ingersoll assumed the presidency of Colorado Agricultural College and became director of its experiment station. Eight years later he moved to the University of Nebraska, where he held three appointments as professor of agriculture, dean of the industrial college, and director of the experiment station. After Frank A. Gulley graduated in 1880, he became professor of agriculture at the Mississippi

Figure 6. Selected Graduates of Michigan Agricultural College Holding Professorial Appointments in 1900 in Colleges, Universities, and Experiment Stations

William Willard Daniells, '64, *professor of chemistry, University of Wisconsin*

Alfred Gurdon Gulley, '68, *professor of horticulture and superintendent of the gardens and grounds, Connecticut Agricultural College; previously horticulturist, Vermont Experiment Station*

Charles Spencer Crandall, '73, *professor of horticulture and botany, Colorado Agricultural College*

Ransom H. McDowell, '74, *professor of agriculture, horticulture, and forestry, and horticulturist and agriculturalist in the experiment station, University of Nevada; previously assistant in agriculture, Colorado Agricultural College*

William Carroll Latta, '77, *professor of agriculture, agriculturalist in the experiment station, and superintendent of farmers' institutes, Purdue University*

Louis George Carpenter, '79, *professor of civil and irrigation engineering and director of the Colorado Experiment Station, Colorado Agricultural College*

Clarence Moores Weed, '83, *professor of zoology and entomology and entomologist in the experiment station, New Hampshire College of Agriculture and Mechanic Arts; previously entomologist and botanist, Ohio Experiment Station*

Edward Ralph Lake, '85, *professor of botany and horticulture, Oregon Agricultural College*

Hiram Taylor French, '85, *professor of agriculture, agriculturalist in the experiment station, and superintendent of farmers' institutes, University of Idaho*

Edgar Albert Burnett, '87, *professor of animal husbandry and superintendent of farmers' institutes, University of Nebraska; previously professor of animal husbandry, South Dakota Agricultural College*

Clare Bailey Waldron, '87, *professor of horticulture and forestry and horticulturist in the experiment station, North Dakota Agricultural College*

Agricultural College. He directed the Texas Agricultural Experiment Station from 1888 until 1890. Gulley then assumed the position of dean of the school of agriculture and the first paid faculty member in the fledgling University of Arizona. He employed a faculty (including Charles Barnard Collingwood, '85, and James William Toumey, '89), published bulletins, equipped the experiment station, created a curriculum, and laid the groundwork for field stations. Gulley resigned in

Arthur Burton Cordley, '88, *professor of zoology and entomology and entomologist in the experiment station, Oregon Agricultural College*

Fred Hebard Hillman, '88, *entomologist and botanist in the experiment station, University of Nevada*

Nelson Slater Mayo, '88, *professor of veterinary science, Connecticut Agricultural College; previously professor of veterinary science and veterinarian in the experiment station, Kansas Agricultural College*

Welton Marks Munson, '88, *professor of horticulture, University of Maine*

Charles Hulbert Alvord, '95, *assistant professor of agriculture, Texas Agricultural College*

James William Toumey, '89, *assistant professor of forestry, Yale University; previously professor of biology and botanist in the experiment station, University of Arizona*

Louis Adelbert Clinton, '89, *assistant agriculturalist in the experiment station, Cornell University; previously assistant professor of agriculture and assistant agriculturalist in the experiment station at South Carolina Agricultural Station*

Vincent Hunt Lowe, '91, *entomologist in the New York Agricultural Experiment Station*

Charles Phillip Chase, '95, *professor of horticulture and botany, Utah Agricultural College; previously assistant horticulturist, New York Agricultural Experiment Station*

Howard Remus Smith, '95, *acting professor of agriculture, Missouri Agricultural College*

Guy L. Stewart, '95, *assistant in botany and vegetable pathology in the agricultural college and pathologist in the experiment station, Maryland Agricultural College*

Hugh Elmer Ward, '95, *instructor in soil physics, University of Illinois*

Ezra Dwight Sanderson, '97, *entomologist in the Delaware Experiment Station; previously assistant entomologist in the Maryland Agricultural College and Experiment Station*

Clift Fay Austin, '99, *assistant horticulturist in the Alabama Experiment Station; previously assistant horticulturist in the Montana Experiment Station*

Source: *Michigan State Agricultural College General Catalogue of Officers and Students, 1857–1900* (Agricultural College, Michigan, 1900), 31–91.

1893, after he lost a power struggle within the university.[32] In 1900 he was working in private industry.

Another group of alumni contributed to the expansion of agricultural and technological research and its application to the public good through their service with the USDA and the U.S. Patent Office. The listing of selected M.A.C. graduates employed by the USDA in 1900 provided in figure 7 gives insights into their work as

Figure 7. Selected Graduates of Michigan Agricultural College Employed by the United States Department of Agriculture in 1900

Charles Christian Georgeson, '78, *special agent in charge of Alaska investigation, Sitka*

Frank Benton, '79, *assistant entomologist, Washington, D.C.*

Lyster Hoxie Dewey, '88, *assistant botanist, Washington, D.C.*

Lyman James Briggs, '93, *assistant chief of Division of Soils and physicist, Washington, D.C.*

William Woodbridge Tracy, '93, *horticulturist, Washington, D.C.*

Mary Lilian Wheeler, '93, *scientific aid in Division of Vegetable Physiology and Pathology, Washington, D.C.*

Maurice Grenville Kains, '95, *special crops culturist, Washington, D.C.*

Huron Willis Lawson, '95, *assistant in the Office of Experiment Stations, Washington, D.C.*

Lewis Storms Munson, '97, *scientific aid, Division of Chemistry, Washington, D.C.*

Macy Harvey Lapham, '99, *scientific aid, Division of Soils, Washington, D.C.*

Source: *Michigan State Agricultural College General Catalogue of Officers and Students, 1857–1900* (Agricultural College, Michigan, 1900), 31–91.

---

well as the scope of the department's ever expanding efforts to improve American agriculture. Several alumni served as examiners in the U.S. Patent Office in Washington, D.C. including Edward Noah Pagelsen and Joseph Harlan Freeman, '90, both of whom pursued careers in private practice as patent attorneys after a few years with the federal government.

FOREIGN SERVICE

The United States extended its influence and presence deeper into the world as the nineteenth century drew to a close. It sent its soldiers, knowledge, and capital to distant places, to build an empire and to help people improve their lives. In the wake of the Spanish American War, alumni served their country in the Philippines and Cuba: John Park Finley, '73, Captain, 9th U.S. Infantry, Manila; Dale Afton Smith,

'88, Sergeant, Company B, 35th U.S. Infantry, Manila; Oliver Cary Hollister, '89, Chief Clerk, Department Headquarters of the Adjutant-General's Office, Matanzas, Cuba; John Weldon O'Bannon, '89, Chief Yeoman, U.S.S. *Glacier,* Manila; and Robert Sylvester Welsh, '94, 1st Lieutenant, Company M, 39th U.S. Volunteers, Manila. (In 1904, Yasuma Ishikawa, '88–'91, commanded a Japanese boat during a Japanese attack on the Russian fleet at Port Arthur.[33])

Other alumni had taken postings overseas for private enterprises or foreign governments. Edward Mason Shelton, '71, spent a year as agriculturalist to the Japanese government and nine years teaching agriculture in Queensland, Australia. Shelton's work foreshadowed a much greater involvement by M.A.C.–trained agriculturalists, who would introduce the findings of scientific research at the college and their practical applications in the field to people living in other countries. One way that this worked was illustrated by Shelton's request to President Clute to send him another copy of Professor A. J. Cook's bulletin dealing with insecticides and the use of carbon sulphide. He needed a second copy because he had loaned his copy to editors in Australia who had "cut it up for use."[34] Frank Paul Davis, '68, died in 1900 at Guayaquil, Ecuador while working as assistant chief engineer for the Guayaquil & Quito Railway. Prior to that he had been an engineer for the Nicaragua Ship Canal project and the Canadian Pacific Railway. George Edward Kedzie, '73, lived in Calle Coliseo, Durango, Mexico, where he was a consulting miner and metallurgical engineer.

Alumni Serving on the M.A.C. Faculty

Serving on M.A.C.'s faculty gave graduates an opportunity to teach and to begin their research careers before they moved on to other institutions. Forty-four men and one woman held the faculty rank of professor before 1900, and nine of them, all male, had earned their bachelor of science degree at M.A.C. Eight of the nine had also earned their masters of science degrees from the college. Albert N. Prentiss, '61 (M.S. '64), left his professorship in botany and horticulture in 1868 for a long tenure at Cornell. Liberty Hyde Bailey, '82 (M.S '86), joined Prentiss at Cornell in 1888 after four years on the M.A.C. faculty. After farming for a few years, Eugene Davenport, '78, was professor of practical

LEFT: Frank S. Kedzie came to M.A.C. in 1863, when he was six years old, by virtue of being the son of Robert C. Kedzie, the newly appointed professor of chemistry. Frank devoted the rest of his life to serving students, faculty, and the public until his death in 1935. He followed in his father's footsteps by teaching chemistry, for over forty years. In 1916 "Uncle Frank," as he was affectionately known by students, took the reins of the presidency from President Snyder. Kedzie resigned as president in 1921, and he served as dean of Applied Science for the next six years. As his career drew to close, Kedzie functioned as the college historian. He gathered much information from former students and others and began writing a history of the college that was never finished. Courtesy of Michigan State University Archives and Historical Collections (People, Frank S. Kedzie, no. 1621).

RIGHT: Kate Marvin, of Lansing, married Frank S. Kedzie on 30 December 1885 in Plymouth Congregational Church, located on the corner of Washtenaw Street and Capitol Avenue in Lansing. A musician, Mrs. Kedzie entertained many audiences and taught music to many students over the years. She and her husband did not have any children of their own, but Kate befriended M.A.C. students and earned their respect and devotion. When Mabel Lucas, '27, received a fellowship to do graduate work at the University of Chicago in 1930, Kate and Frank sent their hearty congratulations. She, like other women connected to the college, remained interested in alumni after they found their places in the world after their student days. Courtesy of Michigan State University Archives and Historical Collections (Scrapbook 348, UA 3277, UA 2.1.8, Portrait of Kate Marvin).

agriculture and superintendent of the farm from 1889 until 1891. Later he had a distinguished career at the University of Illinois. Rolla Clinton Carpenter, '73, spent fifteen years teaching mathematics and civil engineering at M.A.C. In August 1890 he, too, went to Cornell as associate professor of experimental engineering.

Two other men enjoyed much longer terms on the college's faculty. Albert John Cook, '62 (M.S. '65), moved to Pomona College, Claremont, California, in 1893 after teaching zoology and entomology for twenty-six years. Frank S. Kedzie, '77 (M.S. '82), successively held the ranks of instructor in chemistry, 1880–87; assistant professor, 1887–91; and adjunct professor of chemistry, 1891–1902. Kedzie also worked as assistant chemist at the experiment station. He succeeded his father, Robert Clark Kedzie, as professor of chemistry in 1902.

College administrators appointed male alumni of the agricultural and mechanical courses to fill about 58 percent of the assistant professor and instructor positions on the subfaculty. Twenty-one men (including thirteen alumni) and one woman were assistant professors; incumbents usually taught for one or two years before going elsewhere. Teaching under the direction of the professor of their departments prepared these younger people for careers in college teaching or research. Newly minted scholars left M.A.C. for land-grant colleges in Colorado, Oregon, Nebraska, Illinois, Nevada, and Missouri. Several became professors at M.A.C., including Wilbur Olin Hedrick, '91, and Warren Babcock, '91, who worked their way up the academic hierarchy until they were appointed, respectively, professor of history and economics in 1906 and professor of mathematics in 1909. Another fifty men (including 32 alumni) and six women had served as instructors by 1900.

Prior to 1900 only four female graduates were able to attain appointments to the academic staff of the Michigan Agricultural College. Alice Adelia Johnson, '84, worked as assistant agriculturalist in the experiment station during 1888–89. She was the only woman to hold one of the thirty-five assistant positions, some of which were held by twenty-four male alumni, scattered throughout the departments. Mary Lilian Wheeler, '93, taught mathematics in 1897–98 as an assistant instructor, a classification used for only one other person, also a woman. Mrs. Mary Jane Cliff Merrell, '81, became the first female college librarian in July 1883, and occupied that post until 1888. Teresa Adeline Bristol took a position as assistant chemist following her graduation from the women's course in 1899.

BIOGRAPHIES OF SIX ALUMNI

More detailed biographies of six alums suggest the breadth and depth that M.A.C. graduates brought to their work. The six applied the land-grant philosophy of teaching, research, and service in ways that made a difference in many people's lives. These individuals, like their classmates, used their knowledge and skills to solve problems and improve the quality of life in the world.

Charles E. Bessey, who came from a farm in Wayne County, Ohio, graduated from M.A.C. in 1869 and received his master of science degree in 1872. Bessey dedicated his life to the educational philosophy of Justin Morrill that formed the core of land-grant education in America. Bessey put it in these words: "the *best education* for the industrial classes, [is] that which will enable each man to do his best in the struggle of life."[35] In 1870, Bessey began his distinguished career at the recently opened Iowa Agricultural College in Ames, where he pioneered the instruction of botany using the laboratory method. Bessey moved to the University of Nebraska in 1884, and he remained there until his death in 1915. He built a curriculum in the university's Industrial College and an experiment station that was

Albert J. Cook laid a solid foundation for entomology at M.A.C. during his long tenure as professor of zoology and entomology that began in 1868 and ended in 1893. His groundbreaking research into insects pests and ways to kill them with chemicals proved useful to fruit growers. Cook also taught beekeeping, spoke frequently at farmers' institutes, and worked with the Michigan State Horticultural Society. Courtesy of Michigan State University Archives and Historical Collections (People, Cook, Albert J., no. 687).

deeply rooted in experimental, scientific research that supported the growth and development of agriculture.[36] Bessey embodied the tension between pure science and applied science that gave so much vitality to the land-grant philosophy. Never wavering from his commitment to original, scientific research, Bessey gave leadership to the development of botany as a science, insisting that cultivated plants should have the same standing in botany as wild plants. His contributions to the "new botany," which stressed "that research must replace mere observation and collection,"[37] grew out of his conviction that the study of plant physiology was basic to botanical research. One needs to know a plant's structure in order to study plant pathology, to carry out scientific plant breeding, or to practice scientific agronomy.

Even though Bessey felt that farmers were the most resistant group to scientific agriculture, he always remained committed to them. He believed in the efforts of the USDA and experiment stations (in fact, he contributed to the draft of the Hatch Act that led to the establishment of experiment stations throughout the country), and many of his students found their careers in those agencies. Bessey worked tirelessly to broadcast the results of research in scientific publications, and for many years he edited articles on botany for the *American Naturalist* and for *Science.* He wrote two influential textbooks and *Botany for High Schools and Colleges,* which he intended to serve as "general knowledge of the structure of plants" and as a supplement to laboratory research, was published in 1880.[38] Four years later, an abridged edition titled *Essentials of Botany,* which contained revisions and was put in "considerably less difficult language" was published.[39] He also authored articles for farm publications, spoke at farmers' institutes, and remained connected to the farmer in the field. Bessey committed himself to professional scientific organizations; he was a founder of the Society for the Promotion of Agricultural Science, which elevated the respect for scholarship being offered by agricultural scientists. Bessey left a lasting imprint on the development of the study of botany.[40]

Charles C. Georgeson came from Denmark to study at the State Agricultural College in 1874, and he became the first international student to receive a bachelor of science degree from M.A.C., in 1878, and a masters degree in 1882. Georgeson viewed the world as his workplace. After graduation, he served as an editor for the *Rural New Yorker* for two years before assuming professorships of agriculture in Texas Agricultural College, 1880–83, in the Imperial College of Agriculture at Kowaba, Tokyo, Japan, 1886–89, and in the Kansas State Agricultural College, 1890–97. He took time off from his duties in Kansas in 1893 to carry out an assignment as a special agent for USDA in order to investigate the dairy industry in his native Denmark. All of this prepared him for the longest tour of service in his career.

In 1897 Georgeson went to work full time for the USDA, and he moved to Alaska the next year as special agent in charge of investigations. Many people viewed Alaska as a place that was too cold, had too short a growing season, and had soil that was too inhospitable to support viable agricultural endeavors. Georgeson changed that perception, as he devoted thirty years to developing varieties of grains, fruits, and vegetables that would produce respectable yields for Alaskan settlers, who paid enormous prices for imported foodstuffs. He experimented with over 12,000 hybrids of strawberries to produce a berry that would grow from Sitka to the Arctic Circle. Over time he discovered 170 varieties of potatoes that yielded sturdy crops throughout Alaska. Georgeson also found that the Native people did not view his work in the same way as immigrants. After crossing a native wild crabapple with varieties from the States, he was pleased to see a crop take form. He advised the inhabitants of the village not to pick the fruits that had been carefully labeled and shielded by paper bags. To Georgeson's consternation, just as the apples ripened, the Native women made jelly out of them. He established four experiment stations located in different climatic zones. One year at the Fairbanks station researchers bagged yields of forty-one and seventy-nine bushels of spring wheat and oats respectively in an environment that had only about one hundred days free of frost.[41]

In 1877 Liberty Hyde Bailey left his father's farm in South Haven, Michigan, to take up studies at the State Agricultural College. Bailey had acquired an interest in plants, especially the fruit trees in his father's orchards, and as a boy Liberty had started his own flower garden.[42] His formative years excited his imagination and created a desire to learn more about the plant world. Throughout service spanning three-quarters of a century, Bailey acted on his belief "that horticulture must be an applied science based on pure biology."[43] In 1885, he began his teaching career at M.A.C. having studied under W. J. Beal at the college and, after his graduation, with the renowned Asa Gray at Harvard. While at M.A.C. Bailey placed great emphasis on research into the "fundamental biology of plant life, along with taxonomic study, which help to create a 'new' horticulture."[44] Three years later Bailey went to Cornell, where he achieved greatness as a horticulturist, teacher, writer, and editor. He helped to bring about the creation of a State College of Agriculture there in 1904.[45] Soon after taking his post at Cornell, Bailey and his family traveled to Europe where he observed European agricultural research. After his return, he continued the experimentation with hybridization that he had started while in Michigan and advanced the study of horticulture.[46]

One way to gain an appreciation of the extent of Bailey's influence is to look at some of the books he published before 1900. Six titles demonstrate the scope and focus of Bailey's contributions to the study of horticulture and his commitment to disseminate his knowledge to the public, not just to an academic audience. While on the faculty at M.A.C. he wrote *Field Notes on Apple Culture,* a set of teachings "founded upon successful practice," intended to steer novice apple growers from making mistakes.[47] In 1893 Bailey issued a handbook for people who wished to grow grapes. *American Grape Training* grew out of Bailey's belief that students needed to learn the principles of training grapes—and other principles relative to agriculture as well— by their own experience and experimentation rather than by simply hearing lectures.[48] In *Principles of Fruit-Growing,* he covered a wide range of topics including climate, tillage, fertilization, planting, care of plants, diseases, pests, harvesting, and marketing.[49] The breadth of Bailey's instruction for American fruit growers is demonstrated by *The Pruning-Book* issued in 1898.[50] He left no aspect of pomology unstudied or unreported. Bailey's expertise in horticulture encompassed more than fruit-growing. In 1897 he provided a helpful guide for people who had undertaken the relatively new and very expensive, risky business of growing winter vegetables in greenhouses for the market. *The Forcing-Book* gives information on building and managing forcing-houses, followed by chapters on specific vegetables such as lettuce, cauliflower, radish, asparagus, tomato, and cucumber.[51] Always the consummate educator, Bailey tried to make applications of scientific agriculture understandable to the farming community in *The Principles of Agriculture: A Text-Book for Schools and Rural Societies.*[52]

Katharine Cook Briggs, '93, embodied the land-grant ideal that educated women should view homemaking as a calling of the highest order. She fulfilled the potential of this philosophy in such a way that her life's work outside of her home still influences people in the twenty-first century. Briggs raised the question: "If we define Home as an institution that provides for the health and strength, mental and moral as well as physical, of each and every member of the family, what profession open to man or woman furnishes a better opportunity for self-development than is offered by the business of making a home?"[53] One part of her answer to this question saw her devoting herself to home schooling her daughter, Isabel, who learned how to read from her mother before she was of kindergarten age and knew German, Latin, and French before she enrolled in Swarthmore College.[54] In 1912 Katharine began to put forth her arguments for the benefits of the home as a place of learning when she published two short articles in *Ladies Home Journal* under the pseudonym Elizabeth Childe. Briggs believed that many children soon forgot what they learned in school because their parents did not re-enforce or supplement their lessons at home. Only in the home could a mother focus teaching on the individual needs and capabilities of her daughter or son.[55]

Although Katharine always supported the career of her highly successful husband, Lyman Briggs, '93, she also constructed an identity apart from him.[56] After Lyman completed his doctorate in physics at Johns Hopkins University, Lyman and Katharine moved to Washington, D.C., where they lived for the rest of their lives. He started his career with the federal government at the Bureau of Soils in the USDA. President Franklin D. Roosevelt appointed him director of the National Bureau of Standards in 1933. At the same time, Katharine's keen observations and insights into the education of Isabel stimulated an interest in questions relating to personality and Carl Jung's work on the subject. Writing under her own name in 1926, she published "Meet Yourself: How to Use the Personality Paint Box."[57] In this article she set forth the principles of the Myers-Briggs Type Indicator that Isabel and she developed to put Jung's theory into practice.[58] It is still in use today.

Eugene Davenport, '78, fulfilled the farmers' hope that a young man educated at the State Agricultural College would return to the farm, but he could not be confined there all of the time. His career

Liberty Hyde Bailey Jr. graduated from M.A.C. in 1882, studied at Harvard for two years, and taught horticulture at the college for four years before moving to Cornell University. Louis A. Bregger, one of Professor Bailey's students, held Bailey in high esteem. In April 1888, when it appeared that Bailey was headed to Cornell, Bregger articulated his awe for his professor: "Outside of Europe, he has not his equal. . . . Then think of it, Bailey is only 27 years old. He is a man, however, of indomitable energy and push. Michigan as well as this college cannot afford to lose him but they cannot afford offer him the same inducements. We all like him" (Louis A. Bregger to brother, 3 April 1888, Louis A. Bregger Papers, 1884–1909). Cornell's offer of a higher salary, a new horticulture building, two assistants, and paid sabbaticals to travel and study provided Bailey with opportunities that M.A.C. could not match. Courtesy of Michigan State University Archives and Historical Collections (People, Bailey, Liberty Hyde, no. 133).

The Class of 1900 was the last class to graduate in the nineteenth century. Although thirty people are pictured, only twenty-four actually graduated in 1900. Courtesy of Michigan State University Archives and Historical Collections (*Wolverine*, "Class of 1901," 106–7).

exhibited some of the most far-reaching manifestations of the land-grant philosophy, which encouraged students to teach other people to develop their abilities in order to provide leadership in their families and communities. After his graduation, Eugene worked on his family's farm near Woodland, Michigan, for ten years before answering his alma mater's call to serve as an assistant botanist in the experiment station under the direction of botany professor W. J. Beal. One year later, the college appointed him professor of practical agriculture and superintendent of the farm, a post he held for two years. In 1891 he went to Piracicaba, Sao Paulo, Brazil, to start an agricultural college, but local political conditions prevented him from completing

CLASS 1900.

C. H. Parker, a, Grand Blanc, Mich.

Bronson Barlow, a, Greenville, Mich.   H. L. Chamberlain, m, Lansing Mich.

J. F. Coats, m, Lansing, Mich.   W. T. Parks, a, Pipestone, Mich.

G. B. Fuller, m, Lansing, Mich.   A. F. DeFrenn, a, Owosso, Mich.

C. H. Spring, m, Grand Ledge, Mich.   A. J. Cook, a, Harbor Springs, Mich.

Irma G. Thompson, w, Lansing, Mich.   Bertha Malone, w, Lansing, Mich.

A. G. Bodourian, a, Necomedia, Ismid, Turkey.   L. S. Christensen, m, Marquette, Mich.

J. J. Parker, m,   Wm. Ball, m, Pres.,   F. W. Dodge, m, Vice Pres.,   Paul Thayer, a,
Fowlerville, Mich.   Grand Rapids, Mich.   Lansing, Mich.   Benton Harbor, Mich.

L. L. Appleyard, m, Lansing, Mich.   G. M. Odlum, a, Lake Odessa, Mich.

Harriet Robson, w, Secy., Lansing, Mich.   Alice Cimmer, w, Highland, Mich.

Eugene Price, m, Ithaca, Mich.   E. W. Ranney, a, Belding, Mich.

H. B. Clark, m, Coldwater, Mich.   C. H. Hilton, a, Benton Harbor, Mich.

H. B. Gunnison, m, DeWitt, Mich.   C. W. Bale, m, Vermontville, Mich.

Harry Rupert, m, Fort Wayne, Ind.   J. R. Thompson, m, Treas., Grand Rapids, Mich.

W. B. Nevins, m, Douglas, Mich.

w—Women's Course.   m—Mechanical Course.

a—Agricultural Course.

this mission before his return, via England, to his Michigan farm the next year.[59]

Davenport's stay at his farm was short, for in 1895 he assumed enormous responsibility as dean of the agricultural college at the University of Illinois. When he arrived in Champaign-Urbana, the moribund college had less than ten students, equipment valued at five dollars, and sixty dollars remaining in the budget.[60] Davenport's leadership resulted in the building of a first-rate agricultural college in one of America's richest agricultural states. His work was informed and driven by the land-grant ethic learned at Michigan Agricultural College.[61] A proponent of "industrial education," Davenport argued that it occupied an essential role in "a general scheme of education that aims at a higher efficiency of all classes of people."[62] He expressed his approach to agricultural education in his writings. In 1907 he published a textbook, *Principles of Breeding,* to be used by college students, at experiment stations, and on farms.[63] Realizing that this large text reached a relatively small group, within three years Davenport issued a much shorter *Domesticated Animals and Plants* geared to high school and normal school students.[64] There was little room for elitism or aristocracy in Davenport's system of education.

Kolia S. Thabue, who was from Burma, graduated from M.A.C.'s agricultural program and served as an agent for Michigan business in his home country. Following his graduation in 1891, Kolia worked out agreements with a number of Michigan firms to send their products to Burma. Among the companies Thabue represented were D. M. Ferry and Company, Peninsular Cooking Stove Company, Farrard and Votey Organ Company, and Johnson and Nelson Chemical Company, all of Detroit; Bement and Sons and Company of Lansing; the Sewing Machine Company of Ohio; and the Dry Plate Company of Rochester, New York. Thabue apparently returned to Burma in 1893, then came back to the United States to serve as an assistant in the horticultural department of Massachusetts Agricultural College in Amherst during 1894 and 1895. In 1900 he was back in Burma supporting his country's agricultural and horticultural endeavors as an importer of machinery.[65]

THE NONGRADUATES—MEN

The directions taken by nongraduates after they left M.A.C., like those taken by the graduates, mirrored the growth of the United States and foreshadowed alumni activities in the twentieth century. Nongraduates formed a largely invisible but very significant group of about 3,800 men and women who began their higher education at the State Agricultural College. They saw fit to complete their studies elsewhere or not to finish the requirements for a degree. They made up 82 percent of the total number of students who had enrolled by 1900. Even though they did not graduate, their careers should be recognized when assessing the return of the public's investment in Michigan's land-grant college. Nearly three-fourths of the nongraduates lived in Michigan in 1900.

As a group, nongraduates came closer to meeting the farmers' hopes for students who attended the State Agricultural College than did those who finished their degrees, but they nevertheless followed paths similar to the graduates. The largest number of nongraduates took up farming or entered careers related to agriculture. Approximately 20 percent devoted themselves exclusively to farming, horticulture, or dairying, while another 15 percent mixed farming with teaching, the lumber business, milling, dealing in farm implements, and other occupations. One visible manifestation of the impact of the college's winter short courses was the number of butter makers (22) and cheese makers (10) who appeared on the list. Vocations pursued by nongraduates indicate that their motivations for attending college differed little, if at all, from those of the graduates.

The land-grant college functioned as the point of entry into higher education for the 3,800 nongraduates, many of whom later studied at other institutions to earn the credentials needed to take up a variety of professions. In 1900 eighty-three were enrolled in other colleges, taking both professional and general courses of study. Dental students included Emery R. Austin, '94–'97, at Northwestern University, and Charles Adams Hawley, '97–'99, and Arthur John Norman, '98–'99, at the University of Michigan. Other nongraduates enrolled in medical schools, including Martin H. Coan, '89–'90, and Dirk B. Lanting, '96–'98, at Detroit Medical College; Ralph S. Green, '89–'90, at Chicago College of Physicians and Surgeons; John Ford Gruber, '98–'99, at Saginaw Valley Medical College; and James Shanks, '95–'96, at the University of Michigan Medical School.

It was not unusual for students to spend one or more years at M.A.C. before transferring to the University of Michigan or another college to continue their work towards a bachelor's degree. Frank M. Lamb, '93–'94, Wilbur Judson, '95–'96, and Frank Dennison Longyear, '96–'97, were among the young men who continued their undergraduate studies in Ann Arbor. Wynn A. Bartholomew, '95–'98, went to Houghton to matriculate at the Michigan College of Mines. George Edwin Cryderman, '97–'98, and others moved on to the Michigan State Normal College. Wilbur Judd Cummings, '90–'94, studied architecture at Urbana, Illinois; Stanley E. Marsh, '98–'99, enrolled in

Ferris Institute; and Guy Knox White, '97–'99, became a student of pharmacy at Massachusetts College of Pharmacy in Boston.

Thirty-nine men followed careers in engineering after their stay at M.A.C. Carey L. Alverson, '67–'70, became a civil engineer in San Diego, California. William Adolph Ansorge studied in the mechanical course from 1891 until 1894, before he spent three years as assistant manager with the Elliott Button Fastener Company, and in 1900 he went to work for the Newaygo Cement Company in Newaygo, Michigan. Three years of study in the mechanical course prepared Edward J. Frost, '86–'89, to hold the position as assistant superintendent of the industrial works in Bay City, Michigan. Fred Gallop, '86–'88, was an engineer for the Electric Light and Water Works in Flushing, Michigan. The U.S. government employed Harry Hodgman, '91–'92, as engineering inspector on the Great Lakes. Thatcher A. Parker, '83–'84, applied his skills as a mechanical inventor to manufacturing bridges, tipples, and head frames, among other things, in Terre Haute, Indiana.

Another thirty engaged in a wide variety of tasks for the railroads. Roy B. Fugate, '93–'94, lived in Albuquerque, New Mexico, where he was a locomotive firemen with the Atchison, Topeka, & Santa Fe Railroad. LeRoy D. Goldsworth, '96–'97, traversed the Copper Country in Upper Michigan as a rodman for the Copper Range Railroad Survey in Houghton. Living in Toledo, Ohio, Walter Joseph Graves, '88–'90, a surveyor, worked for the Lake Shore & Michigan Southern Railway as a draftsman. The Cincinnati & Muskingum Valley Railroad hired D. Bird Johnston, '89–'90, an engineer, to maintain their tracks and right of ways. Charles Henry Spencer, '71–'73, became the chief draftsman and designing engineer for the Baltimore & Ohio Railroad's construction department in Zanesville, Ohio.

About sixty nongraduates found jobs as draftsmen or machinists after leaving M.A.C., and it was not uncommon for them to work first as a draftsman and then as a machinist. Clare Larue Bloom, '97–'98, found employment as a machinist with the William Shakespeare, Jr. Company in Kalamazoo, which produced "fine reeds and photograph shutters." The Detroit law firm of Parker & Burton, which specialized in patents, hired Daniel Whitefield Bradford, '86–'88, to be a draftsman. Henry C. Buell, '90–'92, worked in Chicago as a draftsman for the Chicago Telephone Company. Adelbert Dryer, '89–'92, moved to Washington, D.C. to become a machinist at the Navy Yard after he had worked as a draftsman for the Lansing Iron and Engine Works. In 1900 Henry Lavern Heesen, '89–'91, who was both a draftsman and a pattern maker, made acetylene gas machines in Tecumseh, Michigan.

Seneca Taylor, the first student to be admitted to the college in 1857, came from Oakland Township in Oakland County. He attended for three years, but decided to read law in the office of Attorney O. M. Barnes in Mason rather than complete his senior year. In 1861 he finished requirements for his bachelor of science degree at Adrian College. He practiced law in Dexter, Michigan for a few months before raising a company for the Michigan Twelfth Regiment. Illness prevented him from joining his unit, and after his recovery he practiced law in Niles, Michigan, until 1865. Following the Civil War, Seneca moved to St. Louis, Missouri, where he established his permanent law office. He was active in the Methodist Episcopal Church and served on boards of colleges and institutions. The Carlton Institute in Missouri honored him with an LL.D. degree in 1890.[66]

James H. Brown entered the State Agricultural College in 1883 and stayed for only three months, but his experience there showed what it meant to change the farmer rather than the farm. Brown expressed his gratitude for the college and his time there to M.A.C.'s President Lewis G. Gorton in January 1894 after he had attended the farmers' institute in Battle Creek. Brown's words probably echo the feelings of other young men who left M.A.C. to make a living on the farm:

> Was glad to make your acquaintance and hope to meet you occasionally in the future. Although deafness militated against my completing a course at the Agr'l College, I am very grateful for what the college has done and continues to do, for me. It is due to my short, three months study at your school, that I really became interested in farming. During the last five years I have "run" my own and my father's farm. Every spare moment has been spent in studying Agr'l, and Dairy literature, also the bulletins from M.A.C., and department at Washington. Aside from three months at M.A.C., and four or five years in my own district school, my studying has been a "solitary" course in my own house. What would I not give, could I have had four years thorough course of study at your College. Tell the young men on the farms, that they know not what they miss in neglecting a thorough Agricultural course of study.[67]

Brown's words drive home the point that even a short exposure to
knowledge gained from the agricultural college's curriculum changed
a person's life in significant ways. It is worth noting that Brown
assembled the experiment station's bulletins, USDA publications, and
other agricultural literature as a growing repository of information
for informed farmers. He gives evidence that people read the materials written by researchers and professors. Even M.A.C.'s most vehe-
ment opponents would have had a difficult time arguing that it was
not fulfilling its mission in the case of Brown.

Like Brown, Fred Maltby Warner spent only three months at
M.A.C., but his life took a far different course. Warner apparently did
not find college life to be especially agreeable, although he did well
academically. His career indicated that he took to heart his father's
admonition that "brains are to determine the power of individual
influence," and his short stay at M.A.C. during the 1880–81 school
year must have helped him to sharpen his intellectual skills.[68] Warner
reaped great successes as he worked in his father's shop, farmed, and
built cheese factories before he engaged in Republican politics. He
went to the state senate in 1894, had two terms as secretary of state,
and served three terms as governor from 1905 until 1911. As governor, Warner advanced a number of progressive issues, including the
conservation of natural resources, the reform of primary elections,
fairer taxation of railroads, and improvement of roads for automobiles.[69] It is fitting that Warner helped M.A.C. observe its Semi-
Centennial in his official capacity as governor of Michigan.

THE NONGRADUATES—WOMEN

The new century promised more opportunities for college-educated
women in the paid workforce outside of the home. Even though the
women's course did not begin until 1896, by 1900, 324 women had
not completed a degree but had studied at M.A.C. Ninety-three of
them were still enrolled in 1900. M.A.C. women showed a strong
desire to make their own way in the world. Many married and
undoubtedly managed their households, but others also pursued different careers or were preparing to do so. The number and percentage
of women in the overall paid workforce increased during the late

Table 5. Percent of Female Labor Force as Percent of Female Population in the United States

YEAR	TOTAL	SINGLE	MARRIED	WIDOWED OR DIVORCED
1890	18.9	40.5	4.6	29.9
1900	20.6	43.5	5.6	32.5
1910	25.4	51.1	10.7	34.1
1920	23.7	46.4*	9.0	—
1930	24.8	50.5	11.7	34.4

* Single includes widowed or divorced.

Source: "Marital Status of Women in the Civilian Labor Force: 1890 to 1970," *Historical Statistics of the United States: Colonial Times to 1970* (Washington, D.C.: U.S. Department of Commerce, Bureau of the Census, 1975), part 1, 133.

Table 6. Gainful Workers by Age, Sex, and Farm-Nonfarm Occupations, 1870–1930 (in thousands of persons ten years old and older)

YEAR	TOTAL WORKERS	OCCUPATION		SEX	
		FARM	NONFARM	MALE	FEMALE
1870	12,925	6,850	6,07	11,008	1,917
1880	17,392	8,585	8,807	14,745	2,647
1890	23,318	9,938	13,380	19,313	4,006
1900	29,073	10,912	18,161	23,754	5,319
1910	37,371	11,592	25,779	29,926	7,445
1920	42,434	11,449	30,985	33,797	8,637
1930	48,830	10,472	38,358	38,078	10,752

Source: "Gainful Workers, by Age, Sex, and Farm-Nonfarm Occupations, 1820 to 1930," *Historical Statistics of the United States: Colonial Times to 1970* (Washington, D.C.: U.S. Department of Commerce, Bureau of the Census, 1975), part 1, 134.

OPPOSITE: Maud McLeod, who came from Ionia, studied at M.A.C. during the 1897–98 and 1898–99 academic years. She probably lived in Abbot Hall, which housed the women's course until 1900. In a collection of her things in the Michigan State University Museum three interesting handwritten documents are found that give insights into how women felt about college life at the M.A.C. at the close of the nineteenth century. It is not certain if any of these short compositions are in Mc Leod's hand.

This here tale may not seem true,
  Altho its handed down.
The ancient prex he makes the rules,
  He is the ruling clown.
  The matrons all side in
  So you can clearly see
The reasons co-eds ne'er get out,
  At M.A.C.

· · ·

Two more days and we are free
From this school of misery;
No more German no more French
No more sitting on a hardwood bench.

President Snyder is a very nice man
He tries to keep us, long as he can;
He never lets us have our way
And is determined to make us stay.

Our professors are all very smart.

Our professors are nice folks
and they crack us all the jokes
They never forget to give an exam
So every student will have to cram.

· · ·

O boys how can you act up so
Please go away for don't you know
We ancient maidens were at rest and noise at
midnight us distressed
(For we have grown so staid you see)
(Through Prexy's rules at M.A.C.) Repeat.
Tune Doxology   Amen

Courtesy of Michigan State University Museum (Acc. No. 7508.13.1-3, Maud McLeod).

Table 7. Percent of Women in the National Labor Force Who Were Working, between 1890 and 1930

| YEAR | TOTAL | AGE | | | | |
		14–19	20–24	25–44	45–64	65 & OVER
1890	18.2	24.5	30.2	15.1	12.1	7.6
1900	20.0	26.8	31.7	17.5	13.6	8.3
1910			*not available*			
1920	22.7	28.4	37.5	21.7	16.5	7.3
1930	23.6	22.8	41.8	24.6	18.0	7.3

Source: "Labor Force by Age and Sex: 1890–1970," *Historical Statistics of the United States: Colonial Times to 1970* (Washington, D.C.: U.S. Department of Commerce, Bureau of the Census, 1975), part 1, 132.

nineteenth and early twentieth centuries (see tables 5 and 6). To gain a fuller appreciation of the possibilities for undergraduate women in 1900, it is useful to know that the percentage of women between the ages of 20 and 24 in the national work force jumped from 30.2 percent in 1890 to 41.8 percent in 1930 (see table 7). Pearl French, '88–'90, who worked as a pharmacist and manufacturing chemist in Mulliken, Michigan, and Essa Singleton, '95–'96, who set type for the Critic Publishing Company in Caseville, Michigan, exemplified M.A.C. women who took employment away from their home.

As of 1900, thirty-nine nongraduate women had taken up teaching, a profession that had long been open to them throughout Michigan and elsewhere. Jennie S. Bigelow, '73–'74, worked as both a teacher and a farmer in Lansing. Elizabeth M. Broughton, '97–'98 taught fourth and fifth grades in Grand Rapids. Others teaching in Michigan included Etta Isolene Degroat, '98–'99, in Lum; Hattie M. Chase, '96–'97, in Covert; Ina Keillor, '95–'97, in Bear Lake; and Agnes K. Polhemus, '96–'97, in Eaton Rapids. Minnie Katherine McCormick, '94–'95, served as the principal of the Barr School in Escanaba from 1895 until 1897 before moving to Cadillac to teach seventh grade. Several of the female nongraduates moved to other states to assume teaching jobs: Myrta Louise Ely, '98–'99, in St. Croix Falls, Wisconsin; Jeannette Claftin, '89–'90, in Toledo; Elizabeth Jeffreys, '90–'91, in Chicago; and Harriet B. Sargent, '92–'93, in Philadelphia. Bertha

Maud Ronan, '98–'99, instructed young women in calisthenics at the State Normal College in Ypsilanti. Mary H. Watts, '93–'94, served as the principal of the Collegio Americano de Petrópolis in Petrópolis, Brazil.

Twenty-seven young women who had started their college education at M.A.C. attended other colleges in 1900. Some of them clearly intended to seek careers in occupations that had offered few opportunities for women in the past. The University of Michigan welcomed Florence Abbott, '94–'95, June L. Davis, '97–'98, Lucy C. Davis, '94–'95, Florence Hedges, '97–'98, and Lottie Lee Smith, '97–'98, all from Lansing; Emma E. Bach, '97–'98, from Sebewaing; and Lucy Monroe, '96–'98, from South Haven after they departed M.A.C. Hoping to become doctors, Myrtelle May Moore, '98–'99, enrolled in the University of Michigan's medical school, and Mary Emily Green, '95–'96, matriculated in the Chicago College for Physicians and Surgeons. Eva Josephine Gray, '97–'98, worked for a year as a stenographer for B. F. Goodrich and Company in Akron, Ohio, before enrolling in Throop Polytechnic Institute, in California, to study art.

Even a casual reading of the 1900 catalog leaves a strong impression that this group of women easily moved from one place to the next in order to fulfill their vocational aspirations or to be with their husbands. Entries in the catalog for these women show how mobil the nongraduate women could be:

- CLARA SKINNER HINCKLEY, (a) '80–'81. Township Superintendent of Schools, Lansing, 1881–82; Teacher in the Lansing Schools, 1882–85; Principal of Schools at Onekama, 1885–86. (Mrs. Irwin O. Glazier.) Greeley, Colo.
- GERTRUDE HOWE, (a) '70–'71. (Mich. State Normal Coll.) Student at University of Michigan, 1871–72; Preceptress 1st Ward School for two years; Clerk, State Capitol. Missionary to China, 1872–. Kin Kiang, China.
- WINNIFRED JOSEPHINE ROBINSON, (sp.), '95–'96. (Mich. State Normal Coll.) 1892. Student at the University of Michigan, 1898. Teacher of Biology, Vassar College. Poughkeepsie, N.Y.

Higher education inspired women, as well as men, to make their livings in places far beyond the family farm, home, or shop.

Lewis Griffin Gorton's short term as president of M.A.C. started in August 1893 and ended in December 1895. A native of Waterloo in Jackson County, Gorton graduated from the State Normal School in Ypsilanti, taught astronomy, physics, and physiology in high school, and was a high school principal before becoming president of the college. Courtesy of Michigan State University Archives and Historical Collections (People, Gorton, Lewis G., no. 1132).

AFRICAN-AMERICAN STUDENTS

As of 1900, no African Americans had graduated from Michigan Agricultural College and few had even attended. This reality caused some embarrassment for Presidents Oscar Clute, Lewis G. Gorton, and Jonathan L. Snyder. Some leaders of black schools and black land-grant colleges, which came into existence as a result of the second Morrill Act of 1890, looked to M.A.C. as a source of trained instructors. An examination of correspondence relative to this matter provides a few insights into the faint presence of African-American students on campus. In 1893, Booker T. Washington, president of the Tuskegee Institute in Tuskegee, Alabama, wrote to President Clute asking "whether you have at present or have had in the past, in your institution, a colored man competent to take a position in this institution either in mechanical or agricultural work?"[70] Clute answered, "There have been very few colored students at this institution. There has not been one for several years. None of those whom we have had have graduated. . . . My impression is that you would get on the track of a suitable man by addressing the school in Hampton. I think that at the Indian School, at Carlisle, Pa., they have had some negro students."[71] It is noteworthy that Clute did not seem to think any other land-grant college in the North would be able to meet Washington's need either. If they were training black men to be scientific farmers or agricultural scientists, the land-grant institutions evidently did not make it very well known.

The next year Washington asked Gorton, who was Clute's successor, if there was a black man at the college who was qualified to teach at Tuskegee. Gorton passed along the name of a man, but his identity and whether he ever went to Tuskegee can not be ascertained from this correspondence.[72] In 1895, President Thomas Calloway of Alcorn A&M in Alcorn, Mississippi, made a similar request for which there was apparently no candidate.[73] A couple of months before Washington gave the commencement address at M.A.C. in May 1900 he asked Snyder if the college would be graduating a black man from the "departments of Agriculture." Snyder replied that there were no black men in either the junior or senior class. Snyder then told him: "We have one very bright boy in the freshmen class, and I think that he will develop into an exceptionally strong man."[74] This man was in all likelihood William O. Thompson, who had come from Indianapolis to study agriculture and became the first known African-American graduate in 1904. On 1 January 1906, Thompson assumed a position at Tuskegee to teach livestock breeding and to manage the livery and transfer department.[75] Curiously, in 1900 neither Washington nor Snyder mentioned Charles A. Warren, '96–'99, who had left M.A.C. during his senior year in the autumn of 1899 to accept a position as a horticulturist at Tuskegee.[76] In 1908 Snyder responded to Washington's inquiry seeking a black male graduate that the State Agricultural College had graduated "a very capable young woman last year."[77] Myrtle Craig, '07, was teaching domestic science at Western University in Quindaro, Kansas. Washington asked for and Snyder sent him her address.[78] (More on the experience of African-American students after 1900 is discussed in chapter 7.)

Michigan Agricultural College's stature and racial attitudes in America were revealed in the efforts of the North Carolina Agricultural and Mechanical College in Greensboro to recruit a M.A.C. alumnus to create an agricultural program. As a result of the second Morrill Act of 1890, the North Carolina legislature established an A&M college for African Americans in 1891. In 1895 T. B. Keogh, the president of the board of trustees, told Gorton: "From what I have heard, your institution has no superior anywhere." Keogh explained that no agricultural program existed at the college, and it needed a good man to lead "in the right path towards perfection in this department." He assured Gorton that it would be no problem to employ a white man in this position because "In this country there is nothing like social association between the whites and the negroes. The relations between the races are most friendly—but the social line is sharply defined. This however does not interfere with the most friendly business relations."[79] The college had hired a white man from Cornell to run the mechanical department, an arrangement that was working well.

Alva T. Stevens, '93, who had grown up on a farm in Michigan, took the position and introduced M.A.C.'s land-grant approach to practical agriculture to African-American students in North Carolina.[80] Stevens had worked for two years as an instructor and agricultural assistant at M.A.C. and its experiment station, where he had supervised students working on the farm. Gorton found Stevens to be "anxious and enthusiastic about the work with the colored people."[81] Armed with his experience in Michigan agriculture, Stevens headed to a place with a warmer climate that was near tobacco country. Corn, wheat, oats, grass, fruits, and vegetables were the main crops grown in the Greensboro area, and the cotton belt was only seventy-five miles distant.[82] Upon his arrival Stevens discovered that indeed he needed to start at the bottom, since the board had purchased only twenty-five acres of land for his program. His students knew "nothing about good farming tools and as little about good farming." He spent the next three years teaching the business and techniques of farming to young African Americans in North Carolina.[83]

Poverty stood in the way of African Americans who wished to attend Michigan Agricultural College, and the college was not particularly sensitive to their plight. John W. Robinson, who had completed the academic course at Tuskegee, petitioned Snyder: "I now have charge of the Dairy and is able to do profitable work in that dept; but I am anxious to take a college course in Agr. but I am poor and absolutely self dependent. Please write me with advantages to such men, if your college afford any."[84] Snyder offered neither advantages nor encouragement. He stated that the college offered opportunities for students to work, but it was critical to "save up a little money to start with." Although Snyder promised Robinson "to do the best we can" for him, he told Robinson that he was on his own and his success at the college would "almost all depend upon your own energy and ability."[85] Young men like Robinson had virtually no opportunity to save money for college; consequently their poverty prevented them from finding a place in Justin Morrill's industrial classes.

INTERNATIONAL STUDENTS

The presence of international students on the campus of Michigan Agricultural College underscored the increasing significance of the United States in the world and changes occurring in other nations by the end of the nineteenth century. Approximately sixty students from other countries had attended M.A.C. by 1900. Nineteen nongraduates came from Japan, twelve from Canada, three each from Great Britain and Turkey, two each from the Netherlands, Russia, and Bulgaria, and

Figure 8. International Students Who Received a Degree from Michigan Agricultural College by 1900

BACHELOR OF SCIENCE

Charles Christian Georgeson, '78, *Rudkjobing, Denmark. USDA in Sitka, Alaska.*

Michitaro Tsuda, '84, *Tokyo, Japan. Tsuda Bank, Iidamachi, Kudan, Kanda, Tokyo and Secretary in H. I. M.'s Privy Council.*

Kolia San Thabue, '91, *Burma. Agriculturalist, horticulturist, photographer, and agent for the importation of agricultural and horticultural machinery, Thayagon, Wakema, Burma.*

Maurice Grenville Kains, '95, *St. Thomas, Ontario, Canada. USDA, Washington, D.C.*

Shoichi Yebina, '95, *Aomori, Japan. President of Aomori Ken Agricultural School, Sambougi, Aomori Ken, Japan.*

Wahey Matsura, '96, *Tomoika Gumma, Japan. Professor of mechanical engineering, Tokyo College of Technology, Tokyo Kogio Gakko, Tokyo.*

Vadim Sobennikoff, '97, *Kiakhta, Siberia. Installing machinery in Siberian gold mines. Superintending construction of steamships for the Siberian government, Kiakhta, Siberia.*

Joseph Arthur Bulkeley, '99, *Wallerawang, New South Wales, Australia. Professor of science, Department of Agriculture, Hawkesbury Agricultural College, Wallerawang.*

Walter Henry Flynn, '99, *St. Thomas, Ontario, Canada. Draftsman, Cleveland, Lorain & Wheeling Railway, Lorain, Ohio.*

Antranig Garabed Bodourian, '00, *Necomedia, Ismid, Turkey.*

MASTER OF SCIENCE

Kizo Tamari, '86, *B.S. (Imperial College of Agriculture, Japan.) Dr. of Agr. conferred by the Ministry of Education, 1899. Professor of agriculture in Tokyo Imperial University, 1879– . Tokyo, Japan.*

Kumaroka Shoshima, '90, *B.S. (Imperial College of Agriculture, Japan) 1888. Tokyo, Japan.*

Chioji Yoshida, '90, *B.S. (Imperial College of Agriculture, Japan) 1888. Tokyo, Japan.*

---

one each from Cuba, Norway, Armenia, Chile, Puerto Rico, and Australia. Ten students had received bachelor of science degrees and three—who had earned their bachelor's degrees at the Imperial College of Agriculture in Tokyo—had received master of science degrees from the State Agricultural College of Michigan (see figure 8).

Since the largest number of international students came from Japan, it is worthwhile to take a brief look at some of the changes occurring there during the late nineteenth and early twentieth centuries. As Japan underwent major political changes after 1868, it sent some its brightest young men to European and American colleges and universities to acquire knowledge and skills to be used back home. By 1900, nine hundred Japanese students had attended colleges and universities in the United States.[86] They returned to Japan with the expectation that they would promote the "general enlightenment in a wide variety of fields, rather than being restricted to specialized research."[87] As Japan built its educational system, it created technical schools that included agricultural institutions that taught "agriculture proper, agricultural chemistry, sericulture, cattle rearing, veterinary medicine and surgery, forestry, fisheries and marine products, etc."[88] The Agricultural College of the Imperial University of Tokyo emerged as a leading institution of agricultural education. At least twenty of the Japanese at M.A.C., most of whom came from Tokyo, studied agriculture. M.A.C.–trained Japanese students drew upon the knowledge that they had acquired at the college to instruct students in agricultural schools back home and to apply relevant aspects of American agriculture on Japanese farms. For example, Shoichi Yabina, '95, served as the president of "a small agricultural school" in 1915.[89] In 1885 Michitaro Tsuda, '84, had returned to the United States to purchase seeds, plants, cattle (including 100 milk cows for his own farm), and agricultural implements to ship to Japan.[90] Their work was the result of the more expansive effort by the Japanese government to import western technology into the country to spur the growth of industry and agriculture.[91] Yabina and Tsuda were living proof that the land-grant mission extended to the far corners of the world.

Although most international students intended to take their hard-won knowledge at M.A.C. back to their home countries, they also brought their own culture to the college community. They served as ambassadors for their people as they taught Americans about life in their homelands. Students coming from rural Michigan had little idea of what life was like in Russia or Japan or other distant nations. The *Speculum* offered a medium for all students to share their knowledge, experiences, and opinions with their classmates, professors, the alumni, and anyone else who happened to read their essays. Vadim A.

Sobennikoff, who came from Siberia, told his readers in October 1893 about his trip in July from Siberia to Japan. He reported seeing convicts from European-Russia building the Siberian railroad near Vladivostok, the bustling seaport located on the Sea of Japan. Sobennikoff said: "This is our naval station of the Pacific squadron." Ships from many countries, especially Germany, China, and Korea called there.[92] Wahey Matsura shared some of his Buddhist beliefs in a piece that appeared in October 1894. Surely Protestant, Catholic, and Jewish readers had a hard time comprehending Matsura's explanations of the "Hinayana" and "Mahayana" and the contemplative methods used by Buddhists to incorporate these doctrines into their lives. Matsura articulated the significance of Buddhism to the Japanese: "The real foundation of morality in the present Japan is the gift of this wonderful influence of 'Hinayana' planted in the very bottom of the hearts of the people."[93] Students coming from different parts of the world stretched the minds of their American classmates in directions that their professors could not. International students contributed much to the evolution of the land-grant philosophy at the State Agricultural College of Michigan.

6

Student Life

YOUNG MEN AND WOMEN ENROLLED AT MICHIGAN AGRICULTURAL College were stimulated by new, challenging, and competing forces outside of the classroom that influenced their beliefs, their worldviews, and their relationships with other people. Coming from farms and small towns, many students lived in close quarters with more people than they had ever before. Saints' Rest, Williams Hall, and the Women's Building proved to be far different living environments than Bath, Decatur, Hanover, or the farm. College literary and social societies, religious organizations, athletic competition, and student publications all provided opportunities for creative, formative minds to begin to realize their intellectual and emotional potential. Men and women enjoyed dances, concerts, games, plays, and other social engagements that often led to lifelong friendships or marriages. Serious illness and death brought on by communicable diseases frequently struck students, bringing on physical and emotional suffering for them and their friends and families. Sometimes youthful exuberance led students to challenge college rules and to behave in an unruly fashion that resulted in disciplinary action.

The establishment of the women's program in 1896 led to a transformation of campus life and an expansion of M.A.C.'s social environment. Student activities outside of the classroom played a critical role in M.A.C. becoming a truly co-educational institution. From the college's first day, male students welcomed any opportunity to attend functions to which young ladies living in the Lansing area had been invited. On campus, Henry Haigh, '74, remembered sitting across the dinner table from Belle Allen, Mollie Jones, and Libby Sessions, three of first women to attend M.A.C. He expressed strong sentiments about his new classmates: "I promptly fell in love with all of them" and they were "an influence for good."[1] For the college to fulfill its

Some students endured real hardships in order to attend M.A.C. Wallace K. Wonders, who came from Detroit, worked diligently to earn enough money to pay for his education. He enrolled in M.A.C. against the wishes of his father, but with the encouragement of his mother (N.S. Aiken to Jonathan L. Snyder, 18 August 1898 and 4 September 1898, Jonathan L. Snyder Papers, UA 2.1.7, Box 814, Folder 45). Later in life, Wonders wrote the following account of his residence:

This cabin was rented by Marcus B. Stevens '02 and Wallace K. Wonders '02 to be used as a home while they worked at the college during the 1901 summer vacation. It was within easy walking distance of the campus. The drainage was good but the facilities were extremely minimum. The views were 100% rural. The social advantages were 0. If the truth is adhered to not too closely, the clear air of the morning and the peace and quiet of the evening were disturbed only by the activities of the mosquitoes. As I remember it, we had even in those days had a problem to face because our college was obliged to close all its eating clubs because of lack of funds.

The college did not feel it could keep an eating club open for the working students so we had to shift for ourselves. We are happy that we survived.

Courtesy of Michigan State University Archives and Historical Collections (Michigan State University, Students, Off-campus Living, Folder 1)

land-grant mission, women students had to become integral to all parts of the institution's program. The women's course brought with it a recognition that women must occupy an essential place in the life and fabric of the campus.

ROOM AND BOARD

Since Michigan Agricultural College was situated "without the bounds of civilization right in the woods & wolves and bears in the distance,"[2] dormitory rooms became "new homes" for students and the focal point of life while on campus. In 1857, four years before the opening of the Civil War, the first students arrived at the oak opening north of the Red Cedar River. They walked through the doors of the boarding hall, which they soon dubbed Saints' Rest, went up two or three flights of stairs to their rooms and entered a whole new way of life. Saints' Rest was a community in itself. The kitchen, laundry, and "community wash room" were located in the basement. The second floor housed the dining room, parlor, and living quarters for James Shearer, the steward, and his wife, the cook. As many as four young men occupied each of the twenty-eight rooms situated on the third and fourth stories. Two students shared a bed, and each one rented

STUDENTS' ROOM

Interior of a woman's room in Abbot Hall in 1898. After the new Women's Building (Morrill Hall) opened in 1900, Abbot Hall once again housed men. Courtesy of Michigan State University Archives and Historical Collections (Michigan State University, Buildings, Abbot Hall [1], Interiors).

bedding and furniture needed to make his home away from home. They tended the small wood stove that heated their room with fuel retrieved from the common wood box on their floor. Within their newly adopted four walls, each student studied, argued with roommates, slept, plotted pranks, complained, composed essays, read, created disturbances, kept dairies, and wrote letters to friends and family back home.[3] After Saints' Rest burned down in 1876, M.A.C. built (the first) Wells Hall to replace it.

The isolated campus location necessitated the construction of additional dormitories if M.A.C. hoped to integrate women into the student body. When Williams Hall opened in 1870, an opportunity arose that enabled the faculty to admit female students who would live in rooms on the floor where the steward and his family resided.[4]

Although this arrangement did not last long, Isabel Allen, Catherine
C. Bacon, Ella Brock, Mary E. Daniells, Harriet A. Dexter, Gertrude
Howe, Emma H. Hume, Mary L. Jones, Elizabeth E. Sessions, and
Catherine E. Steele demonstrated that women were fully as capable as
men to pursue a land-grant college education.[5] Although the absence
of a dormitory for "ladies" prevented many qualified women from
enrolling in M.A.C, those who were determined to attend found
rooms in faculty houses or commuted from their homes in Lansing.
In 1896 President Jonathan L. Snyder initiated the women's course
and outfitted Abbot Hall, which had been a men's dormitory when it
opened in 1888, to house forty women and other facilities related to
the new program. Finally, with the opening of the Women's Building
in 1900, Michigan Agricultural College had a first rate facility that
housed 120 "co-eds" in double rooms. They slept in single beds and
kept their "voluminous skirts" in large closets.[6]

Believing that his land-grant college should advance democracy,
Snyder became an advocate for the dormitory system and helped to
entrench a tradition at Michigan State that has survived to the present. In 1896, the Edwards, Smith, and Kedzie report called for "the
gradual abolition" of the college's dormitories, but Snyder would have
none of it.[7] He believed that requiring students to live in dormitories
broke down social and class distinctions. While rich students might
find more commodious accommodations off campus, poorer students
living in more affordable dormitories might be stigmatized. After a
streetcar line from Lansing reached the west entrance to campus in
1894, it became easier for students to commute from residences in the
city.[8] When the legislature resisted funding the reconstruction of
Wells Hall after it burned in 1905, Snyder categorically stated: "If we
cannot secure a dormitory it will simply mean that the poor boys cannot come here."[9] The legislature appropriated money for the second
Wells Hall, with a capacity of 156. It welcomed its first residents in
1907.

Since M.A.C.'s residence halls could not accommodate all the college's students, Snyder recognized the need for rooming houses run
by respectable women and men in the emerging community of East
Lansing.[10] But when literary societies and fraternities asked to have
dwellings for their members, Snyder voiced strong opposition, charg-
ing that this was elitist and that it would undermine the democratic

Michigan Avenue in East Lansing, circa 1905. The streetcar line joining M.A.C. to Lansing was completed in 1894. Courtesy of Michigan State University Archives and Historical Collections (Michigan, East Lansing, Street Scenes).

influence of dormitory living where both rich and poor students lived next to each other. He believed that these organizations favored the well-to-do because poorer students could not afford the costs that went with membership. In addition, society houses would erect social class barriers running counter to the democratic spirit that characterized M.A.C.

Despite Snyder's and others' adamant opposition, the societies eventually forced M.A.C. to accept society houses. The Hesperian and Columbian societies sought to establish a house without success in 1905.[11] When the board denied the Olympic Society's petition to establish a house in 1907 Snyder articulated what to him was an integral part of the essence of M.A.C.'s land-grant mission:

This image of Room 1, Williams Hall, shortly after 1900, shows what a man's dormitory room looked like. Courtesy of Michigan State University Archives and Historical Collections (Michigan State University, Students, Dormitory Life, Folder 2).

This College is, first of all, a poor boy's college. The institution stands for the best development of the industrial classes. Its reputation has been made, as far as the alumni are concerned, by the poor boy who came from humble surroundings. This institution is patronized almost entirely by young men and women of moderate means. It is our duty, therefore, to see that as far as possible the expense to students is kept at the minimum; that all tendencies to unnecessary expenses is discouraged.[12]

Henry G. Reynolds, '70, a former secretary to the board, supported Snyder when he called it "monstrous" for the college to foster "the contemptible spirit of caste among its students." M.A.C. needed to make it possible for students of "limited or cramped means" to attend, for after all they supplied the human stuff to build the country. Reynolds urged Snyder to continue promoting the dormitory system.[13] The land-grant mission rested upon the premise that a college education should be available to qualified students from all social and economic classes.

The next year, the State Board of Agriculture's opposition to society houses cracked. The board gave permission to the Eclectics to build a "modern fraternity house" off campus with enough living rooms to lodge most of its members.[14] Even though most of the alumni supported Snyder's position, the board in 1912 refused to stop members of the Olympic Society and three other societies from living in their own houses.[15] Student organizations prevailed for several reasons: an acute need for housing, a national trend for fraternities to have their own houses, and questions over the legality of the board denying enrollment to students who lived in quarters not approved by it. By 1925 the college's dormitories could not accommodate most of its nearly 3,000 students, and living off campus no longer raised the objections that it had twenty years earlier.

Since only Wells Hall housed male students on campus in 1925 (the first Williams Hall had burned in 1919), most men had to find quarters in privately owned facilities. Upperclassmen filled most of the rooms in Wells Hall, forcing freshmen to live off campus in an environment that made it more difficult for them to adjust to college life.[16] By 1923 every female society except one had a house off campus in East Lansing. Women, however, were not permitted to live in their sorority houses until their second term, and they had to have at least a C grade point average to move off campus. The college provided rooms for 235 women in three dormitories: the Women's Building (160), Abbot Hall (50), and the College Residence (25).[17]

After Saints' Rest burned in 1876, M.A.C. adopted a new system for providing meals to students. Several independent boarding clubs were instituted, replacing the steward, who had been employed by the college to purchase and oversee the preparation of food for meals eaten by students. Students had complained that the diet served to them in the dormitories contained too many "canned and prepared goods, which are nearly always unwholesome," and lacked sufficient produce.[18] Leon Drake, '03, soon after he arrived on campus in September 1899, described dinner at Club F to his parents and sister:

We don't get pie very often but when we do I should judge that it was cut in six pieces, we get each a piece in a saucer. That wouldn't be so bad but I guess the tins they make them in are about 4 inches across. We have all the milk we can eat or drink every night, that Smith I spoke of in the other letter is Alvah Browns prototype, he rushes in ahead of the others and grabs the milk pitcher and fills his glass and without letting go of the pitcher he drinks the glass of milk and then fills it up again. Well I must stop and go to dinner if I am 5 minutes late the boys will have the choicest parts of it disposed of. Well, I have been to dinner, we had lemon pie and green peas today and plenty of pretty good beef, tell that to Aunt Et, we don't get pie very often, and as they put the victuals on before we come in, and the fellow who sits on my left wasn't there, I thought I had better make the best of opportunities, I presume the cook wondered whether I or Billy Halleck, who sits next, ate the absent one's pie.[19]

Students managed the clubs, each of which employed its own steward who hired and supervised the staff needed to put food on the tables. The club system offered male students the opportunity to choose eating arrangements that best met their needs. Initially weekly rates charged by the clubs ranged from $2.00 to $2.75. By 1925 the weekly costs varied between $3.50 and $4.00. The clubs incorporated in 1906 for the purpose of purchasing food and supplies in bulk; the secretary to the State Board of Agriculture oversaw the boarding association's finances.[20] But when the second Wells Hall opened in 1907, the new club remained independent of the association. In 1921 the home economics department took over the operation of Club C for women located in the Women's Building and renamed it the Women's Common; this gave female students experience in the management of an institutional food service.[21]

SOCIETIES AND STUDENT PUBLICATIONS

Literary societies stimulated intellectual life outside of the classroom, gave birth to student publications, and staged dances and parties, all of which enriched the lives of M.A.C. students. Beginning with the college's first classes, students fashioned both written and spoken responses to concepts learned from variant political and philosophical viewpoints. Participation in literary societies taught young men and women to discuss difficult matters in challenging, yet civil, ways, preparing them to debate important issues after they left school. They contributed thoughtful essays on scores of topics to the student publication the *Speculum,* from 1881 to 1895. When the *Holcad* was launched in 1909, the new student periodical devoted much more space to M.A.C.'s social life than its predecessor. By the first decade of the twentieth century, the societies were sponsoring many dances and events, which provided opportunities for students to have fun.[22]

From the college's earliest days, the societies met in rooms set aside for them in the residence halls. Members furnished them to create an atmosphere conducive to hosting debates, reading essays, reciting sketches of historical or biographical interest, and presenting declamations. Sometimes a piano or an organ resting on a fine carpet surrounded by book shelves and chairs set the stage for lively discussions and dancing on those Friday evenings when women were invited to attend the all male societies' weekly meetings. One or two violinists, an accordion player, and a pianist might provide the music for dancing after the literary program had been completed.[23]

Ideas and opinions expressed by society members on Friday nights soon led to conflict with President Joseph R. Williams. In late 1858 students served notice that they did not intend to be the lackeys of the administration. Williams asked the Lyceum to amend its constitution to prohibit students from saying or publishing anything "disrespectful to the Faculty or ridiculing any of their sayings or doings." In a "spirit of liberty," the students responded with a resounding No!—they lived in a "Free Country" and would not give up their right to speak their minds.[24] These young men from the farms and small towns of Michigan reminded Williams that they had come to the Agricultural College to learn how to influence their world through democratic means, not to be mere pawns in the hands of authority figures.

Members of the Stoical Pen Yankers' Society addressed numerous issues facing them and their classmates in the *Bubble,* the earliest known student publication of the State Agricultural College, which appeared in 1868 and was edited by Frank S. Burton, '68. A writer for the *Bubble* agonized over the public's lack of interest in the Junior Exhibition that featured "seven orations, a discussion in which four participated, and a poetical prophecy." He expressed the impression that the residents of Lansing and Okemos would have come out in great numbers to see a circus but had little interest in "an intellectual exercise of this kind."[25] He speculated that people had difficulty understanding the purpose of the college and saw academic performances to be of little relevance to them. On the other hand, he noted, local residents entertained society members in their homes, where they enjoyed visiting and dancing with young ladies, eating treats, and listening to women give humorous readings.[26]

Student interest in current events and literary subjects was so pervasive that spirited debates and discussions occurred outside of society meetings. On 18 June 1869 the senior and junior classes held a series of discussions before the public on a number of burning issues:

- Is Self-Support at College Desirable?
- Are Early Marriages Advisable?
- Ought Michigan to Adopt a Compulsory System of Education?
- Shall we have a Protective Tariff?
- Ought Women to be Enfranchised?[27]

Henry Haigh and his classmates valued highly the stimulation of their minds outside of the classroom. He and Arthur Lowell, '73, took great pleasure studying the works of Shakespeare and Sir Walter Scott's *Ivanhoe.* Haigh found his time reading Shakespeare with President T. C. Abbot to be exhilarating: "My! but he can read it O.K." Haigh's love for Shakespeare did not prevent him from being attuned to contemporary events. He worried that his father may have lost money as a result of the closure of the David Preston Bank in Detroit during the financial panic griping the country in September 1873.[28]

Perhaps no topic caused more consternation for some students than Charles Darwin's theory of evolution, which led them to question Christian doctrine regarding the origins of life and the earth. Professor Albert J. Cook convinced many undergraduates in comparative anatomy and entomology that evolution best explained the development of plant, animal, and human life. Although evolutionary theory "did not shatter" the religious beliefs of Henry Haigh and others, it gave them a lot to think about. Cook, R. C. Kedzie, W. J. Beal, and other

H. W. COLLINGWOOD

Dr. W. J. BEAL.

, JR.,

professors had not pushed evolutionary thinking beyond the physical and scientific to encompass its moral, ethical, and religious implications. Nonetheless, M.A.C. students encountered in Darwin one of the most influential thinkers of the second half of the nineteenth century, who challenged them to take a look at the world in a way that differed significantly from what they had learned in their country schools and churches. Lyster H. Dewey, '88, said in 1940, "we emerged from his [Cook's] classes firm believers in evolution."[29]

The publication of the *Bubble* showed that students' participation in literary societies motivated them to move beyond the spoken word to put their thoughts into print and circulate them by means of student periodicals and yearbooks. On 1 August 1881, editor in chief Liberty Hyde Bailey turned out the first quarterly issue of the *College Speculum*. Five societies (Natural History, Delta Tau Delta, Phi Delta Theta, Union Literary, and Eclectic) each chose an editor, and together

The *College Speculum* began publication on 1 August 1881. The Board of Editors included (*bottom row, left to right*) Professor William J. Beal, Science; Liberty Hyde Bailey, Editor in Chief and College News; Professor Samuel Johnson, Treasurer; and Lucius W. Hoyt, '82, Secretary and Correspondence, Exchanges, and Colleges; (*top row, left to right*) Herbert W. Collingwood, '83, Business Manager; Osmond C. Howe[?], '83, Personal and Literary Notes; and John W. Beaumont, '82, Literary Articles. Courtesy of Michigan State University Archives (Michigan State University, Students, Organizations, Publications, Speculum).

NO DEMOCRATIC CLUB AT M. A. C.

ORGANIZATION PREVENTED BY DIPLOMACY OF PRESIDENT REPUBLICAN CLUB.

Through the diplomacy of Charles J. Oviatt of Bay City, president of the M. A. C. Republican club, the organization of a Democratic club at the college was prevented. Oviatt is also president of the athletic association at the college and he selects the dates for all the mass meetings held at the college. Learning that the Democratic students were planning to organize Thursday night, and that they had secured as a speaker a prominent Lansing lawyer, Oviatt got busy with his paste pot and posted notices of a mass meeting in conspicuous places all over the campus. The Bryan men were forced to cancel their meeting and other things interfering, they have been unable to get together for any kind of a meeting.

Although students joined together to cheer on their athletic teams, partisan politics created divisions among them. This undated clipping from an unidentified newspaper, found in the Michigan State University Museum's collections, tells the story of how one Republican student, Charles J. Oviatt, '09, prevented other students from showing their support for William Jennings Bryan, the Democratic Party's candidate for president in 1908. Courtesy of Michigan State University Museum (Newsclippings, Campus Events).

these five formed the board of editors, which controlled the paper together with a student business manager and a faculty member who served as treasurer. Anyone who had ever been a student at M.A.C. was welcome to submit an article to the paper, which aimed to publish those that evinced "originality."[30] For fourteen years, students, faculty, and alumni contributed scores of articles analyzing the college's curriculum, discussing recent scientific advances, commenting on the welfare of students, exposing philosophical tensions experienced by students, and reporting the latest campus news, including athletics. The *Speculum* strengthened ties between alumni, current students, and the college through extensive listings of the activities of members of the extended M.A.C. family. It ceased publication at the end of 1895. During the 1890s, the Union Literary Society issued another student publication, the *Union Lit,* a monthly paper filled with news items, letters to the editor, and short articles.[31] Students also issued the *Harrow,* a yearbook, in 1887, 1888, and 1889, and in 1896 they introduced another annual, the *Wolverine.*[32]

Starting in January 1896 the faculty edited and the college published the weekly *M.A.C. Record.* Unlike the *Speculum,* this paper originated with the faculty, and it communicated M.A.C.'s purpose and programs to students, alumni, and constituencies upon whom the college depended. The underlying assumption that informed the *Record*'s editorial slant was the school's aim "to develop all its pupils into broad-minded men, good citizens and ideal farmers or mechanical engineers."[33] While students contributed to the paper's contents, they did not have editorial oversight, nor did they publish the analytical essays that characterized the *Speculum.* The columns of the *Record* were filled with news from alumni, reports on campus activities, changes in curriculum, athletics, stories about new facilities and buildings, and introductions of new faculty—but it was not a student newspaper. In 1913 the M.A.C. Association, the alumni organization, assumed responsibility for publishing the *Record.*

In response to student demand for a forum "of the student, for the student, by the student" in which they could express their concerns and views, the first issue of the *Holcad* introduced a fresh voice on campus on Wednesday, 10 March 1909. In its first number, the *Holcad* spelled out tensions that characterized the relationship between students and the town. Stung by local journalists' harsh criticisms of stu-

dent behavior, the *Holcad* argued that students occupied a difficult position in the community. On one hand, they had more freedom than other young people, and on the other, they were "held to a much higher code." The press was wrong to charge the school administration with "being lax in discipline" or to call students "a band of hoodlums" whenever some of them engaged in pranks.[34] In essence, the demands, expectations, and regimens of college life forced students to live in two worlds and the space in between them, all at the same time, which resulted in mutual misunderstandings between town and gown.

The *Holcad* reflected the shift in student interest from intellectual to social activities outside of the classroom. Although it lacked the depth of many of the essays found in the *Speculum,* the *Holcad* provided a healthy outlet for student news and views. Extensive coverage of M.A.C.'s social scene informed readers what the societies were doing to entertain themselves. Other stories reported on the content

Members of the Class of 1912 produced the 1911 *Wolverine.* Members of the board included (*seated on floor*) Earl C. Kiefer, Advertising; Earle E. Hotchin, Athletic Editor; Max W. Gardner, Literary Editor; Ralph G. Kirby, Humorous Editor; (*seated on chair*) Otto W. Schleussner, Editor; (*back row*) Rudolph J. Tenkonohy, Business Manager; Ruth Mead, Society Editor; Margaret Logan, Humorous Editor; Durward F. Fisher, Advertising; Philena E. Smith, Art Editor; Alfred Iddles, Associate Editor; Arlie D. Badour, Art Editor. Courtesy of Michigan State University Archives and Historical Collections (Michigan State University, Student Organizations, Publications).

of majors offered at the college and vocational opportunities for students studying forestry, agriculture, domestic science, horticulture, and other subjects. Readers learned about the victories and defeats of the Aggies' football, basketball, rifle, and other athletic teams. Local merchants recognized that students were potential customers, and they advertised the goods and services available in their stores and shops in the paper. When the institution's name was changed to Michigan State College of Agriculture and Applied Science in 1925, the *Holcad* became the *Michigan State News.*

SOCIETY LIFE

With few exceptions, the societies at Michigan Agricultural College were merely local organizations until the 1920s. However, two national fraternities, Delta Tau Delta and Phi Delta Theta, established chapters on campus in 1872 and 1873 respectively. The faculty terminated the Iota Chapter of Delta Tau Delta in 1896, and the Michigan Beta Chapter of Phi Delta Theta reconstituted itself as a local entity called the Phi Delta Society in 1898. Between 1898 and 1921 board policy prohibited national fraternities from organizing chapters at M.A.C. Regardless, many new (nonfraternity) local societies formed as the number of students increased around the turn of the century. The variety of societies on campus between 1872 and 1925 reflects the growing diversity of the student body (see figure 9). For instance, the Feronian Society became the first organization for women in 1891—five years before the beginning of the women's course. The Cosmopolitan Club worked to integrate international students into college life, and various music groups came into being to bring together men and women who wished to give public concerts. Beginning in the 1890s, the intellectual and literary pursuits of the societies gave way to more elaborate and extravagant social events. In the early 1920s national organizations assumed a new presence on campus— the Chi Chapter of Alpha Gamma Delta Sorority became the first national Greek letter society after the board lifted the ban against national organizations in 1921. In 1922, the Forensic Society became the Gamma Omicron Chapter of Lambda Chi Alpha Fraternity and Alpha Gamma Rho founded the Tau Chapter.[35]

Students who were left out of existing societies took the initiative to form new ones. Many years after she had graduated from M.A.C., Irma Thompson Ireland, '00, told how the Themian Society came about and how it gently challenged the social structure of the student body. The influx of women into the college after 1896 naturally produced a demand for more associations to fulfill their social needs. The Feronian Society's membership included the daughters of faculty and women living in Abbot Hall, but the society had no room for the "day students [about 20] . . . who came from Lansing on cranky old street cars—or drove in from farms with 'Dobbin' [a horse] and something in the way of transportation on four wheels." Thompson and ten of her classmates "slipped by twos and threes into a thicket of trees and shrubs that forms a sylvan 'hideway' almost under the noses of College officialdom, and there in solemn conclave agreed that there would be a new society called THEMIAN from 'Themis,' of the race of Titans who sat upon the throne of Jove to give him counsel!" President Snyder, with reservations, allowed them to organize and to meet

Clara Snyder played an active role in the life of Michigan Agricultural College and the Lansing community. She entertained women from the college and the town at many receptions, meetings, and parties. Her most memorable luncheon took place on 31 May 1907, when she and her husband hosted a luncheon for 135 guests in honor of President Theodore Roosevelt before he delivered the commencement address at the college's Semi-Centennial celebration. Active in the Presbyterian Church, Mrs. Snyder taught a Sunday School class for college women and assisted Edith McDermott establish the Young Women's Christian Association at M.A.C. She was a founding member of the East Lansing Women's Club and served two terms as its president. In this photograph, college women have gathered in 1903 for one of her garden parties. Courtesy of Michigan State University Archives and Historical Collections (Michigan State University, Students, Activities, Folder 1).

Figure 9. Literary and Related Societies, 1862–1925

NAME	YEAR
Cincinnatus Lyceum	1862
Sons of Demeter	1862
Agricultural College Lyceum	1868
Stoical Pen Yanker's Society	1868
Delta Tau Delta, Iota Chapter, National Fraternity	1872
Phi Delta Theta, Michigan Beta Chapter, National Fraternity	1873
Philomathesian Society	1873
Excelsior Society	1876
Union Literary Society	1876
Eclectic Literary Society	1877
Olympic Literary Society	1885
Hesperian Literary Society	1889
Feronian Society*	1891
Columbian Literary Society	1892
Phi Delta Society (reorganization of Michigan Beta Chapter as a local literary society)	1898
Themian Society*	1898
Shakespeare Club**	1898
Debating Society**	1901
Prohibition League	1902
Sororian Society*	1902
Eunomian Literary Society (organized under name Sigma Mu Beta in 1903, named changed in 1904)	1903
M.A.C. Women's Club*	1903
Ero Alphian Society*	1904
Aurorean Literary Society	1905
Forensic Literary Society	1907
Idlers*	1908
Delphic Literary Society	1908
Athenaeum Literary Society	1909
Ionian Literary Society	1909
Glee Club	1909

NAME	YEAR
Sesame Society*	1910
Phylean	1910
Cosmopolitan Club**	1910
Delta Club	1910
Dramatic Club**	1910
Round Table Club	1910
Sociological Club	1910
Alpha Freshman Society	1910
Beta Freshman Society	1910
Women's Glee Club*	1912
Mandolin Club	1912
Trimoira Literary Society	1913
Hermian Literary Society	1915
Ae-Theon Literary Society	1915
Dorian Literary Society	1915
Orphic Literary Society	1916
Alpha Gamma Delta Sorority*, Chi Chapter	1921
Alpha Gamma Rho Fraternity, Tau Chapter	1922
Lambda Chi Alpha Fraternity, Gamma Omicron Chapter, founded from Forensic Literary Society	1922
Ulyssian Literary Society	1922
Alpha Phi Sorority*, founded from the Feronian Society	1922?
Delta Sigma Phi Fraternity, Alpha Pi Chapter, founded from the Aurorean Literary Society	1923
Phi Kappa Tau Fraternity, Alpha Alpha Chapter, founded from the Dorian Literary Society	1924
Phi Chi Alpha Fraternity	1925
Pi Kappa Phi Fraternity, Alpha Theta Chapter, founded from the Orphic Literary Society	1925

*Organizations for women. **Organizations for men and women.

Sources: Ben Westrate, *Greek History, Michigan State College, 1939* (East Lansing: Michigan State College, Interfraternity Council, 1939); W. J. Beal, *History of the Michigan Agricultural College and Biographical Sketches of Trustees and Professors* (East Lansing: The Agricultural College, 1915), 205–13.

on Saturday evenings in the chapel, where they presented "crude essays, hackneyed recitations, and jangled piano tunes." Before long, the men of Phi Delta invited them to use their room in Wells Hall that included space for dancing. After recruiting members from the class of 1898, the Themians fielded a basketball team that defeated the Feronians.[36]

In order for a young woman to gain admission to a society she had to subject herself to a rigid discipline designed to impress upon her the importance of living up to the group's ideals. The Themians, like other societies, required pledges to show proper respect to older members and to submit to various indignities intended to keep them in their place. A page from a notebook of Helen M. Sheldon, '12, president of the Themians, contains a list of instructions for a prospective member:

1. Can't speak to a man
2. "   go fussing
3. Must make sign of T when they meet old girls except in dining room and classes
4. Must do anything that the old girls ask (i.e. making beds, carrying books, mending etc.)
5. Must get off of walk & out of the way when they meet old girls
6. Two pledges can't be seen together on the campus
7. Must stay in their rooms after 7 P.M.
8. Must only speak to the old girls when spoken to
9. Every pledge *must* appear at breakfast
10. Every girl must supply herself with a pair of pajamas. If they are to be borrowed it must be borrowed outside the society.[37]

Once a woman passed through initiation, she made friendships that lasted for the rest of her life. Of more immediate importance, however, she could participate fully in the life of her society.

Parties and dances with all of their attendant features formed the heart of the societies' entertainment calendar, but a faculty committee supervised activities closely. When a society applied to the committee for permission to hold a dance, it supplied the date and identified the chaperons. If there were no conflicts with other campus functions, the committee approved the event. Although each society was allowed to

Irma G. Thompson, '00, drew this sketch showing men, dressed in their military uniforms, and
women, attired in beautiful dresses, dancing
away the evening at the Military Hop held on
campus. Attached to the sketch is this poem:

Oh that the caller might
go on calling—
The music go on falling—
and no one go away

Courtesy of Michigan State University Archives and Historical Collections (Irma Thompson Papers, 1896–1904, UA 10.3.35, Box 761,
Folder 72).

stage two parties per term, they usually put on no more than one.
Strict curfews were enforced: parties without dances ended at 10 P.M.,
dancing parties, known as eleven o'clocks, concluded at 11 P.M., and
the Junior Hop and commencement parties lasted until 3 A.M. Women
were not allowed to attend more than four dancing parties a term, and
if their academic performances was sub par, their dancing privileges
were revoked at midterm.[38]

Dances wove female and male students and the faculty and their
spouses into the social fiber that both shaped and perpetuated the
spirit of the college. For several hours, all participants, no matter what
their status, danced to the same music and enjoyed each other's company freed from the demands of the classroom or workplace. Cards
listing each dance not only laid out the program, but also enabled
young women to preserve their precious memories of the occasion.
Ruth E. Carrel, '08, saved a published description in her scrapbook of
the Union Literary Society's eleven o'clock held on 14 March 1908.
Outfitted in their prettiest dresses and finest suits, women and men
danced into the night surrounded by a setting that took them far away
from the cold of Michigan:

The decorations were something unique and caused much
favorable comment. The west end of the armory represented the
side of a steamer, with its ports, paddlebox, gangplank and life
buoys. The gallery made an excellent "upper deck" and could be

reached by two "ship's ladders." A lighthouse in one corner of the building burned brightly and just across from it the College Orchestra discoursed sweet music from "a green, rock bound island."

Two numbers on the program were 'Search Light' dances[:] some of the lights were turned out and a search from the 'upper deck' was thrown upon the dancers.

Athletic Director and Mrs. C. L. Brewer, Dr. and Mrs. A. F. Gordon, Prof. and Mrs. J. Fred Baker and Prof. and Mrs. C. D. Smith served as chaperones and helped all present to enjoy the short "Ocean voyage."[39]

Students invested much time and energy building the sets for dances and making the arrangements needed to bring together all of their components for an enjoyable evening.

A steady diet of parties made up an important piece of student life and filled many weekend evenings on the social calendar. The Idlers, for example, held a Halloween party in 1909 followed by a dance several weeks later. Glowing candles and jack-o-lanterns lit the way for women, decked out "in the most fantastic costumes" on their way to the dining room in the Women's Building. After eating, they played games in the parlor and danced in the gymnasium until nine.[40] Their next entertainment took the Idlers to the third floor of the Engineering Building for an eleven o'clock, where they and their guests danced to music played by the college orchestra. Dean George W. and Mrs. Bissell and Professor Victor T. and Mrs. Wilson chaperoned the affair.[41] The following Saturday night the Military Department brought Fisher's Orchestra from Detroit to the Armory to lift ecstatic couples as they floated across the dance floor beneath American flags and bunting draped along the walls. Members of each class attended, and all upperclassmen donned their military uniforms.[42] The next week sixty couples attended the Sororians annual fall eleven o'clock, and were surrounded by "dainty arrangements of pillows, pennants, screens, and rugs" in the Armory.[43] These delightful evenings enriched the lives of many, but no other dance compared in either the excitement or splendor of the J-Hop—the social event of the year.

The Junior Class spared no effort when it staged the annual J-Hop. From its beginning on campus in 1888, the gala evolved into an extravaganza held in Lansing. A rendering of the glamour of the J-Hop of 1914, put on by the Class of 1915, demonstrated why it was

The gala-goers from the Class of '14 paused long enough to pose for this photograph during the J-Hop of 1913 at the Masonic Temple in Lansing.

Courtesy of Michigan State University Archives and Historical Collections (*Wolverine*, 1913, 240).

"written indelibly in the memories of some three hundred and twenty Juniors and their guests." At 4:30 in the afternoon of 13 February a group of young men and women, with their hearts thumping, boarded special streetcars and headed to the Masonic Temple in downtown Lansing, where a sumptuous banquet and an elegant night awaited them. Upon arrival, President and Mrs. Snyder and other dignitaries greeted each guest at the reception. Women from the Universalist Church in Lansing served a feast in the banquet hall, which was outfitted in a scheme of pink. The menu read as follows:

Cocktail aux Fruits

Consomme

Gaufres

Batons de Pain

Celeri

Poulet Roti

Pommes de Terre en Puree

Many of Donald W. Francisco's sketches appeared in the *Holcad*. He drew this cover to highlight the upcoming J-Hop in 1914. Courtesy of Michigan State University Archives and Historical Collections (*Holcad* 6, no. 18 [16 February 1914]: 1).

Pettits Pains Chaud

Punch Glace   *Petits Pois*

Olives   *Cornichons*

Salade   *Gaufres*

Glaces a la Crème   *Gateaux*

Pastilles de Menthe

Raisins   *Noix*

Cafe.

Toasts by class president Arthur L. Bibbins and others commemorated the occasion before the dance began. Hanging baskets of nasturtiums, Art Nouveau decorations, and white cloth-covered walls created the "summer garden" setting for the grand march and twenty-eight "syncopated selections" played by Finsel's orchestra from Detroit. The programs served as mementos of the evening, along with "folding leather picture frames of dark brown leather, with the class seal in sterling silver" for women and "brown leather cigarette cases, mounted in silver and with the class seal in sterling silver on the side" for men. The music stopped at 2:30 A.M., and the gala goers headed back to campus.[44]

The J-Hop of 1914 generated controversy that evinced stress and growing pains resulting from changes at M.A.C. and differences in generational perspectives on the content and pace of life in American society. Criticisms of the cost and extravagance of the J-Hop that had been publicized in newspapers attracted the attention of Louis A. Bregger, '88, who lived on his "Outlook Farm" near Bangor, Michigan. Bregger wrote to Snyder, raising several significant concerns, shared by other people, relative to changed ways and perceptions at M.A.C. since his student days more than twenty-five years before:

> Is it true as our papers have it that the late "Junior Hop" cost $6000 to $7500? (My paper had it $7500 and I promptly rejected such a story.) Is it also true that the "favors" at the banquet were "cigarette cases" and "powder pencils"? Ye Gods—is *that possible.*
>
> If only half true, what has come over the old, sturdy, virile spirit of M.A.C.? Is such luxurious expenditure, vulgar in its magnitude and rivalling that of the so-called functions of the idle, parasitic rich of large cities in any way compatible with the families the students are from, and, is it indicative of present day life and work at M.A.C.?

> Do the anaemic elements of society represented by "cigarette cases"
> and "powder pencils" voice the sentiments and control the classes to
> such an extent that such lavish and unholy expenditure are
> accepted without protests and general ridicule by the student body?
> If so, God save the M.A.C. All the more sorry and pitiable a specta-
> cle after such magnificent recent football history. It must not be. I
> can not and will not believe it.[45]

M.A.C. had changed from Bregger's student days. The student body
had grown from 340, of whom 21 were women, in 1888 to 1,655 in
1914, of whom 321 were women. Just as students prepared to enter
vocations in the industries, businesses, and professions of pre–World
War I America, they also wished to participate in the entertainments
of the era. Bregger seemed bothered that values espoused by a class of
people wealthier than most M.A.C. students seemed to be driving the
social life of the campus. He failed to consider that some of the sons
and daughters from his "families" wanted to live a life of better
means. On the other hand, Bregger took great pleasure in another
development that appeared to him to transcend social class lines—
M.A.C.'s success on the football field in 1913. The Class of 1915 felt no
pangs of guilt as they embraced evolving social values of the contemporary world. Tacitly agreeing with Bregger, Snyder lamented that
students spent more money and were "breaking away from the simpler customs with which we in our younger days were familiar."[46] The
young men and women who whiled away the evening at the Masonic
Temple understood the world far differently than did the generation
who came before them. Their education at the land-grant institution
stretched their comprehension of the world in directions that Bregger
did not grasp.

PROFESSIONAL SOCIETIES, DEBATE, AND THE PERFORMING ARTS

Students formed other kinds of organizations to stimulate and satisfy
their intellectual curiosities and to broaden their professional horizons. The Natural History Society, founded in 1872, created a forum
for students and faculty to report on their scientific experiments and
observations of the physical world. Students interacted with their professors in a less formal environment than the classroom and had

opportunities to share their findings as fellow, albeit junior, scholars. Between 50 and 150 people attended monthly meetings in the chapel, where they heard papers on such topics as birds eating insects or cherries, characteristics of local Indian mounds, grasses, ensilage in England, and the effect on a bee after it had stung a person. The society maintained its own library, financed by donations, and made its holdings accessible to everyone affiliated with M.A.C. Members contributed geological, biological, and cultural specimens to the museum, which could be used for individual study.[47] By the mid 1880s, other clubs appeared that reflected the expanding curriculum and specialized interests of students. Within another twenty years chemistry, zoology, forestry, horticulture, engineering, and farmers' clubs, among others, had supplanted the Natural History Society.[48]

Instrumental music has a magical way of bringing people together for a common cause, and the college band provided entertainment that helped to unify different segments of the campus community. Richard H. Gulley, '78, Dean F. Griswold, '75, and Evart S. Dyckman, '75–'77, appear to have organized the first band in 1875. Within two years, the college had acquired seven instruments, and students added seven of their own to give the band a respectable number of performers. The band supplied music for events at M.A.C., without charge, and at times played for hire at functions off campus for non-college entities, earning money needed to purchase instruments and music.[49] When the Military Department was established in 1884, the student band was put under the department's jurisdiction, and it became known as the cadet band. M.A.C.'s administration took pride in the band's performance and appropriated $400 to upgrade its instruments, enabling it to "furnish good music" for the Semi-Centennial celebration in 1907.[50] Arthur J. Clark, instructor in chemistry, assumed the directorship in 1907 and turned the cadet band into a highly respected unit that had fifty-one members by 1913. The next year the State Board of Agriculture made funds available for more instruments so that the band could spice up "Get together Parties" for students.[51] The band enlivened many campus activities, arousing people at patriotic meetings, nourishing students' spiritual needs by playing hymns at religious concerts, or escorting the senior women to their seats to watch the Aggies whip Olivet 62-0 on the gridiron in 1910.[52]

Arthur John Clark, who rose from instructor of chemistry in 1907 to professor in 1916, made major contributions to the development of instrumental music at M.A.C. He served three terms as director of the band, 1906–16, 1918, 1922–25. He is shown with the Michigan Agricultural College Military Band and the College Orchestra in 1915. In November 1914 the band accompanied the football team on its trip to Pennsylvania, where it defeated Penn State. Of particular interest in both of these photographs is the presence of Everett Claudius Yates, '16, (*first row, fourth from left, in top photo; fourth from left, in bottom photo*) who was probably the first African American to play in either the band or the orchestra. Courtesy of Michigan State University Archives and Historical Collections (*Wolverine*, 1915, 178, 256).

Debate and oration occupied a prominent place in student life. Their development as organized activities illustrated M.A.C.'s growing relationships with other colleges and the slow integration of women into campus life. M.A.C. joined the Michigan Oratorical League in 1887, and sent representatives to its competition until 1914. Elva Davis, '05, became the first woman from M.A.C. to participate in the annual contest at Adrian in March 1904. Four years later the league held separate competitions for men and women.[53]

The M.A.C. Debating Society began holding weekly meetings in 1901 and presented arguments on such topics as "Mechanical students have greater chances of success than Agricultural students" and "Dancing should be prohibited at ten o'clock parties."[54] Mixed teams debated "*Resolved,* that women should be given the right of suffrage in the United States" in November 1904.[55] The society scheduled ten debates in 1906 and selected the squad for each event by drawing names from a hat. The four members of each team determined among themselves who would take the affirmative and negative sides on a topic of their choice. Mary Allen, Helen Ashley, Bertha Lunn, and Alleen B. Raynor formed the teams for the ninth debate, the only debate with female participants.[56] In 1912 the Public Speaking Association urged each men's and women's society to choose a team to enter into a series of debates with their counterparts in the society rooms, culminating in a public debate between the top female and the top male teams.[57]

Debate drew Michigan Agricultural College into intercollegiate competitions that eventually included a women's team. For years, the annual debate with the Normal School at Ypsilanti energized a sizable portion of the student body. On 4 June 1910, 265 Normal students came to East Lansing to cheer their baseball team in the afternoon during a 5-1 loss to the Aggies. After dinner, Normal's debate team defeated M.A.C in the ninth annual renewal of their intercollegiate rivalry, an event held in the pavilion of the new Agriculture Hall.[58] Four years later the cadet band entertained an audience packed into the Armory to watch M.A.C.'s debaters, R. W. Snyder, G. H. Meyers, and F. B. Meisenheimer, lose to a team from Iowa State College.[59] Debates with Iowa State and Purdue made up the annual Tri-State contests. On 5 April 1922 the English department sent a three-man team west for two weeks to debate at "Western State Normal, Iowa

Harry L. Kempster, '09, William E. Piper, '07, and Charles C. Taylor, '09, comprised the debate team that defeated its counterpart at the State Normal College in Ypsilanti on 17 May 1907. They successfully argued the negative side to the question "Resolved, that a progressive inheritance tax should be levied by the Federal government." In a front-page story the *M.A.C. Record* proudly proclaimed: "The work of the team was such as to give pride to every loyal alumnus, student and friend of M.A.C." This achievement helped to prepare the campus for the college's Semi-Centennial celebration, which was less than two weeks away. Courtesy of Michigan State University Archives and Historical Collections (Michigan State University, Students, Organization, Debate Team).

State, Denver University, Colorado Agricultural College, Montana Wesleyan, Montana State College, North Dakota State College, South Dakota State College, and Gustavus Adolphus College." Back home another team debated squads from Purdue and Colorado Agricultural College.[60]

The participation of women in intercollegiate debate competition moved forward in December 1922 with the formation of a women's team. Dorothy Snyder, '25, Marion Harper, '25, Elsie Gelinas, '23, Mrs. Susie B. Emshwiller, '25, Alberta Bates, '25, Corlan Lyman, '25, Marion Stein, '24, and Lillian Lawton, '24, made up the teams to argue the proposition: "Resolved, that the United States government should own and operate coal mines within a period of five years." On the night of 7 March 1923, half of the squad held forth at M.A.C. against a visiting team from Western State Normal School, while the other half traveled to Kalamazoo to engage another team of debaters at Western.[61] The audience expanded on 16 May, when the women's team staged a debate for listeners to WKAR radio. The men had done likewise a week earlier.[62]

Michigan Agricultural College's emphasis on scientific and technical study created a need for more opportunities for students to learn about the arts and humanities. The Liberal Arts Union, established in 1909, made it possible for students to see and hear artists perform and intellectuals share their knowledge with the public. The union's purpose was "to encourage and stimulate a greater interest in those matters that make a broader social and cultural life in the college, and to this end shall foster and support all organizations that tend to bring about such conditions."[63] The attitude behind the Liberal Arts Union (renamed Liberal Arts Council in 1912) led to the expansion of opportunities for the creative and performing arts and public discourse on matters of the mind and society. In its first year the union sponsored a series of public events including performances by Mme. Rita Fornia, soprano from the Metropolitan Opera, and the Kaltenborn String Quartette, and lectures by the American sculptor Lorado Taft and by E. W. Hock, former governor of Kansas.[64] The council enthusiastically encouraged students to attend its 1914–15 lecture series at the Armory with these words at the bottom of the handbill listing the year's schedule:

> This is a part of your education
> Your money is invested
> Yours to have and to hold[65]

The performing arts at M.A.C. forged ties between the college and the public during the fifteen years preceding World War I. Students and faculty planned, produced, and executed a wide range of programs for audiences made up of students and guests from off campus. During her tenure between 1896 and 1902, Mrs. Maud Marshall, instructor in music, organized the chorus, which began the rich tradition of choral music at Michigan Agricultural College. Louise Freyhofer, Marshall's successor in 1902, provided stellar and creative leadership for music programs at M.A.C. until 1919. Freyhofer invited all students, faculty, and community residents who had a talent for singing to join together for "an hour of recreation each Monday evening."[66] Freyhofer directed the annual May Festival, first presented in 1904, which became a highlight of the cultural year at M.A.C. and in the Lansing community. Among the works performed were two Mendelsohn oratorios, *Elijah* in 1907 and *St. Paul* in 1908, and Handel's *Messiah* in 1911.[67] Four soloists and an orchestra accompanied the chorus for each of these presentations. In 1913 an audience of nearly 1,000 people

watched the 200-member chorus of women, dressed in white, and men, dressed in black, present a concert of selections from a number of composers including Wagner, Dvorak, and Verdi.[68]

Men's and women's glee clubs, formed in 1909 and 1912 respectively, made it possible for students to improve their vocal skills and entertain the public. Nineteen men comprised the M.A.C. glee club in 1914, under the direction of B. E. Hartsuch. They sang "selections from grand opera, medleys and humorous numbers" on a tour across Lower Michigan that took them to Portland, Detroit, Fenton, Saginaw, Bay City, St. Louis, Belding, Grand Rapids, Allegan, Battle Creek, and Lansing. The eight members of the mandolin club accompanied the singers and also played selections of their own.[69] Lacking the means to go on tour, the "Girls' Glee Club," with sixteen members, stayed on campus and gave its concerts in Agriculture Hall. Louise Freyhofer's group also entertained at local events including "the bankers' banquet, the round-up and the state pioneer meeting."[70]

Performances by students and others enhanced the significance of music in the cultural life of M.A.C. and the greater Lansing community. The parlor in the Women's Building hosted piano recitals, with an event in January 1904 featuring the school's "new Grand piano."[71]

Director Louise Freyhofer is pictured here with the Women's Glee Club for the 1915–16 academic year. Freyhofer was a gifted piano teacher who also taught voice and theory in the music department of the women's course from 1902 until 1919. The members of the Glee Club are (*top row*) Mrs. Paul Rood, Katheryn Smith, Louise Clemens, Marion Pratt, Louise Freyhofer, Marion Cameron; (*middle row*) Louise Smith, Mildred Mead, Carolyn Wagner, Mildred Coors, Olive Cole; (*bottom row*) Agnes MacIntyre, Murial Dundas, and Marguerite Ryan. Courtesy of Michigan State University Archives and Historical Collections (*Wolverine*, 1916, 276).

Cast picture for *A Midsummer Night's Dream*, presented on 10 June 1914 in the "Forest of Arden." Courtesy of Michigan State University Archives and Historical Collections (Michigan State University, Theatre, Folder 2).

Although most of the pianists were women, in February 1911, Samuel Langdon, '11, and Lee N. Hutchins, '13, joined twelve women students on a program that featured works of Liszt, Haydn, Schubert, Schumann, Grieg, and Rachmaninoff, among others.[72] Music brought people from off campus to perform as well as to listen, which strengthened the ties between the consumers of the arts at M.A.C. and in the community. On a cold Saturday night in January 1913, Mrs. Della Knight Boice and Mrs. Kate Marvin Kedzie, two of Lansing's most accomplished entertainers, presented a program of music and story before nearly 250 people in the large lecture room of the Chemical Laboratory.[73]

Presentations by the dramatic club found students and faculty working together to put on elaborate functions for the entertainment of the public. On Wednesday evening, 10 June 1914, a forty-person cast put on Shakespeare's *A Midsummer Night's Dream* outdoors in front of College Hall for an audience of 1,500. Arthur J. Clark, associate professor of chemistry, directed an eighteen-member orchestra, and Edith W. Casho, instructor in physical culture, coordinated the singing and "fairy dances" with the scenery provided by "the natural slope of the ground and the somber background of massed trees."[74] In a most pleasing manner, the performers used what came to be known as the "Sacred Space" of the land-grant campus as the stage to present their interpretation of a work that had little resemblance to practical farming. T. C. Abbot would have been pleased.[75]

STUDENT GOVERNANCE

Michigan Agricultural College expected students to follow its rules and to behave in responsible ways that contributed to the well-being of the college community. Although students may have been away from the watchful eyes of their parents, college officials supervised their behavior and administered discipline when they went astray. President Snyder and the faculty acted in loco parentis, assuming parental responsibilities for students who had not yet reached their majority. Some struggled to discover the appropriate line between acceptable behavior and violations of common decency. Alcohol, class and organizational rivalries, seemingly repressive rules, and some-

The student cast that presented the third annual campus night musical revue on 13 February 1923. The *Lansing State Journal* reported that the J. W. Knapp Company of Lansing supplied many of "latest spring styles in suits, evening dresses and sports wear" worn by the participants. The program was "a unique revue of characteristic styles and fashions in Michigan over a period of 80 years." Wishing to wear authentic historic dress, female performers made use of the campus library to research "the various fashions and dresses of these quaint periods" (*Lansing State Journal*, 9 February 1923, 24). Courtesy of Michigan State University Archives and Historical Collections (Michigan State University, Theatre, Folder 6).

times even institutional traditions could lead students into harming themselves, others, or property.

From the time students first came to campus, they played pranks and engaged in noisy disturbances irritating others. For instance, on the night of 7 February 1859, residents of Saints' Rest "made all sorts of noise around the hall[,] threw the ash pail down stairs twice and put Lane's buggy on top of the shed." The faculty suspended George W. Ingersoll, '58–'59, for his part in this mischief.[76] Two days later Edward G. Granger and his roommates discovered that someone had tied their doorknob to another door, imprisoning them in their room. In March 1863 young men dancing to their fiddle music shook Saints' Rest late into the night.[77]

Beginning in 1876, the college introduced a form of student government for dormitory life that, over time, changed from a democracy to a paternalistic way for maintaining discipline. Men living in ten-room "districts" elected captains and lieutenants to oversee the behavior of their floor mates and to pass judgment on students who violated rules pertaining to noise, cleanliness of rooms, garbage disposal, hazing, and other things that disrupted an orderly existence in the hall.[78] Unfortunately, students often elected friends to office principally because they would not enforce the rules.[79] This contributed to the system being abandoned soon after Jonathan L. Snyder became president of M.A.C. in 1896. He did not believe that self-government was in the students' best interest: it was up to the faculty and administration to rule the roost. Snyder did "not think that parents who let children run the home have the best home or the best children."[80]

The administration and the faculty had their hands full much of the time filling in for mothers and fathers. Three incidents demonstrate that some students had little regard for the serious harm that they might cause to themselves, other people, or property. In June 1901 a group of men enacted a custom called "a night-shirt parade." Dressed in night-shirts or in "other hideous ways," they carried torches and set off fireworks as they visited homes of the faculty, who were expected to respond with a speech. That night, others dumped water out of dormitory windows onto students when they returned from the march. Then things got out of hand when students began smashing electric lights, windows, and doors. The next day, a faculty committee expelled three students and suspended several others.[81] The

annual rush or "scrap," intended to show which class was superior, brought together freshmen and sophomore men in a confrontation that could result in serious injury to participants. On the night of 13 October 1902, freshmen and sophomores gathered to challenge each other in the annual scrap. After disregarding the faculty's directive and upperclassmen's admonition not to hold the event, men from the two classes went off of campus to engage in a "brutal and fierce" battle. Eventually the upperclassmen convinced each side to choose a few men to settle the scrap with a series of wrestling matches. In the wake of the freshmen and sophomore classes' breach of discipline, the faculty suspended seventeen.[82] Several sophomores continued a class tradition by disrupting the 1906 J-Hop at the Armory when they tampered with electrical wires to short out the lights. Only backup

"Scraps" between sophomore and freshmen classes could turn quite ugly. Many male students participated with considerable enthusiasm in this rush between the Classes of 1903 and 1904. Understandably, the faculty viewed these encounters with apprehension and concern that students could suffer serious injuries. Courtesy of Michigan State University Archives and Historical Collections (Michigan State University, Student Activities, Class rivalry, Folder 2).

Male freshmen suffered through a number of indignities, including the "stacking" of their dormitory rooms, as part of their initiation into college life. Gordon Stuart gave this account, that appeared in an unidentified publication, after his room was stacked in 1902.

I send you a photograph of my room as it looked one night when I came back from class. Every Freshman must have his room "stacked" by the sophomores, so my turn came in due course. They climbed over the transom and literally stacked everything in one corner of the room. Every garment had at least one hard knot in it, and some of them two or three. Over a thousand stamps I had collected, which were loose in a box, were scattered over the whole room. Six packs of playing cards were also thrown in the "stack." My tooth-brush was put in the water-pitcher and coal oil was poured over it. Nothing but the map on the wall was left in its place.

The stackers hung out a sign from the window, "Stack." Of course, every student saw the sign and came up to see how the room looked. It was past twelve o'clock that night before I got my bed down so as to sleep on it. The "stacking" is not done with any malicious intention; only for fun and pastime.

Courtesy of Michigan State University Archives and Historical Collections (Michigan State University, Students, Dormitory Life, Folder 1).

gasoline lamps prevented a cancellation of the event. But even more serious, someone else had constructed an apparatus under the Armory to produce enough hydrogen sulfide gas, in chemistry professor Frank Kedzie's opinion, to have killed everyone in the building if the machine had been activated. The faculty expelled two men from school permanently and suspended another.[83]

Hazing frequently went beyond the boundaries of harmless harassment to produce terrifying, if not life-threatening, situations that challenged the administration's and faculty's capacity to keep order. Class rivalries, particularly between freshmen and sophomores, instigated most hazing. Sophomores yanked freshmen from their beds, doused them with water, made them run about campus blindfolded, or stacked their furniture in the center of their rooms with little regard for the freshmen's dignity.[84] In late 1906 several students grabbed a freshman and rubbed broken eggs into his scalp before cutting off his hair, tying a sack over his head, and holding him in the river before he escaped his captors' grasp and swam to safety.[85] Other victims had their hands tied behind their backs, were blindfolded, and had soap stuffed into their mouths.[86] Serious injuries resulted from some of these attacks. Over the years, incidents like these were common, and the school's efforts to control their frequency and severity often met with little success. The 1908 regulation prohibiting hazing and class scraps did not end the practices.[87] Even agreements between the freshman and sophomore classes to cease hostilities in 1913 and join together for a barbecue did not prevent hazing. In fact, that year Snyder lamented that the fall term saw "the worst epidemic of hazing" since he first came to M.A.C. in 1896.[88] Hazing persisted right up to the outbreak of World War I.[89]

In 1908 students organized a council or union that, in conjunction with the faculty, defined relationships between classes and established regulations for individual behavior on campus. The union had six members elected by the classes—three seniors, two juniors, and one sophomore. Rules passed by the council called for the abolishment of hazing and class scraps and required each freshman to wear "an official brown cap with a small vizor and a gray button" during the fall and spring terms. In the years that followed, the council enacted other measures affecting student life.[90] For example, in 1912 it proposed a tax, approved by a vote of the students, to support athletics, oratory and debate, the *Holcad,* and the Liberal Arts Council.[91] In 1913 the *Holcad* asked the council to investigate an incident that occurred at the Women's Building or the "coop," as it was commonly called. Some sophomore men had insulted their female classmates by singing "suggestive songs" in front of the women's residence and had forced a cow into temporary residence in the vestibule. Finding little humor in these acts, the women called upon the upperclassmen to "impress the new men with the real meaning of the old M.A.C. spirit toward the girls."[92] The council provided a mechanism for students to take part in their governance, but the final authority remained with the administration and faculty.

Although the student council codified the college tradition of not smoking on campus, many students still smoked. They frequently smoked in their rooms, although the practice was prohibited on the

The Holcad carried many advertisements from local merchants and national manufacturers trying to interest students in their products. Some of the most alluring advertisements were placed by tobacco companies. This notice presents cigarette smoking in an appealing way. Courtesy of Michigan State University Archives and Historical Collections (*The Holcad* 5, no. 23 [24 March 1913]: 19).

grounds—even on the athletic field and on the farm. Most students willingly observed the ban on outdoor smoking. On occasion faculty members advised young smokers to break their cigarette habits because they believed it interfered with their social or academic growth.[93] Enticements to smoke were real and could be difficult to resist. The *Holcad* published full-page advertisements, like the one in January 1913 that told readers: "The popular [male] Freshman is the one who always has a good supply of Fatimas."[94] Two months later a full-page commercial pictured a young man smoking a Fatima and sitting next to a young woman who clearly found him to be quite appealing.[95] Male students needed something to put in the cigarette cases given to them at the J-Hop in 1914. It does not appear that women indulged in smoking to the same extent as men.

The use of alcohol presented challenges to students and M.A.C. officials alike, as student drinking increased with the simultaneous growth of Lansing and enrollment in the college. In 1879, 92 (out of a total of 232) students endorsed the temperance movement when they joined the Red Ribbon Club and pledged "to abstain from buying, selling, or using Alcoholic or Malt Beverages, Wine and Cider included."[96] In the years that followed, increasing numbers of students subscribed to neither the letter nor the spirit of temperance, notwithstanding the fact that men caught drinking in dorm rooms faced suspension or expulsion.[97] President Snyder pleaded with Lansing saloon keepers not to sell liquor to his students, either by the drink or by the bottle.[98]

By 1909 alcohol abuse reached such an alarming level that M.A.C. hired a Pinkerton agent to identify men guilty of excessive drinking and of committing other breaches of discipline. Mr. J. E. Spencer enrolled, under an alias, as a special forestry student, and took up residence in Wells Hall. Spencer gambled at cards with his "classmates," went with them to the Lansing saloons, and became privy to their plans to steal fruit from college boarding clubs and to grease the streetcar track after swiping a barrel of oil. Spencer's reports contain disturbing accounts of young men drinking to excess, not studying, and living their lives out of control.[99] He also noted their contacts with women: "Huber, whom I believe is a freshman, had Beulah Rose, a divorced woman of questionable character, to Waverly Park to the dance."[100] Incredibly, Spencer's boss back in Chicago complained that the students Spencer was running around with were so "prone to spend liberally upon their vices" that Spencer needed more money to keep up with them![101] Spencer's three-week undercover operation revealed clearly that critics of M.A.C. had a point when they castigated the administration for not keeping all of the students under control. The *Detroit Free Press* reported that on the basis of Spencer's findings, M.A.C. booted eleven offenders out of school. Readers applauded.[102]

M.A.C.'s ability to govern the behavior of female students rested on a common assumption that young women could not be trusted in their relationships with men. Of course college officials did not trust the men either, but they chose to regulate women more closely. Administration and faculty often acted in lieu of a student's mother

Sr. Sleigh
Ride
Class '10

The Class of 1910 enjoyed this sleigh ride in the snow and brisk air of an East Lansing winter.
Courtesy of Michigan State University Archives and Historical Collections (Michigan State University, Students, Activities, Folder 2).

and father—a common practice in nineteenth- and early twentieth-century America. The college feared that parents would not send their daughters to M.A.C. if the school did not enforce vigorously its rules requiring women to seek permission to take part in activities off campus and to associate with men socially only when properly chaperoned.[103] Consequently, the faculty handed out harsh penalties to women who took part in social outings without permission from the dean of women, especially if men participated in the activities. Some errant young women suffered the indignity of having their rights to attend social functions on or off campus revoked. Others packed their bags and headed to the train station bruised by the blow of expulsion.

In January 1903 Gertrude Peters, '06, from Springport, Michigan, committed a host of violations after getting permission to spend a night with a friend in Lansing, but she did not acquiesce in her punishment. Peters never intended to stay with her friend. Instead she had supper with a man in a restaurant, joined a sleigh-ride party before heading east to Williamston for a dance, and returned to campus between seven and eight in the morning. When she appeared before a faculty committee, Peters admitted that she had also attended one more party during the term than she was allowed to by the regulations. She saw no reason why she should be limited to four dances a term, and according to Snyder she seemed to "chafe under such restrictions as we are required to place upon young women for their own good." Her actions and attitude resulted in her loss of "all social privileges" during the winter term and one half of the spring term. Her future as a student at M.A.C. depended upon her capacity to knuckle under to college rules. Apparently Peters towed the line after this incident, for she not only graduated in 1906, but she also went on to earn a master of arts degree from Columbia University in 1909.[104]

Other women faced harsher penalties. Expulsion awaited two who clandestinely met their dates for a sleigh ride in January 1905. Without permission from the dean of women, they sneaked out of the back door of the Women's Building, went for their ride, and spent the night in a nearby rooming house after discovering the back door had been locked. Since they were already on probation, the faculty did not deliberate long before rendering their sentence.[105] When Kittie Huckins was threatened with expulsion in 1903, she expressed remorse to Dean Maude Gilchrist: "Now Miss Gilchrist I can be good and do what is right and I promise you faithfully that if you will try me again I will do what is right and you will have no cause for complaint."[106] Unfortunately for Kittie, her transgressions were too numerous in the eyes of the faculty, and she was not allowed to return in the fall of 1903 to study cooking before going on to attend the Boston Cooking School.[107]

On other occasions, Snyder played a paternalistic role: "caring father." In April 1904 he informed Mrs. Melissa Adams that her daughter, Ethel, had gone to two parties and the theater in Lansing without permission. Even though Ethel easily could have been removed from M.A.C., her mother was not happy with her daughter's lesser punishment that "cut off" all of "her social privileges" for the rest of the year, including attendance at the commencement party. Snyder gave this reason for what he saw as leniency: "But, believing her to be a good girl at heart, we were willing to go to the limit in order to save her."[108] Clearly M.A.C. and Mrs. Adams drew their boundaries at different places. At another time, Snyder wrote to C. H. Hall that his daughter, who lived in Mrs. Cannel's boarding house, had been eating in public restaurants—sometimes with men. Furthermore, men as well as women lived in the rooming house, which made it difficult for Mrs. Cannel to supervise Hall's daughter. Snyder told Hall that he believed it was his responsibility "to place the facts before" him.[109]

Students faced disciplinary action because they did not make the grade academically or because of cheating. Too much interest in socializing caused some to neglect their studies to such an extent that their behavior was seen as "neither conducive to [their] own welfare nor to the good of the institution." Once college officials came to such a conclusion, they sent the student home.[110] For a time during the first two decades of the twentieth century, the Committee on Doubtful Cases determined whether or not students who were doing poorly in the classroom could re-enroll at the beginning of a new term.[111] A student returned home to Detroit in 1907 after he had been found guilty of copying another student's work into his chemistry notebook and passing it off as his own to Professor Frank Kedzie.[112] The faculty suspended a young man for using notes while writing an exam in his botany class in April 1911.[113] All of these students suffered embarrassment or shame when their offenses became known publicly. From the college's perspective, this was not necessarily bad, for the faculty hoped that disciplined students might learn from their errant ways and make amends.

Students stumbled in their studies for a number of reasons, some of which were beyond their control. Sometimes poor teaching led to lack of success in the classroom. After 77 out of 693 students failed courses in the English department during winter term 1905, Snyder concluded "that the instruction is not what it should be or that the standard maintained is too high for the class."[114] Mrs. Minnie Hendrick, an instructor in history, provided a variety of explanations to the Committee on Doubtful Cases for some of her students receiving grades of D or F during winter term 1908: "poor recitations, poor examinations," "long absence and poor work," "unable to work on account of eyes, consequently-poor work," and "absent nearly half the term, failure in consequence."[115] Illness, jobs, and lack of academic preparation also accounted for failures.[116] Frank Kedzie blamed a failure rate of 15.7 percent among women students in his Chemistry I class, fall term 1911, on poor attendance.[117] Fortunately, many more students succeeded academically than failed.

RELIGION

The practice of religion at Michigan Agricultural College fits into the context of religious life in the United States during the second half of the nineteenth and early twentieth centuries. Evangelical Protestant Christianity had enormous influence in American society throughout the nineteenth century, but science, new social thinking, and immigration led to increasing religious diversity before the outbreak of World War I. New ideas coming from scientists and clergymen gave impetus to liberal Protestant Christianity. For instance, many college men and women across the country—faculty and students—found Charles Darwin's theory of evolution and Walter Rauschenbusch's social gospel to be more compatible with their beliefs, studies, research, and aspirations than evangelical Protestant thinking. Liberals questioned, among other things, the inerrancy of the Bible and its account of creation, and they put forth a theology that allowed for the perfectibility of human beings in this world—not just the next world. Immigration also expanded religious pluralism in the United States. After the Civil War, Roman Catholics, Jews, and Orthodox Christians came in great numbers to America from eastern and southern Europe. During M.A.C.'s first seventy years, relatively few Catholics, Orthodox Christians, or Jews enrolled in the college. Although there seems to have been no overt prejudice against believers of non-Protestant faiths, M.A.C. made little provision for them to practice their religion in organized ways on campus. Religious life at M.A.C. reflected the large presence of Protestant Christianity in America.[118]

Protestant Christianity shaped religious life at Michigan Agricultural College. Professors led daily chapel exercises until 1911, and local clergy conducted Sunday worship services on campus. The Young Men's Christian Association (organized in 1881, succeeding the College Christian Union, organized in 1871) and the Young Women's Christian Association (organized in 1896) nurtured students in their faith, connected them to the world mission movement, and challenged them to put their beliefs into practice. The YMCA specified that its policy was not "towards evangelistic work" on campus, but the YMCA actively supported foreign missions.[119] The formation of Peoples Church in 1907 relieved the college of responsibility for holding formal services and brought together a group of energetic Christians dedicated to serving both M.A.C. and the growing community of East Lansing. The YMCA, YWCA, and Peoples Church sponsored social activities encouraging the integration of students into the life of the town.

M.A.C.'s remote location forced early presidents and faculty to provide opportunities for worship in particular for Protestant and unchurched students who made up the majority of the student body. Although they were free to attend churches in Lansing, the three-mile walk kept most students on campus on Sundays, where they listened to sermons preached by local ministers or, on occasion, by faculty members, some of whom had studied theology. Starting in 1894, streetcar service to Lansing made it much easier for students to attend worship services in the city. Daily chapel exercises, which were required until about 1885, included scripture reading and prayers and announcements from the faculty.[120] By the 1870s many students had grown lax in attending both chapel and Sunday services. Hoping to revitalize religious functions, the *College Speculum* in August 1885 called for the creation of a "faithful choir to lead in the singing" and the employment of an "inspiring preacher."[121] Neither suggestion seems to have come to fruition. Starting in 1894, a "College Sunday School" focusing on Bible study, enabled Professors Robert C. Kedzie, Clinton D. Smith, Walter B. Barrows, and Wilbur O. Hedrick to teach "the most elementary facts connected with or contained in the Bible" to students wishing to learn about Biblical teachings.[122] For many students these classes may have been their introduction to the Bible, for few students appear to have had been affiliated with any church. In 1906, out of 766 full-time students only 130 claimed a church membership, with another 115 stating a denominational preference. In 1915, out 1,655 full-time students, 264 were church members and 165 indicated a preference (see table 8).[123] Roman Catholics went to St. Mary's Church, which had been established in 1866, in Lansing.[124] A group of Christian Science believers, both men and women, met weekly in the Women's Building parlor on Thursday nights.[125]

Table 8. Church Membership, 1915

DENOMINATION	MEMBERS	PREFERENCE	TOTAL
Baptist	27	11	38
Catholic	40	2	42
Christian	3	—	3
Christian Science	1	5	6
Church of God	1	—	1
Congregational	36	36	72
Episcopal	22	13	35
Evangelical	1	2	3
German Evangelical	1	—	1
Hebrew	2	4	6
Lutheran	16	1	17
Methodist Episcopal	55	57	112
Methodist Protestant	1	1	2
Presbyterian	52	30	82
Quaker	1	—	1
Reformed	1	—	1
United Brethren	3	1	4
Unitarian	1	1	2
Universalist	—	1	1

The remainder gave no preference.

Source: *Fifty-Fourth Annual Report of the Secretary of the State Board of Agriculture of the State of Michigan . . . from July 1, 1914, to June 30, 1915* (Lansing: Wynkoop, Hallenbeck Crawford Co., 1915), 150.

The YMCA and YWCA worked together, stimulating interest in Christianity and encouraging the spiritual growth of Protestant students.[126] These two organizations, which were part of large, active national movements, used Bible studies as a primary vehicle to teach young people Protestant Christianity. Although liberal and fundamentalist branches of Protestantism fought intense battles over the inerrancy and divine inspiration of the Bible, they all tended to believe that the scriptures contained the essence (although they did not often agree on the substance) of Christianity. Examining the contents of the Bible fit in nicely with the non-denominational Protestant

Many female students listened to speakers and participated in Bible studies and social gatherings sponsored by the Young Women's Christian Association. The YWCA cabinet posed for this picture on the steps of the Women's Building in 1921. Courtesy of Michigan State University Archives and Historical Collections (College of Human Ecology, Young Women's Christian Association).

tradition that had existed at M.A.C. since 1857. Christian students could search their scriptures for spiritual inspiration and edification together even if they differed on interpretations of the history and meaning of the text. Both the YWCA and the YMCA held classes on campus. During winter term in 1901 the YWCA offered "Women of the Bible" and "Christ in Old Testament" and the YMCA offered "Studies in the Acts and Epistles" and four sections of "Studies in the Life of Christ."[127]

The YMCA and the YWCA encouraged students to become part of the worldwide missionary movement and not rest comfortably with their Bibles on their laps. Their 1903 missionary committees broadcast activities of missions, gave financial support for missions, and generally stimulated interest in missions.[128] Missionaries and their supporters spoke to separate and union meetings of the YMCA and YWCA, discussing mission operations and opportunities in different countries and encouraging students to consider committing themselves to missionary work. M.A.C. students, along with students from other colleges, also attended missionary conferences out of town.[129]

Students who became missionaries brought together their commitment to follow the teachings of Jesus Christ with the land-grant philosophy of serving people where they lived. The social gospel of liberal Protestantism admonished Christians to work diligently to improve living conditions of other people. Missionary work enabled Protestant Christians to teach people in distant lands to better their lives, to try to convert them to Protestant Christianity, and to apply in practical ways things they learned at M.A.C. Graduates of Michigan Agricultural College traveled to the four corners of the earth serving their Lord and their fellow human beings.

Following his graduation in 1902, Dillman S. Bullock taught agriculture, horticulture, and carpentry to the Patagonian Indians of southern Chile under the auspices of the American Missionary Society of the Church of England. After seven or eight years in Chile, he served as principal of an agricultural school in Marinette, Wisconsin, and he later worked for the Bureau of Markets, USDA. In 1923 Bullock was back in South America, where he studied the stock, sugar, and cotton industries in Peru before resigning his position with the USDA. He and his wife then returned to Chile to teach in the Methodist Episcopal Church's agricultural mission in Angol.[130] Ralph W. Powell, '11, and his wife, Maude E. Nason, '13, went to China, where he taught physics and engineering at the mission campus of Yale-in-China.[131] His sister, Alice Powell, '17, followed them to China, where she served as a teacher of English, American history, and nature study at the Union Girls' High School in Hangchow, Cheking, in 1922.[132] Alice Smallegan, '16, went as a missionary to Rannipettai, South India, and in 1921 she studied "the Indian language" at a school in Kodia Kanal.[133] Missionaries Henry A. Jessop, '16, and Bernice Hales Jessop, '17, introduced some contemporary technology to their Zulu students in South Africa when they played a gramophone for them in their "rude school rooms."[134]

Peoples Church, dedicated in 1911, stood east of Abbott Road on Grand River Avenue. A few years later an auditorium was built next to the church to accommodate many of the church's activities as the congregation ministered to students and the growing community of East Lansing. It was dubbed "McCune's Garage" in honor of Pastor Newell A. McCune, '01, who served the church from 1917 until 1949. In 1926 Peoples Church moved into a new structure west of Abbott Road on Grand River Avenue. Courtesy of Michigan State University Archives and Historical Collections (Michigan, East Lansing, Churches).

PEOPLES CHURCH

A momentous event occurred at the college armory on Sunday, 8 December 1907, when Baptists, Congregationalists, Methodists, and Presbyterians organized Peoples Church in East Lansing (which had been incorporated as a city in the same year). They chose to unite because no denomination had enough adherents to form a congregation, and no group had sufficient funds to go its own way. Ray Stannard Baker, who grew up a Presbyterian but preferred Quaker worship services, recalled a meeting that paved the way for the church's establishment: "Baptists, Congregationalists, Methodists, Presbyterians, Unitarians, and others attended our meeting and of those present all seem satisfied on one point: that the town was too small for a church for each denomination. We must all get together on a community church: a kind of old fashioned meetinghouse, with emphasis upon its social aspects."[135] On 22 October 1911, the Bemis Violin Quartette and the College Chorus performed at the dedication of the

church's edifice located across the street from campus on Grand River Avenue a few doors east of Abbott Road.[136]

Seven days a week, students and nonstudents used the facility. The 28 January 1913 issue of the *M.A.C. Record* reported: "The total number of persons visiting the People's Church at East Lansing for the week ending Jan. 18 was 2080. In the Bible School more than 100 short course men have reported each Sunday and on Tuesday evenings the men enjoy basketball and other games. The church as a social center is certainly fulfilling its mission in E. Lansing."[137] The town and the college met at Peoples Church every day.

The church hosted YMCA functions, including activities designed to integrate several hundred short course students into the community during their eight weeks on campus. Most of them were young men who came from farms in January to learn crafts such as making butter or cheese or to improve their general knowledge of agriculture. Without the effort put forth by the community's religious organizations, these students faced a potentially lonely stay in East Lansing. On Saturday evenings as many as two hundred of them came to the church to sing, hear speakers, and socialize, and many of them came back on Sunday mornings to attend Sunday school. Throughout the week students studied in the church's reading rooms.[138]

The church's commitment to student ministry deepened in April 1919 when it employed the Presbyterian Rev. O. B. Behrens as pastor for college students. A graduate of McCormick Theological Seminary,

Many students found spiritual and social fellowship at the Peoples Church. Several hundred people were in attendance on Wednesday, 5 October 1921 for one of the church's frequent functions held in "McCune's Garage." Courtesy of Michigan State University Archives and Historical Collections (Student Activities, Folder 1).

Behrens had been a minister since 1912, had served as a chaplain in the navy during World War I, and he and his wife, Margaret, had two young daughters and a son. He worked closely with the YMCA and the YWCA to introduce students into the college community through a range of ministries. The YMCA sent upperclassmen to meet new students when they arrived by train in Lansing, and it helped many find part-time jobs. The church aided students in finding places to live, a valuable service given the dearth of dormitory rooms on campus after fire consumed Williams Hall on 1 January 1919. At the start of the school year a thousand or more students gathered in the college gymnasium to drink cider, eat doughnuts, and make new friends at a reception. The church welcomed many students into its respective organizations for men and women that met together socially every other Saturday night. Residents of East Lansing and students watched "moving pictures" or rubbed shoulders at Community Night in the church auditorium. An association for married students enabled them to help each other as they faced challenges that differed from their single classmates. The Cosmopolitan Club, an organization sponsored by M.A.C., served a similar function for international students. Behrens organized a host of Sunday School classes, Bible studies, and special courses that focused on Christianity. Peoples Church, the YMCA, and the YWCA worked tirelessly to meet spiritual and social needs of M.A.C. students. In May 1926 the congregation dedicated its new building situated a block west of its first structure.[139]

HEALTH

From its first day, Michigan Agricultural College struggled to provide health care for students and to devise strategies to prevent the outbreak and spread of communicable diseases, which were frequent causes of serious illness and death. Mosquitoes breeding in swamps in the Lansing area infected many people with malaria, causing them to suffer chills and high fevers that could result in death. Initially, the closest doctor lived in Lansing, so students suffering from malaria or ague, as it was commonly called, received medicine from chemistry professor Lewis R. Fisk. Manly Miles, a trained physician who joined the faculty in 1861, was the first college doctor. The young students lived in very close quarters, which facilitated the spread of the microorganisms that caused mumps, scarlet fever, measles, smallpox, and diphtheria. Primitive sanitation facilities and careless disposal of human and other organic wastes (throwing food scraps out of dormitory windows, for example) created breeding grounds for potentially deadly microbes, especially those causing typhoid fever. Recurrent epidemics challenged administrators to improve water purity and sewage systems and to build and staff health care facilities. On 1 October 1920 M.A.C. established a health service to provide comprehensive medical care to the student body.[140]

Outbreaks of diphtheria and measles in the spring of 1892 illustrated the complexity of the prevention, containment, and treatment of disease on campus. After Cecil John Barnum, '94, complained of a sore throat, Dr. Watson from Lansing examined him in his dormitory room on two successive days, determining that Barnum did not have diphtheria. Barnum had been exposed to sore throats and other ailments while he taught school during the winter before returning to campus for the spring term. Confident that he was suffering from nothing more serious than a sore throat, Barnum attended the Hesperian Society's dance. Four days later two of his dancing partners came down with diphtheria. One of these young women then transmitted the contagion to her family in Lansing. Fortunately, it appears that all of these people recovered. Within a few days three male students, who lived in Barnum's dormitory, also fell prey to diphtheria. The attending physician sent them into quarantine, where they recuperated in a rented farm house about one-half mile east of campus that served as the college hospital.[141] Soon after diphtheria released its hold, a measles epidemic took the life of Albert Blanding, '91–'92, and sent twenty-five students to isolation rooms in the Veterinary Laboratory and Abbot Hall.[142]

Controversy that arose over the effectiveness of the college's efforts to curb the spread of diphtheria led to the construction of a hospital on campus. Barnum, it appears, never had diphtheria, but he lived in a room that had been occupied by a student who had died from the disease late in the fall term. Dr. J. H. Wellings of Lansing had supervised the efforts to rid the dormitory of germs that had caused diphtheria. He disinfected rooms by fumigating them with sulphur, cleaning, scraping and whitewashing the walls, and repainting the

Professor Charles E. Marshall strengthened research in bacteriology during his sixteen-year tenure at M.A.C. He came in 1896 as an assistant bacteriologist in the Experiment Station, received appointment as assistant professor of bacteriology and hygiene in 1900, and assumed the professorship of bacteriology and hygiene in 1902. In 1912 Marshall moved to the Massachusetts Agricultural College, where he worked until his death in 1927. Courtesy of Michigan State University Archives and Historical Collections (Michigan State University, People, Marshall, Charles E., no. 1863).

woodwork. Highly critical stories in the *Lansing Republican* and the *Detroit Journal* castigated college officials for letting Barnum attend the dance rather than isolating him from other people. Calls for improvements in health care at M.A.C. grew out of the debate between several prominent doctors over the merits of Barnum's case. President Oscar Clute bemoaned the fact M.A.C. did not have a hospital building except a rented house off campus, which was staffed by a married couple who had no medical training. Doctors Henry B. Baker, secretary of the state board of health, Rush J. Shank, and J. H. Wellings all recommended that a hospital, outfitted with proper equipment and supplies, be erected to facilitate care for sick students. Patients afflicted with contagious diseases could be isolated from others and treated there by medical professionals. Consequently, M.A.C. built a seven-room hospital with two bathrooms in 1894. In 1909, four cottages with isolation wards were put up behind the Bacteriology Laboratory for patients afflicted with contagious diseases.[143]

Typhoid fever afflicted students frequently. Efforts to combat it required a constant vigilance over the college's and community's water supply and methods of sewage disposal, and close communication between officials and patients' family members. Bacteriology professor Charles E. Marshall ran scores of tests annually on the institution's water system between 1896 and 1912, trying to ensure the water's purity.[144] In Collegeville, the developing community north of campus, residents discharged sewage into drains designed to carry off surface water. The resulting pollution was believed to have caused a case of typhoid fever in 1904. M.A.C. demanded that property owners clean up the mess.[145] Lydia Zae Northrup led an investigation to find the source of sewage that polluted the water supply in 1913 that forced many students to drink "distilled aqua."[146] The dread of typhoid fever instigated the construction of new and improved sewage disposal systems.

Another outbreak of typhoid fever showed the disease to be insidious, with deadly consequences. Keeping the campus water pure did not prevent contagions from penetrating M.A.C.'s food supply. On 12 April 1911, Snyder wrote these painful words to Kilian Klement in Detroit: "We were very sorry indeed to hear of your son's death. He was a good, earnest boy, and had he been spared, no doubt would have made a very useful citizen."[147] Three days later Snyder informed James Knowles that his son, Bernard, along with six or seven others,

A student conducting an experiment in the Bacteriology Laboratory (opened in 1903 and later renamed Marshall Hall) in 1905. Courtesy of Michigan State University Archives and Historical Collections (Michigan State University, Buildings, Marshall Hall, Interior).

had become ill with typhoid fever and was recuperating in the college hospital under the watchful eye of nurse Rowena Ketchum and Dr. Oscar Bruegel.[148] Snyder assured Fred Post's father that he had visited his son, who was also suffering from typhoid fever, daily, and found him "to be in good spirits and getting along alright" even though occasionally he ran a high fever.[149] These young men had all got sick after drinking milk produced by cows that had consumed well water contaminated by sewage on a farm south of campus.[150]

Fortunately, smallpox could be prevented by inoculation. In 1902, fearing that an epidemic of it in Michigan had reached East Lansing, Dr. Henry B. Baker encouraged college officials to be careful not to misdiagnose a sick man's fever and aches as being chicken pox when it was more likely to be smallpox. Baker also told them to vaccinate everyone on campus and not rely on isolation as a means to prevent the spread of the disease.[151] As smallpox raged across Michigan in November 1910, the faculty cancelled Thanksgiving recess and ordered all students to remain at the college during the holiday and to be vaccinated immediately. The faculty, in conjunction with state and local health officials, hoped to keep students from picking up the disease while at home and bringing it back to East Lansing, where no cases of smallpox had been reported yet.[152]

Effective control of infectious diseases on campus required sick students and those who thought they had been exposed to follow the prescribed procedure of remaining in their rooms until examined by a physician. Self initiated quarantines thus reduced the opportunities for ill students to infect others. In March 1917 President Kedzie became alarmed when two students who were already exhibiting the symptoms of scarlet fever went to the office of the city health officer, Dr. Bruegel, for examination rather than stay in their rooms. Kedzie issued a "Scarlet Fever Bulletin" ordering students who thought that they had come down with scarlet fever not to leave their rooms, or go to Cottage D, which had been set up to receive patients who suspected that they had a communicable disease. In order for a quarantine to be effective, patients, medical workers, and unexposed persons all had to obey the rules.[153]

On 10 October 1918, Spanish influenza, which American soldiers returning from the war in Europe had brought to the United States, swept across the country, striking M.A.C. with such intensity that it necessitated a mobilization of most college and community resources to combat its deadly impact. The campus overflowed with more than 1,000 members of the Student Army Training Corps, who had come to receive training as auto mechanics, truck drivers, and in other skills needed by the military for the war. They lived in temporary barracks constructed east of the Horticultural Laboratory and in Agriculture Hall and Olds Hall of Engineering. During the epidemic, college officials set up makeshift hospital wards in one of the barracks for between 230 and 240 patients. Before the scourge abated, eighteen men died, but a six-week long quarantine kept the disease from infecting most women students, with only two mild cases being reported. At least three more deaths occurred in East Lansing. Senior domestic science students played a key role in the battle against the flu by operating a "dietetic kitchen" in the basement of the Horticultural Laboratory. They made meals for recovering patients under the supervision of Professors Mary E. Edmonds, Louise Clemens, and Edna Garvin.[154] In December, a new bout of influenza sent twenty-six women to bed and took the life of Rose M. Taylor, instructor in botany.[155] The epidemic also caused the closure of public schools in East Lansing.[156]

Women living in East Lansing worked tirelessly to ease the pain and suffering of patients and their families during the influenza epidemic. Mrs. Margaret Holt, hostess for Peoples Church, found rooms for family members who came to East Lansing to be near ailing students. The East Lansing Red Cross set up a sewing room in the East Lansing State Bank, where women from the community and college sewed an incredible list of items including "315 sheets, 293 pillow cases, 106 pajama suits, 72 pneumonia jackets, 623 handkerchiefs, 128 cubicals, 45 surgeons' robes, 324 masks, 15 surgeons' coats and 262 utility bags." The Red Cross also supplied pillows, towels, jelly, grape juice, and other furnishings and supplies to the hospital.[157]

M.A.C. took a major step forward to ensure the well-being of its students by creating the Health Service on 1 October 1920. Prior to this date the bacteriology department had assumed the role of health protector on campus. For a long time Dr. Oscar Bruegel, in his role as the health officer for the City of East Lansing, had provided many services to students, especially during epidemics.[158] But a more proactive approach to student health was needed. Finally, the State Board of Agriculture acted upon a M.A.C. faculty committee report recommending the establishment of an "adequate health service."[159] The new entity, in conjunction with the Department of Physical Training, arranged for physical examinations of all incoming students and conducted lectures on "personal hygiene, first aid, rural sanitation, and sex hygiene." Each student paid a fee of $1.50 per term, and received in return medical treatment and advice, including hospital care, all on campus, for no further cost. Services were available to both men and women, although the lack of a female doctor limited women's access. All students received smallpox vaccinations unless they had been inoculated previously. Dr. Clyde Reynolds served as the first director of the Health Service, and Dr. Richard M. Olin assumed that position in 1925.[160]

THE ALUMNI ASSOCIATION

Once a young woman or man enrolled in Michigan Agricultural College, she or he became a lifetime member of the institution, and the

The *Wolverine*, the college yearbook, frequently recognized "prominent alumni." Three of the alumni honored in the 1926 Wolverine were Eduard Christian Lindeman, Catherine E. Koch, and Anne (Anna) Cowles Herr. Courtesy of Michigan State University Archives and Historical Collections (*Wolverine*, 1926, 307).

PROMINENT ALUMNI OF M. S. C.

EDUARD CHRISTIAN LINDEMAN, '11, has become one of the prominent sociologists and educators of the United States today. Among his many and varied activities are included editorships on the New Republic and the Journal of Social Forces, as well as acting as contributor to the Survey, Saturday Review of Literature, Review of Reviews, etc., a trusteeship on the National Child Labor Committee, member of the Research Committee of the Federal Council of Churches and acting as Field Secretary for the American Country Life Association, along with active membership in a number of clubs and associations dealing with the furtherance of science and education. He is the author of several books and pamphlets and is known as a lecturer on social psychology and philosophy at New York schools of social work and research, besides having lectured at such colleges as Harvard, Columbia University, Bryn Mawr, Ohio State and others.

CATHERINE E. KOCH, of the class of '09, took graduate work at Cornell and later attended Vassar. Her position as Assistant Professor of Landscape Architecture in the Botany Department of Smith College testifies to the success of her work there.

ANNE COWLES HERR was graduated from Michigan State in '15, and was made assistant in the Home Economics Extension Department. Later she became state leader of the Boys' and Girls' Club work. In 1919 she entered community center work and was an organizer for community councils in New York. Then she became assistant to the National Director of the Red Cross with headquarters in Washington, and did Red Cross work in the Philippine Islands.

Page Three Hundred Seven

college worked hard to create a spirit of kinship and unity among alumni to promote the school. At commencement in November 1868 old and new graduates formed an alumni organization, enabling them and future graduates and nongraduates to participate in M.A.C.'s expanding presence in Michigan, the United States, and the world. The alumni association held reunions for all alumni every three years from 1870 through 1913.[161] In addition, the college sent out letters asking alumni for information about their vocations and whereabouts so that this data could be included in issues of the *Catalogue of Officers and Graduates* that were published periodically. In 1913 the Michigan Agricultural College Association took over the editing and production of the *M.A.C. Record* under the direction of the association's secretary, George C. Sheffield, '12.[162] The association began hosting annual get-togethers, beginning on 23 June 1914, when over 200 alumni and their families returned to campus.[163]

The secretary took on responsibility for nurturing and forming new local associations to strengthen existing bonds and to build new relationships between alumni and M.A.C. Organizations thrived in some major cities outside of Michigan. Fifty-one graduates put their names on the roll for the Washington, D.C., association in 1902.[164] Alumni attending the 7 March 1914 meeting of the Chicago association shook hands with President Snyder, Librarian Linda Landon, Dean George Bissell, Grounds Superintendent Thomas Gunson, and Professors Frank Kedzie, Herman Vedder, W. O. Hedrick, and Harry J. Eustace before they sat down to dinner at the Hamilton Club. Next, Coach John Macklin waived the green and white as he relived the triumphs of his undefeated football team, especially the victories over Michigan and Wisconsin. With their spirits lifted, the alumni danced away the evening.[165] In 1914 approximately 700 alumni took part in association meetings in Lansing, Flint, Saginaw, Bay City, Grand Rapids, Jackson, and Detroit.[166] The organization of alumni never stopped. For instance, readers of the *Record* were instructed to send the names of alumnae living in the Detroit area to Julia Grant, '05, who was putting together a women's club of alumnae for Detroit in 1924.[167]

M.A.C. courted its alumni, hoping to enlist their financial support by playing on their fond recollections of their days on campus and the value they placed on their education, their classmates, and the professors who had taught them. Since the *M.A.C. Record* kept them informed of changes and advances in courses of study, victories in debate and athletics, construction of new buildings, and achievements of M.A.C. graduates, the alumni felt a special affinity with the college long after they had graduated, often for the rest of their lives. So it appeared only natural to ask alumni to encourage young people to attend M.A.C., to cheer on the Aggies, or to make a financial contribution to their alma mater.

The M.A.C. Association made the construction of the Union Memorial Building its first major funding effort. The project required alumni, students, faculty, staff, administration, and the governor (an alumnus of the University of Michigan) to work together to bring the Union into reality. Initially, the association had hoped to restore College Hall for the Union, but that plan went up in dust when its plank footings, hollow walls, and soft bricks gave way and the grand dame of M.A.C. collapsed in August 1918. Five years later, on 16 June 1923, W. K. Prudden, '78, broke ground for the new structure, an honor that was accorded to him because he was the largest single donor to the project. The association's president, E. W. Ranney, '00, summed up the significance of the Union Memorial, designed by Pond and Pond of Chicago, in his remarks before an audience of about 1,000 people gathered for the ceremony on Alumni Day: "This is an occasion toward which the M.A.C. association has been looking over a period of five years. . . . It will be a memorial to the M.A.C. soldiers and sailors who fought for their country, it will be a mark of alumni loyalty and appreciation for their college and it will symbolize our sentiment by serving constantly."[168] Lack of money threatened the project's completion even though $130,000 was given by faculty, students, and alumni.

Students, with some help from the faculty, dug the hole for the Union's foundation during one week in November 1923. Professor W. O. Hedrick and Robert J. McCarthy, alumni secretary, instigated this endeavor to keep up interest in the Union and to save some money. Writing in the *Record,* a reporter for the *Holcad* gave the following analysis of this monumental effort:

> Although it required 67,500,000 pounds of energy to remove the 5,000 yards of dirt for the foundation of the M.A.C. Union Memorial Building nearly two per cent of this work was made possible through the efforts of the co-eds. Figures compiled on the activities of Excavation Week show that the girls were responsible for 1,049,000 foot pounds of work produced. Not because they wielded shovels but because they supplied nourishment for the workers.
>
> · · ·
>
> In all, 2000 frankfurts and rolls, one barrel of cider, 14 bushels of apples, 150 gallons of coffee, besides the 70 quarts of cream and 30 pounds of sugar essential to making up the coffee were served to the diggers.
>
> The 5,000 yards of dirt removed weighed approximately 13,500,000 pounds. If there was a team for every load of dirt removed and these were placed end to end they would make a line over eleven miles long.[169]

Sweat, sugar, and hot dogs did not pay the bills. But Governor Alex J. Groesbeck came to the rescue by working with the legislature to authorize the sale of bonds from funds in the state's sinking fund to pay construction costs of the building.[170]

The construction of the Memorial Union symbolized the evolution of Michigan Agricultural College into Michigan State College of Agriculture and Applied Science in 1925. The sons and daughters of the college had spent several of the most formative years of their lives on its campus. They had trained to take up a vast array of vocations in a world that faced challenges and crises presented by new, often dangerous, political philosophies, changing technology, and severe economic disruptions. Even as the Union memorialized those who had served and died in the Great War, it looked to the future by providing a facility where students could learn more about the world and its problems—and to help them discern their role in it.

The construction of the Union Memorial building saw the student body, faculty, and alumni work together in a way that not only produced a facility, but that also strengthened the bonds between the different elements of the M.A.C community. In this picture students are digging up the soil to prepare the ground for the Union Memorial's foundation. The *M.A.C. Record*, published on 25 February 1924, set forth a lofty goal for the Union: "The building is designed with the thought of serving men and women alike. The Union includes all students in its membership and the alumna has contributed generously to the fund. It is important that this section of the college and the alumni body be made as welcome as the men." Courtesy of Michigan State University Archives and Historical Collections (Michigan State University, Buildings, Union Building-Construction).

African-American Students

THE STORY OF THE EARLY AFRICAN-AMERICAN STUDENTS AT MICHI-
gan Agricultural College reveals courageous efforts by young
men and women, burdened by the brutal legacy of slavery, to
find a place in American society where racial discrimination
and segregation put severe limitations on their liberties and opportunities. Jim Crow laws forced the physical separation of African Americans from other Americans by determining where they could live,
work, worship, eat, sleep, and even be buried. Laws forbidding racial
intermarriage institutionalized racial separation, and other statutes
codified the principle that any person having any African ancestry
was black, even if he or she had one white parent. No matter how
benign racial attitudes at M.A.C. appear to have been, it is important
to remember that most white Americans' attitudes toward African
Americans were informed and shaped by laws made by white legislators and the prejudices that led to their enactment.[1] By 1930 only
eleven African Americans held degrees from Michigan State College,
but the small size of this number should not be allowed to obscure the
significance of what it measures. Evidence suggests that the number
of black students at the college was greater than first meets the eye,
though the records unveil their identity and the details of their lives
only grudgingly.

Even though the college admitted African-American students, it
also quietly embodied the attitudes about race that permeated
American society—North and South. M.A.C.'s inclusion of a few
African-American students was premised on a paradoxical concept—segregated integration. This contradiction in terms both
reflected and reinforced the prevailing public attitudes in Michigan
and the United States, which allowed, even encouraged, racial integration in some spheres of public and private life, but not in others.

Black students attended classes with white students, sat on the same platforms at graduation, and found their pictures and names included alongside their classmates' in the yearbooks. But black women did not live in college housing, and it is uncertain whether the black men were welcome in the dormitories. Literary societies, sororities, and fraternities were closed to African Americans, although black male students did take part in athletics and in some social and religious organizations.

The college, nevertheless, did provide the opportunity for a few African Americans to receive land-grant educations, preparing them for a life of service, most often teaching at schools attended only by African Americans. In a small way, M.A.C. fulfilled the intention of the Second Morrill Act (1890) by training young African-American men and women to teach in the growing system of black schools scattered throughout the South and elsewhere. Positions held by black M.A.C. alumni in 1930 graphically illustrate the extent of racial segregation in American education. The commitment of the black graduates to educate other African Americans no doubt contributed to the betterment of many lives, but the relegation of black alumni to teaching in institutions reserved only for African Americans shows how little American democracy extended to black citizens. Black students at M.A.C. simply did not have the same opportunities as their white classmates, even after they completed their studies.

In order to appreciate fully the African-American experience at M.A.C., it is essential to recognize and understand the racial climate on campus. Beneath the official veneer of equality, it is clear that M.A.C., an institution populated primarily by European Americans and run by them, exhibited the racial prejudice that saturated American life. President Snyder's assertion in 1905 that "there is no antipathy in this institution against colored students" seems to have been accurate, but he viewed African Americans differently than others because of their race.[2] When he hired Mrs. Evans, a black woman, to manage the kitchen in the Women's Building, he made it a point to call her "a colored woman," and went on to say: "She has a husband who will work with her, and they will hire all the other help; it is all to be *chocolate* [emphasis added] in color."[3] A few African Americans worked for M.A.C. in service positions at the time, but there is no evidence of any black faculty members or administrators. The absence of

overt antipathy did not prevent the evils of prejudice and social segregation from occupying prominent places on campus.

The memory and pain of slavery, segregation, and the need to
find ways to improve their place in American life motivated African
Americans to build up their own communities. Although blacks
formed a small minority in Michigan, their numbers expanded
significantly between 1890 and 1920. The population of African
Americans in the state grew slowly from 15,000 in 1890 to 17,000 in
1910, but it leapt to 60,000 by 1920, at which point it represented
about 1.64 percent of the total number of people living in Michigan.
As the twentieth century began to unfold, African Americans in
Michigan, especially in Detroit, experienced the hardening of racial
prejudice. European immigrants took jobs in manufacturing that had
once been filled by blacks, and racial segregation confined African
Americans to separate neighborhoods. Racism and the decline of the
influence of the black elite in white society in Detroit led to the rise of
a black middle class, which worked to create a viable African-American society.[4] In Lansing the African Methodist Episcopal (AME)
Church brought together blacks who had come from many different
places. Pastor Joseph W. Jarvis articulated the challenge faced by "the
colored man": "he should embrace all the opportunities that were
open to him; that he should help mould sentiment for the betterment
of more than three million of souls—every little bit helps."[5] Education
of black youth was essential to the development of black communities
throughout the country. Principles espoused at the local AME Church
undoubtedly motivated William O. Thompson, '04, Myrtle Craig, '07,
Margaret E. Collins, '24, and Mabel Jewel Lucas, '27, all of whom are
known to have been part of this church, and other African-American
students at M.A.C. to use their college education for the advancement
of their own people. In response to the overwhelming need for black
instructors, at least nine M.A.C. alumni answered the call to teach in
schools that had been established for black students elsewhere in the
United States.

William O. Thompson's life after leaving M.A.C. demonstrates
that a college education did not open doors for a black man in a society controlled by whites. His experience also says something about
the racial discrimination in Lansing during the early decades of the
twentieth century. Born in Kentucky in 1875 or 1876, Thompson

came to M.A.C. in 1899 from Indianapolis and graduated with the class of 1904. Neither his name nor his picture appears with the senior class in the college yearbook for 1904 (several white students were also not included). Furthermore, his name does not appear on the membership lists of any student organizations. His name does appear with his freshmen class in the 1900 edition of the *Wolverine*.[6]

On 1 January 1906 Thompson started teaching stock breeding at Tuskegee Institute in Alabama; he also managed the school's livery stable of 112 horses and mules.[7] It is not known how long he remained at Tuskegee, but he moved back to Lansing by 1911, when he took a job at Wynkoop Hallenbeck Crawford Co., the state printer.[8] Thompson had been active in the Lansing AME Church during his student days, and in all likelihood his involvement with the Lansing black community influenced his decision to return. Thompson's employment record exhibited a lack of permanency until about 1916, when the Reo auto manufacturing company hired him to be a "ganitor."[9] Thompson, even though he held a bachelor of science degree from M.A.C., could only land a menial job in a firm that hired almost exclusively white workers to make its automobiles.[10] From the time he returned to Lansing and up to 1916 he had also worked a year at M.A.C., in an unknown capacity, and at the Seager Engine Works.[11]

Thompson held leadership positions in the African Methodist Episcopal Church and served the people of that congregation until his death on 7 February 1923.[12] He married Elsie Merchant, a cook at Scheidt's Restaurant in Lansing, on 30 November 1916.[13] The couple lived on West Ionia Street, in a racially mixed neighborhood in Lansing.[14] They had three children [William?] Horton, Milton, and Violet. The 1920 census identifies both William's and Elsie's race as "Mu" (mulatto) with "B" (black) having been written in and then crossed out by the census taker.[15] The *M.A.C. Record*'s obituary for Thompson gives us a tantalizing insight into his demeanor with the words "and his employers found his marked ability handicapped by a reserved attitude."[16] Thompson found himself at home among African Americans. In the workplace his white classmates and their friends did not make room for him in front of one of their machines on the shop floor. His role was to sweep up the debris.

Preparing herself for a teaching career, Myrtle Craig proved that an African-American woman could earn a degree at M.A.C. In 1902 her father, Stephen, insisted that Myrtle, originally from Adrian, Michigan, enroll at M.A.C., rather than at the University of Michigan. She had graduated from George R. Smith College, a small school for black youth in Sedalia, Missouri, and had taught for a year in an elementary school in Missouri.[17] Since her father worked as a messenger for the secretary of state at the Capitol in Lansing, Myrtle kept in touch with him during her time at M.A.C. Like many of her white classmates, Craig had to work her way through college, but she either was not allowed or could not afford to live in the Women's Building. First she lived with the family of Addison M. Brown, secretary to the State Board of Agriculture, where she earned her keep by cooking. During her second year, she boarded with Assistant Professor of Drawing Chace Newman and his family; thereafter she lived in Lansing. She held jobs as a sales clerk in a clothing store and waiting tables. When Craig reminisced in 1972 about her time at M.A.C., she said: "I didn't feel any different because of my race. There were sororities for girls, but I was so poor, I didn't have the time or money for those things."[18] She worshipped and socialized with African Americans at the AME Church in Lansing. Craig and William O. Thompson, along with several others, signed the contract for the church when it purchased its first piano.[19] On Friday, 31 May 1907, Myrtle Craig, a poor, black woman received her diploma from the hand of Theodore Roosevelt, president of the United States—a fitting tribute to this gifted young woman.[20]

Craig devoted her next forty years to educating African-American youth in Kansas and Missouri. She lived up to the assessment of President Snyder that her "dignity" and "good sense" together with her solid academic training prepared her to "exert a most excellent influence upon young women."[21] From 1907 until 1910 she taught at Western University in Quindaro, Kansas, a school of about two hundred students from Kansas, Missouri, and Oklahoma. The state of Kansas provided funding for the institution's industrial program, and the AME Church paid for the literary department.[22] After leaving Western, Craig taught domestic art for two years at Lincoln Institute, a land-grant school created under the provisions of the 1890 Morrill Act, located in Jefferson City, Missouri.[23] While at Lincoln, she won a blue ribbon and other prizes for work in domestic arts that she exhibited at the Missouri State Fair in 1911.[24] After teaching for three years

Myrtle Craig, '07. Courtesy of Michigan State University Archives and Historical Collections (*Jubilee Wolverine*, 1907, n.p. [n-749]).

at Sumner High School in Kansas City, Kansas, she married the Reverend W. H. Bowen in 1915. Craig sandwiched two more tours of duty at Lincoln (1917–1922 and 1934–1947) around twelve years of teaching elementary school in Fulton, Missouri. She married George H. Mowbray in 1951 and lived in Kansas City, Missouri, at the time of her death on 8 November 1974.[25]

Gideon Edward Smith broke through racial barriers on campus in a highly visible way, but even so, his participation in campus life appears to have been limited by his race. Born in Virginia on 13 July 1889, Smith graduated from Hampton Institute in 1910. Before coming to M.A.C. he studied at Ferris Institute in Big Rapids, Michigan. The first African American to play football at M.A.C., he starred at left tackle for the 1913, 1914, and 1915 Aggies. His intensity magnified his great strength as he blocked and tackled his opponents with a determination that few had ever seen.[26] Like all other black students at the college before 1930, Smith's name and picture are not included in the men's societies' pages in the yearbooks. He belonged, nevertheless, to the Cosmopolitan Club, serving as recording secretary during 1914–15 and as treasurer for the winter term 1916. This club drew together students from other countries and a few Americans committed to the motto "Above all Nations is Humanity."[27] Gideon Smith was a charter member of the M.A.C. Varsity Club when it was organized in 1915 for all upperclassmen who had earned a monogram, or varsity letter. Also in 1915 he was in the Farmers' Club. He received his bachelor of science degree in agriculture in 1916.[28]

Smith applied his education and experience on the gridiron to a lifetime of teaching and coaching African-American youth. Following his last season with the Aggies in 1915, he became one of the first African-American professional football players. He competed for the Canton Bulldogs; Jim Thorpe was one of his teammates, and Knute Rochne one of his opponents.[29] Along with many of his classmates, Smith served in the military during World War I. After his discharge, he taught agriculture at West Virginia Collegiate Institute, in Institute, West Virginia, then at Virginia Normal and Industrial Institute in Petersburg, Virginia, and also at Princess Anne Academy in Princess Anne, Maryland.[30] He went back to Hampton in 1921, to begin his 34-year career there coaching football, teaching physical education, and serving as assistant athletic director.[31] Smith spent a sabbatical as a graduate student at Michigan State College in 1947. He died in Salem, Virginia, on 6 May 1968.

Everett Claudius Yates was the first known African-American student to play in the college's cadet band and orchestra. Yates was born in Lawrenceville, Virginia, on 3 February 1892. Sometime before he enrolled at M.A.C., Yates moved to the Boston area. In 1912 he came from Roxbury, Massachusetts, to study agriculture at M.A.C., and he graduated with Smith in the Class of 1916, with a major in horticulture. Yates, a percussionist, performed in the cadet band, and in September 1915 he played the band's new set of cathedral chimes.[32] Yates rose to the rank of Second Lieutenant in the Corps of Cadets. He was also a member of the orchestra, a small group with about fifteen members who provided music for dances on campus. During his senior year, the orchestra joined with the college chorus to put on productions of Cowen's *Rose and Maiden* for the mid-winter concert and Haydn's *The Seasons* for the commencement concert. Yates also participated in the Horticultural Club and the New England Club.[33]

After graduation, Yates went back to Boston for a year, before going to Nashville, Tennessee, to teach at the Tennessee Agricultural and Industrial State Normal School. This institution, now known as Tennessee State University, was founded in 1912, as a land-grant institution under the 1890 Morrill Act. Yates joined an all-black faculty at the normal school, which enrolled African-American students in grades five through twelve. In late 1922 Yates returned to Boston and began teaching at Rice School, a public elementary school for boys and girls. He earned a M.E. degree from Boston Teachers College in 1931.[34]

It is of interest to note that although Smith and Yates graduated together in the Class of 1916, the school has remembered and celebrated Smith's presence on campus but has virtually forgotten Yates. Until research for this book "discovered" William O. Thompson, Smith alone was recognized as the first black male to have earned a degree at M.A.C.[35] Although one can only speculate as to why Yates has not been accorded the same honor, such consideration can enhance our understanding of the college's perceptions of its early African-American students. Available evidence suggests that both Smith and Yates broke racial barriers on campus when they took part in highly visible activities—Smith in athletics and Yates in music. There is little doubt that when Smith's athletic performances were publicized through newspapers and word of mouth they created a public persona that Yates never had. Being a star tackle on a football team that beat the University of Michigan twice drew far more attention than suiting up with the band to play songs before and after the game. Furthermore, the memory of the Smith's athletic achievements became part of an institutional athletic heritage that has continued to grow by leaps and bounds long after his graduation. Over time his contributions to the college—as an African American—became a great asset to the university as it sought to demonstrate its acceptance of African Americans. Gideon Smith could not be forgotten. On the other hand, Everett Yates never attracted the public's attention, so he faded into the institution's past. Playing the cathedral chimes simply did not make as lasting an impression as tackling one of Fielding Yost's powerful running backs.

Delbert McCulloch Prillerman came to M.A.C. from Institute, West Virginia, in September 1914 to study horticulture. Although the literary societies were closed to him, he joined the Cosmopolitan Club. Prillerman and Smith both appear in the club's 1914–15 group photo, which may be the first time that more than one African American appears in a published photograph of M.A.C. students. One of the stated purposes of the club was "to satisfy the same social need among the foreign students that the various societies and fraternities satisfy among native students."[36] Prillerman and Smith were not the only American students in the club, but their presence indicates that a perception existed at M.A.C. that African Americans, like international students, were different from European American students. It was more acceptable for Smith and Prillerman to meet their "social need" among students from other countries than in a society or fraternity apparently reserved for white Americans only. A good tennis player, Prillerman competed successfully on his class team alongside three white teammates.[37] During his senior year he participated in the Horticulture Club. He graduated in 1917 and served in World War I. Prillerman moved back to West Virginia where he taught at Bluefield Colored Institute in Bluefield.[38]

Margaret Elizabeth Collins, a Michigan native, renewed the link between the African-American community in Lansing and M.A.C. Collins commuted from her home on West Main Street beginning in the fall of 1920 and until her graduation in 1924.[39] Her stepfather,

Charles A. Campbell, had worked with William O. Thompson in 1913 and 1914 at the Seager Engine Works, and in 1920 the Michigan Department of Labor employed him as a janitor.[40] Collins was secretary of the Sunday School at the AME Church, where she worshipped with Thompson, who also held leadership positions in the Sunday School.[41] There is no evidence that Collins participated in any social or professional organizations at the college.[42] Her social life apparently centered around her church, family, and friends. After she received her bachelor of science degree in home economics, Collins taught for a year in Lansing. In 1930 she was on the faculty at the Georgia Normal and Agricultural College (now Albany State University) in Albany, Georgia.[43]

Mabel Jewel Lucas followed Collins to M.A.C. in 1923. She lived with her father, Stephen, and stepmother, Rosetta, on South Sycamore Street in Lansing.[44] Over the years, Stephen worked as a waiter at the Downey Hotel (one of Lansing's best), a self-employed window washer, and an employee at a barber shop.[45] The Lucas family worshipped at the AME Church, where Stephen held the office of trustee between 1912 and 1917.[46] Mabel's name is not found on rosters of student organizations at the college, and her picture does not appear in group photos. Lucas broke new ground for African-American women at Michigan State College when she received her bachelor of science degree in applied science in 1927. After graduation, she moved to Prairie View, Texas, to teach chemistry and biology at Prairie View State Normal and Industrial College, now Prairie View A&M University.[47] While there, she sought Professor Frank Kedzie's advice about going on to graduate school. With his encouragement, Lucas entered the University of Chicago in 1930 to take up graduate studies in chemistry.[48]

Benjamin L. Goode and Clarence E. Banks earned degrees in agriculture and competed in athletics, but they, too, were not accepted completely into the social life of the college. Goode came to M.A.C. in 1921 from Charleston, West Virginia, and graduated in 1925. He played halfback on the football team, winning a monogram in 1924, and he participated in track and inter-class sports.[49] Banks arrived in 1922 from Evansville, Indiana. A strong runner, he earned minor sports monograms in cross country in 1923, 1924, and 1925. He had less success in track, winning only a "Reserve Award" in 1926.[50] Banks

Margaret Collins, '24. Courtesy of Michigan State University Archives and Historical Collections (*Wolverine*, 1924, 34).

Mabel J. Lucas, '27. Courtesy of Michigan State University Archives and Historical Collections (*Wolverine*, 1927, 59, [n-3633]).

GOODE
Half Back

The 1925 *Wolverine* featured Benjamin L. Goode with this photograph in the section reliving the 1924 football season. Courtesy of Michigan State University Archives and Historical Collections (*Wolverine*, 1925, 134).

found fellowship in the YMCA and the Student Citizenship League, a religious group. He joined the Dairy Club, and both he and Goode studied apiaries in the Bee Seminar.[51] Yet, there is no evidence that fraternities welcomed either of these men into their organizations. Goode and Banks, like Gideon Smith and Delbert Prillerman before them, took teaching positions at black schools. In 1930 Goode taught agriculture and coached at West Virginia Collegiate Institute, and Banks worked in the Dairy Department at the Manual Training Industrial School for Colored Youths in Bordentown, New Jersey.[52]

Aeolian J. F. Lockert and Clarice A. Pretlow left fainter footprints than the other nine pre-1930 African-American graduates. Lockert, who was from Clarksville, Tennessee, received a bachelor of science degree in agriculture. His photograph is included with his graduating class in 1926, but neither his name nor his picture show up anywhere else in the yearbooks.[53] Possibly he came to Michigan State College as a transfer student for his senior year only. Coming from Smithfield, Virginia, Pretlow enrolled in the college about 1925 and graduated in 1930 with a bachelor of science degree in home economics. Her photograph appears with her class in the *Wolverine* of 1927, but she is not pictured with her class in 1930.[54] Apparently of little means, Pretlow worked as a live-in housemaid in 1928 for the Owen Knapp family in East Lansing. Knapp was the assistant manager of the Reo Service Station in Lansing.[55]

This group of eleven determined men and women opened the way at Michigan Agricultural College for others to follow in their footsteps. They embodied the land-grant ideal of service by dedicating their lives to teaching black youth to make their way in a society that kept them, because of their race, on the margins. Motivated by a strong will to grow intellectually in a society that kept them apart, the early African-American alumni showed that it was possible to take advantage of the educational opportunities offered at M.A.C. in spite of the racial discrimination. Whether or not anyone really noticed at the time, they helped to enlarge Justin Morrill's "industrial classes" to include the descendants of slavery. That reality, in itself, pushed out the perimeters of democracy just a little farther.

Clarence E. Banks (*front row, second from the right*) is pictured with the Dairy Club in the
1925 Wolverine. The *Wolverine* described the club:

> The Dairy Club is a professional club organized on the campus for the primary
> purpose of promoting interest in dairying and dairy questions. One of the
> definite things promoted and financed by the club is the annual trip of the Dairy
> Judging Team which is completed by the national intercollegiate competition at
> the National Dairy Show. The Club is also actively demonstrated during Farmers'
> Week when it carries on such functions as dairy demonstrations, an eating cafe-
> teria, booths, etc. The club shows the activity of one of the largest departmental
> divisions of the study of agriculture on the campus. They point with pride to the
> enviable record of their judging team annually at the National Dairy Show.

Courtesy of Michigan State University Archives and Historical Collections (*Wolverine*, 1925, 345).

8

International Students

DURING THE EARLY TWENTIETH CENTURY, A SMALL NUMBER OF international students came to Michigan Agricultural College to receive a land-grant education. The college welcomed young men from all over the world who wished to study scientific agriculture, and in some cases engineering, most of whom intended to return home and teach others ways to improve agricultural practices and production. A few remained in the United States and made significant contributions to the development of agriculture in America and abroad. Although M.A.C. and its American students generally accepted students from other countries, at times international students faced racial or ethnocentric prejudice. Some struggled with the English language, and some had to deal with family or political crises at home while living in East Lansing.

Only a tiny percentage of students from other countries studying in the United States attended M.A.C. prior to the 1920s. In 1904–5, 16 of 2,673 international students were enrolled at M.A.C.; in 1911–12 the numbers were 17 of 4,856, and in 1920–21 the totals were 20 of 6,901. During this sixteen-year period the number of Chinese students studying in America grew from 93 to 1,443, the number of Japanese students rose from 105 to 525, Canadians studying in the United States increased from 614 to 1,294, and Mexican enrollees dropped to 282 from 308.[1] Men from Armenia, Bulgaria, Canada, China, England, Japan, Mexico, Norway, the Philippine Islands, Puerto Rico, Romania, Russia, Scotland, and Turkey earned degrees at M.A.C. between 1904 and 1925 (see figure 10). It appears that M.A.C. did not enroll a woman international student until 1924.

Though faculty and administrators tried to integrate international students into campus life, they were not always successful in maintaining harmony between American students and their classmates from

Figure 10. International Students Who Received a Degree from Michigan Agricultural College between 1901 and 1925

BACHELOR OF SCIENCE

Henry W. Geller, '04, *Focşani, Romania*

Alfonso G. Palacio, '07, *Durango, Mexico*

Francis O'Gara, '08, *Ottawa, Ontario, Canada*

Joseph A. Rosen, '08, *Tula, Russia*

Alfonso Garcinava, '09, *Durango, Mexico*

Arao Itano, '11, *Okayamaken, Japan*

Charles Junzo Okada, '11, *Iwo, Japan*

Yoshio Kawada, '12, *Kuré, Japan*

Morris Crasniansky Ellman, '13, *Smiela, Russia*

Pow Kwang Fu, '14, *Swatow, China*

Ming Sear Lowe, '15, *Sunn Ning, China*

Michael Ivanovitch Wolkoff, '15, *Zoubtsov, Tver, Russia*

James Alexander Berry, '16, *Aberdeen, Scotland*

Dimitar Atanasoff, '17, *Gramada, Bulgaria*

Mehmed Ali Mehmedoff, '17, *Constantinople, Turkey*

Christo Sardjoff, '19, *Roustchouk, Bulgaria*

Ludovico Hidrosollo, '20, *Manila, Philippine Islands*

Santiago Iledan, '20, *Capiz, Philippine Islands*

Christo Demetrius Christoulias, '21, *Kotalis, Turkey*

Jen Han Shu, '21, *Kiang-Su, China*

Sen Yu, '21, *Canton, China*

Panos Demetrius Caldis, '22, *Smyrna, Asia Minor*

John Der Hovhannesian, '23, *Harpoot, Armenia*

Felix Aquino Pineda, '23, *Concepcion, Philippine Islands*

---

other nations. At times prejudice bubbled up and M.A.C.'s American students demeaned and hurt students who came from places that had different cultures. One Saturday night in late April 1888, eight or ten American-born M.A.C. students harassed three Japanese students physically and verbally and destroyed their property. Although the Japanese students resisted, they lacked the strength to overcome their attackers, who terrorized them for a couple of hours. After conducting an investigation, the faculty suspended four American students for

Ming Tat Young, '23, *Hong Kong, China*

Clato Rich Coe, '24, *Moose Jaw, Saskatchewan, Canada*

George Wesley Greene, '25, *London, England*

MASTER OF SCIENCE

Konstantin N. Svetlikoff, '13, *(Mannsky Agricultural College), Suzran, Russia*

Hisata Ogiwara, '15, B.S., (Tokyo Agricultural College), '06, *Chikujen, Japan*

Michael Ivanovitch Wolkoff, '16, B.S., '15 *Zoubtsov, Tver, Russia*

Pow Kwang Fu, '16, B.S., '14, *Swatow, China*

James Alexander Berry, '17, B.S., '16, *Aberdeen, Scotland*

Manuel Justo, '17 (University of Puerto Rico), '15, *Juncas, Puerto Rico*

William Kia-shen Sie, '17, (University of Nanking), '14, *Wuhu, China*

Hung Chung Chang, '22, A.B., (University of Peking), *Mukden, China*

Christo Demetrius Christoulias, '22, B.S. '21, *Kotalisk, Turkey*

Sarkis Garabed Der Sarkissia, '22, B.A. (International College, Smyrna), '20,
  Smyrna, Asia Minor

Daniel K. Ming Lee, '25, (Morningside College), '24, *Foochow, China*

DOCTOR OF VETERINARY MEDICINE

Morris Jaque Sisley, '15, *Koono, Russia*

Henrik Jordan Stafseth, '17, B.S. (North Dakota Agricultural College), '15,
  Arlesund, Norway

Schwen Yuing Chen, '25, *Changchow, Ku, China*

Source: Compiled from student lists in M.A.C. annual catalogs, 1900–1925, and annual reports of the State Board of Agriculture.

---

one year, and President Edwin Willits gave a private reprimand to one of the Japanese students.[2] All three Japanese students left the college. K. G. Minakata explained his reasons for dropping out of M.A.C.:

> But, now I must go out this college, because I suspect that we may not be hereafter in this college in comfortable condition. Of course, I have done everything possible in Last Saturday nights case, and they have nothing to say about their punishment. Yet, I fear that, all the men in this college, except a few, will constantly display their

Alfonso Garcinava, '09, was the second student from Mexico to earn a degree at Michigan Agricultural College. A member of the Hesperian Society, Garcinava was quite active in the college's social life. He is pictured (*on the left*) with two friends while sailing on the Detroit River on 20 September 1908. In 1930 he lived in Mexico City and worked in the insurance business. Courtesy of Michigan State University Archives and Historical Collections (Alfonso Garcinava Papers, 1906–09, UA 10.3.14, Scrapbook 52).

worst feelings upon us; or, at least, some bodies will be in utmost animosity against us, for they naturally like their own countrymen further than a stranger. I am very unhappily to go out, but it must be done so.[3]

In March 1900, an American student made prejudicial and derogatory remarks about a student from Puerto Rico in a letter to his parents. In all likelihood, attitudes revealed by these comments were held by many others, for the Puerto Rican student intended to transfer to another college.[4]

Even though some students held ethnocentric attitudes, President Snyder took great pains to assure prospective students from other parts of the world that they would experience no racial prejudice at M.A.C. When S. L. Maddox, director of the Department of Land

Records and Agriculture, Bengal, India, made inquiry in May 1905 as
to the suitability of Indian students studying at the college, he wanted
to know "if race antipathy or any other difficulty" would keep them
from being accepted as equals.[5] Snyder responded: "There is no
antipathy in this institution against colored students or students from
foreign countries. They would receive fair and cordial treatment from
every one."[6] Three years later he expressed a similar attitude relative
to the possible enrollment of men and women from Ceylon.[7] W. J.
McGee, '96, affirmed the longstanding acceptance of international students at M.A.C. when he suggested that a friend from India might
study at the college in 1908. Remembering that in his time M.A.C.
"was very kindly towards foreign students," McGee drew a contrast to
colleges in the American South, where he knew of prejudice directed
against students of "dark complexion" from Cuba and Brazil.[8] Snyder
assured McGee that he did "not think the feeling could be more democratic anywhere" than at M.A.C.[9]

   In 1906, H. Caramanian, who had come from Turkey in 1896 to
attend M.A.C., recalled with fondness the kind treatment that he
received upon his arrival and during his stay at the college and its
impact on him:

> A period of twelve years seperates me from that happy time when
> I put myself under your [Snyder's] and Prof. [Clinton D.] Smith's
> care: The first day I was on the campus I was picked up by him and,
> on my request, was carried to his office to write a letter. On the same
> day I met you in your office, and introduced you myself. One of the
> first questions you asked me was "How much money have you." I
> answered "Five Dollars only," and you laughed heartily, since you
> had suggested me to come with $50 for the first year. From that day
> on, with your guidance and under Prof. Smith's care, I have never
> been misfortunate until some family misfortunes called me home
> on the autumn of 1897.[10]

At the time he wrote this reminiscence, Caramanian contemplated
moving back to Michigan to take up farming and escape "the life of a
common worker" in Turkey. His experience hints at the tensions created within him (and no doubt in the minds of other international
alumni) from the contrast between the pleasant memories of his experience at M.A.C. and the challenges brought on by family responsibilities after returning home.

At times, foreign governments sent students to study at colleges and universities in the United States. Joseph A. Belkeley, a student at Hawkesbury Agricultural College in Australia, found his way to M.A.C. having won a two-year scholarship from the Department of Mines and Agriculture in New South Wales, Australia, to study agriculture abroad. The colonial government there chose M.A.C. because of its "high standing," even though agricultural colleges in the South enjoyed a climate more similar to that of New South Wales. Belkeley fulfilled his benefactors' expectation that he teach agriculture and share his expanded knowledge with farmers in New South Wales by returning home after his graduation in 1899.[11]

In 1904 the government of the Philippine Islands enrolled six students in M.A.C. Ramon J. Alvarez, Ludovico Hidrosollo, Rafael F. Montenegro, Balbino Palmares, Andres Sevilla, and Pastor Airsado were part of a 110-member contingent of Filipinos, supported by their government, studying at various American colleges and universities. Upon their arrival in Lansing, the administration looked after their needs. The students spent their first night in a hotel and the next two weeks in a boarding house, until the construction of their permanent residence was completed. None of them initially stayed long enough to earn a degree, since their government intended that each one learn as much as possible during their years in the United States. After two years at M.A.C., four of them moved on to other land-grant institutions: Palmares studied irrigation at Colorado Agricultural College in Fort Collins; Sevilla took farm mechanics at the University of Wisconsin in Madison; Hidrosollo learned about harvesting and marketing agricultural products grown in the South at Louisiana State University in Baton Rouge; and Airsado received an introduction to engineering at Ohio State University in Columbus.[12] Hidrosollo came back to M.A.C. in 1919 and received his bachelor of science degree the next year.

Young people living in distant lands expressed other reasons for considering studying at M.A.C. One man in Uruguay thought that the agricultural course at M.A.C. might prepare him to start the first agricultural college in his country.[13] J. R. Maurer, living in Zurich, Switzerland, recognized a need for his son to study scientific agriculture at an American college to prepare him to run his own farm or to manage a larger farm in the United States. However, it does not appear that his

Many potential students requested college catalogs by sending postcards to the president of M.A.C. Christo Sardjoff sent this card from Roustchouk (Rustchuk), Bulgaria, in 1913. The "recent war" that Sardjoff mentions is the Balkan Wars of 1912–13 that brought considerable turmoil to his homeland on the eve of the First World War. The *Alumni Catalogue* issued in 1931 reported that Sardjoff was "Director, Agr. School Nadarevo, Eski-Djoina, Bulgaria." Courtesy of Michigan State University Archives and Historical Collections (J. L. Snyder Papers, UA 2.1.7, Box 820, Folder 160).

son enrolled, although the college was willing to accept him.[14] Others expressed a desire to expand the formal education and practical experience in agriculture they received in their home countries without stating a specific career objective. For example, A. S. Onischick had graduated from a Russian teachers' institute, had repeated course work in English in the United States, and claimed a "wide experience in horticulture, in gardening, in the cultivation of bees and with little experience in floriculture."[15] He, too, never came to M.A.C.

M.A.C. offered no financial assistance to prospective students from foreign countries. Consequently, expenses prevented some aspirants from attending the college. International students had to pay out-of-state tuition that amounted to $15 per term in 1925, in addition to other fees charged to all students. Travel costs were prohibitive for individuals such as a young Peruvian boy who promised to bring "relics" for the museum if M.A.C. sent him enough money to pay for his passage to East Lansing in 1907—a request that went unmet.[16] Potential students received advice to come with enough money in hand to get them through their first year. Although President Snyder and professors helped them find jobs as waiters and janitors and doing "odd jobs for faculty members," their earnings were usually insufficient to pay all of their bills.[17]

Alumni living abroad encouraged young people to attend M.A.C. Although the link cannot be established with certainty, George E. Kedzie, '73, appears to have induced the first students from Mexico to study agriculture at M.A.C. Kedzie worked as a consulting mining and metallurgical engineer in Durango, where he made known educational opportunities available at his alma mater.[18] In 1903, two students from Mexico, both from Durango, are included on lists of students. Alfonso G. Palacio entered as a freshman and graduated with the Class of 1907, and Alfonso Garcinava enrolled as a sub-freshman and graduated with the Class of 1909. Over the next several years, five more men from Durango came to M.A.C., although none of them earned a degree.

Limited ability with the English language proved to be a significant obstacle for international students, and M.A.C. offered little help to improve their reading, writing, and speaking.[19] When Osorio Susano Sanjurjo arrived from Puerto Rico in September 1906, his command of English was so poor that the college sent him to Ferris Institute in Big Rapids for language study.[20] Three years earlier, Victor C. Gonzalez, from Arequipa, Peru, who spoke some English, and Dimos Burbano Bowen, from Ste. Ana, Ecuador, who spoke no English, became discouraged and did not stay at M.A.C. beyond their first year.[21] In 1914, disruptions resulting from the outbreak of World War I led the Bulgarian government to send three students, only one of whom spoke any English, to study agriculture at M.A.C. Professor Ernst A. Bessey, who had spent many years in Europe and knew their language, helped Dimitar Atanasoff, '17, from Gramada, Koula, Bulgaria, and Christo Sardjoff, '19, from Roustchouk, Bulgaria, to complete their degrees. Ivan Zwetan withdrew from M.A.C. when his government failed to forward money needed to pay his college expenses.[22]

The Cosmopolitan Club from the 1914–15 academic year as it appeared in the 1915 *Wolverine*. This photograph is noteworthy because it may be the earliest extant picture that has two African-American students attending M.A.C. in it. Delbert Prillerman, '17, is in the front row, fourth from the left, and Gideon E. Smith, '16, is in the top row, first from the left. Christo Sardjoff, '19, appears in the center of the top row. Courtesy of Michigan State University Archives and Historical Collections (*Wolverine*, 1915, 241).

The handful of international students matriculating at M.A.C. helped communicate the troubled state of world affairs to men and women at the college, most of whom had lived on farms or in small towns and had little firsthand knowledge of people in other countries. The world in 1910 stood on the verge of the Great War. Competition between Japan, China, and Russia had led to the start of the Sino-Japanese War in 1894 and the Russo-Japanese War in 1904. European imperialism created tensions in Asia and Africa. Instability in Russia, the Balkans, and the Ottoman Empire contributed to global uncertainty. In the United States, many Americans had strong anti-Japanese and anti-Chinese feelings that strained U.S. relations with the two countries.

The Cosmopolitan Club created a forum where students from different nations could meet, socialize, exchange views, and connect with international students at other colleges and universities. The club's constitution set out a lofty goal for the organization: "to study the conditions of, and to promote a better understanding of the political, social, and cultural problems of various countries, and thereby foster the spirit of universal brotherhood."[23] Over two hundred people from the college and the community attended the first open meeting on 26 February 1910. They heard Mexican and Japanese students sing their national hymns, Jewish students sing the "Jewish Song," Osman Abdel Rasik, an Egyptian student give a talk entitled "The Culture of Egypt," and Julius Kaplan speak about the political situation in his native Russia.[24] Soon after Sotin Alan Chow, of Shanghai, China, enrolled in the agricultural course, he addressed the club in January 1911. Chow had attended a meeting of the Cosmopolitan Club at Cornell University while en route to East Lansing from London. Members of the club at Cornell passed along their greetings to their counterparts at M.A.C. through Chow.[25] "Russo-Japanese Night" on 10 February 1911 featured a showing of Japanese lantern slides and a fencing demonstration.[26] Later in 1911 the club raised money for famine relief in China, hoping to match contributions given at the University of Wisconsin.[27] Ghose Makham, a Hindu from Bengal, India, shared insights about his country and its people with an audience in November 1912.[28]

The club never lost sight of its commitment to bring together members from its diverse constituency and to address issues relevant to them. Professor William W. Johnston, along with Herman E. Segelin, '22, led a discussion on Hebrew literature in May 1922.[29] The schedule for the weekly meetings for fall quarter 1925 included programs that focused on China, South Africa, Japan, and the Philippines. Other subjects covered were "General Conditions in Europe," "The Historical Development of International Institutions," "Economics and Internationalism," and "Contributions of Religion to International Good Will."[30] Knowledge of other people's ways made possible a greater understanding of them, which in turn encouraged acceptance and tolerance—entities in very short supply during the early twentieth century.

International students reached larger audiences through the thoughtful articles they published in the *M.A.C. Record* and the *Holcad.* Students, faculty, alumni, and others learned about conditions in Russia, Japan, Egypt, and other places from writers who had grown up in these countries. Vadim Sobennikoff, '97, enlightened readers of the *M.A.C. Record* in 1896 on the state of affairs in Russia in two articles "Russian Peasant Life" and "Occupations and Systems of Land Ownership."[31] A few issues later Shoichi Yebina, '95, wrote about agriculture in Japan.[32] In 1909, Gikan Fujimura wrote a series of pieces for the *Holcad,* including a sensitive, prophetic article entitled "Relations between Japan and the United States" that contains insights that could be articulated only by a person who had lived in both countries.

He believed that ties between Japan and the United States were much stronger than the ill will caused by a few rabble rousers in California and Nevada. Fujimura predicted that if an armed conflict were to occur between the two nations, it "would be the most inhuman and hellish event in the world's history."[33] Time proved him correct.

Two other international students brought the Middle East to the attention of the M.A.C. community. Osman Abdel Razik, a graduate of the University of Cairo, used some of his time at M.A.C. in 1909 to educate Americans about his native Egypt. In the *Holcad* he discussed Egyptian crop rotation and how his country's climate divided the year into "three seasons for agriculture, summer, winter, and flood." At a lecture to the Lansing Women's Club at a meeting that was gathered in the parlor of the Women's Building, Razik made a special reference to the role women played in Egyptian history from ancient times to the present.[34] Morris Crasniansky Ellman, '13, a Jew from Smiela, Russia, portrayed the last four centuries of Jewish suffering and shared his dream of a restored Jewish state in Palestine. In "The Wandering Jew," Ellman introduced Theodor Herzl and the Zionist movement to readers of the *Holcad* in 1910.[35]

Chinese students (all men) began arriving in East Lansing in 1910, and approximately twenty-five of them studied at M.A.C. during the next fifteen years. Some of them had done coursework at other institutions in the United States or China before enrolling at M.A.C. Pow Kwang Fu, '14, from Swatow, and Kang Ching Luke, from Hong Kong, came for the spring term in 1911 from Valparaiso University in Valparaiso, Indiana.[36] Following his graduation, Fu earned a master's degree in plant breeding with a minor in chemistry. Returning to China, he taught at the Canton Christian College. Surrounded by "political troubles" in 1923, Fu looked forward to receiving the *M.A.C. Record,* which kept him in touch with his beloved alma mater and offered him an avenue for communicating with other alumni.[37] During fall term 1911, Ming Sear Lowe, '15, from Sunn Ning, took up the agricultural course; he had lived in the United States for ten years and had graduated from a high school in Berkeley, California.[38] Lowe was working for Metro-Goldwyn-Mayer of China in Tientsin in 1930.[39] Shwen Yuing Chen had earned his bachelor's degree at Nanking University in 1922 before starting work on his doctor of veterinary medicine degree at M.A.C. in 1923, which he obtained in two years. He then returned to his home in Changchow Ku.[40]

Fewer Japanese attended M.A.C. after 1900 than before; only four Japanese students earned degrees between 1900 and 1925. From 1897 until the end of the Russo-Japanese War in 1905, no Japanese names can be found on student lists. The freshman class in 1907 boasted six Japanese, including Arao Itano, '11, from Okayama, and Charles J. Okada, '11, from Iyo. After 1910 there were never more than two Japanese on the annual rolls. Itano played an active role on campus and ultimately earned a doctorate from Massachusetts Agricultural College in 1916, where he studied under Professor Charles E. Marshall, formerly of M.A.C. After serving on the faculty at Massachusetts, Itano went back to Japan in 1924 to be head of the division of microbiology and chemistry at the Ohara Institute for Agricultural Research in Kurashiki, Okayama.[41] Yoshio Kawada, '12, taught English, natural science, and drawing after returning to Japan before he became a patent attorney in Tokyo.[42]

Henry W. Geller emigrated from Focşani, Romania, to M.A.C. in 1900 to begin his studies for a career in service to agriculture. He quickly learned English, and he became energized by the college's land-grant philosophy and graduated in 1904.[43] After completing a master's degree at the University of Illinois, he assumed the superintendency of the Baron de Hirsch Agricultural School in Woodbine, New Jersey, an institution where young Jewish men learned practical farming. Geller summed up his work there in a letter to Snyder:

> We teach practical agriculture, take in boys of 18 years of age, young men who have never seen the country, never handled horses or cattle, boys who all their life has been spent in the city or even their fathers and ancestors since the destruction of the Jewish Kingdom, have never seen a farm and farm work; and teach them the art of farming by making the boys do the work.[44]

Bliss B. Brown, '03, and Robert D. Maltby, '04, taught at the same school, where the attendance had risen from eighteen to eighty during Geller's first two years.[45] Geller served as a conduit to M.A.C. for Moses Levine, a young Jewish man from Lutzin, Vitebsk gov., Russia, who came to Woodbine in August 1910. Levine spent a few weeks

there acclimating himself to the United States before he headed to East Lansing to enroll in M.A.C.[46] In 1912 Geller was back in Michigan working for the Duluth, South Shore & Atlantic Railroad as its agriculturalist in the Upper Peninsula.[47] He died in 1918 in Texas.

Joseph A. Rosen, '08, a Russian Jew, epitomized the land-grant educated man who devoted his life to improving a world that for him was characterized by a continuously expanding horizon. Born in Moscow, Russia, in 1876, Rosen attended Moscow University before being sent to prison in Siberia for his opposition to the tzar. In 1903 he made his way to the United States, and he became a student in the agricultural course at M.A.C. in 1905. Following graduation, he went to work for the Russian province of Ekaterinoslav, with an office in Minneapolis, Minnesota. Rosen wrote a series of ten reports analyzing and discussing American agriculture for the benefit of Russia. A listing of some of his titles reveals the extent of Rosen's survey: *Agricultural Experiment Stations in the United States of North America; Questions of Dry Land Farming in the United States of North America; Methods of Investigating Seeds in Experiment Stations of America; How the American Farmer Fattens Cattle; The Organization of the Bread Trade in the United States;* and *Agricultural Extension Work in the United States of North America.* In addition he purchased seeds and equipment to ship to Russian farmers. After he left his position with the Russian government, he served as superintendent of the Baron de Hirsch Agricultural School from 1914 until 1916.[48]

In 1921, Rosen's life entered a new phase when he joined the U.S. government's relief work in Russia by directing the Joint Distribution Service. One benefit of this job was that Rosen could visit his aging parents and relatives in Russia, whom he had not seen for ten years. While en route to Russia on the high seas, Rosen expressed how the land-grant philosophy influenced his upcoming assignment. He wrote to his mentor and old friend Frank Kedzie:

> The night is pitch dark. For an hour I was looking into the black abyss of the sea. My inner-self watching the undulations of my thoughts and the memories of my three years came back to me as the happiest recollections of my American life. These were the years that prepared me for the work that I am taking up at present. I am to be the Agricultural Advisor on the Commission and I hope to be of real service to both my native and adopted countries.[49]

While living in Russia, he helped to resettle Jews in the Crimea and the Ukrainian region of Krivog-Rog, encouraging them to plant corn and sorghum.[50]

Rosen also made important contributions to agriculture in Michigan and other states. In 1909, he gave to the college a pound of rye seed grown in Russia. M.A.C. researchers found it produced higher yields than varieties then being sown by Michigan farmers. Wartime demands for higher yields from American grain fields encouraged Michigan farmers to commit 500,000 acres of their land to "Rosen Rye" by 1919.[51] In 1917 Rosen, who at that time represented the State Agricultural Society of Charkow, Russia, sold 3,600,000 pounds of sugar beet seed to American farmers who were struggling to meet the wartime demand for sugar. This quantity made up over one-sixth of the seed planted in the United States in that year.[52]

Maurice G. Kains came to M.A.C. from St. Thomas, Ontario, in 1891, and after graduating in 1895, he had a long career as an agricultural writer. Kains received his master's degree from Cornell University in 1897, where Liberty Hyde Bailey was one of his professors. He spent three years doing "investigational and crop culture work" with

M.A.C. alumni gathered together as part of local associations or made special efforts to arrange meetings wherever they happened to find themselves. The *M.A.C. Record* published this photograph of "M.A.C.'s Outpost in the Pacific" when this group of alumni, who were working in the Far East, held an impromptu gathering and luncheon in Manila in February 1923: (*back row*) Leroy Thompson, '13, supervising mechanical engineer for the bureau of public works, Manila; Ludivico Hidrosillo, '20, director of agricultural extension for the Philippine Islands; C. W. Edwards, '09, in charge of Experiment Station in Guam, USDA; D. D. Wood, '11, conservator of forests for the British North Borneo company; E. W. Brandes, '13, bacteriologist, USDA, making a world-wide study of plant, especially sugar cane, diseases; J. L. Myers, '07, owner Myers-Buck Co., engineers and contractors, Manila; (*front row*) Santiago Iledan, '20, instructor of agriculture at the Central Luzon Agricultural School at Munoz, N. Ecija; Mrs. C. W. Edwards; Anna B. Cowles, '15, director of Junior Red Cross for the Philippine Islands; Zella Kimmel (Mrs. D. D.) Wood, '11, E. G. Hoffman, '10, manager of the Binalbagan Estate and Refinery, Island of Negros. Courtesy of Michigan State University Libraries, Special Collections (*The M.A.C. Record* 28, no. 30 [21 May 1923]: 5).

the USDA. Next, he taught horticulture and science for two years before assuming the editorship for agriculture, horticulture, and botany for the twenty-volume *New International Encyclopedia* issued in 1905.[53] Kains then went to work for the Orange Judd Company, a publisher of five regional agricultural periodicals and books. For five years he produced or edited all the stories relating to horticulture that appeared in the *American Agriculturalist,* in addition to supplying articles on poultry and editorials for all five weekly Orange Judd papers. The *American Agriculturalist* served a region comprising Ohio, New York, Pennsylvania, New Jersey, Maryland, Delaware, West Virginia, and southern Ontario. Kains also edited more than twenty books for Orange Judd. In the preface to *Making Horticulture Pay* Kains set forth the land-grant emphasis upon the practical by drawing upon the work of "actual farmers rather than of specialists, and especially of women, because on so many farms the women are the gardeners and small fruit growers" for ideas that could be implemented to improve the lives of farmers.[54] Kains left the company in 1914 to head the Department of Horticulture at Pennsylvania State College, a position he held for two years before he resumed his career in publishing. While at Penn State, Kains wrote *Plant Propagation,* a textbook that was widely used by plant propagators and nurserymen for many years.[55]

International students made their presence felt at M.A.C. by stretching the worldviews of their American classmates, many of whom had never been out of the state of Michigan. As they shared their beliefs and practices, and the histories of their countries with the college community, they created an environment that opened up opportunities for the people of M.A.C. to extend the land-grant way to other parts of the world. International students not only became part of the M.A.C. family, but they developed strong attachments to their college that they took with them when they returned home. Although small in numbers, they opened the way for the institution's future efforts to meet the needs of ordinary life for people living throughout the world.

9

Athletics

THE ORGANIZATION OF ATHLETIC TEAMS BY STUDENTS, AND THEIR gradual acceptance by Michigan Agricultural College's administration made it possible for student athletes to compete against teams from other institutions, injected energy into campus life, and generated public interest in the college. After M.A.C. overcame its early reluctance to support athletics financially, it built facilities, hired coaches, and made institutional changes permitting its teams to take their place in the growing business of American intercollegiate competition. Like their counterparts at other colleges, M.A.C. administrators, alumni, and students came to believe that success on the gridiron was an important asset to developing a school identity. Sports, especially football, proved to be an effective way to bring alumni back to campus and to stimulate their interest in the future of their alma mater. M.A.C. students and alumni, even though they may have held conflicting worldviews and had different academic and vocational interests, could join together as they rooted for the Aggies from the grandstand, in the newspaper, or in front of their radios. By 1925 M.A.C.'s men's teams competed at the intercollegiate level in football, basketball, ice hockey, wrestling, track and field, swimming and diving, baseball, tennis, cross country, and rifle shooting. Women's teams included tennis, swimming, and rifle shooting.

Growing student support for and participation in intercollegiate athletics at M.A.C. and other colleges was primarily a male phenomenon, and it generated passionate debates over the place of athletics in American higher education. Donald P. Yerkes, '85–'87, expressed one view in the *College Speculum* in 1887 when he argued that people need a balance between physical exercise and mental studies in order to reach their potential.[1] Five years later the *Speculum* moved beyond Yerkes's philosophical apologetic by stating plainly the benefits of athletics: "They stimulate a healthy interest, they form an outlet for innocent enthusiasm, they cultivate a most necessary *esprit de corps;* and who will say that these are not valuable elements of college life?"[2] Some members of college faculties in the East and the Midwest, however, felt that intercollegiate athletics, especially football, led to excesses that distracted students from their studies and encouraged both players and fans to commit acts of violence. Critics, such as history professor Frederick Jackson Turner at the University of Wisconsin, charged that the lure of football revenues and glory blinded universities to such evils as team supporters secretly paying athletes and to the game's brutality, which in turn corrupted student ethics. Denunciations of football during the first decade of the twentieth century did not cause its demise, as some hoped it would, but rather instigated changes in rules that spurred its continued growth in popularity.[3] Intercollegiate athletics thus became a permanent feature of campus life across the United States, including in East Lansing.

Gender inequity in intercollegiate athletics might obscure the fact that M.A.C.'s position was that physical education was for all students. M.A.C.'s commitment to teaching physical education created an environment that encouraged both sexes to take part in athletics. Therefore, in 1898 a women's team played M.A.C.'s first interscholastic basketball game. The women's basketball team soon faded away, but women's interest and desire to engage in athletics did not abate. While newspapers reported the victories and defeats of men's intercollegiate football, baseball, basketball, and track and field teams, women staged a host of competitions, often out of public view, in the gymnasium in the Women's Building. Contests between class teams as well as activities required for physical education courses provided opportunities for women to play sports.

EARLY TEAMS, COMPETITION, AND ISSUES

Organized student athletics at M.A.C. started with baseball and gained impetus through class competition and field days. Alfred G. Gulley recalled that students in the 1860s, caught up in the national interest in baseball, formed the college's first club, The Stars, in 1865, to play teams from surrounding communities.[4] The next year a college "nine" continued to meet challenges from neighboring towns; among its victories was a 37-21 triumph over the "Sheridan club of Mason." The first game against a team from the University of Michigan resulted in a 2-1 loss for M.A.C. in 1868.[5] Participation in baseball grew as classes fielded their own teams and engaged in spirited intramural contests. The best intramural players made up the "first nine" to take on opponents from other colleges. In May 1889, for example, the M.A.C. varsity humbled Olivet by a score of 17-6.[6] M.A.C. athletes faced their competition wearing green and white uniforms; the 1899 team wore "new suits" that were "similar to the old—white shirts and green caps and pantaloons."[7]

The formation of the M.A.C baseball club spurred on the development of that sport and started an effort within the college to properly supervise athletics. Expressing a wish for M.A.C. to catch up with other institutions, in 1882 the *College Speculum* called for the establishment of an athletic association to organize teams and contests for the students' enjoyment.[8] The baseball team heeded this call in 1886 when it created the M.A.C. club and put a manager in charge of selecting players, arranging practices, and disciplining team members.[9] Professor Rolla C. Carpenter, '73, served as the first manager, but the organizational structure that allowed the manager to be either a faculty member or a student made it possible for students to exercise significant control over athletics.

The Athletic Association, which originated with the baseball team's codification of its procedures in 1886, provided financial and organizational support for the college's embryonic athletic program.[10] Students comprised its membership and played a large role in forming and overseeing M.A.C.'s teams. In 1897 Henry Keep, '97–'99, enrolled as a student and coached the football team to a record of four wins, two losses, and one tie.[11] The next year Walter H. Flynn, '99, arranged a seven game schedule for Keep's Aggies that featured two victories over Ypsilanti, and one each over Albion and Olivet.[12] The association looked after the needs of other sports by buying such things as a wrestling mat and blank cartridges.[13] When the association's officers failed to make the effort needed to get the freshmen to join in 1901, it faced near bankruptcy for want of their dues.[14] Some years later the college assessed each student $2.00 a term for membership in the association, which entitled them to free admission to all

Students at Olivet College challenged M.A.C. to a football game during Olivet's field day scheduled for 19 May 1884. Lack of time prevented the game from being played. M.A.C.'s football team, which may have been the school's first "college" team (as opposed to class teams), is shown here. Courtesy of Michigan State University Archives and Historical Collections (Michigan State University, Athletics, Football, 1884).

home athletic events, debates, lectures, and the May Festival. Since all gate receipts went to the association, it assumed the responsibility of meeting guarantees to visiting teams and paying wages to game officials and student workers. All monies were deposited with the college treasurer, who paid the bills upon receiving proper receipts and invoices from the association.[15]

The first local field day, held in 1884, set the stage for M.A.C. to send its athletes into competition against Olivet and Albion in 1886. Albion and Olivet each had approximately 360 students, and M.A.C.'s student body numbered 295.[16] Field days provided opportunities for male athletes to participate in a wide range of contests, including wrestling, baseball, running, jumping, boxing, kicking, shot putting, and even football. The tennis team included women, but it is unclear when women first competed. On 14 and 15 May 1886 M.A.C. hosted thirty athletes from Olivet and another thirty-five from Albion for "friendly but spirited contests" that fostered "a feeling of brotherhood" between the three colleges. Bad weather curtailed some outdoor events, but baseball and football squads completed their games. Boxing, wrestling, and field events entertained fans from each college in the Armory all day Saturday. Three weeks later M.A.C.'s team traveled to Albion to take part in that college's field day.[17]

Field days drew M.A.C. into a formal arrangement with Olivet, Albion, and Hillsdale colleges when they formed the Michigan

Inter-Collegiate Athletic Association (MIAA) in 1888. Each institution agreed to host the annual field day, on a rotating basis, under the provisions of the MIAA's constitution and by-laws. During the next few years, Kalamazoo, Alma, and the Normal School at Ypsilanti joined the association. Adrian became a member in 1908, the year M.A.C. was asked to leave the MIAA.[18]

Athletics at M.A.C. entered a new era when it hosted the MIAA's first annual field day on 31 May and 1 and 2 June 1888. The college put forth its best efforts as it opened its facilities to house, feed, and entertain more than 150 visiting students. Once classes ended on Thursday morning, the competition commenced with the running of the 100-yard dash and the 50-yard backward run, a baseball game between Hillsdale and Olivet, and tennis matches. In the evening, jovial students from the four colleges enjoyed music and talk in the literary societies' rooms. After the second day's events, the cadet companies put on an exhibition, President Edwin Willits welcomed his guests, and the women students served refreshments in the Armory. Leander Burnett, '92, who may have been M.A.C.'s first Native American student, led his team by taking first place in four events: standing broad jump; backward jump; running hop, step, and jump; and high kick with both feet. While resting between contests, Burnett played third base and batted lead-off for the baseball team that beat Olivet, 12-2, and Albion, 10-8, for the association championship. An estimated crowd of 1,500 to 2,000 people attended the affair.[19]

Before Henry Keep brought respectability to football at Michigan Agricultural College, it had an inauspicious beginning. Although the college prepared a team to play against Olivet in June 1884, it appears that M.A.C.'s first games were losses to Olivet (0-8) and Albion (0-79) during field days in 1886.[20] In 1888 "the committee on football" outfitted thirty players in "new canvass suits," and in April, 1894 the Class of 1896 lost to a team from Ann Arbor High School.[21] The M.A.C. school calendar, with its long vacation from early November until February, prevented fielding a team in fall. After the long vacation was moved to summer in 1896, a football team could be assembled to represent the college late into November, when many schools played their most important games. The 1896 team struggled without a coach and finished last in the MIAA with a record of no wins and two losses in conference play and an overall record of one victory, two

Track and field competition dominated field days that began in the mid 1880s. In 1886 M.A.C. hosted teams from Albion and Olivet for an intercollegiate field day. Two years later these three schools along with Hillsdale College formed the Michigan Inter-Collegiate Athletic Association (MIAA), which spurred the growth of athletics at M.A.C. After the college left the MIAA following the 1907 competition, M.A.C. track teams ran against teams from larger institutions. The 1899 team posed for its team picture in this photograph. Courtesy of Michigan State University Archives and Historical Collections (Michigan State University, Athletics, Track, 1890–1899).

defeats, and one tie. The *M.A.C. Record,* however, pledged that "a competent coach" would be in place for the next season.[22] Henry Keep, the last student football coach at M.A.C., fulfilled that promise.

The college's administration soon recognized that competent coaching was an essential ingredient to a successful physical education program that encompassed athletics, and that the only way to make coaches answerable to the administration was to keep them on the M.A.C. payroll. At the insistence of President Snyder and board member L. Whitney Watkins, '93, the college hired the Reverend Charles O. Bemies to be director of physical culture and coach of M.A.C.'s men's intercollegiate teams in 1899.[23] With Bemies's appointment the college integrated organized athletics into a larger physical education program benefiting all students. Bemies summed this up very well in his first annual report: "The work of this department has been along the three-fold line of athletics and gymnastic work for young men, and calisthenics for young women. The ultimate object is to give systematic physical training to all the students, to correct defective positions of the body, and to stimulate general athletics in such a way as to actively engage a larger number of students."[24] During his two-year tenure, Bemies's football and baseball teams enjoyed little success, but the 1901 men's basketball team won all three of its games, and the track and field team took first place in the MIAA field day held at Hillsdale College. The football team's poor performance in 1900, which ended with a 23-0 defeat by Alma, led the athletic association to ask for Bemies's resignation. Although the players saw Bemies as a "good man," they refused to play for him after the season ended.[25] Of particular note, Marguerite A. Nolan, '02, became the first M.A.C. woman to score a point in women's tennis singles in a field day competition in June 1901, where she took third place.[26]

A host of thorny issues cropped up in the early 1900s that forced M.A.C.'s administration to pay more attention to its athletic program as the institution struggled to build a winning football team. A Hillsdale fan, disappointed over his school's 22-0 football loss to M.A.C. in October 1901, charged that three members of the Lansing Athletic Club had taken the place of three students on M.A.C.'s team. Although the accusation proved false, it raised the larger question of student eligibility for intercollegiate athletics, which was being faced across the country.[27] President Snyder, who liked football and was "anxious" that M.A.C. "have a winning team," took great interest in the athletic department and gave it wholehearted support, but as much as Snyder wanted his football team to win, he would not countenance non-students masquerading as Aggies. Always sensitive to allegations that some M.A.C. football players were being paid, Snyder held that even receiving a salary to teach high school should render students ineligible to play on college teams.[28]

The faculty fixed the minimum academic standards that athletes had to meet in order to retain their eligibility and set parameters for team activities. But rules governing athletics were rarely cast in bronze. In 1905 any student who was not doing passing work in two or more subjects was declared ineligible. But it appears that this restriction was not followed very closely in subsequent years, since Snyder, uncharacteristically, seemed not to remember if the college had such a rule when questions arose over the status of athletes carrying "conditions" in February 1910.[29] In May of that year the faculty approved an extensive set of regulations that limited a student's eligibility to three years, prevented freshmen from participating in intercollegiate athletics, outlawed pay for playing on college teams, and required athletes to compete under their own names. The football team could not begin practice until the new school year had started,

The first organized physical education class for women might have happened in 1887 or 1888 when female students took daily exercise under the direction of Second Lieutenant Wendell Lee Simpson, professor of military science and tactics (Louis A. Bregger to brother, 3 April 1888, Louis A. Bregger Papers, 1884–1909). Here, women enrolled in the Women's course (circa 1896–99) are doing calisthenics in the Armory.
Courtesy of Michigan State University Archives and Historical Collections (Michigan State University, Health, Physical Education and Recreation, Women).

and could only play teams from other colleges and schedule no more than nine games. Sixteen games were allowed for both the baseball and basketball teams.[30] But only one year later, in an effort to enable John Macklin, the newly appointed athletic director, to build his teams quickly, the rules were revised to allow freshmen to compete.[31] Snyder acknowledged the growing significance of athletics at M.A.C. telling Macklin: "Your department is just as important as any other department in this institution and in some respects even more so. There is no one connected with the institution whose personal influence will count for more than your own."[32] Ever anxious to keep players on the gridiron, in 1914 Snyder encouraged Macklin to hire a tutor for a football player who was having academic problems.[33]

As early as 1902, Athletic Director George E. Denman actively recruited young men to play football at M.A.C., initiating practices that brought the college into competition with other institutions for the services of talented athletes. After graduation, Denman rushed from campus to New York State in order to see a young man who played football and baseball. Denman hoped to bring him to M.A.C. for the 1902 season against his father's wish for him to delay college for another year. He also wrote letters to "promising athletes" extolling the virtues of M.A.C. and urging them to attend the college.

Denman lamented that few people in the East had ever heard of M.A.C. This put him at a severe disadvantage when he recruited against better-known eastern football powers. In late August, Denman happily reported to Snyder that George W. Talladay would be attending M.A.C., and he asked him to find a room in a dormitory or near campus for his new recruit. Confident that his efforts to field a good football team would succeed, Denman assured Snyder that "unless something very unexpected happens this fall, we will have a great football team."[34] Unfortunately, the Aggies won only four of nine games and suffered a humiliating 119-0 thrashing at the hands of Fielding Yost's Michigan Wolverines.

For M.A.C. to field competitive teams, the institution had to have adequate facilities. After the Armory was completed in 1886, it served as an all-purpose indoor gymnasium where students played basketball and baseball, wrestled, boxed, performed gymnastics, and practiced track and field events. The baseball diamond moved about campus until it found a permanent home on the south side of the Red Cedar River on land (now known as Old College Field) that M.A.C. had purchased for athletics in 1900. A quarter-mile track, grandstand, bleachers, and a football field were also constructed to accommodate both performers and observers.[35] The gymnasium in the Women's Building, which opened in 1900, provided a facility for women to participate in a range of organized physical activities, including basketball, calisthenics, and gymnastics. When the Gymnasium, now called IM Circle, was dedicated in 1918, M.A.C. finally possessed an indoor facility devoted exclusively to physical education and athletics. The new pool in the Gymnasium made it possible for men and women to learn to swim and to engage in swimming as a competitive sport.

Although Bemies and Denman pioneered institutionally paid coaching at M.A.C, Chester L. Brewer took the athletic program into the twentieth century in a way that won the respect of students, faculty, and administrators. He showed what a director of athletics could do. During his first tour of duty at M.A.C. from 1903 until 1910, he led the Aggies to such dominance in the MIAA that the conference expelled the college (in 1908) after they had won ten consecutive field days.[36] The expulsion freed M.A.C., which had grown larger than the other colleges in the conference, to schedule stronger opponents, but M.A.C. continued to play some games against MIAA schools.[37] Brewer's football teams scored some notable successes against big name foes, including a 0-0 tie with Michigan in 1908, a 17-0 upset of Notre Dame in 1910, and victories over Marquette in 1909 and 1910. His basketball teams had a record of seventy wins and twenty-five losses over seven seasons. Brewer took pride in his 1908 track team that won a dual meet with Notre Dame.[38] More important, Brewer carried out the faculty's desire that all athletes playing for M.A.C. be "bona fide students."[39] He negotiated an end to the freshmen-sophomore scraps that had caused so many injuries and such bad publicity for M.A.C., and he supervised the men's dormitories in a way that engendered student cooperation. Snyder captured the students' esteem for Brewer when he told board member William H. Wallace "There is no man in the institution today who has the influence with the boys that he has."[40] In addition, Brewer managed monies appropriated by the college for athletics efficiently and used his influence to direct the athletic association to expend its funds responsibly.[41] It was a sad day in East Lansing when Brewer's wife's poor health compelled his family to move to the University of Missouri in 1910. (Brewer returned in 1917 and left in 1921.) The couple enriched the lives of many people on campus and in the community.[42]

Football's Rise to Prominence

Between 1900 and 1915 intercollegiate football at Michigan Agricultural College became more than just contests on the field, as the role of the crowd at games took shape. In October 1902 an editorialist for the *M.A.C. Record,* concerned about the lack of student interest in the football team, expressed unequivocally the college's attitude regarding attendance at home games:

> Football is the logical evolution of the forces that have been at work in college life for the past two decades. . . .
>
> The College environment should represent a social unit and it is wrong for any individual in this environment to appropriate continually individual gain without ever making a contribution. All the members of the faculty, every student at the College, and all the Lansing alumni should patronize the home games. A large and appreciative crowd of spectators is an absolute necessity for the success of a game. We have an example worthy of imitation from the large Universities in many of which faculties and students turn out en mass for intercollegiate contests. Why can this not be the condition at M.A.C.?[43]

Fan interest varied with the opponent, and the *M.A.C. Record* pulled no punches when it evaluated each game's following. "The crowd went wild" when the Aggies tied the hated Michigan Wolverines in 1908, but a few weeks later 1,200 students watched their team beat a good Wabash team, 6-0, as if they were "witnessing a funeral procession."[44] The different responses to these two games exposed a potential weakness in the athletic program—too much emphasis placed on defeating Michigan and too little interest in contests against other teams.[45] Students and others attending a game were expected to cheer the home team and make a lot of noise in the process.

Women marching in the 1921 Homecoming Parade call attention to the history of women at M.A.C. Alumni enjoyed returning to campus as early as the 1860s, and their initial organization was formalized in 1868. Homecoming became an annual event shortly before the beginning of World War I. Graduates enjoyed visiting with classmates and old friends, admiring improvements to campus, and opportunities to express opinions. Courtesy of Michigan State University Archives and Historical Collections (Michigan State University, Students, Activities, Homecoming, 1921, Folder 1).

Competing against strong opponents called for pageantry before the game and celebration afterwards that went beyond the grandstands, instilling a spirit of pride in M.A.C. for everyone who came. These activities enabled men and women who were not actually playing the game to expand the contest beyond the gridiron so that they, too, could participate in the event. On 10 November 1909, students, alumni, friends, and "gaily bedecked autos and ladies from Lansing" formed a colorful entourage as they made their way to the football field. Packed into every seat, they cheered and yelled as the Aggies upset Marquette, 10-0.[46] On Saturday afternoon, 29 October 1910, the Aggies stunned Notre Dame, 17-0, before 4,000 excited fans. That night "several hundred students painted Lansing red." They invited Lansing residents to join them in a parade, built a bonfire in town, and called upon local businessmen to make speeches extolling the virtues of M.A.C. and its great football team.[47] Three weeks later, the band met the senior women at the Women's Building and squired them to their seats in the bleachers to root for M.A.C. during their 62-0 rout of Olivet.[48]

On two successive Saturdays in October 1913 the football team lit up the college community with its stunning victories over the University of Michigan in Ann Arbor and the University of Wisconsin in

FOOTBALL
MASS MEETING

ON

TUESDAY NIGHT,
6:30 IN THE ARMORY

to practice yells and songs for the

Wabash Game

SPEAKERS

Pres. Snyder, Capt. Small, Coach Brewer, Dean Bissel,
Dr. Blaisdell, Ass't Coach Halligan

YELLS

(Locomotive yell.)

I. Rah!--Rah!--Rah!
Uz, Uz, Uz.
M. A. C.--Tiger.

II. Osky-wow-wow- -Skinny-wow-
wow.
Skinny-wow-wow- - wow-wow-
wow.

III. Goodness,-Gracious,-Mercy me.
Wabash and old M. A. C,
Willy Willy Wad! Hitem with a
Cod!
Wabash! Wabash! Oh my
Gracious!

IV. Rat-a-ta-thrat! ta-thrat!ta-thrat!
Terrors to lick! to lick! to lick!
Kick-a-ba-ba! Kick-a-ba-ba!
M. A. C.! M. A. C.! Rah! - Rah-
Rah!

V. Ripity rip! Ripity row!
They're up against the real thing
now!
Wabash wants to win renown,
She better go way-back-and sit
down!

VI. Hold! M. A. C. Hold! M. A. C.!
(repeat.)

VII. Block that kick! Block that kick!
(repeat.)

SONGS

I. They say that our team, they cannot play ball,
Play ball all the while, play ball all the while,
They say that our team, they cannot play ball,
Play ball all the while, all the while.--Rah! rah!

II. Down before the farmers, down before the farmers,
Down before the farmers Wabash goes.
Um!--ah!--Tiger!

III. Cheer! boys, cheer! our team has got the ball,
My! oh, my! but won't they take a fall,
And when we hit their line, they'll have no line at all,
There'll be a hot time on the old farm tonight.
M. A. C. (Repeat all.)

IV. We are, we are, we are, we are the M. A. C.
We are, we are, we are, we are the M. A. C.
And when we get to heaven, we'll give the good old yell.
And those who are not so fortunate in Albion will dwell.
Cheer up boys--There is no Albion.

V. Are we all dead yet? Are we all dead yet?
No! by golly! There's eleven left yet.
Come, get your Quinine.

Get wise to these yells and songs before you come
And save these bills for future use

THE M·A·C·
RECORD

VOL. XX   TUESDAY, MARCH 2, 1915.   NO. 21

ALMA MATER.

Close beside the winding Cedar's
Sloping banks of green
Spreads thy campus, Alma Mater
Fairest ever seen.

Chorus
Swell the chorus! Let it echo
Over hill and vale;
Hail to thee, our loving mother,
M. A. C., all hail.

First of all thy race, fond mother,
Thus we greet thee now,
While with loving hand the laurel
Twine we o'er thy brow.

Backward through the hazy distance
Troops the days of yore
Scenes and faces float before us,
Cherished more and more.

College Hall and Wells and Williams,
Abbot and the rest,
Halcyon days were those spent with you,
Days of all the best.

Fold us fondly to your bosom,
Alma Mater, dear,
Sing we now thy endless praises,
Sounding cheer on cheer.

Published by
MICHIGAN AGRICULTURAL COLLEGE ASSOCIATION
East Lansing, Michigan

Music linked together students and alumni separated by age, scholarly interests, and distance. Addison M. Brown, secretary to State Board of Agriculture, wrote "Close Beside the Winding Cedars" for the Semi-Centennial Celebration in 1907. It served as the Alma Mater until "M.S.C. Shadows" replaced it in 1949. Courtesy of Michigan State University Libraries, Special Collections (The M.A.C. Record 20, no. 21 [2 March 1915]: 1).

Many people on campus believed that student enthusiasm spurred the football team on to victory. In anticipation of a tough game against Wabash College, the Athletic Association staged a mass meeting on Tuesday, 22 October 1907, to stir up interest in the following Saturday's game. Included among the speakers were President Snyder, Dean of Engineering George W. Bissell, and English Professor Thomas C. Blaisdell. After the Aggies defeated Wabash, 15-6, before what the M.A.C. Record described as "the biggest crowd that ever attended a football game here," the students celebrated with a "night shirt parade in Lansing." It also appears that President Snyder, who had gone downtown to buy a new hat, was prevailed upon to give a speech standing on a trunk near the corner of Washington and Michigan Avenues. Courtesy of Michigan State University Museum, framed collection.

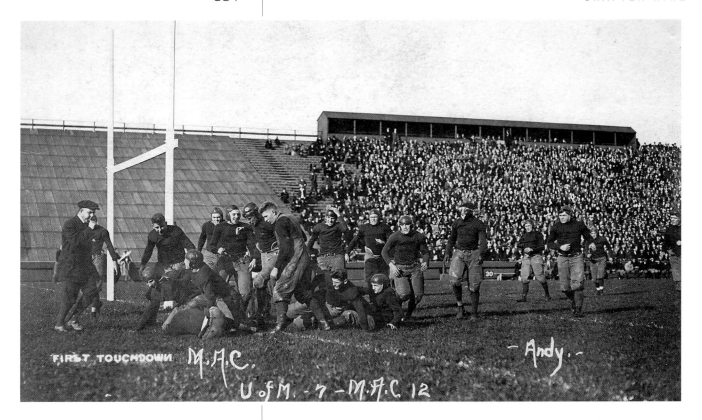

FIRST TOUCHDOWN M.A.C.
U. of M. - 7 - M.A.C. 12
— Andy. —

Carp Julian scores M.A.C.'s first touchdown, less than ten minutes into the game against Michigan, on a three-yard headlong dive. After time expired, M.A.C. fans celebrated like they had never done before. The *M.A.C. Record* for Tuesday, 21 October 1913, captured the exhilaration for its readers:

A thousand joyously raving students climbed, squirmed, wriggled and otherwise proceeded over and through the fence surrounding the field and pounced upon the team. Willing hands quickly grasped the twenty-odd members of the team and lifted them to shoulders. Like the conquering heroes they were, the Aggie team was borne in triumph to the dressing room. And then the rooters hastened pell mell back to the gridiron, where a scene, the like of which no M.A.C. man, past or present, had ever taken part in, was being enacted. Up and down the field marched the band, and in its wake writhed and twisted a human chain, composed of more than a thousand shrieking, crying, absolutely insane fans. Gray haired professors linked arms with the youngest preps, staid alumni cast dignity to the winds, forgot the cares of business, and impartially hammered the backs and shoulders nearest them. For half an hour, the cup of victory was drained by the mob, thirsty after years of disappointment. And then it was not ended, for the scene of operation simply shifted from Ferry Field to the streets of Ann Arbor. From the time the game was over until the special trains began to depart, there was not a single dull moment. On Monday, the band and the team led the student body on a noisy parade to the Capitol to the delight of the residents of Lansing. The joy of victory melted away "the old spirit of antagonism between the town and college forces."

Courtesy of Michigan State University Archives and Historical Collections (Michigan State University, Athletics, Football 1910–1919).

Madison. The Aggies raced onto Ferry Field to the howls of 1,500 students and alumni and salutes by the M.A.C. cadet band to face Fielding Yost's powerhouse on 18 October. In a rugged contest, Gideon Smith, the first African American to play for M.A.C., and Hugh Blacklock dominated play on the line. George Gauthier's passing, and touchdowns by Carp Julian and Hewitt Miller provided points needed to shock the favored Wolverines, 12-7. Halfback Blake Miller sustained a head injury and had to be carried off the field, only to return to play again the next week. When the game ended, Aggie fans carried their "conquering heroes" to their locker room. The fans then returned to the field where the cadet band marched and "in its wake writhed and twisted a human chain, composed of more than a thousand shrieking crying, absolutely insane fans." After sipping "the cup of victory" for half an hour, the crowd and the band took their celebration to the streets of Ann Arbor until they boarded their special trains back to Lansing. When they arrived home, the M.A.C. faithful treated downtown Lansing to another celebration, and the faculty declared a campus holiday on Monday.[49]

The next Saturday the "farmers" headed to Camp Randall on the campus of the University of Wisconsin to take on the Badgers, who had posted a 7-0 record in 1912 and were undefeated thus far in 1913. Many fans went by train, stopping first at St. Johns on the way to Grand Haven, where a ferry took them across Lake Michigan to Milwaukee. The M.A.C. enthusiasts then took another train from Milwaukee to Madison. The band led Aggie fans past the state capitol and finally into the stadium, where they yelled themselves hoarse as the Aggies stopped the Badgers, 12-7. Delirious with joy, the M.A.C. faithful celebrated on the field, marched back to the station, and returned to their excited campus in East Lansing. The *Holcad* complimented both Michigan and Wisconsin supporters for taking their teams' unexpected defeats in a spirit of good sportsmanship.[50]

As the M.A.C. family basked in the glory of Coach Macklin's masterful victories, the celebrants, caught up in the excitement of the moment, seemed to think that M.A.C. had entered a new era. Snyder gloated that it had "taken a good many years of patient, conscientious work to get the team up to this time."[51] During the campus holiday celebration following the Wisconsin game, speakers, including Snyder and Professor Frank Kedzie, exhorted their audience "that with added

Blake Miller makes 10   U. of W. 7 - M.A.C. 12. Sat.

football glory came added responsibilities to the student body." Putting their new obligation into a perspective of Herculean proportions, the *Holcad* went on to say: "The eyes of the entire country are turned upon this hitherto unheard of 'farmer's school,' and our reputation from now on depends upon the way in which the students here carry themselves."[52] Fortunately, Snyder kept his feet on the ground and recognized that it could be a "long time" before the Aggies pulled off a repeat performance.[53] The team finished the season with a perfect record of seven wins and no losses or ties and rightfully claimed the title "Champions of the Midwest." There would not be another undefeated, untied season until 1951.

KEEPING FOOTBALL UNDER CONTROL

The huge football success of 1913 challenged M.A.C. to figure out how to handle the large sums of money generated by the team in the following years. The victory over Michigan guaranteed that twelve to fifteen thousand spectators would show up to see if the Aggies could repeat the feat when their arch rival took the field in East Lansing. A big crowd would spend a lot of money for admission to the game, and Snyder wanted tight control over these funds. Since Macklin had his hands full coaching the team, Snyder urged Jacob Schepers, the college cashier, to handle the sale of tickets and the accounting of the receipts, all of which were to be credited to the athletic association.[54] Tickets, of course, had to be passed along to people whose support the college valued, including Governor Woodbridge N. Ferris.[55] The frenzy created by the "big time" game with Michigan caused Snyder to question, if only for a moment, whether "the emphasis is not placed on the wrong track." In his next breath he reassured himself that M.A.C. was doing "better work" with its current 2,000 students than it had done with 200 a few years before.[56] Big-time football, or at least the desire for it, had become a permanent part of M.A.C.'s psyche.

In 1915, hoping to gain more institutional control over intercollegiate sports, the college strengthened the Board in Control of Athletics that had been established a few years earlier to allow for more faculty supervision. Student oversight of athletics virtually disappeared as faculty and administrators now held all of the positions of power.

OPPOSITE, TOP: Like the week before, the Aggies simply outplayed a more experienced opponent to the absolute delight of their supporters. Bursting with pride, the *M.A.C. Record* issued on Tuesday, 28 October 1913, reported to Aggies everywhere that "the Green and White machine had been much under-estimated." The paper also noted:

> The new style of football never showed to better advantage than in last Saturday's game. While a majority of the substantial gains were made on the line, there were forward passes and end runs in profusion, with several exhibitions of broken field running which brought the 8,000 spectators to their feet. Blake Miller is pictured gaining ten yards; he also scored the first touchdown.

Courtesy of Michigan State University Archives and Historical Collections (Michigan State University, Athletics, Football 1910–1919).

OPPOSITE, BOTTOM: One hundred and seventy-five students, alumni, and supporters of the M.A.C. football team traveled to Madison to cheer their heroes. They rode on this Grand Trunk coach from Grand Haven on their way back to campus. Note the members of the band in the lower left hand corner, posing with the rest of the contingent at St. Johns, Michigan. Alumni gloated over the Aggies' success in 1913 long after the season had ended. In early December the band played "Hail, Hail, The Gang's All Here" and other songs and marches as it escorted the team (riding in automobiles) down Fort Street in Detroit for a banquet to honor M.A.C.'s shining stars at the Hotel Tuller, hosted by the Detroit alumni club. Courtesy of Michigan State University Archives and Historical Collections (Michigan State University, Athletics, Football, 1910–1919, Fans).

The board consisted of two faculty members, one of whom had to be the chairman, two alumni, four undergraduates, and four ex-officio members—the president and secretary of the college and the director and assistant director of athletics. The chairman and the director of athletics comprised the finance committee, who had to approve expenditures of funds received by the athletic association or the board. The two faculty members and the director of athletics certified the eligibility of athletes. If a student wished to participate on an M.A.C. team, he had to fulfill the following requirements: be a bona fide student for at least one year, never have played professionally, have academic deficiencies or failures in no more than ten credits, play under his own name, and receive no remuneration or pay for playing. The athletic director and two of the undergraduates looked after buildings and grounds, and the chairman appointed three board members to make sure that there was appropriate music at athletic contests. By 1925 the board had shrunk to seven members: the president and one member of the State Board of Agriculture as ex-officio members, two faculty, the secretary of the Michigan State College Association, a member of the Alumni Varsity Club, and one student.[57]

Success on the gridiron led boosters to support the athletic program in ways that created headaches for college administrators. A flap in 1916 over M.A.C.'s efforts to recruit the best athletes in Michigan reveals problems related to athletics that arose off the field. In November, the Board in Control of Athletics decided to write letters to alumni living throughout Michigan urging them "to use every legitimate means" to encourage star athletes to enroll in M.A.C. The Lansing Automobile Club, an organization of prominent Lansing businessmen, immediately offered to help in this endeavor. The club had previously raised funds to send the football team and the band to play at Wisconsin, Penn State, Nebraska, and Ohio State. In addition, for several years C. P. Downey, one of Lansing's foremost businessmen and owner of the Downey Hotel, had turned his cottage at Pine Lake (Lake Lansing) into the football team's training camp before the start of classes. Downey hired a chef to make meals (from food that had been purchased by the automobile club) for the team. Club members also promised to help newly arrived athletes find good summer jobs, the only financial inducement allowed by the college. After the *Detroit Free Press* reported this scheme, President Kedzie told Downey and

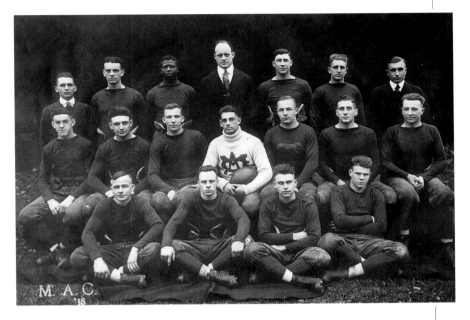

The 1915 football team pummeled Michigan, 24-0, in Ann Arbor, only to lose their next game to Oregon Agricultural College, 20-0, in East Lansing. Although that loss cost the Aggies a perfect season, they rebounded to close the season by whipping Marquette, 68-6. The team won five games and lost one. Team members: (*top row*) George E. Gauthier, assistant coach; Herbert Day Straight; Gideon Edward Smith; John Macklin, head coach; Hugh McNeil Blacklock; Hiram Hewitt Miller; M. S. Fuller, manager; (*center row*) Howard E. Beatty; C. O. Butler; Lyman L. Frimodig; Blake Miller; Ralph B. Henning; Jerry De Prato; Del Van Dervoort; (*bottom row*) Clarence R. Oviatt; Harold A. Springer; Robert Heubel; Hilmar Andreas Fick. Courtesy of Michigan State University Archives and Historical Collections (Michigan State University, Athletics, Football, 1910–1919, Team Members, Varsity Team, 1915).

his friends that he appreciated their support, but he reminded them that the college controlled athletics, not them. He reiterated his intention that athletics "were going to be kept free from the taint of professionalism as they had been in the past." Tension between outside supporters of athletics and institutional jurisdiction became one aspect of the growing sports program.[58]

INTERSECTIONAL COMPETITION AND A NEW STADIUM

In 1914 M.A.C. began to play intersectional football games against larger institutions, which enabled it to connect with some of its graduates and gave it exposure beyond Ohio, Indiana, Illinois, and Wisconsin. The University of South Dakota became the most distant team to play M.A.C. when it came to East Lansing in 1913. The next year, Coach Macklin took his Aggies on the road, losing to the University of Nebraska in Lincoln, 24-0, and defeating Pennsylvania State College in State College, 6-3. The visit to Nebraska created an opportunity to bring together alumni living in the area to meet the team, share a meal, and exchange memories from student days at M.A.C. In this way, the college administration began using football as an instrument to cultivate relations with the alumni.[59] The triumph over Penn State demonstrated that M.A.C. belonged on the same field as an Eastern

M.A.C. Field became the home of Aggie football with the dawn of the 1923 season. On 10 October 1924, M.A.C. dedicated the stadium with a game against the University of Michigan. Nearly 20,000 fans packed the stands to watch the Wolverines spoil the Aggies' party, 7-0. In 1935 the facility was renamed Macklin Field in honor of Coach John F. Macklin, and in 1957 it became Spartan Stadium. This picture shows the largest crowd ever to have attended an athletic event at the college up to 1924. Courtesy of Michigan State University Archives and Historical Collections (Michigan State University, Athletics, Football, 1920–1929, Action Shots).

power. Oregon Agricultural College traveled from Corvallis to East Lansing in 1915 to spoil M.A.C.'s hopes for a perfect season by trouncing the Aggies 20-0 a week after Jerry DaPrato, Gideon Smith, Hugh Blacklock, and the brothers Blake and Hewitt Miller had led M.A.C. to a 24-0 pounding of Michigan.[60] Losing, however, was part of the price to be paid for playing first-rate competition. The 1913 and 1915 seasons proved to be the high water mark for football until strong teams were built in the 1930s.

Sagging fortunes in football after World War I prompted a movement to build a stadium large enough to lure young men to play for the Aggies and to entice major university teams to put M.A.C. on their schedules. Board member L. Whitney Watkins and Governor Alex J.

AGGIES			LINE-UPS	ALBION		
Name	Weight	Number	Position	Number	Weight	Name
Kipke	142	28	L.E.	15	173	Upton
Taylor (Capt.)	177	2	L.T.	13	175	Eastman
Eckert	187	4	L.T.		178	Cretcher
Hultman	174	15	L.G.	9	170	Coin
Eckerman	172	3	C.	19	185	Olson
Hackett	190	29	R.G.	12	178	Smith (Capt.)
Haskins	167	7	R.T.	20	185	Preshaw
Edmonds	160	14	R.E.	21	185	Lightbody
Richards	160	25	Q.	3	150	Williams, V.
Lloret	170	22	L.H.	10	160	Tumblyn
Schmyser	105	27	R.H.	5	155	Williams, H.
Neller	170	24	F.B.	22	185	

SUBSTITUTES—
Aggies: Thayer (5), Vogel (8), Mortensen (9), Place (10), Speikerman (11), H. Smith (11), Goode (13), Robinson (16), Schultz (17), McInnis (18), Burris (19), Frank (20), Gasser (21), Beckley (22), Crane (28), Boehringer (30), Lyman (31), Gofton (32), Thompson (33), Pearl (34), Vogelsang (35), Elliot (36), Dobben (37), D. Smith (38), Anderson (39), Caywood (40), Murray (42), Thomas (44).
Albion: Spannenburg (2), Griffin (4), Fisher (6), Barlow (8), Pahl (11), Bedient (14), Carcarelli (16), Cooper (18), Price (23), Boldt (24), Voelker ( ), Ship ( ).
OFFICIALS—
Costello, (Georgetown)—Referee. Roper, (West. State Normal)—Umpire. Dalrymple, (Knox)—Linesman.

Football Program

ALBION COLLEGE
vs.
MICHIGAN AGGIES

M. R. TAYLOR   Captain, 1923

Saturday, October 20, 1923
New Stadium

Program from the 20 October 1923 football game, the second game played at Michigan Agricultural College's new stadium. Courtesy of Michigan State University Museum Collections.

Groesbeck led the charge that convinced the legislature to loan the college the money needed to construct a facility that seated 15,000 people.[61] With the 1923 season opener fast approaching, the *M.A.C. Record* bubbled with pride as it described M.A.C. Field, renamed Macklin Field in 1935, for its readers:

> Early fall finds the stadium practically completed and ready for shouts of partisans such as this great game only knows, ready for the thud of pigskin and the whistle of the referee. The great concrete stands are gaunt in their gigantic outlines, the green sod offers a welcome and seats are being placed for the first crowds of the season. Goal posts of iron pipe are in place, the track is being filled with cinders and a fence is being constructed around the field with a brick gateway at the southern end and brick ticket offices.[62]

New coach Ralph H. Young's team beat Lake Forest College and Albion College and lost to Ohio Wesleyan University and Creighton University on the new "green sod."

COALESCENCE OF AN ATHLETIC PROGRAM

Baseball

Although football garnered the most attention, M.A.C. athletes competed in other sports throughout the academic year, which led to the coalescence of a well-rounded athletic program. Success on the baseball diamond instilled a sense of pride in the college. The 1901 student yearbook, the *Wolverine,* boasted that since 1888 M.A.C. had won nine MIAA baseball championships.[63] In a game played at M.A.C. in April 1903 the Aggies nipped Michigan, 10-9, before 1,000 fans. The *M.A.C. Record* joyfully reported:

> The ninth came and Millar, first man up, drove the ball through the infield, Tower got out but Pinance sent a line drive over Roch's head to the river [,] Millar came in with the winning run and the crowd swarmed onto the field, the players being carried off on the shoulders of the enthusiastic onlookers.
>
> The campus could not hold anyone after the game and a crowd of six hundred paraded the streets of Lansing during the evening. The procession was headed by the band followed by the players hauled by the students. Several Lansing business firms burned fireworks along the line of march and cannon crackers helped on the celebration. When the crowd returned to the campus bonfires and speeches by the members of the team ended the events of the day.[64]

Nothing tasted better to Aggie supporters than a victory over Michigan.

The baseball team pioneered M.A.C.'s entry into intersectional competition, and the institution came to see this as an important means to promote itself before an ever larger public. John Macklin's 1911 team's 11 and 5 record was highlighted by victories at home over Ohio State University and Syracuse University.[65] A trip east came to be a regular part of the schedule each season. In 1916, the team, coached by John Morrisey, claimed victories on their road trip over the universities of Buffalo and Rochester, and lost to Syracuse.[66] In 1926, the year after Michigan Agricultural College became Michigan State College, second-year coach John H. Kobs took the team on its first trip into the South between the winter and spring terms. The college hoped that this endeavor would "increase interest in [M.S.C.] college baseball by developing a real diamond machine." Another sign

Baseball was the first sport in which the State Agricultural College competed against teams from outside of the college. The Stars played their first game in 1865, and it is possible that the college's men had organized a team even earlier. The squad pictured here is from the 1890s. During that decade, M.A.C. played a total of 67 games—winning 32, losing 33, and tying 2. The length of the early schedules ranged widely, from the 1892 team that won the only two games it played to the 1898 team that had a 6-5-1 record. Courtesy of Michigan State University Archives and Historical Collections (Michigan State University, Athletics, Baseball, 1890–1899, Team members).

Monogram Men in College

A. L. Alderman—Track
H. E. Beatty—Track
A. L. Bibbins—Baseball
C. F. Barnett—Track
H. M. Blacklock—Football
W. W. Blue—Track
L. A. Cobb—Football
F. G. Chaddock—Football
C. T. Dendel—Football
N. J. Daprato—Football, Basketball
R. E. Dinan—Track
L. L. Frimodig—Basketball, Baseball
H. A. Fick—Baseball
M. S. Fuller—Baseball
R. B. Henning—Football
K. W. Hutton—Football
C. R. Herr—Track
C. C. Hood—Baseball, Basketball

D. F. Jones—Track
G. E. Julian—Football, Track
C. W. Loveland—Football, Track
A. M. La Fever—Baseball
A. L. McClellan—Basketball
H. H. Miller—Football, Basketball
W. B. Miller—Football, Basketball Baseball.
O. R. Miller—Football, Basketball.
C. H. Peterson—Baseball
D. L. Peppard—Track
F. G. Ricker—Basketball
G. E. Smith—Football
H. D. Straight—Football
Miss Ethel Taft—Tennis
L. F. Vaughn—Football
A. D. Van Dervoort—Football
M. G. Weeder—Baseball

Female tennis players earned monograms as did male athletes who participated in intercollegiate tennis and other sports. The list of "Monogram Men in College" that appeared in the 1915 edition of the *Wolverine* included Miss Ethel Taft, '16. Courtesy of Michigan State University Archives and Historical Collections (*Wolverine*, 1915, 163).

that M.A.C. had grown into the "State College" was that the 1926 *Wolverine* referred to the team as the "Spartans."[67] George Alderton, of the *Lansing State Journal,* first used "Spartons" [*sic*] to refer to an M.S.C. team when he wrote a story about the baseball team's southern trip in 1926.[68]

Michigan Aggie
Schedules

BASKETBALL
INDOOR TRACK
WRESTLING
SWIMMING

1925

M. E. NUTTILA
Basketball Captain
1925

Michigan Aggie
Schedules

SPRING, 1925

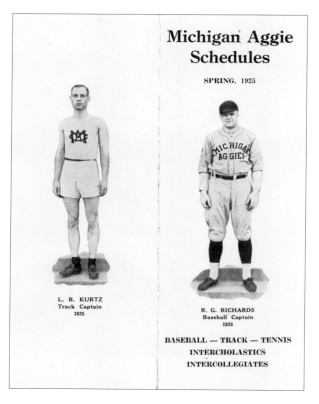

L. B. KURTZ
Track Captain
1925

R. G. RICHARDS
Baseball Captain
1925

BASEBALL — TRACK — TENNIS
INTERCHOLASTICS
INTERCOLLEGIATES

This series of printed athletic schedules shows the uniforms worn by the college's athletes in 1925. It also illustrates a change in the college's logo during the course of the year to reflect the institution's new name—Michigan State College of Agriculture and Applied Science. The 1925 football and cross country schedule may have been the first appearance of the block "S" on a college publication. Courtesy of Michigan State University Museum Collections.

DON HASKINS
Captain, 1925

Michigan State
College

S

FOOTBALL
CROSS-COUNTRY
1925

Basketball

Women fielded the first basketball team at M.A.C. to play against an off-campus rival. In late January 1898, Lansing High School's women's team defeated M.A.C. 16-2. A week later the two teams met again in the Armory, where the Lansing High Schoolers wore down their opponent en route to a 26-6 triumph. The women played four ten-minute quarters. Sara G. Lowe, '97–'00, scored two baskets and Katherine Clute, '97–'98, added another for the losing team. Readers of the *M.A.C. Record* learned that although their team showed promise of improvement, it needed more practice:

> Since the game a week ago, our girls have improved much in defensive work, but are still weak in the offensive. They lack team work, pass the ball without looking to see where it will go, and are not aggressive enough in their playing. Miss Birdie Dean is easily the strongest and most darling player on our team, but there are several others who will make good players when they get more practice. The weakening of our team toward the end of the game shows that they need more vigorous practice to develop endurance.[69]

For the next few years, the women's team continued to compete and with more success. In 1900 they took revenge on Lansing High School, 16-4, and three years later they defeated the teachers from the Flint School for the Deaf, 17-15.[70] In February 1901 women students at M.A.C. entertained the Normal School team from Ypsilanti before the "Normalists" embarrassed their hosts 26-0. Hoping to soften the blow, the *M.A.C. Record* pointed out that some of Ypsi's players were "specialists" in their institution's physical training department.[71]

After the early years of the new century, intercollegiate women's basketball at M.A.C. faded away, but the competitive spirit among M.A.C. women never wavered as they formed class teams for intramural play. In 1899 freshmen organized two teams and the sophomores one to compete against each other.[72] Two years later, the top players from six intramural teams made up the varsity, who stood ready to answer "challenges" from other colleges.[73] Two elite teams, also drawn from the six teams—the Reds and the Blacks—played "a very interesting game of basketball" in March 1902. The Reds defeated the Blacks by a score of 37-21.[74] The seniors' team in 1913 called upon Norman

The 1903 men's basketball team: Identification of team members is made possible because their last names are written on the back of the photograph and first names have been taken from student lists in the 1903–04 catalog: (*top row*) Robert A. Bauld, Wilson Floyd Millar, William E. Morgan; (*middle row*) H. Foley Tuttle, Edward Balbach, John Philip Haftenkamp, Roy R. Tower; (*front row*) John Edward Schaefer. The team won all six of its games: M.A.C 43, Detroit Y.M.C.A. 8; M.A.C. 49, Hillsdale 2; M.A.C. 23, State Normal 7; M.A.C. 19, Governor's Guard 7; M.A.C 49, State Normal 5; and M.A.C. 42, Grand Rapids Y.M.C.A. 7. Courtesy of Michigan State University Archives and Historical Collections (Michigan State University, Athletics, Basketball, 1900–1909).

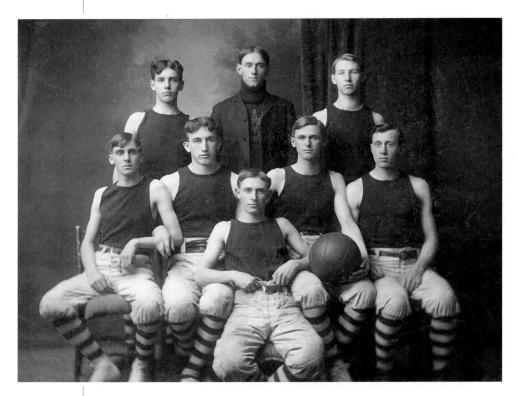

M. "Baldy" Spencer, '14, a member of the men's varsity, to coach them in preparation "to meet any other ladies competition."[75]

After weeks of practice, a team of M.A.C. men traveled to Olivet College in late February 1899 to play the college's first men's intercollegiate basketball game. Olivet prevailed 15-6. A rematch in the Armory at M.A.C. saw the teams tied after two twenty-minute halves; Olivet scored on a free throw eight minutes into a ten-minute overtime to notch a 7-6 win. Criticism of officiating by M.A.C. fans originated in the institution's first men's home basketball game. According to the *M.A.C. Record:* "The game was hard and fast from start to finish but was characterized by entirely too many fouls that were not noticed by the referees."[76] The next season, the men's team took pride in its 25-8 victory over a visiting team from the State Normal School in Ypsilanti as the highlight of its 2-3 season.[77] The installation of a new court "in the center of the armory" enabled all of the fans to see every play of M.A.C.'s win over the Detroit YMCA when the two teams started the 1903 season.[78] The "bird cage armory" proved to be a friendly home for the Aggies and a nightmarish den for opposing teams until the Gymnasium opened in 1918. M.A.C.'s men's basketball teams posted a composite record of 143 wins and only 68 losses beginning with the 1900–01 season through the 1916–17 season.

The men's basketball team's schedule usually did not include many large schools. Conscious of this reality, M.A.C. highlighted the strength of some opponents. Coach George E. Gauthier's 1917 squad (11 and 5) won victories over the Detroit YMCA, Wabash, Illinois Athletic Club, Northwestern, and Buffalo teams, who were described as "the best in this part of the country," whatever that meant.[79] The Aggies won all of their home games in the "bird cage" during the 1917 season. Lyman M. Frimodig, who won ten monograms in football, baseball, and basketball from 1914 until 1917, coached the 1921 team to a 13 and 8 record. The season's results provide a representative example of an M.A.C. schedule during the first thirty years of men's basketball.[80]

M.A.C.		OPPONENT	M.A.C.		OPPONENT
26	Mt. Pleasant Normal	21	21	Oberlin	23
30	Kalamazoo College	18	37	Notre Dame	25
31	Hope	17	24	Michigan	37
22	Western State Normal	16	23	De Pauw	26
40	Mt. Union	26	26	Mich. College of Mines	19
19	De Pauw	39	29	Oberlin	37
23	Notre Dame	36	10	Michigan	17
29	Western State Normal	19	27	Hope	23
20	St. John's Univ.	21	20	Grand Rapids Jr. Coll.	11
30	Mt. Union	28	41	Bethany	18
27	Creighton	20			

The mid 1920s saw the Aggies' and Spartans' fortunes sink as they lost most of their games during the 1924–25 and 1925–26 seasons prior to Benjamin F. Van Alstyne becoming the coach.

Tennis

Tennis made a special contribution to intercollegiate athletics at M.A.C. because it was the first sport that both women and men participated in. Chippie Harrison, '88, and Mamie Smith, '89, served on the "committee of control" for the "Racket," an association formed by students in April 1888 "for the purpose of getting a reduction on tennis

Women and men formed a tennis team that competed with teams representing other schools in field days and in individual matches. This is believed to be M.A.C.'s team for 1905. Courtesy of Michigan State University Archives and Historical Collections (Michigan State University, Athletics, Tennis).

supplies, to have a system in games, and to select and train contestants for the annual field day."[81] Apparently no women competed in the 1888 field day, but women played both singles and doubles matches in 1896 and subsequent years.[82] After 1908, when M.A.C. no longer took part in the MIAA field day, a tennis team comprised of women and men engaged in matches against such old foes as Olivet. The first known contest took place in May 1909, when Hazel Taft, '10, and Margaret Kedzie, '11, each won their singles match and together won a doubles match against opponents from Olivet. The men won two of their three matches.[83] After 1916 the women's tennis team played a schedule separate from the men's team.[84] The 1920 *Wolverine* celebrated their status on campus: "Tennis is the only branch of inter-collegiate athletics in which the co-eds take part, consequently their activities are followed with great interest." The 1920 *Wolverine* also claimed that 1919 team members were the first women to win monograms for their athletic excellence.[85] But three years earlier the *Wolverine* had credited Haidee F. Judson, '17, with winning monograms in 1914, 1915, and 1916, and Julia M. Rigterink, '18, with winning one in 1916.[86] (The 1915 *Wolverine* lists Ethel Taft, '16, as a monogram winner.) By the mid 1920s the men's team was no longer playing MIAA teams, and it scheduled Michigan, Central State Normal, Western State Normal, Detroit City College, University of Detroit, and Oberlin in 1925.[87]

Track and Field

Beginning in the 1880s, track and field became an integral part of M.A.C.'s athletic program. The desire of male students to compete against each other in running, jumping, and throwing motivated them to organize track and field events, which in turn became one of the driving forces behind the establishment of the MIAA. Students put on meets in the Armory during winter to help determine who would represent the college at the conference's annual field day in the spring.[88] M.A.C.'s teams did very well in these games; in 1900, for example, the Aggies won twelve gold medals, leaving only eighteen for the other six schools.[89] M.A.C.'s dominance in track and field was one of the reasons the college's membership was terminated in 1908.[90] M.A.C., however, continued to face MIAA teams in some sports during the second decade of the twentieth century, and as late as 1916 they were "not quite ready" to step up to the level of the University of Michigan.[91] M.A.C. athletes competed in the Western Conference (Big Ten) annual championship meet, which was open to teams from all over the country, from 1901 until 1926. In 1908 Ralph J. Carr, '08, won the two-mile race at the Big Ten meet in Chicago.[92] Track and field at M.A.C. suffered after World War I called athletes to service, but a period of reconstruction elevated the college to a higher level in the 1920s. M.S.C. took part in the Ohio Relays in Columbus, the Drake Relays in Chicago, and the Big Ten meet in 1926. Bohn Grim, '27, and Frederick P. Alderman, '27, both entered the upper stratum of college sprinters after their performances at the Illinois Relays on 27 February 1926. Grim tied the world record for the 75-yard run (:07.6), and Alderman did the same in the 300-yard dash (:31.2). Alderman had previously won the Big Ten championship in the 220-yard dash (:21.12) at Ann Arbor in 1925.[93]

Cross Country

Cross country has occupied a niche in the college's athletic program ever since it debuted in 1907. The first meet saw two teams of M.A.C. students run a course that followed both sides of the Red Cedar River.[94] In 1910 the Aggies competed in an intercollegiate meet for the first time. They won the Hope College Invitational, defeating runners from Olivet, Hope, the Grand Rapids YMCA, and Muskegon High School.[95] The next year M.A.C. harriers took second place in the same event.[96] Michigan, Albion, and Western State Normal traveled to M.A.C. for the college's only meet in 1916, which Michigan won.[97] The 1920 team beat Michigan, won the state intercollegiate cup, and placed eighth (out of fourteen teams) in the Big Ten run at Urbana, Illinois.[98] Five years later M.A.C. engaged in seven meets, highlighted by wins over Michigan and Marquette.[99]

Swimming and Diving

The natatorium in the Gymnasium enabled M.A.C. students to take up swimming and diving as a course in physical education and as a competitive sport. If women wanted to receive credit for the swimming course first offered in 1919, they had to pass a final examination that included satisfactory execution of a face float, back float, back stroke, and a preliminary dive.[100] Two years later a swimming course was added for men.[101] Intercollegiate competition soon evolved, and in 1923 a team of eight women traveled to Detroit to vie with a squad from the Detroit Junior College at the Detroit Athletic Club. Their male counterparts, who were in their second season of competition, won their first meet at home against Grand Rapids Junior College before losing home and home meets to Michigan and a meet to Indiana University in Bloomington.[102] Spectators in attendance at the East Lansing pool on 1 February 1924 watched the men's varsity defeat Ypsilanti Normal, 44-24, and the women's team lose 39-11.[103] Swimming, as a team sport, was still in its infancy.

Rifle Shooting

Male sharpshooters engaged in intercollegiate rifle competition for many years before women took their places at the firing line against the University of Illinois in 1920.[104] The Military Department organized, coached, and supervised this sport. The men won a national

GIRLS RIFLE TEAM   M.A.C 1923   PHOTO BY HARVEY PHOTO SHOP

Rifle shooting proved to be quite popular among female students in the 1920s. On 25 February 1924 the *M.A.C. Record* reported: "Nearly forty co-eds qualified for the boxes of candy offered by the military department to the sharpshooters among the girls who could register five consecutive bull's eyes." Having acquired refined skills in shooting, women fielded a varsity rifle team, trained by staff from the military department, that competed at the intercollegiate level. The women's team for 1923 appears in this photograph. *Courtesy of Michigan State University Archives and Historical Collections (Michigan State University, Athletics, Rifle Team).*

championship in 1916, but Washington State College prevailed over M.A.C. the next year.[105] In 1922 the college military staff began training women shooters to compete against other teams. In 1924 the women's team took on Illinois, beating them 491-489 in an engagement where scores were exchanged by telegraph.[106] During the 1926 season the men won twenty-eight matches with colleges located across the country, which were also conducted by telegraph. At the end of that season the women's team had a record of ten wins, ten loses, and one tie.[107]

Ice Hockey

Though ice hockey did not have a lasting presence on campus until the mid 1920s, in 1906 M.A.C. students from northern Michigan, undaunted by the cold of January, formed a hockey team that practiced "on a sheet of ice near Cushman's farm" and played two games against Lansing High School above Piatt's dam. M.A.C lost that first encounter 2-1, but came back to take the second by a score of 3-1.[108] Thirty-five students, many of them from the Upper Peninsula, answered Athletic Director Chester Brewer's invitation to put together a hockey team in January 1922. One day after the squad organized they lost 4-1 to the University of Michigan in Ann Arbor. The next week the team suffered a 3-1 defeat at the hands of the University of

Notre Dame on a rink constructed above the dam on the Red Cedar.[109] If hockey was to survive and thrive it needed a better facility, so the college built a new rink behind the Women's Building in 1926.[110] John H. Kobs, who had played hockey at Hamline College, coached the 1926 varsity team, which lost all four of its games: two to Michigan and one each to the Fisher Body Company Team of Battle Creek and the University of Minnesota.[111]

Wrestling and Fencing

Modern wrestling and fencing teams competed for the first time in 1922 and 1925 respectively. From its beginning, the M.A.C wrestling team tangled with topflight competition. In their first season, Coach Jimmy Devers's grapplers lost to Iowa State, Indiana, and Michigan, but they defeated Michigan 24-20 in Ann Arbor.[112] The 1925–26 team did not shy away from stiff competition either, although they lost all five meets to Indiana, Purdue, Cornell College (Iowa), Ohio State, and Michigan.[113] O. M. Lebel, an instructor in French, organized a fencing team that engaged in only one contest during the spring of 1925, an 8-1 loss at Michigan. Yussef Waffa, '27, a student from Egypt, coached a team for 1925–26, and led them to a victory over Michigan by a score of 10-6. There had been at least one fencing match previous to that, when Shoichi Yabina, '95, defeated an opponent from the State Normal School during the MIAA field day in 1895.[114]

Formation of the Women's Athletic Association

Though opportunities for women to participate in intercollegiate athletic competitions at the varsity level were limited to tennis, swimming, and rifle shooting in 1925–26, the Department of Physical Education for Women offered courses in field hockey and soccer, basketball, baseball and track, tennis, dancing, swimming, gymnastics, campcraft, and playground management.[115] But women were not satisfied with instruction in a sport for long; they often formed teams in order to compete against each other. The *M.A.C. Record* reported how quickly baseball caught on after it was taught in physical education classes in 1919:

> That college girls can play baseball and become regular fans is being demonstrated on the drill grounds this week during the girls'

morning gymnasium classes. For the first time in the history of the
college, the drill ground, that has been sanctuary of the art of war
and other sports purely masculine was invaded by feminine athletes. The spring schedule of out-of-door exercises is being applied
to the women as well as men with the result that girls are playing
baseball these nice spring mornings and so rapid has been their
progress in learning the game that already they are arranging an
interclass schedule.[116]

Another step forward came in 1925 with the birth of the
Women's Athletic Association, a student group formed to oversee
women's sports. Its president, Margaret Foote, '26, officiated over a
membership of ninety. The association sponsored competition
between teams representing each college class to determine class
champions in baseball, field hockey, soccer, basketball, volleyball, and
rifle. It also put on an annual women's field day comprised of a 50-
yard dash, 60-yard low hurdles, a running high jump, javelin throw,
discus throw, basketball throw, baseball throw, and sack race. Accomplished female swimmers made up the Women's Life Saving Corps of
Michigan State College.[117]

Women learned how to play field hockey in
courses offered by the Department of Physical
Education. The game in progress in this picture,
circa 1922, may have been between teams representing two different classes as part of the
annual competition to determine class
supremacy in athletics. Courtesy of Michigan State University Archives and Historical Collections (Michigan State University,
Athletics, Field Hockey).

Looking to the Future

Although sports at Michigan Agricultural College had grown into a highly organized program by 1925, intercollegiate athletics at Michigan State College had a long way to go to reach the upper levels in a very competitive business. Several of the men who joined the athletic department in the mid 1920s helped prepare M.S.C. for its eventual entrance into the Big Ten in 1949. Ralph H. Young provided leadership as athletic director from 1923 until 1954. He also coached men's track until 1940. Serving with him for many of those years were coaches John H. Kobs, baseball; Benjamin F. Van Alstyne, basketball; and Charles D. Ball Jr., men's tennis; and trainer Jack Heppinstall. The new football stadium made it possible for M.S.C. to host the big-time opponents that gave the college the exposure necessary to be considered for Big Ten membership. Perhaps activity in the Gymnasium in 1925 foretold the future better than the promising coaching staff that was then arriving on campus. Students kept the gym humming from 6:30 A.M. until 10:00 P.M. Four hundred and thirty-five women participated in intramural soccer, field hockey, basketball, swimming, bowling, tennis, indoor baseball, rifle, canoeing, and track, and 1,374 men played intramural horseshoes, basketball, indoor baseball, indoor track, bowling, swimming, baseball, and outdoor relays in 1924–25.[118]

When college officials looked to the future in the mid 1920s, what they wanted most was membership in the Big Ten conference. To that end, the Board in Control of Athletics voted on 8 February 1923 to send its chairman, Professor Ralph Chase Huston, to Chicago to attend the Big Ten's spring meeting "with the view of determining the status of the Michigan Aggies, and if possible, to lay the grounds for reception of the school into the conference."[119]

The First World War Turns M.A.C. into a Military Camp

THE FIRST WORLD WAR SERIOUSLY DISRUPTED THE LIVES OF M.A.C. students and faculty, changed the physical and social landscape of the campus, and brought on new opportunities for the college to carry out its land-grant mission. When the war erupted in Europe in August 1914 most Americans hoped that their country's soldiers would not have to fight in it. German advances on the continent and its submarines' deadly attacks on Allied and American ships forced the United States to declare war on Germany on 6 April 1917. The United States committed the nation's resources to defeat the Central Powers, and it called upon the M.A.C. community to devote the next year and a half to do all it could to aid the Allied cause. Students, alumni, faculty, staff, and administrators fought and died on the battlefields of Europe, worked with Michigan farmers and homemakers to bolster the supply of food, and saw their campus turned into a virtual military camp. M.A.C. people joined government efforts on federal, state, and local levels to meet needs generated by the war. Even as programs related to the war disrupted campus life, M.A.C. expanded the way it trained young people to serve the public.

It is remarkable how little anti-German rhetoric appeared in the *M.A.C. Record* and the *Holcad,* M.A.C.'s two weekly newspapers during the war. The papers reported changing circumstances on the campus, emphasizing M.A.C.'s contributions to the war effort. In many parts of the country, people directed strong anti-German feelings against Americans of German descent along with a condemnation of Germany for causing and prosecuting a brutal war. No doubt there were people at the college who were prejudiced toward ethnic Germans, but the administration and faculty neither encouraged nor promoted such opinions. The college rallied its students, faculty, alumni, and staff to support the war because the defeat of Germany was in the best interests of the United States. Professor of History and Political Science E. H. Ryder might have captured the mood of the campus when he wrote a three-part series of articles titled "The World Conflict with Militarism," published in the *M.A.C. Record* in April and May 1918.[1] Ryder gave a balanced account of the reasons for "each nation's participation in the struggle." But he blamed Germany for the war when he concluded: "It seems apparent that the claims of Germany in calling it a war of defense are absolutely absurd. Rather her appearance as a leader of a great Pan-German ambition with Austria and Turkey as sworn allies, stands forth in uncontrovertible evidence."[2]

Ryder reinforced President Wilson's condemnation of "Germanism," which had appeared in the *Holcad* on 8 January 1918. Wilson considered the leaders of Germany the real culprits, not the German people. The *Holcad* carried the short article quoted below:[3]

THE PRESIDENT'S DEFINITION OF GERMANISM

Innumerable articles and many books have been written to define "Germanism" and show to the world what it means.

In his message to Congress December 4, President Wilson defines it as follows:

> "This intolerable Thing of which the masters of Germany have shown us the ugly face, this menace of combined intrigue and force which we now see so clearly as the German power, a Thing without conscience or honor or capacity for covenanted peace."

This thing must be crushed, and if not truly brought to an end, at least shut out from the friendly intercourse of the nations, says the President, and it is only when this Thing and its power are indeed defeated that the time can come when we can discuss peace with the German people.

Although the *Holcad* offered no commentary on Wilson's remarks, most people at M.A.C. undoubtedly shared their president's views. America was fighting the war to defeat the "masters of Germany."

The war and its impact upon the people of Michigan caused M.A.C. president Frank Kedzie to reflect upon his institution's history. Remarkably, the land-grant ethos that had generated and sustained M.A.C. had put in place a network of trained men and women to enable

CLASS OF 1917 EXPECTS TO DO ITS BIT.

The U.S. entry into the Great War in 1917 profoundly shaped the immediate future of the Class of 1917. This sketch, which appeared in the 15 May 1917 issue of the *M.A.C. Record*, captures the seriousness of the times and the expectation that M.A.C. graduates would do their part to move the nation's war effort forward. Courtesy of Michigan State University Archives and Historical Collections (*M.A.C. Record* 22, no. 29 [15 May 1917]: 7).

Michigan to meet its wartime obligations in an expeditious manner. Kedzie, who had spent his life at the college, understood the unique capacity of land-grant-educated men and women to serve their country in this time of great uncertainty and need. In a letter, dated 26 October 1917, to Ellwood C. Perisho, who was at the South Dakota Agricultural College, Kedzie, in a reflective mood, captured the moment:

> Unconsciously this college has been preparing for years for this war. Its training of the men of Michigan who come to us as four year and short course students has been to make them more efficient as are also our young women trained for efficiency. It is through them as they are scattered throughout the state that the college is enabled to contribute a wonderful stimulating influence favorable to the progress of the war.[4]

The extension service provided the structure through which M.A.C. quickly put men and women to work in the nation's effort to reduce the domestic consumption of food and increase production of crops needed for the war. As extension workers brought together the kitchen and the field, they put into practice the intent of land-grant education to improve the overall quality of life of American families. Federal funds granted to M.A.C. to make emergency appointments swelled the ranks of county agents, home demonstration agents, and leaders for boys' and girls' clubs. By June 1918 sixty-eight agricultural agents, thirty-seven of whom had been hired in the past year with federal funds, were helping farmers increase their yields by procuring good seeds, working closely with local farm bureaus, and encouraging the use of tractors in seventy-one of Michigan's eighty-three counties. Twenty-four newly employed home demonstration agents, assisted by food specialists from extension, showed Michigan women how to conserve wheat, meat, fat, and sugar, and how to can perishable foods for home use. Extension added five full-time people to work with boys' and girls' clubs teaching young people how to participate in food programs.[5]

The campus evolved into a large classroom, preparing students, farmers, and draftees to become more proficient in their use of new technology and machines, while at the same time creating opportunities for civilian and military personnel to work together. The presence of men in uniform was not a new sight, because male students had been required to spend three hours a week on the drill field in uniform during their first three years at M.A.C. But this activity had not prepared the campus community for the scale of military operations that took place in 1918. At the beginning of the year, the U.S. Signal Corps offered an intensive course in radio technology. Senior engineering students who were registered for the draft enrolled in this full-time program that taught them to operate, build, and repair all sorts of "electrical machinery," skills that they could use in a variety of wartime activities.[6] In March, the Department of Farm Mechanics put on a special two-week short course for farmers to learn to "intelligently purchase, operate, and care for the tractor."[7] Five-hundred draftees arrived from Wisconsin on 15 May to begin an eight-week course in auto mechanics overseen and taught by military officers and instructors who came from off campus. Three Lansing companies—Reo, Oldsmobile, and Duplex—loaned instructional materials that supplemented the ten "Class B" army trucks brought to campus for use in the course.[8]

By October 1918 M.A.C. had taken on the appearance and the function of a military camp. The War Department established a Student Army Training Corps (SATC) unit on campus. Eight temporary barracks that could house 1,400 men and two mess halls were built east of the Horticultural Laboratory to accommodate the 500 men of "Class B" enrolled in the auto mechanics course. Another 500 men of

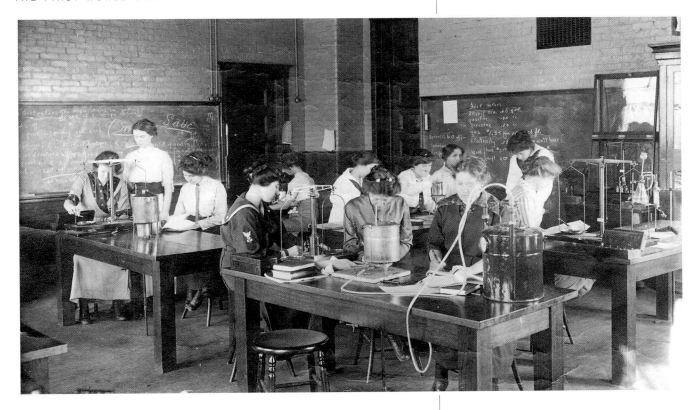

Members of the Class of 1916 getting ready to perform experiments in the Old Physics Lab.
Courtesy of Michigan State University Archives and Historical Collections (College of Human Ecology Collection, Physics).

"Class A," who met the requirements to be regular college students, arrived in October to study subjects useful in the military such as mathematics, chemistry, war history, drawing, and French. They lived in the men's dormitories and on the top floors of Agriculture Hall and the Olds Hall of Engineering. A contingent of fifty from the navy, who were taking a naval training course, joined the SATC men.[9] Less than two weeks after the official opening of the camp, an outbreak of Spanish influenza rendered the program inoperative until the war ended on 11 November. The camp closed without any Class A members completing their studies.

The war profoundly affected life at M.A.C. in other ways, as well. Beginning in November 1917, five mounted members of the Michigan State Constabulary protected vital facilities as they patrolled the grounds encompassing "the power house, and the Engineering, Agricultural, and farm buildings" each night beginning at 8:00 P.M. and lasting until 6:00 A.M.[10] The *Holcad* ceased publishing for a couple of months, and the men's literary societies shut down in fall 1918, as the enrollment of men in the regular curriculum declined.[11] Although the women's societies kept functioning, women directed much of their

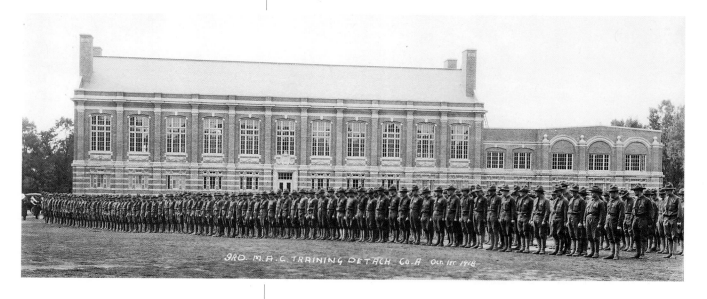

3RD. M.A.C. TRAINING DETACH CO. A Oct 1st 1918

Several significant occurrences converge in this photograph taken on 1 October 1918. Within two weeks, many of the healthy young members of the Student Army Training Corps (SATC) standing at attention would be stricken with a virulent outbreak of the Spanish flu that claimed eighteen lives and sent hundreds more to bed. Behind them stands the new Gymnasium, which opened in the spring of 1918. The Gymnasium made possible a fuller integration of women into campus life. Beginning in the fall term 1918 women used the building for physical education classes and athletics, and they had access to the swimming pool. The college intended to use this new structure as "a center of recreational, athletics, and social activity for the whole campus" ("Gymnasium Open to Girls," *M.A.C. Record* 24, no. 1 [24–30 September 1918]: 3). Courtesy of Michigan State University Archives and Historical Collections (Michigan State University, Military Science, 1900–1919, Drills).

energy to wartime activities. They joined the Red Cross, knitted "helmets, wristers, and scarfs for the Navy," put on canning demonstrations, and operated a kitchen for men suffering from Spanish influenza.[12] School started two weeks late in 1917 and closed a month early in 1918 to allow students to help with the fall harvest and the spring planting.[13] During the winter of 1918, forestry students surveyed wood lots on 185 farms near campus to provide information to the Lansing fuel committee as it strived to supply enough wood to keep local homes warm.[14]

Faculty and staff took war-related positions in government and civilian service and joined the military in significant numbers. For instance, in September 1917 Herbert Hoover tapped Harry J. Eustace, professor of horticulture, to oversee the Food Administration's work with perishable foods.[15] Others who left campus included Henrik J. Stafseth, L. H. Cooledge, J. Frank Morgan, Charles G. Nobles, I. F. Huddleson, and F. W. Fabian, all members of the bacteriology department who received commissions in the army.[16] Athletic Director Chester Brewer served as athletic director at Camp Sherman in Ohio. Ralph C. Huston, associate professor of chemistry, was commissioned captain in the Sanitary Corps, and Joseph H. Cox, associate professor of farm crops, took a captaincy in the army's airplane production program.[17] Elizabeth M. Palm, '11, left her post as assistant librarian in September 1918 to undertake nurse's training at Camp Custer, near Battle Creek, Michigan.[18]

Even more dramatic was the departure of students from campus to serve in the army and the navy as part of the national fighting force. M.A.C. men who had taken courses in military science and had been members of the cadet corps now saw their training transformed into the harsh reality of service in wartime. On 7 December 1917 the *M.A.C. Record* reported that six or seven men either enlisted or were drafted each week.[19] Twenty engineering students signed up for the engineer's branch of the Enlisted Reserve Corps in January 1918; this act allowed them to finish their studies before going on active duty. Veterinary medicine students entered a similar program.[20] Because of shrinking male enrollments, Abbot Hall closed during the winter term, and Wells Hall and Williams Hall had unoccupied rooms during the spring term.[21] The formation of a Reserve Officers Training Corp on campus in spring 1917 was another sign of the military's dominating presence on campus.[22]

Reports in the *M.A.C. Record* of M.A.C. men wounded or killed in battle added a somber note all too soon to the patriotic enthusiasm within the M.A.C. community. An account of the first M.A.C. casualty appeared in the 14 September 1917 issue of the *Record:*

THE FIRST

A casualty notice, believed to be the first one of the death of anyone connected in any way with M.A.C. in the Great War, is stated simply: "John Woodbridge killed in action at Vimy Ridge, France, April 9, 1917. A member of the 72^d Highlanders, Canadian Infantry. He died A Christian and a hero."

John Woodbridge was a short course student at the college in 1915. While this is the first such notice, it is the hope, though possibly a vain one, that this may also be the last.[23]

Tragically, many more solemn casualty reports followed. William R. Johnson, '12, perished when his ship the *Tuscania* went down off the Irish coast on 6 February 1918. Philip M. Hodgkins, '17, Frank E. Hausherr, '17, and Stanley F. Wellman, '18, all survived this tragedy.[24] Gordon Webster Cooper, '18, died on 13 June 1918, in an airplane accident while serving at Taliaferro Field in Fort Worth, Texas.[25] On 30 September the *Record* listed four more M.A.C. deaths: Frank Huston Esselstyn, '18, Lester P. Harris, '17, George S. Monroe, '18, and Cosmer M. Leveaux, who was engaged to Hildah Cummings of the

To honor the men who had died in the Civil War, the Spanish American War, and World War I, the Washington, D.C., M.A.C. Association gave this plaque to be hung in the new Union Memorial in 1925. It now hangs in the lower level of Alumni Memorial Chapel. Courtesy of Michigan State University Archives and Historical Collections (Michigan State University, Buildings, Union Building).

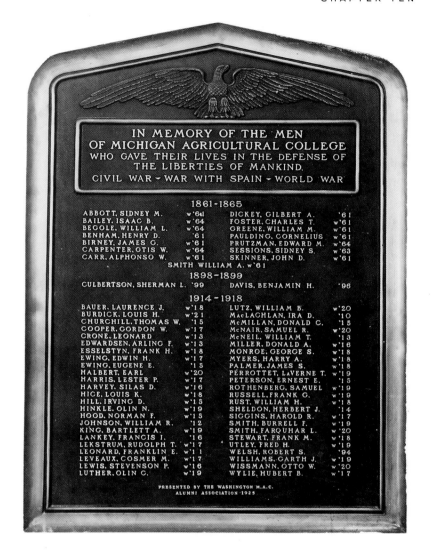

bacteriology department, all lost their lives in France.[26] Before the guns were silenced, the *M.A.C. Record* carried many more grim accounts of M.A.C. men giving their lives in the Great War.

The unpredictability of war brought together M.A.C. men in unexpected ways. Ralph J. Johnson, '16, weathered a hail of bullets that tore through his forearm, ear, and ammunition belt on 31 March 1918, while fighting on the western front in France. He described his frightful experience in a letter to his forestry professor, Alfred K. Chittenden: "The pain of being wounded is one of the real trials of this modern war." While convalescing in the hospital, Johnson was cheered when he met Paul Ginter, '21, another forestry student at

M.A.C. Ginter had sustained bullet wounds to his hand and jaw in a different battle. Johnson and Ginter took walks together each day, enjoying a special camaraderie as "brother forester[s]" from M.A.C.[27]

Not all news from the war front was bad, however. Some soldiers shared what they had learned from their noncombat experiences with readers of the *Record*. The hardships and horror of war did not prevent fighting men from investigating new places as they followed their curiosities. Censorship rules prevented Russel A. Warner, '12, from telling much about military operations, but he reported, with great satisfaction, on his visit to the French patent office in Paris. Warner delighted in seeing "the first steam engine that ran out of Paris about 1770" and other machines and models chronicling the history of "French inventions." Warner found this visit to be especially instructive since he had made many visits to the U.S. Patent Office as part of his job with General Electric.[28] Roger W. Billings, '22, noted that the "French peasants do without many things that the poorest American considers necessities." They went without coal, electricity, and theater, and the women prepared meals over an open fireplace. Bicycles were the primary means of transportation.[29]

The *Record* brightened the day for O. A. Taylor, '15, who received an issue in December 1917, while stationed "somewhere in France." Upon reading it, he passed on his sentiments about his alma mater, especially his memory of the football team, and how they shaped his view of the war in a letter to the editor:

> Like a ray of sunshine, the RECORD came drifting into camp and you can just bet that it was mighty welcome. I read it through and then vice versa until even the advertisements had been carefully surveyed.
>
> .   .   .
>
> I was glad to see the long list of names of the men who are already in the fight. Can't beat the old M.A.C. spirit, whether on the gridiron or in the trenches.
>
> When the right time comes we'll do our bit to beat "Kaiser Bill," just the way we "got" Michigan and Wisconsin back in '13.[30]

Former student W. J. MacKenzie, while serving with the British in Belgium in 1916, also made reference to football. He told Frank Kedzie: "I firmly believe that when the allied armies make their attack, they

Michigan Agricultural College rejoiced when Calvin J. Overmyer won a coveted Rhodes Scholarship in 1920 to study chemistry at Oxford University in England. Overmyer graduated in 1918 and served in the Great War as a second lieutenant in the army. After leaving the military, he taught high school in Arizona before coming back to campus to teach chemistry. Upon receipt of the scholarship, he went to Oxford, where in three years he completed his doctorate in organic chemistry. In 1930 he lived in Louisville, Kentucky, where he worked as a chemist for Peaslee Gaulbert Paint and Varnish Company.
Courtesy of Michigan State University Archives and Historical Collections (*Wolverine*, 1926).

In 1925 Douglas V. Steere, '23, became the second M.A.C. graduate to receive a Rhodes Scholarship. Steere was an associate professor of philosophy at Haverford College in Haverford, Pennsylvania, in 1930. Courtesy of Michigan State University Archives and Historical Collections (*Wolverine*, 1926).

are going through, just as M.A.C. football team used to do, but the cost in men is going to be very great."[31]

Alice Latson, '09, nursed soldiers while they recuperated from their wounds. She volunteered for service in July 1918, and was sent to Asbury Hospital in Minneapolis to train for work in a military hospital. From there she went to the army base hospital at Camp Gordon, Georgia, before joining the American Expeditionary Forces in France. Latson was a dietician for the Red Cross at Base Hospital No. 77 in Beaume, France, after the war.[32]

World War I ended on 11 November 1918, and M.A.C. faced an enormous challenge to return to normal operations. Nearly 1,200 students, faculty, staff, and alumni had joined the various services with 425 serving as army officers and 18 as naval officers.[33] Since most undergraduate men on campus were part of the SATC that awaited disbandment, the college needed to integrate returning and new students into academic and social programs that had been either curtailed or shut down. Faculty and staff had to adjust to an institution that had changed from what they had known just a year earlier. Fire destroyed Williams Hall on New Year's Day, 1919, significantly reducing living quarters for returning veterans. On top of this, the cost of living in East Lansing had risen more than 60 percent between July 1914 and March 1919.[34] But these problems were small in comparison to the sacrifices made during 1917 and 1918. Michigan Agricultural College had demonstrated by its contribution to the allied cause in the Great War that it was an institution composed of men and women who put the public good ahead of their own self-interest, just as millions of other Americans had done. The children and grandchildren of Justin Morrill's industrial classes had done their part, and now they needed to put their college back together to continue its mission of public service.

Not only had the college changed, but the world in 1919 was a different place as well—a reality driven home by former President William Howard Taft's visit to campus on 2 April. Taft came to M.A.C. to promote President Woodrow Wilson's proposed League of Nations. After Taft's arrival in Lansing, the M.A.C. band escorted him to the Capitol, where he addressed a joint session of the Michigan legislature. In the evening, Taft spoke before an enthusiastic audience of 1,700 people from campus, East Lansing, and Lansing in the Gymnasium.[35] The outcome of the war meant that the United States would

play an increasing role in the affairs of the world. This, in turn, would open up many opportunities in the future for M.A.C. men and women to apply their land-grant training to help meet human needs throughout the world.

World War I spurred another transformation of Michigan Agricultural College in 1925. As the nations of the world struggled to bring a new order out of the destruction wrought by the war, M.A.C. faced the challenge of reorganizing itself to better prepare students to take their place in the post-war world. To help accomplish this objective, the college introduced new courses of study in applied science, liberal arts, and business administration. While M.A.C. had always recognized that it was more than simply an agricultural college, the public finally acknowledged this reality when the Michigan legislature passed a measure in 1925 that officially changed the institution's name to Michigan State College of Agriculture and Applied Science. The contributions of M.A.C. and its alumni, students, and faculty during the war proved beyond all doubt that the institution was an essential and integral part of all of Michigan. The "practical farmer" that had bedeviled M.A.C. from its inception became a casualty of the Great War.

Looking east down Grand River Avenue from the junction of Michigan and Grand River Avenues, circa 1925. Courtesy of Michigan State University Archives and Historical Collections (Michigan, East Lansing, Street Scenes).

Epilogue

ICHIGAN AGRICULTURAL COLLEGE'S STUDENTS ULTIMATELY determined the success or failure of Michigan's great experiment in higher education. Throughout the college's first seven decades, alumni used their education to help people meet their daily needs. Attention to the "ordinary" became a hallmark of the land-grant way, but the sons and daughters of M.A.C. did not stop with the ordinary. They expanded the intellectual, scientific, cultural, and commercial horizons of people at local, state, national, and international levels. Their ideas, inventions, and hard work generated new ways of thinking and doing that challenged the conventional, and they championed causes that empowered people to contribute to the economic, political, and social development of the United States. Perhaps the greatest achievement of Michigan's land-grant college—and of land-grant institutions across the country—is that its alumni everywhere expanded the base of American democracy (see appendix 5 for places of residence for graduates in 1930). Brief accounts of the careers of six alumni of M.A.C. sum up some of the ways they and their classmates made a difference in the world.

These six people are examples of how M.A.C. men and women used their land-grant education to enrich the lives of people in Michigan, the United States, and the world. They are not intended to be representative of M.A.C. alumni—for no such distinction can be applied to any student of the college. But Helen R. Hull, Louis Guy Michael, Mary Allen, Paulina E. Raven, Donald W. Francisco, and Ray Stannard Baker show us that a land-grant education empowered women and men to work for the betterment of the world, which was Justin Morrill's intent. Their lives are an appropriate conclusion to this history of Michigan Agricultural College.

417

Helen R. Hull, '05–'07

Helen R. Hull, an early feminist writer began her college education at M.A.C., but like many others, she completed her studies elsewhere. She graduated from Lansing High School in 1905 and commuted to M.A.C. for two years before she left the college to teach in local schools for three years. After completing her degree at the University of Chicago in 1912, Hull taught English at Wellesley College. While at Wellesley she met Mabel Louise Robinson, who became her lifelong partner. In 1916 Hull joined the faculty at Columbia University where she remained until retirement. She published her first novel, *Quest,* in 1922. The novel is a fictionalized account of her difficulties with her troubled family, her attempts to come to grips with her sexuality, and her efforts to realize her potential free from the demands of her mother. Hull's fictionalized self, Jean Winthrop, provides Hull's readers the opportunity to empathize with young women in the early twentieth century who discovered that their innermost desires ran counter to what was expected of them by others. Hull published seventeen novels and sixty-five short stories before her death in 1971.[1]

Louis Guy ("Mike") Michael, '03

Louis Guy Michael followed in the footsteps of earlier M.A.C. graduates who used their land-grant education to improve agricultural practices in other countries. After teaching at Iowa State College, in 1910 Michael went to Bessarabia, then part of Tzarist Russia, to teach peasants how to test seed corn before planting it in order to increase yields. Michael worked among people from several ethnic groups, including Germans, Bulgarians, and Moldavians, all of whom had their own methods of farming. Large landowners who did not want to see peasants empowered in any way opposed him and his work. He also ran into resistance from corrupt government officials. Drawing upon the example of boys' and girls' clubs in the United States, Michael formed numerous corn clubs among the youth and successfully introduced methods of seed testing to them. In November 1917 he witnessed the Bolshevik revolution as it unfolded in Petrograd (St. Petersburg).[2] By the early 1930s Michael had moved to become the agricultural attaché for the USDA Bureau of Agriculture's office in Belgrade, Yugoslavia. His office analyzed and reported on agricultural matters in Bulgaria, Greece, Hungary, Romania, Turkey, and Yugoslavia that were relevant to production, exportation, and American agriculture.[3]

Mary Allen, '09

Pluck and persistence pushed Mary Allen, a contralto, to pursue her dream of entertaining people as a professional singer rather than continue her career as a home economics instructor. After teaching for one year in Detroit schools and for three years in Cape Girardeau, Missouri, Allen took voice lessons and sang with the St. Louis Symphony as soloist. Next, she traveled for a year with a concert company, performed in vaudeville, and for two summers worked the Chautauqua circuits. Stranded in New York after a vaudeville troupe disbanded, she began singing in large churches in the city and with the New York Symphony and New York Oratorio Society. In late 1922, Mary Allen, a daughter of Lansing, Michigan, stood before a packed house in Carnegie Hall to sing Handel's *Messiah* with the Symphony and Oratorio Society.[4] On 17 May 1923, Mary Allen returned to campus to thrill more than five hundred friends and music lovers by singing a concert (which included compositions by Schubert, Brahms, Debussy, and Rachmaninoff) in the Gymnasium to raise funds for the construction of the Union Memorial Building.[5]

Paulina E. Raven, '05

Paulina E. Raven had three careers after she graduated from M.A.C. In 1905, Raven, a new graduate of the women's program, headed to South Dakota to teach at the St. Elizabeth's Indian Mission in Wakpala. Two years later she moved on to the Northern Normal and Industrial School in Aberdeen, South Dakota, to direct its domestic art and science department and to serve as dean of women for two years. Raven's next job put her in charge of the Department of Domestic Science and Art at the State Normal School in Warrensburg, Missouri. Ill-

CONCERT

May 17

8:00 P. M.

College Gymnasium

Mary Allen

'09

Auspices
The Alumnae Council

Benefit
Union Memorial Building Fund

Mary Allen, '09, raised funds for the Union Memorial Building through her benefit concert given on 17 May 1923. *The M.A.C. Record*, in its 28 May 1923 edition, reported: "More than 500 of her friends and admirers made up the crowd which comfortably filled the seating capacity of the gym; a half dozen organizations sent flowers and a large portion of the audience paid homage to her achievements at the reception after the concert." Courtesy of Michigan State University Libraries, Special Collections (*The M.A.C. Record* 28, no. 29 [14 May 1923]: 13).

ness prevented her from completing graduate studies that she started in the summer of 1909 and had continued during the second semester of 1911 at Columbia University.[6] On 1 September 1914, Raven started M.A.C.'s extension work in home economics. Assisted by several seniors, she offered a series of one-week courses on diet, textiles, and home management in different communities throughout Michigan. She resigned her position in April 1917 in order to marry Fred C. Morse and to go live with him on his "Grant Elm Farm" near Jasper in Lenawee County, Michigan.[7] Twelve years later Raven was an active homemaker, raising children, and participating in programs for home economics created by Lenawee County's extension service and the Grange. She rounded out her community service through her membership in the Michigan Federated Women's Clubs.[8]

DONALD W. FRANCISCO, '14

Living in Los Angeles in 1930, Donald W. Francisco held the title vice president and Pacific Coast manager for the advertising firm Lord & Thomas and Logan. Francisco had studied agriculture, with an emphasis on horticulture, at M.A.C. During his college years, he drew many cartoon figures that appeared in the *Holcad.* He clearly understood the land-grant concept of agriculture as a business that started with the seed, vine, or tree in the ground and ended with the product on the consumer's table. Advertising agricultural products formed one link in the chain from the soil to the dinner plate, and Francisco devoted his career to devising strategies to increase people's demand for oranges, lemons, walnuts, rice, apples, and other produce. Francisco's business was just one of many that had to be created and developed to enable American agriculture to improve yields, move products from farm to market, generate profit, and grow food to sustain healthy diets. Even though he may never have picked an apple, planted a seed, or sprayed a tree after he graduated from M.A.C., Francisco contributed to the well-being of American farmers and the public.[9]

Ray Stannard Baker, '89

Ray Stannard Baker embodied the breadth and depth of M.A.C.'s land-grant philosophy and its commitment to expanding the boundaries of American democracy. Baker was born in Lansing and grew up in St. Croix Falls, Wisconsin. He earned his bachelor of science degree in agriculture at M.A.C. in 1889. Married to Jessie Beal, '90, the daughter of Professor W. J. Beal, Baker remained close to his alma mater throughout his life. The land-grant goal of empowering young men and women to change the world by overcoming obstacles created by ignorance, corruption, abuse of power, and just plain human inertia formed a fundamental principle of Baker's career and writings. In his *American Chronicle* he spelled out why reform was so difficult in the United States:

> It seems to be a failing peculiarly American to begin dosing before the diagnosis is complete; we dislike to be quiet and slow; we hate to think things through. We believe in formulas and slogans, we like "drives." We have a pathetic faith in legal enactments such as the 18th Amendment, and changes in the "system" such as the initiative and referendum. There may be considerable education of individual minds in the process of campaigning for such reforms, but it is astonishing how little, how very little, they change actual conditions.[10]

Baker made every effort to be thorough when doing research needed for his writing. He marched with William Coxey's "army" in 1894, as it walked from Massilon, Ohio, to Washington, D.C., hoping to convince Congress to pass legislation to provide jobs for people put out of work by the depression of the mid 1890s. He traveled to Newfoundland to be present when Guglielmo Marconi received the first transAtlantic wireless transmission from England in 1895. Baker wrote insightful articles critical of American industry, labor, and race relations among other topics for *McClure's* and *American Magazine,* two leading periodicals of the time.

Baker's reputation as a reformer who wished to see barriers to American democracy knocked down brought him to the White House to consult with three presidents: Theodore Roosevelt, William Howard Taft, and Woodrow Wilson. Baker became a devoted supporter of President Wilson's demand that *"more democracy, real*

Ray Stannard Baker, '89, was remembered as "State's most distinguished man of letters" by the *Record* after his death on 12 July 1946. The same obituary reprinted Baker's reflections on his college days after attending his class's fiftieth reunion in 1939. He paid homage to Professors Beal, Kedzie, Cook, Bailey, and Edwards as "first-class teachers." And the impact of his classmates on the world drew this comment: "I think the greatest enjoyment I had at the reunion was in meeting a number of men I knew at the college fifty years ago and in hearing something of their careers. Good men, whose success is the best evidence of the virility of that old Michigan institution which in later days came to occupy such a place in the educational procession of the nation." Courtesy of Michigan State University Archives and Historical Collections (People, Baker, Ray Stannard, no. 170).

democracy, [be] applied in all aspects of the 'common interest.'"[11] He served in the Wilson Administration after World War I during the peace conference in Paris, and earned such respect that the president gave him special access to his papers. The crowning achievement of Baker's career occurred in 1940 when he received a Pulitzer Prize for the final volumes of his eight-volume biography *Woodrow Wilson: Life and Letters.*[12]

The "experiment, wholly new and untried" undertaken in 1855 at the Agricultural College of the State of Michigan succeeded because its students went from the college to make a difference in the lives of thousands, even millions, of people. M.A.C. equipped the youth of Justin Morrill's "industrial classes" to serve people everywhere using "the several pursuits and professions of life" learned on the banks of the Red Cedar River. Helen R. Hull, Louis Guy Michael, Mary Allen, Paulina E. Raven, Donald W. Francisco, and Ray Stannard Baker, along with thousands of other alumni, strengthened democracy in America by enabling more people, including themselves, to participate more fully in the economic, political, social, and intellectual life of the country. Their service to their communities, scattered across the continent and the world, helped people enjoy healthier, safer, and more informed lives. The land-grant experiment that started in Michigan spread to every state in the Union transforming and empowering millions of women and men into well-informed citizens who have taken an active part in the affairs of their world, which is the bedrock of a healthy democracy.

The Curriculum, 1925–26

T HE 1925–26 CATALOG OFFERED EIGHT COURSES OF STUDY, WITH A number of specializations, groups, and majors available for students to chose from after taking prerequisite courses as freshmen and/or sophomores. The courses were offered by the departments to students following any one of the courses of study.

COURSES OF STUDY

Agricultural Course

- General Agricultural Series
- Horticultural Series
- Landscape Architecture Series
- Agricultural Business and Marketing
- Six Year Course in Agriculture and Veterinary Medicine (leading to the degree of B.S. at the end of four years and D.V.M. at the end of six years)

At the beginning of the sophomore year, specialization begins in General Agriculture, Horticulture, Forestry, Landscape Architecture, Agricultural Economics, and Agricultural Engineering. Further specialization was also possible in Dairy Husbandry, Animal Husbandry, Poultry Husbandry, Soils, Farm Crops and Apiculture.

Engineering Courses

- Chemical Engineering Group
- Civil Engineering Group
- Electrical Engineering Group
- Mechanical Engineering Group
- Engineering Administration Group

Home Economics Course

- Major in General Course
- Major in Foods and Nutrition
- Major in Institutional Management
- Major in Clothing
- Major in Textiles
- Major in Related Arts
- Major in Vocational Education in Home Making
- Five-Year Course in Home Economics and Nursing

Forestry Course

Veterinary Course

Applied Science Course

- Pre-Medical Training (two-year course)
- Pre-Dental Course
- Major Subjects (Bacteriology, Botany, Chemistry, Economics, Entomology, Geology, History and Political Science, Mathematics, Physics, Zoology)
- Six-Year Course in Applied Science and Veterinary Medicine (leading to the degree of B.S. at the end of four years and D.V.M. at the end of six years)

Liberal Arts Course

- Majors (English, Economics, Drawing and Design, Music, History and Political Science, Mathematics, Modern Language, Sociology)

Business Administration Course

- Personnel, Secretarial and Welfare Management groupings
- Accountancy grouping
- Production or General Business Management groupings

DEPARTMENTS OF INSTRUCTION

Agricultural Engineering

Power Machinery; Farm Machinery; Farm Structures; Poultry House Construction; Farm Conveniences; Farm Drainage and Land Clearing; Concrete and Materials; Farm Equipment; *Graduate Courses:* Farm Structures; Farm Conveniences; Power and Farm Machinery.

Anatomy

Osteology; Arthrology, Myology, and Splanchnology; Angiology, Neurology, Esthesiology, Regional or Operative, followed by Comparative; Histology; Embryology; Embryology and Histology; Mammalian Anatomy; Avian Anatomy; Avian Embryology; *Graduate Courses:* Advanced Gross Anatomy; Advanced Embryology; Advanced Histology.

Animal Husbandry

Study of Breeds; Chemistry of Nutrition; Applied Feeding; Stock Judging; Meat Production; Livestock Production and Management; The Livestock Industry; Types of Livestock; *Graduate Courses:* Swine Husbandry; Sheep Husbandry; Beef Cattle Management; Horse Breeding or Management; *Agriculture:* Development of Agriculture; Farm Management.

Animal Pathology

General and Systemic Pathology; Pathology of Infectious Diseases; Meat Inspection; General and Special Pathology; *Graduate Course:* Advanced Pathology.

Bacteriology and Hygiene

General Bacteriology; Morphological and Cultural; Physiological; Applied; Antiseptics and Disinfectants; Water and Sewage; Food Preservation and Decomposition; Dairy Bacteriology; Fermentation Bacteriology; Pathogenic Bacteriology; Plant Bacteriology; Soil Microbiology; History of Bacteriology; Immunology and Serology; Public Health Bacteriology; Bacteriology of Digestive Tube; Apiarial Bacteriology; Systematic Bacteriology; Thesis; Parasitology; Parasitic Protozoa; Parasitic Helminths; Sanitary Science; Hygiene of Foods; Dairy

Hygiene; Pathological Bacteriology; Industrial Hygiene; Personal Hygiene; Hygiene of Fur-Bearing Animals; *Graduate Courses:* Bacteriological Studies of Antiseptics and Disinfectants; Bacteriological Studies of Water and Water Supplies, of Sewage and Sewage Disposal; Bacteriological Studies of Foods, Food Poisoning, Food Preservation and Decomposition; Bacteriology of Milk and Dairy Manufactures; Fermentation Bacteriology and Enzyme Studies; Pathological Bacteriology; Plant Bacteriology; Soil Bacteriology; History of Bacteriology; Research in Animal Parasitology.

Botany

Plant Anatomy and Physiology; Morphology of the Lower Plants; Systematic Botany; General Botany; Poisonous Plants; Mycology; Forest Pathology; Plant Pathology; Fruit Diseases; Diseases of Field and Garden Crops; Methods in Plant Pathology; Plant Nutrition; General Plant Physiology; Plant Physics; Weeds; Economic Plants; Woody Vegetation; Grasses; Botanical Technique; Cytology of Heredity; Plant Reproduction; The Teaching of Botany; Advanced Plant Physiology; Introductory Botany; Plant Disease Control Methods; Plant Anatomy; Advanced Technique; Physiology of Fungi; Phyto-Chemistry; Ecology; Advanced Mycology; Botanical Readings; Phytogeography; Morphology of the Lower Plants; Elementary Genetics; *Graduate Courses:* Systematic Botany; Anatomy and Morphology; Pathological Investigations; Investigations in Plant Physiology; Seminar.

Chemistry

General Chemistry; Qualitative Analysis; Organic Chemistry; Physical Chemistry; Quantitative Analysis; Food Analysis; Technical Problems and Reports; Physiological Organic Chemistry; Inorganic Technology; Organic Analysis; Organic Technology; Advanced Technology; Metallography of the Common Metals; Physiological Chemistry; Elementary Textile Chemistry; Pyrometry; Chemistry for Civil Engineers; Industrial Chemistry Laboratory; Power Plant Chemistry; Inorganic Chemistry; History of Chemistry; Biological Chemistry; Biological Chemical Preparations; Biological Analytical Chemistry; Advanced Plant Chemistry; Fertilizer Chemistry; Technical Analysis; Advanced Quantitative Analysis; Applied Electro Chemistry; Problems and Reports; Plant Chemistry; Colloid Chemistry; Special Topics in

Organic Chemistry; Organic Chemistry Laboratory; Dairy Chemistry; Electro Chemistry; Gas and Electric Furnace Chemistry; Quantitative Textile Analysis; Bleaching and Laundering; Application of Dyes; Textile Design and Weaving; Sizing and Finishing of Textiles; Textile Chemistry Problem; Chemistry of Nutrition; Chemical Thermodynamics; *Graduate Courses:* Agricultural and Biological; Inorganic and Electro-Chemistry; Physical Chemistry; Industrial Chemistry.

Civil Engineering

Surveying and Leveling; Surveying Methods; Topographic Surveying; Elements of Structural Design; Mechanics of Engineering; Mechanics of Materials; Graphics of Framed Structures; Hydraulics; Hydraulic Laboratory; Hydraulic Motors; Advanced Surveying; Topographic Surveying; Geodesy; Railroad Surveying; Topographic Mapping; Bridge Stresses; Bridge Analysis and Design; Masonry and Arches; Pavements; Technical Problems and Reports; Contracts and Specifications; Astronomy; Water Supply; Sand, Cement and Concrete; Reinforced Concrete; Road Construction; Drainage; Sewerage; State Highway Construction and Law; Highway Design; Advanced Concrete Design; Seminar; *Graduate Courses:* Structural Engineering; Concrete Structures; Steel Structures; Structural Materials; Hydraulic Engineering; River and Harbor Improvement; Hydraulic Motors; Water Power Plants; Highway Engineering; Transportation; Highway Transportation; Highway Traffic Regulation and Safety; Railroad Engineering; Railroad Practice; Block Signals and Automatic Train Control; Railroad Terminals; Sanitary Engineering; Sewerage Collecting System; Public Health Engineering; Sewage Disposal Plants.

Dairy Husbandry

Buttermaking; Market Milk and Condensed Milk; Ice Cream Making and Cheesemaking; Advanced Dairy Judging; Advanced Testing; Dairy Barn Practice; Dairy Farm Management; General Dairy Production; Dairy Seminar; Advanced Dairy Products Judging; Farm Dairying; Elements of Dairying; *Graduate Courses:* Dairy Production; Dairy Manufacture.

Drawing and Design

Mechanical Drawing: Foundation Course in Mechanical and Topographical Drawing; Mechanical Drawing; Architectural and Mechanical Drawing; Descriptive Geometry; Kinematics; Machine Design; Mechanics of Machinery; Shades, Shadow and Perspective; Advanced Machine Design; House Architecture; Architectural Drafting; Advanced Mechanical Drawing and Empirical Machine Design.

Freehand Drawing: Short History of Art; History of Art; Perspective and Freehand Sketching; History of American Art; History of Painting; History of Architecture; Freehand Drawing and Painting; Drawing and Painting; Advanced Drawing and Painting; Decorative Composition and Design; Advanced Composition and Illustration; Color Composition; Advanced Color Composition; Advertising Arrangement; Costume Fashion Drawing; Advanced Costume Fashion Drawing; Advanced Poster; Professional Problems; Modeling; Sketching; Advanced Design and Lettering; *Graduate Courses:* Advanced Drawing and Painting; Advanced Color Problems; Advanced Design and Lettering.

Economics

Economic Geography; Principles of Co-operation; Accounting Technique; Principles of Accounting; Labor Conditions and Labor Problems; Personnel Management; Industrial Government; Public Finance; Elements of Economics. Production and the Economic Agencies; Corporation Finance; Agricultural Economics; Principles of Marketing; Marketing Organizations and Methods; Elements of Economics; Occupations, Wages and Unions; Market Technique; Business Organization and Management; Money and Banking; Railroad Transportation; Elements of Economics. Value and the Economic Shares; Economic Theory; Marketing of Special Products; Agricultural Economic Research; Institutional Accounting; History of Co-operation and Farmers' Movements; Extension Organizations and Methods; Agricultural Prices; Foreign Trade; Rural Credit and Finance; Elements of Accounting; Banking, Securities and Investments; Transportation; Business Law; Business Statistics; Industrial Relations; Business Cycles; *Graduate Courses:* The Economics of Food Supply; Economic Research; Land Economics; State and Local Taxation; Seminar in Agricultural Economics; Economics of Agricultural Production.

Education

Psychology; Special Methods in Home Economics; Science of Education; History of Education; Practice Teaching in Home Economics; Agricultural Methods; Practice Teaching in Agriculture; Educational Psychology; School Administration; Practice Teaching; *Graduate Courses:* Problems in Secondary Agricultural Education; The Economic Background of Agricultural Education; Philosophy of Education; Seminar in Rural Education; History of Rural Education; Seminar. Curriculum Construction; Educational Surveys; *Life Planning:* Principles and Art of Life-Planning.

Electrical Engineering

Direct Current Circuits; Magnetic Circuits; Electric and Magnetic Circuits; Direct Current Generators and Motors; Direct Current Dynamo Laboratory; Dynamo Laboratory; Principles of Alternating Currents; Alternating Current Circuits; Alternating Current Circuits Laboratory; Alternating Current Machines; Alternating Current Engineering; Alternating Current Machinery Laboratory; Direct Current Circuits and Apparatus; Direct Current Circuits and Apparatus Lab; Alternating Current Circuits and Machines; Alternating Current Circuits and Machines Lab; Technical Problems and Reports; Design of Electrical Apparatus; Illumination; Batteries and Ignition; Batteries and Ignition Lab; Electrical Measurements; Electrical Measurements Laboratory; Alternating Current Measurements; Alternating Current Measurements Laboratory; Electric Power Transmission; Electrical Distribution; Electrical Railways; Electric Communication; Electric Communication Laboratory; Radio Communication; Radio Communication Laboratory; Hydro-Electric Power Development; Switchboard Design; Motor Control Systems; Distribution of Electric Power; Advanced Electrical Design; Advanced Electric Circuits.

English and Modern Languages

English: Composition; Argumentation and Public Addresses; Writing for the Press; Writing for Farm Journals; Advertising Copy; Agricultural Publicity; Writing and Editing Farm Bulletins; Public Speaking; Parliamentary Law and Organization; Dramatics; Advanced Public Speaking; Argumentation; Technical Writing for Engineers; Engineering Prose; The Tragedies of Shakespeare; Revolutionary Era in

English Literature; Tennyson; Greek Literature in Translation; Development of the Drama; Victorian and Contemporary Prose; The English Novel; Milton and his Age; Browning; The Pedagogy of English; The Continental Novel; Modern English for Engineers; Development of the Drama; The English Old Testament as Literature; American Literature; Literature of Country Life; Current Tendencies in Literature; The English New Testament; Development of the Drama; English. Continental Short Story; Survey of English Literature; From Ibsen to Hamsun; Advanced Composition. Creative Writing; Advanced Composition. Expository Writing; Writing the Short Story; News Writing; Reporting and Correspondence; Editing; Special Articles; Articles for Specialized Journals; Editorial Writing; Critical Writing; *Graduate Courses:* Anglo-Saxon (Old English); Chaucer; Seminar in Bibliography and Methods of Investigation; Seminar in the Theory and History of Criticism; Seminar in the Development of the English Drama; Seminar in Regional Literature in the United States; Seminar in Present Tendencies in English and American Poetry.

Modern Languages: French: Elementary French; Second-Year French; Introduction to French Literature; Study of Selected Authors. *German:* Elementary German; Second-Year German; Third-Year German. *Spanish:* Elementary Spanish; Second-Year Spanish; Elementary Spanish for Engineering Students.

Entomology

Introductory; Fruit Insects; Farm Crop Insects; Forest Insects; Applied Entomology; Apiculture; Introductory and Household Insects; Systematic Entomology; Medical Entomology; Comparative Anatomy of Insects; Microscopic Technology for Insect Tissues; *Graduate Courses:* Investigation in Life Histories and Insect Control; Systematic Studies in Some Limited Groups of Insects.

Farm Crops

Field Crop Industry; Cereals, Grain Grading, and Marketing; Forage Crops; Advanced Genetics in Relation to Plant Breeding; Plant Breeding; Seed Production and Marketing, and Crop Improvement; Special Crops; Advanced Crop Project and Seminar; Agronomic Seminar; *Graduate Courses:* Plant Breeding; Crop Ecology; Cultural Investigations.

Forestry

Farm Forestry; Dendrology; Forest Mensuration; Lumbering Methods; Silviculture; Forest Protection; Forest Physiography; Wood Technology; Lumbering; Forest Policy; Forest Products; Forest Valuation; Timber Seasoning and Wood Preservation; Forest Regulation and Working Plans; Municipal Forestry; Forestry Seminar; Forest Conservation; Forest Management; *Graduate Courses:* Dendrological Research; Research in Silviculture; Wood Technology and Forest Products; Forest Management.

History and Political Science

History: General History of Europe From the Reformation to the Congress of Vienna; General History of Europe From the Congress of Vienna to 1870; General History of Europe From 1870 to 1926; History of England From Early Beginning to the Stuart Period; History of England From the Stuart Period to the 20th Century; History of England-British Colonization; American History, Including the Colonization and the Revolutionary Periods; American History, From the War of 1812 to the Civil War; American History From the Civil War to the Present Time; Industrial History—Rise of Industry in England; Industrial History—Development of Industry in America; History of Agriculture; Reformation; French Revolution; National Unity; Constitutional History of England; *Graduate Courses:* Seminar in American History; Seminar in Political Science; Seminar in Industrial History.

Political Science: National Government; State Government; International Organization; Local Government; Comparative Government; Theory of Government.

Home Economics

Foods; Experimental Foods; Fancy Cookery and Meal Service; Problems in Foods; Selection and Preparation of Food; Camp Cookery; Dietetics; Dietetics in Abnormal Conditions; Field Work in Child Nutrition; Field Work in Social Service; Nutrition; Advanced Nutrition; Problems in Nutrition; Large Quantity Cookery; Institutional Management Practice; Tea Room Practice; Food Service and Sales; Institutional Laundry; Home Care of the Sick; Household Administration; Home Management Practice; Home Care of the Child; Experimental Study of Physical and Mental Development of Children; The

Modern Family; Elementary Nutrition and Dietetics; Orientation Course for Home Economics Freshmen; Clothing; Principles of Clothing Selection; Advance Clothing; Millinery; Commercial Millinery; Study of Costume; Field Problems; Selection and Construction of Clothing; Textile and Clothing Economics; Textiles; Buying of Textiles and Clothing; Salesmanship; Salesmanship Practice; Exhibits and Display; Design and Color; Applied Design; History of Costume; Advanced Design; The House; House Furnishing; *Graduate Courses:* Seminar; Research in Nutrition; Research in Experimental Foods.

Horticulture

Fruit Production; Systematic Pomology; Systematic Olericulture; Applied Plant Breeding; Vegetable Forcing; Commercial Floriculture; Herbaceous Crops; Advanced Pomology; Geography of Horticulture; Commercial Horticulture; Vegetable Gardening; Small Fruits and Grapes; Plant Propagation; Truck Farming; Spraying; Fruit Handling; *Graduate Courses:* Special Investigations; Methods of Horticultural Research; History and Literature of Horticulture.

Landscape Architecture

Elements of Landscape Architecture; Landscape Architecture; Rural Landscape Design; Landscape Art; Elements of Civic Design; Landscape Design.

Mathematics

Algebra; Agricultural Mathematics; Mathematics for Business Administration; Trigonometry; Plane Trigonometry; Analytic Geometry; Calculus; Survey of Elementary Mathematics; Statistical Methods; Applied Statistics; Theory of Equations; Differential Equations; Analytic Mechanics; Advanced Calculus; Solid Analytic Geometry; Modern Geometry; Graphical Methods and Interpolation; Mathematics of Finance and Investment; Vector Calculus; Statistics; Analytic Projective Geometry; *Graduate Courses:* Theory of Functions of a Real Variable; Theory of Functions of a Complex Variable; Theory of Differential Equations with Applications; Elliptic Integrals with Applications; Vector Analysis; Finite Differences; Modern Geometry; Fourier's Series and Spherical Harmonics; Statistics; Kinetic Theory of Gases; Theory of Relativity.

Mechanical Engineering

Elements of Engineering; Carpentry; Forge; Industrial Management; Human Engineering; Steam Power; Gas Power; Heat Engines; Boiler Room Economy; Gas Engines; Thermodynamics; Metallurgy; Timber Testing; Metallurgy of Non-Ferrous Metals; Metallurgical Calculations; Engineering Laboratory; Automobile and Truck Engine Design; Automobile and Truck Chassis Design; Heating and Ventilation; Power Plant Design; Costs, Accounting; Hydraulic Power Plants; Technical Problems and Reports; Pattern Work; Forge Work; Foundry; Machine Shop; Seminar; Heat Treatment of Carbon Steels; Heat Treatment of Alloy Steels; Heat Treatment of Non-Ferrous Metals and Alloys.

Meteorology

Meteorology

Military Science

Freshman Training; Sophomore Training; Elective Courses; Equitation.

Music

Piano; Violin; Voice.

Physics

Mechanics and Heat; Magnetism and Electricity; Sound and Light; Mechanics and Electricity; Heat, Electricity; Optics; Advanced Course in Heat; General Physics; Heat, Magnetism, and Electricity; Electron Theory; Electrical Discharge Through Gases; Heat Measurements; The Vacuum Tube—Its Characteristics and Applications; Experiments on Present Problems in Physics; Geometrical Optics; Physical Optics; Electricity and Magnetism; Current Problems; *Graduate Courses:* Light; Heat; Molecular Physics; Electrical Physics.

Physical Education

Physical Education for Men: Physical Education; Football; Basketball; Track and Field Athletics; Baseball; Athletic Training; Wrestling and Boxing; Organization and Administration; Swimming; Theories and Systems of Physical Education; Calisthenics; Mass Physical Activities; Gymnastics; Anthropometry and Physical Examination; Intramural Athletics.

Physical Education for Women: Physical Education; Hockey and Soccer; Basketball; Baseball and Track; Tennis; Dancing; Swimming; Individual Gymnastics; Organization and Administration; Health Education; Playground; Campcraft; Principles and Methods of Physical Education.

Physiology and Pharmacology

Physiology: Elementary Physiology; Veterinary Physiology; The Nervous System; Nutrition; Muscular Exercise; Advanced Physiology; Internal Secretions; Abnormal Physiology; Circulation; Special Senses.

Pharmacology: Materia Medica and Pharmacy; Pharmacy; Toxicology; Therapeutics; Pharmacology.

Poultry Husbandry

Farm Poultry; Poultry Judging; Marketing, Breeding and Diseases of Poultry; Incubation and Brooding; Farm Records; Advanced Poultry.

Religious Education

Fundamental Moral and Religious Concepts; Christianity and Social Problems; Science and Christ; Religion in the Home; Woman's Place in Modern Religious Movements; Religious Callings.

Sociology

Principles of Sociology; Social Attitudes and Values; Social Evolution; Rural Sociology; Principles of Community Development; Historical and Comparative Rural Life; The Family and Child Welfare; Rural Population; Educational Sociology; Sociological Phases of the Extension Movement; Principles of Leadership; Public Opinion and Social Control; Modern Social Problems; Social Progress and Democracy; Rural Life Values and Tendencies; The Rural Church as a Social Institution; Rural Recreation and Art; The Field of Social Work; *Graduate Courses:* Methods of Research in Sociology; Educational Sociology; Rural Social Problems; Field Studies and Surveys; Community Problems; Social Theory.

Soils

Soils; Soil Physics; Soil Fertility; Soil Classification and Management; Fertilizers; Muck Soils; *Graduate Courses:* Soil Physics; Soil Fertility;

Soil Chemistry; Origin of Soils and Principles of Soil Classification—Podology.

Veterinary Medicine, Surgery and Clinic

Medicine: Diagnosis; Theory and Practice; Jurisprudence.

Surgery and Clinic: General Surgery; Principles of Horseshoeing; Special Surgery; Clinic, Medical and Surgical; Lameness and Soundness; Obstetrics.

Veterinary Science and Zootechnics: Types and Breeds, and Care of Small Animals; Veterinary Science; Hippology.

Zoology and Geology

Zoology: Introductory; Ornithology; Forest Zoology; Aquatic Zoology; General Zoology; Human Heredity and Eugenics; Animal Ecology; Systematic Zoology; Genetics; Genetics and Eugenics; Advanced Ornithology; *Graduate Courses:* Economic Zoology; Genetics and Eugenics.

Geology: General Geology; Soil Geology; Engineering Geology; Physiography; Industrial Geology; Mineral Deposits; Dynamic Geology; Structural Geology; Historical Geology; Interpretations of Geologic Maps; Stratigraphy; Geography; Petrology; Field Geology; Advanced Petrology; Mineralogy and Crystallography; Advanced Geology; Invertebrate Paleontology; Determinative Mineralogy.

Winter Short Courses

Two-Year Courses in Agriculture: Sixteen Weeks' Course: First Year; Second Year; Eight Weeks' Course.

Dairy Courses: Dairy Production Course (ten weeks); Dairy Manufacturers' Course (eight weeks); Ice Cream Making Course (two weeks).

Horticultural Courses: Eight Weeks' Course; Market Gardeners' Course; Fruit Growers' and Nurserymen's Short Course; Ornamental Horticulture Short Course.

Poultry Course: Eight Weeks.

Agricultural Engineering Courses: Farm Mechanics (four weeks); Truck and Tractor Course (four weeks).

Post Graduate Veterinary Short Course: One week for graduate veterinarians.

Home Economics: One week Farm Women's Institute in July.

Farmers' Week.

Comparison of Leading Industries in Michigan, 1900 and 1904

INDUSTRY	1900		1904	
	NUMBER OF ESTABLISHMENTS	CAPITAL	NUMBER OF ESTABLISHMENTS	CAPITAL
Agricultural implements	59	$8,932.344	42	14,342,106
Beet sugar	9	4,013,743	19	12,989,630
Bread and other bakery products	455	1,920,818	614	2,829,028
Carriages and wagons[1]	299	7,935,269	216	13,610,900
Cars and general shop construction and repairs by steam railroad companies	42	2,527,256	34	2,462,881
Cars, steam railroad, not including operations of railroad companies	4	6,693,209	4	7,783,077
Cheese, butter and condensed milk[2]	286	1,250,897	371	1,888,385
Chemicals[3]	25	7,481,873	23	16,130,766

INDUSTRY	1900		1904	
	NUMBER OF ESTABLISHMENTS	CAPITAL	NUMBER OF ESTABLISHMENTS	CAPITAL
Druggists' preparations	10	2,951,503	20	7,650,266
Flour and grist mill products	393	6,755,237	405	7,654,270
Food preparations	25	793,742	55	3,445,786
Foundry and machine shop products[4]	364	19,595,771	419	24,290,075
Furniture	124	13,900,789	134	15,797,769
Iron and steel[5]	10	3,859,050	15	5,795,551
Leather tanned, curried, and finished	27	5,214,042	25	6,860,797
Liquors, malt	77	6,235,484	86	8,951,185
Lumber and timber products	1,391	66,489,960	766	38,507,207
Lumber, planing mill products, including sash, doors, and blinds	235	8,571,453	246	8,770,537
Paper and wood pulp	27	4,505,741	30	8,397,576

INDUSTRY	1900		1904	
	NUMBER OF ESTABLISHMENTS	CAPITAL	NUMBER OF ESTABLISHMENTS	CAPITAL
Printing and publishing, newspapers and periodicals	591	4,114,760	663	4,725,303
Salt	53	4,759,865	41	3,538,149
Slaughtering and meat packing, wholesale	8	1,265,151	8	1,252,767
Smelting and refining copper	3	1,523,407	3	2,378,315
Tobacco, cigars and cigarettes	599	1,931,635	696	2,462,314

1. 1904 total establishments includes: carriages and wagons, 183; automobiles, 22; and automobile bodies and parts,11.

2. 1904 total establishments includes: butter, 203; cheese, 162; and condensed milk, 6.

3. 1904 total establishments includes: chemicals, 14; and wood distillation 9.

4. 1904 total establishments includes: foundry and machine shop products, 382; metal working machinery, 13; steam pumps and pumping machinery, 3; and stove sand furnaces, 21.

5. 1904 total establishments includes: iron and steel, blast furnaces, 11; and iron and steel, steel works and rolling mills, 4.

Source: "Comparative summary of twenty-four leading industries—1904 and 1900," *Census of the State of Michigan, 1904: Agriculture, Manufactures and Mines* (Lansing: Wynkoop Hallenbeck Crawford Co., State Printers, 1905), 2: 620–21.

Places of Residence or Death by State or Country in 1900 of M.A.C. Graduates

Alabama	2	New Mexico Territory	2	
Arizona Territory	5	New York	37	
Arkansas	1	North Carolina	1	
California	26	North Dakota	6	
Colorado	20	Ohio	20	
Connecticut	6	Oklahoma Territory	3	
Delaware	1	Oregon	5	
District of Columbia	22	Pennsylvania	17	
Florida	2	Rhode Island	0	
Georgia	5	South Carolina	2	
Idaho	6	South Dakota	4	
Illinois	51	Tennessee	5	
Indiana	17	Texas	2	
Iowa	5	Utah	2	
Kansas	4	Vermont	0	
Kentucky	1	Virginia	2	
Louisiana	1	Washington	6	
Maine	3	West Virginia	1	
Maryland	3	Wisconsin	7	
Massachusetts	1	Wyoming	2	
Michigan	415	Alaska	4	
Minnesota	11	Hawaiian Islands	0	
Mississippi	3	Philippine Islands	4	
Missouri	2	Australia	1	
Montana	7	Burma	1	
Nebraska	11	Ecuador	1	
Nevada	4	Japan	3	
New Hampshire	1	Mexico	1	
New Jersey	4	Siberia	1	
		TOTAL	789	

Source: Michigan Agricultural College: *General Catalogue of Officers and Students, 1857–1900* (Agricultural College: The College, 1900).

Places of Residence or Death in Michigan of M.A.C. Graduates in 1900

COMMUNITY	POPULATION IN 1900	RESIDENTS
Adrian, Lenawee Co.	9,654	2
Agricultural College, Ingham Co.	P.O.*	27
Albion, Calhoun Co.	4,519	3
Allegan, Allegan Co.	2,667	3
Alma, Gratiot Co.	2,047	1
Ann Arbor, Washtenaw Co.	14,509	9
Armada, Macomb Co.	863	1
Bangor, Van Buren Co.	P.O.	3
Baraga, Baraga Co.	1,185	1
Bath, Clinton Co.	P.O.	1
Battle Creek, Calhoun Co.	18,563	4
Bay City, Bay Co.	27,628	1
Belding, Ionia Co.	3,282	2
Bellaire, Antrim Co.	1,157	2
Bellevue, Eaton Co.	1,074	1
Benton Harbor, Berrien Co.	6,562	2
Big Rapids, Mecosta Co.	4,686	3
Birmingham, Oakland Co.	1,170	2
Bloomingdale, Van Buren Co.	379	1
Boyne Falls, Charlevoix Co.	431	1
Bridgman, Berrien Co.	P.O.	1
Buchanan, Berrien Co.	1708	1
Byron Center, Kent Co.	P.O.	1
Cadillac, Wexford Co.	5,997	1
Calumet, Houghton Co.	P.O.	1
Capac, St. Clair Co.	547	1
Caro, Tuscola Co.	2,006	1

COMMUNITY	POPULATION IN 1900	RESIDENTS
Carson City, Montcalm Co.	906	2
Caseville, Huron Co.	507	3
Cass City, Tuscola Co.	1,113	1
Cassopolis, Cass Co.	1,330	1
Central Lake, Antrim Co.	1,307	1
Centreville, St. Joseph Co.	645	2
Climax, Kalamazoo Co.	398	1
Clio, Genesee Co.	640	1
Coldwater, Branch Co.	6,216	1
Coloma, Berrien Co.	687	1
Colon, St. Joseph Co.	P.O.	1
Commerce, Oakland Co.	P.O.	1
Constantine, St. Joseph Co.	1,226	1
Coopersville, Ottawa Co.	660	2
Corunna, Shiawassee Co.	1,510	1
Covert, Van Buren Co.	P.O.	1
Custer, Mason Co.	269	1
Dearborn, Wayne Co.	844	2
Decatur, Van Buren Co.	1,356	2
Dennison, Ottawa Co.	P.O.	1
Detroit, Wayne Co.	285,704	39
Dimondale, Eaton Co.	P.O.	1
Douglas, Allegan Co.	444	1
Dowagiac, Cass Co.	4,151	2
Durand, Shiawassee Co.	2,134	1
Eagle, Clinton Co.	142	1
East Cohoctah, Livingston Co.	P.O.	1
East Jordan, Charlevoix Co.	1,205	1
Eastport, Antrim Co.	P.O.	1
Eaton Rapids, Eaton Co.	2,103	1
Edenville, Midland Co.	P.O.	1
Elmdale, Ionia Co.	P.O.	1
Elsie, Clinton Co.	576	1
Essexville, Bay Co.	1,639	1
Evart, Osceola Co.	1,360	1
Fisher's Station, Kent Co.	P.O.	1

COMMUNITY	POPULATION IN 1900	RESIDENTS
Flint, Genesee Co.	13,103	3
Fowler, Clinton Co.	426	1
Frankfort, Benzie Co.	1,465	1
Galesburg, Kalamazoo Co.	689	1
Ganges, Allegan Co.	P.O.	1
Gaylord, Otsego Co.	1,561	2
Gobleville, Van Buren Co.	505	1
Grand Haven, Ottawa Co.	4,743	3
Grand Ledge, Eaton Co.	2,161	2
Grand Rapids, Kent Co.	87,565	19
Grattan, Kent Co.	P.O.	1
Greenville, Montcalm Co.	3,381	2
Grindstone City, Huron Co.	P.O.	1
Hamburg, Livingston Co.	P.O.	1
Hanover, Jackson Co.	378	1
Harbor Beach, Huron Co.	1,149	2
Harbor Springs, Emmet Co.	1,643	3
Harrietta, Wexford Co.	419	1
Hart, Oceana Co.	1,134	3
Hastings, Barry Co.	3,172	1
Hetherton, Otsego Co.	P.O.	1
Hickory Corners, Barry Co.	P.O.	1
Highland, Oakland Co.	P.O.	2
Highland Park, Wayne Co.	427	1
Hillsdale, Hillsdale Co.	4,151	1
Holland, Ottawa Co.	7,790	1
Holly, Oakland Co.	1,419	1
Houghton, Houghton Co.	3,359	2
Howell, Livingston Co.	2,518	2
Hudson, Lenawee Co.	2,403	1
Hudsonville, Ottawa Co.	P.O.	1
Hunter's Creek, Lapeer Co.	P.O.	1
Imlay City, Lapeer Co.	1,122	1
Ionia, Ionia Co.	5,209	11
Iron Mountain, Dickinson Co.	9,242	1
Ithaca, Gratiot Co.	2,020	1

COMMUNITY	POPULATION IN 1900	RESIDENTS
Jackson, Jackson Co.	25,180	5
Kalamazoo, Kalamazoo Co.	24,404	3
Keeler, Van Buren Co.	P.O.	1
Laingsburg, Shiawassee Co.	690	2
Lake Odessa, Ionia Co.	1,037	1
Lansing, Ingham Co.	16,485	36
Lawrence, Van Buren Co.	598	1
Leland, Leelanau Co.	P.O.	2
Lewiston, Montmorency Co.	P.O.	1
Ludington, Mason Co.	7,166	1
Mancelona, Antrim Co.	1,209	1
Manchester, Washtenaw Co.	1,209	2
Manistee, Manistee Co.	14,260	1
Manistique, Schoolcraft Co.	4,126	1
Manton, Wexford Co.	579	2
Maple Rapids, Clinton Co.	579	3
Marcellus, Cass Co.	1,025	1
Marlette, Sanilac Co.	996	1
Marshall, Calhoun Co.	4,370	1
Mason, Ingham Co.	1,828	3
McBain, Missaukee Co.	709	1
Medina, Lenawee Co.	P.O.	1
Mendon, St. Joseph Co.	777	1
Menominee, Menominee Co.	12,818	2
Meridan, Ingham Co.	P.O.	1
Midland, Midland Co.	2,363	1
Moore Park, St. Joseph Co.	P.O.	1
Morenci, Lenawee Co.	1,334	2
Mount Clemens, Macomb Co.	6,576	1
Mount Pleasant, Isabella Co.	3,662	2
Munising, Alger Co.	2,014	1
Muskegon, Muskegon Co.	20,818	3
Napoleon, Jackson Co.	P.O.	1
New Baltimore, Macomb and St. Clair Cos.	922	1
North Aurelius, Ingham Co.	P.O.	1
North Lansing, Ingham Co.	P.O.	3

COMMUNITY	POPULATION IN 1900	RESIDENTS
Northville, Wayne Co.	1,755	1
Nunica, Ottawa Co.	P.O.	1
Oak Grove, Livingston Co.	P.O.	1
Old Mission, Grand Traverse Co.	P.O.	1
Onondaga, Ingham Co.	P.O.	1
Orion, Oakland Co.	756	2
Otsego, Allegan Co.	2,073	2
Overisel, Allegan Co.	P.O.	1
Owosso, Shiawassee Co.	8,696	6
Paw Paw, Van Buren Co.	1,465	6
Pentwater, Oceana Co.	1,061	1
Perkins, Delta Co.	P.O.	1
Perry, Shiawassee Co.	641	1
Petoskey, Emmet Co.	5,285	3
Pewamo, Ionia Co.	446	1
Pinckney, Livingston Co.	500	1
Plainwell, Allegan Co.	1,318	2
Pontiac, Oakland Co.	9,769	4
Port Huron, St. Clair Co.	19,158	5
Portland, Ionia Co.	1,874	1
Prairieville, Barry Co.	P.O.	1
Saginaw, Saginaw Co.	42,345	1
St. Johns, Clinton Co.	3,388	1
St. Joseph, Berrien Co.	5,155	1
St. Louis, Gratiot Co.	1,989	1
Sanilac Center, Sanilac Co.	578	1
Saranac, Ionia Co.	768	1
Somerset, Hillsdale Co.	P.O.	1
South Haven, Van Buren Co.	4,009	1
Springport, Jackson Co.	559	1
Stanton, Montcalm Co.	1,234	1
Sturgis, St. Joseph Co.	2,465	2
Swartz Creek, Genesee Co.	P.O.	1
Tawas City, Iosco Co.	1,228	1
Tekonsha, Calhoun Co.	573	1
Three Oaks, Berrien Co.	994	3

COMMUNITY	POPULATION IN 1900	RESIDENTS
Traverse City, Grand Traverse Co.	9,407	2
Tuscola, Tuscola Co.	P.O.	1
Unadilla, Livingston Co.	P.O.	1
Union City, Branch Co.	1,514	1
Union Pier, Berrien Co.	P.O.	1
Unionville, Tuscola Co.	427	1
Wacousta, Clinton Co.	P.O.	2
Washington, Macomb Co.	P.O.	1
Wayland, Allegan Co.	P.O.	1
Webberville, Ingham Co.	346	2
White Pigeon, St. Joseph Co.	705	1
Whitford Centre, Monroe Co.	P.O.	1
Willis, Washtenaw Co.	P.O.	1
Woodland, Barry Co.	319	1
Ypsilanti, Washtenaw Co.	7,378	3

* Post Office

Sources: *Michigan State Agricultural College: General Catalogue of Officers and Students, 1857–1900* (Agricultural College, The College, 1900); *Michigan Official Directory and Legislative Manual for the Years 1901–1902* (Lansing: Wynkoop Hallenbeck and Crawford Co., 1901), 210–22, 281–304.

Places of Residence or Death of Graduates in 1930

I N 1931 MICHIGAN STATE COLLEGE PUBLISHED *Alumni Catalogue Number: Michigan State College Bulletin: List of Graduates, Officers and Professors of the Faculties 1857–1930* (June 1931, vol. 25, no. 11). Two paragraphs from the Foreword are quoted below to explain the composition of the catalog and how the class lists were compiled.

> The chief items of information are presented in the following parts: (1) A list of all officers of the College and members of the faculty who from 1857 to 1931 have held the title of professor; (2) The names, occupations, and addresses of all graduates to the same date; (3) A geographical index of the graduates; (4) An alphabetical index of officers, professors and graduates.
>
> For the class lists an effort has been made during the past year to obtain the official position and address of every living graduate. These lists give the year of graduation according to the official college records. Necessarily on this account, some individuals who were irregular will find their names with classes other than the ones to which they would naturally belong. For those graduates deceased the date and place of death is given wherever known.
>
> The numbers of graduates that appear in each list include living and deceased graduates.

PLACES OF RESIDENCE OR DEATH, BY STATE, OF 6,076 GRADUATES IN 1930

Alabama	7	Nebraska	12
Arizona	9	Nevada	1
Arkansas	5	New Hampshire	8

California	195	New Jersey	58
Colorado	27	New Mexico	1
Connecticut	21	New York	213
Delaware	5	North Carolina	6
District of Columbia	75	North Dakota	11
Florida	36	Ohio	205
Georgia	13	Oklahoma	15
Idaho	12	Oregon	32
Illinois	283	Pennsylvania	102
Indiana	96	Rhode Island	1
Iowa	23	South Carolina	1
Kansas	16	South Dakota	12
Kentucky	16	Tennessee	21
Louisiana	4	Texas	44
Maine	7	Utah	4
Maryland	32	Vermont	7
Massachusetts	43	Virginia	24
Michigan	4,132	Washington	40
Minnesota	29	West Virginia	25
Mississippi	1	Wisconsin	67
Missouri	44	Wyoming	14
Montana	21		

PLACES OF RESIDENCE OR DEATH OF 82 GRADUATES OUTSIDE OF THE UNITED STATES IN 1930

Alaska	1	Hawaii Territory	3
Argentine Republic	1	Holland	1
Australia	1	India	3
Bolivia	1	Japan	1
British North Borneo	1	Java	1
Bulgaria	1	Jugoslavia	1
Canada	22	Liberia	1
Chile	3	Lower Burma	1
China	7	Mexico	5
Cuba	3	Panama	1
Dominican Republic	1	Philippine Islands	7

Egypt	1	Puerto Rico	1
England	1	Russia	1
Germany	1	Siberia	1
Greece	1	South Rhodesia	2
Guam	1	Union of South Africa	3
Haiti	1	Venezuela	1

Notes

FOREWORD

1. Brooks Mather Kelley, *Yale: A History* (New Haven: Yale University Press, 1974), 161–64.
2. Willis Frederick Dunbar, *Michigan Through the Centuries* (New York: Lewis Historical Pub. Co., 1955), 1:250.
3. William S. Tyler, *History of Amherst College during Its First Half Century, 1821–1871* (Springfield, Mass.: C. W. Bryan, 1873), 325.
4. *Revised Constitution of the State of Michigan* (1850).
5. Quoted in George M. Marsden, *The Soul of the American University: From Protestant Establishment to Established Nonbelief* (New York: Oxford University Press, 1994), 107.
6. Louis Menand, *The Metaphysical Club* (New York: Farrar, Straus, and Giroux, 2001).

INTRODUCTION

1. Moses Wisner, "Governor's Inaugural Address," 5 January 1859, *State of Michigan, No. 2 Legislature,* 1859, 15–16, Moses Wisner Correspondence, 1859–60, Madison Kuhn Collection (hereafter Kuhn Collection), UA 17.107, Box 1140, Folder 88, Michigan State University Archives and Historical Collections (hereafter MSU Archives).
2. Alonzo Sessions, "Report of the Joint Committee on Education and Agriculture," *Michigan Farmer* 1 (new series), no. 12, 19 March 1859, 89, from Herbert Andrew Berg, "The Pioneer Land-Grant College: The Formative Years of Michigan State University as Revealed by Excerpts from Issues of *The Michigan Farmer,* January 1, 1845–June 1864," bound typescript collection, Michigan State University Libraries, Special Collections, East Lansing, Mich., 1965, 249.
3. "The First Morrill Act, 1862," in *Agriculture in the United States: A Documentary History,* ed. Wayne D. Rasmussen (New York: Random House, 1975), 1:616–18.
4. Carl E. Kaestle, "The Development of Common School Systems in the States of the Old Northwest," in " . . . *Schools and the Means of Education Forever Be Encouraged": A History of Education in the Old Northwest, 1787–1880,* ed. Paul H. Mattingly and Edward W. Stevens, Jr. (Athens: Ohio University Libraries, 1987), 31–44.

5. Calculated from data taken from *Historical Statistics of the United States: Colonial Times to 1970* ([Washington, D.C.]: U.S. Department of Commerce, 1975), part 1, 11–12, 29, 139, 457.

6. Richard Hofstadter, *The Age of Reform from Bryan to F.D.R.* (New York: Vintage, 1955), 39.

7. Clark L. Brody, *In the Service of the Farmer: My Life in the Michigan Farm Bureau* (East Lansing: Michigan State University Press, 1959), 12–19.

8. "The Case of the M.A.C.," *Adrian Daily Telegram,* 1 June 1914: 4.

9. "What the Attack on the M.A.C. Means," *Detroit Tribune,* 2 April 1914: 4.

10. "M.A.C.'s New Troubles," *Detroit Free Press,* 9 March 1916: 4.

11. Ray Stannard Baker, *American Chronicle* (New York: Charles Scribner's Sons, 1945), 514.

CHAPTER 1. THE BEGINNING

1. Joseph R. Williams, "Address to the Board of Education," in *The Agricultural College of the State of Michigan* (Lansing, Mich.: Hosmer and Fitch, 1857), 33.

2. Robert F. Johnstone, "The Agricultural College," *Michigan Farmer* 15, no. 5 (1857): 129–31, from Berg, "The Pioneer Land-Grant College," 142–43.

3. Kenneth E. Lewis, *West to Far Michigan: Settling the Lower Peninsula, 1815-1860* (East Lansing: Michigan State University Press, 2002), 284, 287.

4. Justin L. Kestenbaum, *Out of a Wilderness: An Illustrated History of Greater Lansing* (Woodland Hills, California: Windsor Publications, 1981), 24–27; Ford Stevens Ceasar, *The Bicentennial History of Ingham County, Michigan* ([Lansing: The Author, 1976]), 50–60.

5. *Manual, Containing the Rules of the Senate and House of Representatives of the State of Michigan, with the Joint Rules of the Two Houses, and Other Matter* (Lansing: By Authority, 1861), 186, 215, 218.

6. Kestenbaum, *Out of a Wilderness,* 29–43.

7. Willis Frederick Dunbar, *All Aboard!: A History of Railroads in Michigan* (Grand Rapids: Eerdmans, 1969), 97.

8. Charles Sellers, *The Market Revolution: Jacksonian America, 1815–1846* (New York: Oxford University Press, 1991), 4.

9. Lewis, *West to Far Michigan,* 235–300.

10. Alice Felt Tyler, *Freedom's Ferment: Phases of American Social History from the Colonial Period to the Outbreak of the Civil War* (New York: Harper & Row, 1944), 227–64.

11. Eric Foner, *The Story of American Freedom* (New York: Norton, 1998), xvii.

12. Charles R. Starring and James O. Knauss, *The Michigan Search for Educational Standards* (Lansing: Michigan Historical Commission, 1969), 3.

13. Bernard Bailyn, *Education in the Forming of American Society: Needs and Opportunities for Study* (New York: Vintage Books, 1960), 48.

14. J. Hector St. John de Crèvecoeur, *Letters from an American Farmer and*

Sketches of Eighteenth-Century America (New York: The New American Library, 1963), 64.

15. Horace Mann, *Lectures on Education* (New York: Arno Press, 1969), 117.

16. Michael Bezilla, *The College of Agriculture at Penn State: A Tradition of Excellence* (University Park: Pennsylvania State University Press, 1987), 3.

17. Margaret W. Rossiter, *The Emergence of Agricultural Science: Justus Liebig and the Americans, 1840–1880* (New Haven: Yale University Press, 1975), xiii. Rossiter provides an insightful analysis of the how agricultural science evolved during the middle of the nineteenth century.

18. "Executive Meeting," *Transactions of the State Agricultural Society, with Reports of County Agricultural Societies for 1850* (Lansing: Published by Order of the Legislature, 1851), 18.

19. Edward Danforth Eddy, Jr., *Colleges for Our Land and Time: The Land-Grant Idea in American Education* (New York: Harper and Brothers, 1957), 10–11.

20. Earle D. Ross, *Democracy's College: The Land-Grant Movement in the Formative State* (Ames: Iowa State College Press, 1942), 39–43.

21. Eddy, *Colleges for Our Land*, 12–20. For informative discussions of developments in higher education before the passage of the Morrill Act in 1862, see Ross, *Democracy's College*, 14–45 and J. B. Edmond, *The Magnificent Charter: The Origin and Role of the Morrill Land-Grant Colleges and Universities* (Hicksville, N..Y.: Exposition Press, 1978), 3–14. See also Bezilla, *The College of Agriculture at Penn State,* 12–15, and Saul Sack, *History of Higher Education in Pennsylvania* (Harrisburg: Pennsylvania Historical and Museum Commission, 1963), 2:478–88.

22. Elizabeth M. Farrand, *History of the University of Michigan* (Ann Arbor: Register Publishing House, 1885), 15–25.

23. Howard H. Peckham, *The Making of the University of Michigan, 1817–1992* (Ann Arbor: University of Michigan, Bentley Library, 1994), 17–32, and Willis F. Dunbar, *The Michigan Record in Higher Education* (Detroit: Wayne State University Press, 1963), 58–68.

24. Warren Isham, ed., *Michigan Farmer* 6, no. 16 (15 August 1848): 252, from Berg, "The Pioneer Land-Grant College," 20.

25. E. H. Lothrop, Address, 26 September 1849, in *Transactions of the Michigan State Agricultural Society with Reports of County Agricultural Societies [1849]* ([Lansing, N.p.: 1850]), 102.

26. "Joint Resolution Relative to a Donation of Land by the General Government, for Establishing Agricultural Schools, and for Other Purposes," *Transactions of the State Agricultural Society . . . 1850,* 258.

27. Bela Hubbard, *Michigan Farmer* 8 (1850): 18–19, from Berg, "The Pioneer Land-Grant College," 35–39.

28. "Constitutional Provision," *The Agricultural College of the State of Michigan,* 9.

29. J. W. Scott to Isham, 29 January 1851, *Michigan Farmer* 9 (1851): 76–77, from Berg, "The Pioneer Land-Grant College," 55–56.

30. *Michigan Farmer* 9, no. 1 (January 1853): 18, from Berg, "The Pioneer Land-Grant College," 74.

31. A. Winchell, "On the location of the Agricultural College," 10 March 1855, Kuhn Collection, UA 17.107, Box 1142, Folder 32.

32. Henry P. Tappan to J. C. Holmes, 3 December 1852, Kuhn Collection, UA 17.107, Box 1139, Folder 65.

33. Ibid. See also "Proceedings of the Board of Regents," 1837–1854, University of Michigan, Board of Regents, Bentley Library, University of Michigan; and Tappan, "Report of the President of the University," 18 December 1855, "Proceedings of the Board of Regents, 1855–1870, 13–16, University of Michigan, Board of Regents. See also Madison Kuhn, *Michigan State: The First Hundred Years* (East Lansing: Michigan State University Press, 1955), 6–8.

34. "Dr. H. L. Tappan's Address before the Michigan State Agricultural Society, Sept. 1853," typescript, 14–16, Kuhn Collection, UA 17.107, Box 1143, Folder 67.

35. Ibid., 23. For discussions of Tappan and his contributions to the University of Michigan see Peckham, *The Making of the University of Michigan,* 35–57. Tappan was greatly influenced by German higher education. For an interesting analysis of the influence of the German model on graduate training at the University of Michigan see James Turner and Paul Barnard, "The German Model and the Graduate School: The University of Michigan and the Origin Myth of the American University," in Roger L. Geiger, ed., *The American College in the Nineteenth Century* (Nashville: Vanderbilt University Press, 2000), 221–41.

36. "Dr. H. L. Tappan's Address," 16.

37. "Mr. Tappan's Address," *Albany Evening Journal,* 7 August 1857, Henry Lewis Tappan Papers, Box 1278, Folder 15 [miscellaneous], Bentley Library, University of Michigan.

38. Ibid.

39. J. C. Holmes to Justus Gage, 15 October 1861, State Board of Agriculture Correspondence, UA 1, Box 864, Folder 2, MSU Archives.

40. Philo Cultus to editor, *Michigan Farmer* 12, no. 2 (February 1854): 43–44, from Berg, "The Pioneer Land-Grant College," 108–9.

41. J. S. Tibbits, *Michigan Farmer* 12, no. 8 (August 1854): 241, from Berg, "The Pioneer Land-Grant College," 115.

42. Warren Isham, *American Citizen* 4, no. 25 (14 February 1853).

43. "Act of Organization," *The Agricultural College of the State of Michigan,* 12–13. W. J. Beal gives voice to many of the arguments put forth during the debates over the creation of an agricultural college in *History of the Michigan Agricultural College and Biographical Sketches of Trustees and Professors* (East Lansing: The Agricultural College, 1915), 6–15.

44. J. Wilkins to J. R. Williams, 24 May 1856, Joseph R. Williams Papers, UA 2.1.1, Box 871, Folder 30, MSU Archives.

45. C. Beckington to Williams, 17 May 1857, Williams Papers, UA 2.1.1, Box 871, Folder 26.

46. H. K. Moss to Williams, July 1857, Williams Papers, UA 2.1.1, Box 871, Folder 28.

47. [Warren Ibo] to Williams, 4 August 1857, Williams Papers, UA 2.1.1, Box 871, Folder 26.

48. J. W. Hoyt to Williams, 12 September 1857, Williams Papers, UA 2.1.1, Box 871, Folder 42.

49. [D.C. Linsley] to Williams, 27 August 1857, Williams Papers, UA 2.1.1, Box 871, Folder 42.

50. Theodore Brown to Williams, 16 June 1857, Williams Papers, UA 2.1.1, Box 871, Folder 18.

51. Williams, "Report of the President of the Agricultural College," 1 December 1858, *Michigan Farmer* 1 (new series), no. 4 (22 January 1859): 26, from Berg, "The Pioneer Land-Grant College," 238.

52. For Holmes's recollection of his efforts to have the college created as a separate entity see his letter to T. C. Abbot, 24 August 1876, in Theophilus Capon Abbot Papers, UA 2.3.1, Box 861, Folder 37, MSU Archives.

53. Kuhn, *Michigan State,* 11–17.

54. Ibid, 13–18, 25–33, 58–67; Beal, *History,* 385–406.

55. Ibid, 17.

56. "A farmer's wife," *Michigan Farmer* 3 (new series), no. 34 (24 August 1861): 270, from Berg, "The Pioneer Land-Grant College," 487.

57. Edward Granger, Diary, 1858–60, Edward Granger Papers, 1858-60, UA 10.3.56, Folder 1, MSU Archives.

58. Charles Jewell to Horace, 4 May [no year], Charles A. Jewell Papers, UA 10.3.5, F.D., Folder 1. MSU Archives.

59. Granger, Scrapbook, 1858, Granger Papers, UA 10.3.56, Folder 2.

60. Henry Graham Reynolds, Diary, 1867, 12–18 March, typescript in the collections of the Michigan State University Museum, original in the Rare Book Room, Library of Michigan.

61. Granger, Scrapbook.

62. Granger, Diary.

63. Eliza C. Smith, "History of the Michigan Female College, and a Sketch of the Life and Work of Miss A. C. Rogers," *Michigan Pioneer and Historical Collections* (Lansing: Michigan Pioneer and Historical Society, 1907), 6:287; Justin L. Kestenbaum, *Out of a Wilderness: An Illustrated History of Greater Lansing* (Woodland Hills, California: Windsor Publications, 1981), 40.

64. Theodore F. Garvin to mother, 21 October 1859, Kuhn Collection, UA 17.107, Box 1140, Folder 3.

65. Sidney S. Sessions to Mary E. Yates, 10 November 1859, Sidney S. Sessions Papers, UA 10.3.9, F. D., Folder 1. MSU Archives.

66. Jewell to Horace, 4 May [no year].

67. Granger, Diary.

68. Dunbar, *The Michigan Record,* 94–97.

69. For a discussion of Gregory's curriculum and its particulars, see Kuhn, *Michigan State,* 52–65. Abbot's new asignment lasted for only one year, 1860–61. In 1861 he was appointed professor of history and English literature, and in 1866, he became professor of mental philosophy and logic.

70. Lewis R. Fisk, Report to the State Board of Agriculture, 4 April 1861, Lewis R. Fisk Papers, UA 2.1.2, Box 862, Folder 4, MSU Archives.

71. T. C. Abbot, Diary, typescript, 150, Abbot Papers, UA 2.1.3, Box 861, Diaries.

72. Fisk, Report.

73. T. T. Lyon, *Michigan Farmer* 3 (new series), no. 2 (12 January 1860): 11, from Berg, "The Pioneer Land-Grant College," 422; James H. Gun-

nison, "The Dawn of the Michigan Agricultural College," n.d., Kuhn Collection, UA 17.107, Box 1140, Folder 8.

74. "An Act to Reorganize the Agricultural College of the State of Michigan, and to Establish a State Board of Agriculture," 15 March 1861, *The Compiled Laws of the State of Michigan*, comp. James S. Dewey (Lansing: W. S. George and Co., 1872), 1:1183. See also Beal, *History*, 37.

75. "An Act to Reorganize the Agricultural College."

76. For discussions of events leading up to the Morrill bill, see Winton U. Solberg, *The University of Illinois 1867–1894: An Intellectual and Cultural History* (Urbana: University of Illinois Press, 1968), 22–58, and Coy F. Cross II, *Justin Smith Morrill: Father of the Land-Grant Colleges* (East Lansing: Michigan State University Press, 1999), 77–84.

77. Justin S. Morrill, "On the Bill Granting Lands for Agricultural Colleges; Delivered in the House of Representatives, April 20, 1858," Washington, D.C.: Printed at the Congregational Globe Office, 1858, Justin S. Morrill Papers, Library of Congress, Container 66, Reel 42, Speeches 1856–93. John H. Florer, "Major Issues in the Congressional Debate of the Morrill Act of 1862," *History of Education Quarterly* 8, no. 4 (winter 1968): 465.

78. Albert E. Macomber, who attended M.A.C. 1857–59, credited Williams for being very influential in the passage of the first Morrill bill in 1859 in Beal, *History*, 36–38. See also Kuhn, *Michigan State*, 49–52.

79. J. R. Williams to J. S. Morrill, 15 November 1858, Morrill Act-Correspondence, Kuhn Collection, UA 17.107, Box 1142, Folder 48.

80. J. R. Williams, "Agricultural Education, An Address Delivered at the State Fair, Syracuse, N.Y., October 8, 1858," Williams Papers, UA 2.1.1, Box 871, Folder 47, MSU Archives. Earle D. Ross, *A History of the Iowa State College of Agriculture and Mechanic Arts* (Ames: Iowa State College Press, 1942), 21.

81. J. R. Williams to J. S. Morrill, 8 December 1858, Kuhn Collection, UA 17.107, Box 1142, Folder 48.

82. See Cross, *Justin Smith Morrill*, 80–83.

83. Paul W. Gates, "Western Opposition to the Agricultural College Act," *Indiana Magazine of History* 37, no. 1 (March 1941): 107–12. Scott Key argues that "in 1858, the debate focused on economics not education. As Morrill presented his case, he emphasized the role that agriculture played in 'the prosperity and happiness' of the nation." ("Economics or Education: The Establishment of American Land-Grant Universities," *Journal of Higher Education* 67, no. 2 [March/April 1996]: 211).

84. Allan Nevins, *The Origins of the Land-Grant Colleges and State Universities: A Brief Account of the Morrill Act of 1862 and Its Results* (Washington, D.C.: Civil War Centennial Commission, 1962).

85. Jonathan B. Turner, *Prairie Farmer*, March 1852, p. 114, quoted in Winton U. Solberg, *The University of Illinois 1867–1894: An Intellectual and Cultural History* (Urbana: University of Illinois Press, 1968), 47.

86. Bezilla, *The College of Agriculture at Penn State*, 12–23.

87. Abraham Lincoln, "Address before the Wisconsin State Agricultural Society, Milwaukee, Wisconsin," 30 September 1859, in *The Collected*

Works of Abraham Lincoln, ed. Roy P. Basler, Marion Dolores Pratt, and Lloyd A. Dunlap (New Brunswick: Rutgers University Press, 1953), 3:480.

88. Ibid. For a provocative discussion of the importance of "free labor," see Eric Foner, "Free Labor and Nineteenth-Century Ideology," in *The Market Revolution in America,* ed. Melvyn Stokes and Stephen Conway (Charlottesville: University Press of Virginia, 1996), 99–127.

89. Louis Ferleger and William Lazonick, "Higher Education for an Innovative Economy: Land-Grant Colleges and the Managerial Revolution in America," *Business and Economic History* 23, no. 1 (fall 1994): 116–28.

90. "The First Morrill Act, 1862," in Rasmussen, *Agriculture in the United States: A Documentary History,* 1:617.

91. Russell Thackrey and Jay Richter, "The Land-Grant Colleges and Universities, 1862–1962: An American Institution," *Higher Education* 16, no. 3 (November 1959): 3–7.

92. See Kuhn, *Michigan State,* 66–70. For information on people associated with M.A.C. who served in the Civil War, see Beal, *History,* 484–95.

93. "Secretary's Report," *Second Annual Report of the Secretary of the State Board of Agriculture of the State of Michigan for the Year 1863* (Lansing: John A. Kerr and Co., Printers to the State, 1863), 28–29.

94. "An Act to Reorganize the Agricultural College of the State of Michigan," 1180.

95. [Faculty Report on the Operation of the Farm, 1862–1864,] State Board of Agriculture Papers, Faculty Reports to State Board [1862–1864], UA 1, Box 865, Folder 66, MSU Archives.

96. In February 1865 the college placed an advertisement in the *Chicago Tribune.* The first two paragraphs clearly spell out the curriculum: "The State Agricultural College, Lansing, Michigan, now offers to students very superior advantages. Four years are required to complete the course of study which is as full in Mathematics, English, Literature, History, Philosophy, &c. as in other Colleges.

"The courses in Chemistry, Meteorology, Botany, Animal Physiology, and other branches of Natural Science are unusually extensive, great attention being paid also to their practical applications" (State Board of Agricultural Papers, Financial Records: 1860–1866, UA 1, Box 865, Folder 74).

CHAPTER 2. DEFINING THE COLLEGE'S MISSION

1. Thomas Jefferson, *Notes on the State of Virginia,* ed. William Peden (New York: Norton, 1982), 164–65.

2. See Leo Marx, *The Machine in the Garden: Technology and the Pastoral Ideal in America* (Oxford: Oxford University Press, 2000). Henry Nash Smith, *Virgin Land: The American West as Symbol and Myth* (New York: Vintage, 1950) presents a classic analysis of the importance of the power of myth and symbol in the development of the United States.

3. Norman Pollack, *The Populist Response to Industrial America: Mid-*

western Populist Thought (Cambridge: Harvard University Press, 1962), 11–12.

4. Lawrence Goodwyn, *The Populist Moment: A Short History of the Agrarian Revolt in America* (Oxford: Oxford University Press, 1981), 39–40. For a thought provoking analysis of Populism, see Norman Pollack, *The Humane Economy: Populism, Capitalism, and Democracy* (New Brunswick: Rutgers University Press, 1990).

5. Willis Frederick Dunbar, *Michigan: A History of the Wolverine State* (Grand Rapids: W. B. Eerdmans, 1965), 526; *Michigan Legislative Manual and Official Directory for the Years 1897–98* (Lansing: Robert Smith Printing Co., 1897), 368–69, 566–67; Richard Jensen, *The Winning of the Midwest: Social and Political Conflict, 1888–1896* (Chicago: University of Chicago Press, 1971), 227–28. Michigan elected Republican governors between 1854 and 1892, except in 1854 and 1892. In every presidential elections between 1856 and 1892, the state went Republican. Also, third party candidates fared poorly.

6. J. S. Mitchell, "The Discontent Among Farmers," *Speculum* 13, no. 7 (15 May 1895): 111–14.

7. Brody, *In the Service of the Farmer*, 9.

8. Fred Trump, *The Grange in Michigan: An Agricultural History of Michigan over the Past Ninety Years* ([Grand Rapids: Fred Trump, 1963]), i–135. Two works that analyze the national granger movement are Solon Justus Buck, *The Granger Movement: A Study of Agricultural Organization and Its Political Economic and Social Manifestations* (Cambridge: Harvard University Press, 1913); and D. Sven Nordin, *Rich Harvest: A History of the Grange, 1867–1900* (Jackson: University Press of Mississippi, 1974).

9. L. H. Bailey, ed., *The Principles of Agriculture: A Text-Book for Schools and Rural Societies* (New York: Macmillan, 1898), v–vi.

10. James H. Gunnison, "The Dawn of the Michigan Agricultural College," Kuhn Collection, UA 17.107, Box 1140, Folder 8.

11. For a summary of this battle see Kuhn, *Michigan State*, 76–82. [J. D. Hawley,] [1867,] "A Bill to Provide for an Agricultural Department in the State University," Kuhn Collection, UA 17.107, Box 1140, Folder 62. Alonzo Sessions to Gov. Baldwin, 15 March 1869, Kuhn Collection, UA 17.107, Box 1140, Folder 32.

12. A concise summary of the case against the college is provided in Beal, *History*, 58.

13. T. C. Abbot, 22 and 27 January 1869, Diary, 451, 452, T. C. Abbot Papers, UA 2.1.3, Box 861, MSU Archives.

14. Richard Haigh, February 1869, notes, Kuhn Collection, UA 17.107, Box 1139, Folder 61; *Detroit Free Press*, 20 February 1869, 2.

15. Haigh, 16 March 1869, notes.

16. House Journal, 1869, pages 774–75, typescript, Frank S. Kedzie Papers, UA 2.1.8, Box 1166, Folder 31, MSU Archives.

17. House Journal, 1869, pages 828–31. T. C. Abbot described some of the events leading up to the 1869 vote in "The Michigan State Agricultural College," *Michigan Pioneer and Historical Collections* (Lansing: Wynkoop Hallenbeck Crawford Co., 1907), 6:123–24, where he says "The discussion of the location of the college did not cease until 1869."

18. T. C. Abbot, "Agricultural College Management," "Letter from President Abbot," Abbot Papers, UA 2.1.3, Box 861, Folder 63.

19. Daniel Strange to W. J. Beal, 29 May 1911, William J. Beal Papers, UA 17.4, Box 891, Folder 2, MSU Archives.

20. Ibid.

21. J. J. Woodman, "State Agricultural College," 25 November 1878, Abbot Papers, UA 2.1.3, Box 861, Folder 51.

22. Beal, *History,* 67.

23. Beal, Address, 9 July 1879, Beal Papers, UA 17.4, Box 891, Folder 11.

24. *Michigan Farmer,* 27 June 1891, copy of article in "Articles on the College" from *Michigan Farmer* 1891, Kuhn Collection, UA 17.107, Box 1141, Folder 66, and L. D. Watkins to Oscar Clute, 31 July 1891, Oscar Clute Papers, UA 2.1.5, Box 863, Folder 52, MSU Archives.

25. I. H. Butterfield to Oscar Clute, 24 January 1891, Clute Papers, UA 2.1.5, Box 863, Folder 39.

26. I. H. Butterfield to Clute, UA 2.1.5, Box 863, Folder 39.

27. W. W. Bemis to *Grange Visitor,* 19 April 1895, Kenyon L. Butterfield Papers, UA 2.1.10, Box 860H, Folder 7, MSU Archives.

28. Alva Sherwood, '81, to Kenyon L. Butterfield, 10 April 1895, , Kenyon L. Butterfield Papers, UA 2.1.10, Box 860H, Folder 7, MSU Archives.

29. E. O. Ladd to K. L. Butterfield, 23 April 1895, , Kenyon L. Butterfield Papers, UA 2.1.10, Box 860H, Folder 7, MSU Archives.

30. Hudson Sheldon to K. L. Butterfield, 3 May 1894, Kenyon L. Butterfield Papers, UA 2.1.10, Box 860H, Folder 7, MSU Archives.

31. A. C. Bird to K. L. Butterfield, 7 May 1894, , Kenyon L. Butterfield Papers, UA 2.1.10, Box 860H, Folder 7, MSU Archives.

32. F. Hodgman, '62, to Editor of the *Visitor,* 8 May 1895, , Kenyon L. Butterfield Papers, UA 2.1.10, Box 860H, Folder 7, MSU Archives.

33. E. D. A. True, '78, to K. L. Butterfield, 20 May 1894, , Kenyon L. Butterfield Papers, UA 2.1.10, Box 860H, Folder 7, MSU Archives.

34. F. N. Clark to K. L. Butterfield, 7 June 1894, , Kenyon L. Butterfield Papers, UA 2.1.10, Box 860H, Folder 7, MSU Archives.

35. *The Grange Visitor* 21, no. 4 (20 February 1896): 1.

36. "Attendance at Agricultural College," *Thirty-Fifth Annual Report of the Secretary of the State Board of Agriculture . . . from July 1, 1895, to June 30, 1896* (Lansing: Robert Smith Printing Co., 1897), 60. The entire report appears on pages 57–70.

37. Ibid., 63–64.

38. Ibid., 68–71.

39. A. H. Zenner to Fred Warner, 25 August 1905, Kuhn Collection, UA 17.107, Box 1140, Folder 85.

40. Leo M. Geismar to J. L. Snyder, 6 December 1911, Jonathan L. Snyder Papers, UA 2.1.7, Box 816, Folder 65, MSU Archives.

41. R. N. Seward to Snyder, 8 February 1915, Snyder Papers, UA 2.1.7, Box 866, Folder 14.

42. Clinton D. Smith to I. H. Butterfield, 1 August 1893, Kuhn Collection, UA 17.107, Box 1139, Folder 95.

43. Smith to Chase S. Osborn, 26 May 1912, Snyder Papers, UA 2.1.7, Box 819, Folder 109.

44. Snyder to R. R. Lyon, 21 November 1912, Snyder Papers, UA 2.1.7, Box 818, Folder 154.

45. Smith to Osborn, 26 May 1912.

46. Snyder, [1905,] Snyder Papers, UA 2.1.7, Box 809, Folder 7; Snyder to A. E. Lawrence, 2 October 1909, Snyder Papers, Box 814, Folder 35.

47. Snyder to Don Dickerson, 4 December 1900, Snyder Papers, UA 2.1.7, Box 814, Folder 13.

48. Snyder to P. Talbot, 24 December 1907, Snyder Papers, UA 2.1.7, Box 812, Folder 9.

49. [W. J. Beal,] "Some Reasons Why M.A.C. should be connected with the U. of M.," [1914,] Kedzie Papers, UA 2.1.8, Box 1166, Folder 32; [Alumni opinions,] Kedzie Papers, UA 2.1.8, Box 1166, Folder 41; Beal, *History,* 496–503.

50. Snyder to Beal, 7 December 1914, Snyder Papers, UA 2.1.7, Box 855, Folder 33.

51. H. B. Hutchins to Beal, 9 December 1914, Snyder Papers, UA 2.17, Box 855, Folder 33.

52. H. B. Hutchins to Snyder, 10 December 1914, Snyder Papers, UA 2.1.7, Box 855, Folder 33.

53. Jason Woodman to Snyder, 12 December 1914, Snyder Papers, UA 2.1.7, Box 855, Folder 185.

54. For a full discussion of Jonathan Le Moyne Snyder's presidency, see Maurice Raymond Cullen Jr., "The Presidency of Jonathan Le Moyne Snyder at Michigan Agricultural College, 1896–1915" (Ph.D. diss., Michigan State University, 1966).

55. In 1890 Congress passed the Second Morrill Act that led to the establishment of Black Land-Grant Colleges. Tuskegee did not receive funding under this law until 1972. For discussions of the Black Land-Grant Colleges see Frederick S. Humphries, "1890 Land-Grant Institutions: Their Struggle for Survival and Equality"; John R. Wennersten, "The Travail of the Black Land Grant Schools in the South, 1890–1917"; Robert L. Jenkins, "The Black Land-Grant Colleges in Their Formative Years, 1890–1920"; and B. D. Mayberry, "The Tuskegee Movable School: A Unique Contribution to National and International Agriculture and Rural Development," all in *Agricultural History* 65, no. 2 (spring 1991). See also B. D. Mayberry, *A Century of Agriculture in the 1890 Land-Grant Institutions and Tuskegee University, 1890–1990* (New York: Vantage Press, 1991).

56. Booker T. Washington, Commencement Address, June 1900, *M.A.C. Record* 5, no. 39 (26 June 1900), 7–8.

57. Theodore Roosevelt, "The Man Who Works with His Hands," *Semi-Centennial Celebration of Michigan State Agricultural College: May Twenty-Sixth, Twenty-Ninth, Thirtieth and Thirty-First, Nineteen Hundred Seven,* ed. Thomas C. Blaisdell (Chicago: University of Chicago Press, 1908), 239–55.

58. Snyder to Don B. Button, 29 January 1905, Snyder Papers, UA 2.1.7, Box 809, Folder 107.

59. Snyder to John J. Coventry, 9 March 1907, Snyder Papers, UA 2.1.7, Box 810, Folder 127; Snyder to R. F. Crane, 27 June 1908, Snyder Papers, UA 2.1.7, Box 812, Folder 77; Snyder to Myra E. Lee, 13 November 1907, Snyder Papers, UA 2.1.7, Box 811, Folder 93.

60. "Agricultural College," *M.A.C. Record* 9, no. 14 (22 December 1903), originally published in *Detroit Tribune,* 10 December 1903.

61. Snyder to W. P. Snyder, 20 February 1911, Snyder Papers, UA 2.1.7, Box 867, Folder 5.

62. E. H. Ryder, "Facts of the New Course [Applied Science], Condensed for the Alumni," *M.A.C. Record* 26, no. 28 (6 May 1911): 8

63. Snyder to A. J. Ladd, 4 October 1912, Snyder Papers, 2.1.7, Box 818, Folder 152.

64. This data has been gathered from the Michigan Agricultural College catalogs for 1885, 1901, 1906, and 1915. When the totals for each division are added together in 1906 and 1915, they do not equal the total given because some students enrolled in more than one program during the year.

65. E. S. Ingalls to Abbot, 22 February 1872, Abbot Papers, UA 2.1.3, Box 861, Folder 28.

66. E. S. Ingalls to Abbot, 9 April 1872, Abbot Papers, UA 2.1.3, Box 861, Folder 29.

67. *College Speculum* 4, no. 2 (15 October 1884): 6.

68. F. F. Rogers, "Desirability of a College Education," *College Speculum* 2, no. 1 (1 August 1882): 2–3.

69. A. C. Redding, "A Knowledge of the Classics Not Necessary to Culture," *College Speculum* 3, no. 1 (1 August 1883): 2–3.

70. Geo. L. Teller, "Science and Practice in Agriculture," *Speculum* 8, no. 3 (10 October 1888): 33.

71. J. W. White, "Farmers Should Have a Liberal Education," *Speculum* 9, no. 3 (10 October 1889): 47–48.

72. O. P. West, "Agricultural Training," *Speculum* 13, no. 9 (15 July 1895): 150–52.

73. Harry J. Eustace to Lewis Griffin Gorton, 6 September 1894, Lewis Griffin Gorton Papers, UA 2.1.6, Box 873, Folder 20, MSU Archives.

74. Hudson Shelton to Kenyon L. Butterfield, 3 May 1894, Butterfield Papers, UA 2.1.10, Box 860H, Folder 7.

75. Mitchell, "The Discontent Among Farmers."

76. Katherine McCurdy, "The Real Student," *M.A.C. Record* 2, no. 8 (23 February 1897): 3.

77. Blanche Clark, "Pedagogical Relation of Domestic Art and Science to the High School," *Holcad* 2, no. 9 (10 February 1910): 2–3.

78. E. V. O'Rourke to Michigan Agricultural College, 28 December 1910, Snyder Papers, UA 2.1.7, Box 815, Folder 83.

79. J. F. Ruff to Snyder, 20 November 1911, Snyder Papers, UA 2.1.7, Box 817, Folder 34.

80. Geo. W. Murrow to Pres., M.A.C., Snyder Papers, UA 2.1.7, Box 815, Folder 57.

81. "Agricultural College," attached to Snyder to J. H. Brown, 3 January 1904, Snyder Papers, UA 2.1.7, Box 806, Folder 142; C. D. Smith to Snyder, 15 October 1907, Snyder Papers, UA 2.1.7, Box 811, Folder 150.

82. Snyder to Mr. Bird, 17 September 1897, Snyder Papers, UA 2.1.7, Box 804, Folder 24; "Report of the President," *Forty-Fifth Annual Report of the Secretary of the State Board of Agriculture of the State of Michigan, . . . from July 1, 1905, to June 30, 1906* (Lansing: Wynkoop Hallenbeck Crawford Co., 1906), 27.

83. Snyder to W. H. French, 5 December 1906, Snyder Papers, UA 2.1.7, Box 809, Folder 157.

84. Snyder to Chas. F. Robinson, 23 February 1903, Snyder Papers, UA 2.1.7, Box 807, Folder 98.

85. Snyder to R. N. Seward, 13 February 1915, Snyder Papers, UA 2.1.7, Box 866, Folder 14.

86. Snyder to R. B. Sleight, 13 March 1909, Snyder Papers, UA 2.1.7, Box 814, Folder 51.

87. Snyder to E. Davenport, 22 October 1904, Snyder Papers, UA 2.1.7, Box 807, Folder 186.

88. Snyder to W. E. Weatherly, 21 April 1896, Snyder Papers, UA 2.1.7, Box 894, Folder 19.

89. Snyder to Mrs. Jas. F. Hancock, 21 August 1897, Snyder Papers, UA 2.1.7, Box 804, Folder 30.

90. Ernest E. Bogue to Gifford Pinchot, 25 October 1904, Kuhn Collection, UA 17.107, Box 1139, Folder 31.

91. E. Wallace Brainard to Snyder, 22 August 1911, Snyder Papers, UA 2.1.7, Box 816, Folder 20; The Tribune Company to [M.A.C.], 10 August 1911, Snyder Papers, UA 2.1.7, Box 817, Folder 50.

92. "The Michigan State Agricultural College," attached to Alf. Washington to Snyder, 15 July 1902, Snyder Papers, UA 2.1.7, Box 806, Folder 111.

93. R. S. Shaw to [Alumni], 20 August 1908, Snyder Papers UA 2.1.7, Box 813, Folder 132.

94. Snyder to Maude Gilchrist, 27 April, Snyder Papers, UA 2.1.7, Box 816, Folder 55.

95. "Results—Advertising, 1912–13," Snyder Papers, UA 2.1.7, Box 820, Folder 80.

96. Charles D. Ellis, 12 October 1903, Snyder Papers, UA 2.1.7, Box 807, Folder 7.

CHAPTER 3. TRANSFORMING KNOWLEDGE

1. *Calendar of the University of Michigan for 1890–91* (Ann Arbor: University of Michigan, 1891), 17

2. *Catalogue of the Officers and Students of the State Agricultural College of Michigan . . . 1890–91* (Agricultural College: The College, 1891), 48.

3. Brian A. Williams, *Thought and Action: John Dewey at the University of Michigan* (Ann Arbor: Bentley Historical Library, 1998), 20.

4. John Dewey, *Lectures in the Philosophy of Education, 1899,* ed. Reginald D. Archambault (New York: Random House, 1966), xx.

5. John Dewey, "Lecture VIII," 18 January 1899, in Archambault, ed. Lectures, 80. Dewey had left Michigan in 1894 to teach at the University of Chicago.

6. Howard H. Peckham, *The Making of the University of Michigan, 1817–1992* (Ann Arbor: Bentley Historical Library, 1994), 96.

7. *Catalogue of the Officers and Students of the State Agricultural College of Michigan, 1866* (Lansing: Jno. A. Kerr and Co., 1866), 12–14, 25.

8. Merle Curti and Vernon Carstensen, *The University of Wisconsin: A History* (Madison: University of Wisconsin Press, 1949), 1: 352–53.

9. *Catalogue,* 1866, 18.

10. Ibid., 16.

11. Oscar Clute, "Education at the Michigan Agricultural College: Its Scope; Its Method; and Its Results," [1890,] 5. Oscar Clute Papers, UA 2.1.5, Box 863, Folder 97, MSU Archives.

12. Theophilus C. Abbot, "Agricultural Education. An Address by President Abbot, of the Michigan State Agricultural College," 4 March 1875. T. C. Abbot Papers, UA 2.3.1, Box 861, Folder 68, MSU Archives.

13. Ibid., 8.

14. Ibid., 14.

15. Eva D. Coryell, "A Practical Education for Women," *Eighteenth Annual Report of the Secretary of the State Board of Agriculture of the State of Michigan, for the Year Ending August 31st, 1879* (Lansing: W. S. George and Co., State Printers, 1880), 173–174.

16. Mrs. Perry Mayo, "Does Education Lead to Extravagance," *Nineteenth Annual Report of the Secretary of the State Board of Agriculture of the State of Michigan for the Year Ending August 31st, 1880* (Lansing: W. S. George and Co., 1880), 194.

17. Mary Mayo, "Practical Education for Young Women," 14 April 1896, Kuhn Collection, UA 17.107, Box 1141, Folder 38.

18. Mary Mayo, "Special Studies for Young Women at M.A.C.," 2 June 1896, Kuhn Collection, UA 17.107, Box 1141, Folder 38.

19. Trump, *The Grange in Michigan,* 34.

20. Charles E. Rosenberg, "Rationalization and Reality in the Shaping of American Agricultural Research, 1875–1914," *Social Studies of Science* 7, no. 4 (November 1977): 405–6.

21. For a good summary of the development of scientific agriculture see Reynold M. Wik, "Science and American Agriculture," in *Science and Society in the United States,* ed. David D. Van Tassel and Michael G. Hall (Homewood, Illinois: Dorsey Press, 1966), 81–106.

22. Gladys L. Baker, Wayne D. Rasmussen, Vivian Wiser, and Jane M. Porter, *Century of Service: The First 100 Years of the United States Department of Agriculture* ([Washington, D.C.]: U.S. Department of Agriculture, 1963), 13–18; Wik, "Science and American Agriculture," 94–95; T. Swann Harding, *Two Blades of Grass: A History of Scientific Development in the U.S. Department of Agriculture* (Norman: University of Oklahoma Press, 1947), 23–33.

23. R. Douglas Hurt, *American Agriculture: A Brief History* (Ames: Iowa State University Press, 1994), 193; "The Hatch Act," from 24 U.S. Statutes at Large 440, in Rasmussen, *Agriculture in the United States,* 2:1232–34.

24. Charles E. Rosenberg, "The Adams Act: Politics and the Cause of Scientific Research," *Agricultural History* 38, no. 1 (January 1964): 8; "Adams' Experiment Station Act, 1906," from 34 U.S. Statutes at Large 63, in Rasmussen, *Agriculture in the United States,* 2:1344–45.

25. Rosenberg, "Rationalization and Reality," 415–18.

26. G. C. Caldwell, "The More Notable Events in the Progress in Agricultural Chemistry, Since 1870," *The Journal of the American Chemical Society* 14 (1892): 83–111.

27. Charles A. Browne, "Agricultural Chemistry," in *A Half-Century of Chemistry in America 1876–1926,* ed. Charles A. Browne (Easton,

Pennsylvania: American Chemical Society, 1926), 187; W. H. Glover, *Farm and College: The College of Agriculture of the University of Wisconsin, A History* (Madison: University of Wisconsin Press, 1952), 118–21; Curti and Carstensen, *The University of Wisconsin,* 2:387–90.

28.   H. W. Wiley, "The Relation of Chemistry to the Progress of Agriculture," in *Yearbook of the United States Department of Agriculture, 1899* (Washington, D.C.: G.P.O., 1900), 253. See Joseph S. Chamberlain, ed., *Chemistry in Agriculture: A Cooperative Work Intended to Give Examples of the Contributions Made to Agriculture by Chemistry* (New York: Chemical Foundation, 1926) for an interesting collection of essays discussing the relationship between chemistry and agriculture as of 1926.

29.   Gustavus A. Weber, *The Bureau of Chemistry and Soils: Its History, Activities and Organization* (Baltimore: Johns Hopkins Press, 1928), 2.

30.   Browne, "Agricultural Chemistry," 191.

31.   L. O. Howard, "Progress in Economic Entomology in the United States," in *Yearbook of the United States, Department of Agriculture, 1899* (Washington, D.C.: G.P.O., 1900), 135–36, 137, 152–153.

32.   See Gustavus A. Weber, *The Bureau of Entomology: Its History, Activities and Organizations* (Washington: Brookings Institution, 1930) for a thorough look at the USDA's work in entomology until 1930.

33.   L. O. Howard, "The Rise of Applied Entomology in the United States," *Agricultural History* 3, no. 3 (July 1929): 135–37.

34.   W. J. Holland, "The Development of Entomology in North America," *Annals of the Entomological Society of America* 12, no. 1 (March 1920): 10–11; Thomas R. Dunlap, "Farmers, Scientists, and Insects," *Agricultural History* 54, no. 1 (January 1980): 93.

35.   G. C. Ainsworth, *Introduction to the History of Plant Pathology* (Cambridge: Cambridge University Press, 1981), 4–5.

36.   Winton U. Solberg, *The University of Illinois: 1867–1894* (Urbana: University of Illinois Press, 1968), 153–55. In June 1885 the name Industrial University of Illinois was changed to the University of Illinois, 226–27.

37.   Ainsworth, *Introduction to the History of Plant Pathology,* 8. See also John A. Stevenson, "The Beginnings of Plant Pathology in North America," in *Plant Pathology: Problems and Progress, 1908–1958,* ed. S. Holton et al. (Madison: University of Wisconsin Press, 1959), 14–23.

38.   B. T. Galloway, "Progress in the Treatment of Plant Diseases in the United States," *Yearbook of the United States Department of Agriculture, 1899* (Washington, D.C.: GPO, 1900), 193–95; Galloway, "Twenty Years Progress in Plant Pathology," *Proceedings of the Twenty-First Annual Meeting of the Society for the Promotion of Agricultural Science,* 1900 (Syracuse: The Society, 1900), 95.

39.   John Charles Walker, *Plant Pathology* (New York: McGraw-Hill, 1969), 40.

40.   Alexius A. J. De'Sigmond, "Development of Soil Science," *Soil Science* 40 (June–December 1935): 77, 80. De'Sigmond gives a very concise, informative survey of the evolution of soil science as a distinctive discipline.

41.   Milton Whitney, "Soil Investigations in the United States," in *Year-*

book of the United States Department of Agriculture, 1899, 342–46.

42. A. G. McCall, "The Development of Soil Science," *Agricultural History* 5, no. 2 (April 1931): 56.

43. D. E. Salmon, "Some Examples of the Development of Knowledge Concerning Animal Diseases," *Yearbook of the United States Department of Agriculture, 1899,* 96–119; T. C. Byerly, "Changes in Animal Science," *Agricultural History* 50, no. 1 (January 1976): 273.

44. Adelynne H. Whitaker, "Pesticide Use in Early Twentieth Century Animal Disease Control," *Agricultural History* 54, no. 1 (January 1980): 72.

45. Salmon, "Some Examples of the Development of Knowledge," 124–34; Whitaker, "Pesticide Use," 72–73.

46. For an interesting discussion of the effort to determine the cause of hog cholera see B. W. Bierer, *A Short History of Veterinary Medicine in America* (East Lansing: Michigan State University Press, 1955), 60–79.

47. U. G. Houck, *The Bureau of Animal Industry of the United States Department of Agriculture: Its Establishment, Achievements and Current Activities* (Washington, D.C.: The Author, 1924), ix.

48. Byerly, "Changes in Animal Science," 258.

49. Bierer, *A Short History of Veterinary Medicine,* 38, 95.

50. Charles Cleon Morrill, *Veterinary Medicine in Michigan: An Illustrated History* (East Lansing: College of Veterinary Medicine, Michigan State University, 1979), 58.

51. Rudolf Alexander Clemen, *The American Livestock and Meat Industry* (New York: Ronald Press, 1923), 211–68, 317–46.

52. Bierer, *A Short History of Veterinary Medicine,* 91, 98.

53. Henry E. Alvord, "Dairy Development in the United States," *Yearbook of the United States Department of Agriculture, 1899,* 391–98; T. R. Pirtle, *History of the Dairy Industry* (Chicago: Mojonnier Bros. Co., 1926), 31–57; John T. Schlebecker, *A History of American Dairying* (Chicago: Rand McNally, 1967), 37.

54. Pirtle, *History of the Dairy Industry,* 60–71.

55. Jenks Cameron, *The Bureau of Dairy Industry* (Baltimore: Johns Hopkins Press, 1929), 17.

56. Pirtle, *History of the Dairy Industry,* 68.

57. Robert K. Winters, "The First Half Century," in *Fifty Years of Forestry in the U.S.A.,* ed. Robert K. Winters (Washington, D.C.: Society of American Foresters, 1950), 3, 5–7, 13.

58. H. T. Gisborne, "Forest Protection," in *Fifty Years of Forestry,* 31.

59. Winters, "The First Half Century," 14.

60. A useful discussion on the use of forest products is Nelson C. Brown, "Forest Utilization," in *Fifty Years of Forestry,* 92–112. A fascinating study of the social implications of cutting over forest land is Robert Gough, *Farming the Cutover: A Social History of Northern Wisconsin, 1900–1940* (Lawrence: University Press of Kansas, 1997).

61. Charles Edwin Bessey, "How They Taught in the Early Years," in *Semi-Centennial Celebration of Michigan State Agricultural College,* ed. Thomas C. Blaisdell, 82–87.

62. W. J. Beal, "Notes for Beal's History of M.A.C., n. d.," William J. Beal Papers, UA 17.4, Box 891, Folder 17, MSU Archives.

63. Eugene Davenport to Elida Yakeley, 23 October 1939, Kuhn Collection, UA 17.107, Box 1139, Folder 39.

64. W. J. Beal, "Annual Report," 11 November 1872, Beal Papers, UA 17.4, Box 891, Folder 16.

65. Beal to Asa Gray, 11 March 1872, Beal Papers, UA 17.4, Box 891, Folder 1; Charles Darwin to Beal, 21 May 1878, Beal Papers, UA 17.4, Box 891, Folder 1.

66. A. N. Prentiss to Beal, 26 September 1876, Beal Papers, UA 17.4, Box 891, Folder 1.

67. C. E. Bessey to Beal, 31 December 1877, Beal Papers, UA 17.4, Box 891, Folder 1.

68. Bessey to Howard Edwards, 11 December 1906, Charles Edwin Bessey Papers (1865–1915), microfilm, roll 16, University of Nebraska Archives, University of Nebraska, Lincoln.

69. W. J. Beal, *The New Botany: A Lecture of the Best Method of Teaching the Science,* 2nd ed., rev. (Philadelphia: C. H. Marot, 1882), 8.

70. Alexander Winchell, 16 May 1884, Beal Papers, UA 17.4, Box 891, Folder 2.

71. Oscar Clute, "The State Agricultural College, Michigan: Report of the President," 19, Kuhn Collection, UA 17.107, Box 1139, Folder 3.

72. *College Speculum* 10, no. 5 (10 April 1891): 76–77.

73. Edwin Phelps to Oscar Clute, 16 July 1892, and Phelps to Beal, 15 July 1892. Clute Papers, UA 2.1.5, Box 863, Folder 68.

74. *College Speculum* 7, no. 1 (1 August 1887): 9.

75. J. W. Toumey, "Our Manual Labor System," *Speculum* 9, no. 1 (10 August 1889): 7.

76. L. J. Briggs, "The New Labor System at the College," *Speculum* 9, no. 6 (10 May 1892): 92–94; Alva Sherwood, "What a Decade Hath Wrought at M.A.C.," *Speculum* 12, no. 1 (10 August 1892): 2. See T. C. Abbot, *Manual Labor: An Address by President Abbot, of the Michigan State Agricultural College* [1874] for a rationale for student labor, Abbot Papers, UA 2.3.1, Box 861, Folder 67.

77. Guy L. Stewart, "Labor in the Agricultural College," *Speculum* 12, no. 15 (11 June 1894): 251–53.

78. Kuhn, *Michigan State,* 173–76.

79. "Student Experiments at the Agricultural College," *Speculum* 13, no. 7 (15 May 1895): 117–18.

80. Clinton D. Smith, "Changes in Student Labor," *Speculum* 14, no. 1 (15 August 1895): 11.

81. Leon Drake reported to his parents and sister, during his years at M.A.C. (1899–1903) how difficult the academic courses were for him and many others. For example, on 11 March 1900, Drake wrote: "There are a lot of fellows who are going to drop out this term . . ." Leon Drake to Mother, father and sister, (Leon Drake Papers, Michigan State University Museum Collections).

82. High schools in the 1880s published brochures listing their courses of study. Examples include: "Course of Study in the High School, Paw Paw, Michigan, Adopted September 3, 1880," "Monroe High School," 1880," Gaylord High School, 1882–3," and "Tabular View of Course of Study for the Public Schools of Menominee, 1884." Kuhn Collection, UA 17.107, Box 1142, Folder 19.

83. Howard Edwards, Clinton D. Smith, and F. S. Kedzie, "Attendance at Agricultural College," *Thirty-Fifth Annual Report of the Secretary of the State Board of Agriculture*, 62, 67.

84. Kuhn, *Michigan State*, 237–38.

85. Curti and Carstensen, *The University of Wisconsin*, 1:473–75.

86. Beal, *History*, 146.

87. Edwards, Smith, and Kedzie, "Attendance at Agricultural College," 64.

88. *Special Short Courses For Busy Men at the Michigan Agricultural College, Winter*, 1909, Kuhn Collection, UA 17.107, Box 1137, Folder 40.

89. Kuhn, *Michigan State*, 274, 288.

90. *Catalogue: Officers and Students of the Michigan State Agricultural College for the Year 1896–97* (Agricultural College: The College, 1897), 27; Beal, History, 141.

91. *Catalog of the Michigan State College of Agriculture and Applied Science for the Year 1925–1926* (East Lansing: The College, 1926), 43.

92. Kuhn, *Michigan State*, 228–30.

93. *Catalogue*, 1866, 22.

94. *Twenty-Third Annual Catalogue of the Officers and Students of the State Agricultural College of Michigan, 1879–80* (Lansing: W. S. George and Co., State Printers, 1879), 28–29.

95. Howard, "Progress in Economic Entomology in the United States," 147–51.

96. R. C. Kedzie to A. B. Copley, [23 February 1871,] Kuhn Collection, UA 17.107, Box 1140, Folder 47.

97. *Catalogue*, 1879–80, 36.

98. Charles E. Browne, "Agricultural Chemistry," in *A Half-Century of Chemistry in America, 1876–1926* ed. Charles E. Browne (Philadelphia: American Chemical Society, 1926), 182.

99. For summaries of Kedzie's work, see Kuhn, *Michigan State*, 105–8, 161–62, and Beal, *History*, 75, 406–07.

100. "Resolutions Passed by the Michigan Dairymen's Association at Their Annual Meeting at Charlotte, Feb. 4th, 1897," Kuhn Collection, UA 17.107, Box 1138, Folder 115.

101. G. A. Bowling, *History of Dairy Development at Michigan State College* (East Lansing: Dairy Department[, 1929].

102. *Catalogue*, 1896–97, 48–51.

103. R. P. Hughes to the Inspector General, 20 November 1889, Clute Papers, UA 2.1.5, Box 863, Folder 6.

104. Kuhn, *Michigan State*, 116–17, 155–56; "Historical Sketch," in *The Michigan State College Band* (East Lansing: Michigan State College, 1931), in Kuhn Collection, UA 17.107, Box 1138, Folder 36; R. H. Gulley, *Agr'l Coll. Cadet Band*, 1877, Kuhn Collection, UA 17.107, Box 1137, Folder 89.

105. Carl A. Wagner to Snyder, 20 May 1907, Snyder Papers, UA 2.1.7, Box 812, Folder 27.

106. *Catalogue of the Officers and Students of the State Agricultural College of Michigan, 1886–7* (Agricultural College: The College, 1887), 30–33.

107. William H. Van Devoort, "Benefit of Mechanical Courses," *College Speculum* 7, no. 2 (1 October 1887): 4–5.

108. E. N. Pagelsen, "Talk—75th Anniversary Program," *Kuhn Collection*, UA 17. 107, Box 1141, Folder 6.

109. H. K. Vedder to Snyder, 22 February 1907, Snyder Papers, UA 2.1.7, Box 812, Folder 12.

110. Snyder to John C. Wanger, 30 March 1907, Snyder Papers, UA 2.1.7, Box 812, Folder 23; Snyder to Mrs. J. Jerome Davis, 24 July 1901, Snyder Papers, Box 805, Folder 57.

111. Snyder to A. E. Lawrence, 2 October 1909, Snyder Papers, UA 2.1.7, Box 814, Folder 35.

112. Undated [1914], unsigned response to an editorial in the [Detroit] *Tribune*, Snyder Papers, UA 2.1.7, Box 855, Folder 33.

113. Philip S. Rose, "The Place and the Field of Agricultural Engineering," [1914,] 2, Snyder Papers, UA 2.1.7, Box 855, Folder 122.

114. Snyder to A. Marston, 2 October 1913, Snyder Papers, UA 2.1.7, Box 820, Folder 102; H. H. Musselman to the State Board of Agriculture, 3 June 1913, Snyder Papers, Box 820, Folder 90; Snyder to R. S. Shaw, 3 December 1914, Snyder Papers, Box 855, Folder 153.

115. Snyder to [State Board of Agriculture], 10 June 1914, Snyder Papers, UA 2.1.7, Box 854, Folder 114.

116. *Catalog of the Michigan Agricultural College for the Year 1918–1919* (East Lansing: The College, 1919), 41–42.

117. James Kip Finch, *The Story of Engineering* (Garden City, N.Y.: Doubleday, 1960), 353–58, 366–67, 383–84, 400–25.

118. *Catalog of the Michigan State College . . . for the Year 1925–1926*, 50–61.

119. Kuhn, *Michigan State*, 302–3.

120. For a discussion of the mill tax see Kuhn, *Michigan State*, 211–17.

121. "M.A.C. Income is Cut off by Auditor-General," *Detroit Tribune*, 26 March 1914: 1; "M.A.C. Can't Borrow Cash, says Fellows," *Detroit Tribune*, 24 May 1914: 2; "M.A.C. Wins in Fight for Tied Funds," *Detroit Tribune*, 30 May 1914: 3.

122. Snyder to J. W. Beaumont, 8 May 1914, Snyder Papers, UA 2.1.7, Box 854, Folder 118.

123. "The Case of the M.A.C.," *Adrian Daily Telegram*, 1 June 1914: 4. In an editorial that was published on 2 April 1914 the *Detroit Tribune* discussed the meaning of the attack on M.A.C. Although the *Tribune* did not necessarily agree with the "practical farmer" argument for the college's existence, it did note: "There is clear evidence it is the agriculturalists of the state who are most displeased with them [President Snyder and the State Board of Agriculture]; that it is for a very considerable constituency back on the farms that the makers of laws have to make laws that clip the wings of the college" (4).

124. "An Explanation," *Fifty-Third Annual Report of the Secretary of the State Board of Agriculture of the State of Michigan . . . July 1, 1913 to June 30, 1914* (Lansing: By Authority, 1914), 38–44.

125. Frank Kedzie to Graduates of M.A.C., 7 March 1917, Kedzie Papers, UA 2.1.8, Box 892, Folder 52; M.A.C. Record 21, no. 28 (25 April 1916): 5. See Kuhn, Michigan State, 266-69, for a discussion of the attempt to move engineering to Ann Arbor and the role played by Kedzie and Olds to save it. "Engineering Over at M.A.C. Belief," *Detroit Free Press*, 7 March 1916: 10.

126. "Report of Dean Bissell of burning of Engineering Buildings," 13 March 1916, Michigan State University Board of Trustee Minutes, UA

1, Book 4, 12 January 1909–16 July 1924, MSU Archives.

127. F. S. Kedzie to Beaumont, 8 March 1916, Kedzie Papers, UA 2.1.8, Box 892, Folder 18.

128. "The Relation Between Agriculture and Engineering: Great Achievements in Agriculture, Science and Engineering," *Holcad* 5, no. 32 (9 June 1913): 2.

129. An interesting interpretation of the women's course at M.A.C. is Carol R. Koch's unpublished paper "Domesticity Institutionalized: The Establishment of the Woman's Course at the Michigan Agricultural College, 1896," Carol Koch Papers, UA 10.3.35, F.D., Folder 1, MSU Archives.

130. William J. Darby, *Nutrition Science: An Overview of American Genius* (Washington D.C.: Agricultural Research Service, USDA, 1976), 6–7.

131. Edward C. Kirkland, "'Scientific Eating': New Englanders Prepare and Promote a Reform, 1873–1907," in *Proceedings of the Massachusetts Historical Society* 86 (Boston: The Society, 1975): 34.

132. Ibid., 29–39; Darby, *Nutrition Science,* 10–11. See Frederic L. Holmes, "The Transformation of the Science of Nutrition," *Journal of the History of Biology* 8, no. 1 (1975): 135–144, for a discussion of the evolution of chemical investigations into nutrition.

133. Leslie J. Harris, "The Discovery of Vitamins," in *The Chemistry of Life: Eight Lectures on the History of Biochemistry,* ed. Joseph Needham (Cambridge: Cambridge University Press, 1970), 156–60.

134. Robert Clarke, *Ellen Swallow: The Woman Who Founded Ecology* (Chicago: Follett, 1973), 121–40. For another account of Swallow's work, see Sarah Stage, "Ellen Richards and the Social Significance of the Home Economics Movement," in *Rethinking Home Economics: Women and the History of a Profession,* ed. Sarah Stage and Virginia B. Vincenti (Ithaca, N.Y.: Cornell University Press, 1997), 17–33.

135. Baker et al., *Century of Service,* 36–37; Paul V. Betters, *The Bureau of Home Economics: Its History, Activities and Organization* (Washington, D.C.: Brookings Institution, 1930), 1.

136. Isabel Bevier, *Home Economics in Education* (Philadelphia: J. B. Lippincott, 1928), 149–53, 158.

137. For a discussion of the society's history and purpose see Mary I. Barber, *History of the American Dietetic Association, 1917–1959* (Philadelphia: J. B. Lippincott, 1959).

138. Jacob Riis, *How the Other Half Lives: Studies Among the Tenements of New York* (New York: Scribner's and Sons, 1890); Jane Addams, *Democracy and Social Ethics* (New York: Macmillan, 1902); John Spargo, *The Bitter Cry of the Children* (New York: Macmillan, 1906).

139. Theodore Roosevelt to L. H. Bailey, 10 August 1908, in *Report of the Commission on Country Life* (New York: Sturgis and Walton, 1911), 41–46, 28–31.

140. Clarke, *Ellen Swallow,* 114.

141. Emma Willard, "An Address to the Public Particularly to the Members of the Legislature of New-York proposing a Plan for Improving Female Education" (1819), in *Pioneers of Women's Education in the United States: Emma Willard, Catherine Beecher, Mary Lyon,* ed. Willystine Goodsell (New York: McGraw-Hill, 1931), 62.

142. Catharine E. Beecher, "An Address on Female Suffrage," 1870, in *Pioneers of Women's Education,* 197.

143. Catharine E. Beecher, *A Treatise on Domestic Economy, for the Use of Young Ladies at School,* rev. ed. (New York: Harper and Brothers, 1859), 5; Catharine E. Beecher and Harriet Beecher Stowe, *The American Woman's Home: Or Principles of Domestic Science; Being a Guide to the Formation and Maintenance of Economical, Healthful, Beautiful, and Christian Homes* (New York: J. B. Ford and Co., 1869), 13. For a full treatment of Catharine Beecher's life see Kathryn Kish Sklar, Catharine Beecher: A Study in American Domesticity (New York: W. W. Norton), 1973.

144. Laura Shapiro, *Perfection Salad: Women and Cooking at the Turn of the Century* (New York: Farrar, Straus and Giroux, 1986), 3–10; Sally I. Halverston and Margaret M. Bubolz, "Home Economics and Home Sewing in the United States 1870–1940," in *The Culture of Sewing: Gender, Consumption and Home Dressmaking,* ed. Barbara Burman (Oxford: Berg, 1999), 303–25.

145. Earle D. Ross, *A History of the Iowa State College of Agriculture and Mechanic Arts* (Ames: Iowa State College Press, 1942), 130–31; James C. Carey, *Kansas State University: The Quest for Identity* (Lawrence: Regents Press of Kansas, 1977), 44–45, 60.

146. Solberg, *The University of Illinois, 1867–1894,* 160–63.

147. James E. Pollard, *History of the Ohio State University: The Story of Its First Seventy-Five Years, 1873–1948* (Columbus: Ohio State University Press, 1952), 143, 149–51; Robert W. Topping, *A Century and Beyond: The History of Purdue University* (West Lafayette: Purdue University Press, 1988), 169; Curti and Carstensen, *The University of Wisconsin* 2:404–6.

148. Ruth Bordin, *Women at Michigan: The "Dangerous Experiment," 1870s to the Present* (Ann Arbor: University of Michigan Press, 1999), 24.

149. Morris Bishop, *A History of Cornell* (Ithaca, N.Y.: Cornell University Press, 1962), 379–81.

150. Charlotte Williams Conable, *Women at Cornell: The Myth of Equal Education* (Ithaca, N.Y.: Cornell University Press, 1977), 113–15.

151. A good treatment of the evolution of co-education in American colleges and universities is the chapter titled "Coeducation," in Thomas Woody, *A History of Women's Education in the United States,* 2 vols. (New York: Octagon Books, 1966), 2:224–303.

152. John Evert, "Co-Education at Our College," *College Speculum* 1, no. 1 (1 August 1881): 3.

153. R. C. Clute, "Co-Education Would be Detrimental to the College," *College Speculum* 7, no. 4 (10 May 1888): 2–3.

154. E. J. Heck, "Education of Women," *Speculum* 14, no. 1 (15 August 1895): 9–10.

155. Edwards, Smith, and Kedzie, "Attendance at Agricultural College," 67.

156. Snyder to R. C. Stevenson, 6 April 1896, Snyder Papers, UA 2.1.7, Box 804, Folder 15.

157. Snyder to Alice M. James, 10 June 1899, Snyder Papers, UA 2.1.7, Box 804, Folder 86.

158. Snyder to C. A. Jewell, Jr., 1 May 1901, Snyder Papers, UA 2.1.7, Box 805, Folder 90.

159. Snyder to Bess Howell, 7 June 1901, Snyder Papers, UA 2.1.7, Box 805, Folder 90.

160. Snyder to W. C. Latta, 29 May 1905, Snyder Papers, UA 2.1.7, Box 809, Folder 9.

161. Margaret W. Rossiter, *Women Scientists in America: Struggles and Strategies to 1940* (Baltimore: John Hopkins University Press, 1982), 51–72.

162. Snyder to Hutchins, 4 May 1912, Snyder Papers, UA 2.1.7, Box 818, Folder 97.

163. *Catalogue,* 1896–97, 28.

164. *Catalogue: Officers and Students of the Michigan State Agricultural College for the Year 1897–8* (Agricultural College: The College, 1898), 110–13. *Catalogue: Officers and Students of the Michigan State Agricultural College, for the Year 1898–1899* (Agricultural College: The College, 1899), 86–87.

165. The best statement of the early years of the home economics program at MSU is Maude Gilchrist, *The First Three Decades of Home Economics at Michigan State College* ([East Lansing]: School of Home Economics, 1947).

166. Snyder to "Madam," 30 May 1899, Snyder Papers, UA 2.1.7, Box 804, Folder 101; Marilyn Culpepper, "Mary Mayo: Leader of Rural Women," in *Historic Women of Michigan: A Sesquicentennial Celebration,* ed. Rosalie Riegle Troester (Lansing: Michigan Women's Studies Association, 1987), 77–79.

167. Gilchrist, *The First Three Decades,* 9–17.

168. "Education for Our Women—Of What Shall It Consist?" *M.A.C. Record* 6, no. 3 (2 October 1900): 1.

169. "The Women's Course at the Agricultural College," *M.A.C. Record* 6, no. 2 (25 September 1900): 2.

170. Hazel Ramsay, "Domestic Science," and Clara Jakway, "Domestic Art," *Holcad* 5, no. 21 (10 March 1913): 12–13.

171. "Battle Creek The Royal Hostess of Senior Girls," *M.A.C. Record* 16, no. 37 (6 June 1911): 1.

172. *Catalog of the Michigan State College . . . for the Year 1925–26,* 62.

173. Snyder to E. A. Bryan, 11 March 1903, Snyder Papers, UA 2.1.7, Box 806, Folder 131.

174. Gilchrist, *The First Three Decades,* 12, 17.

175. *M.A.C. Record* 16, no. 8 (8 November 1910): 3; *M.A.C. Record* 16, no. 34 (16 May 1911): 3.

176. Gilchrist, *The First Three Decades,* 41–43.

177. Ibid., 18–19; Obituary for Jennie L. K. Haner, Kuhn Collection, UA 17.107, Box 1139, Folder 63; Stephanie Perentesis, "'Brain Work With Hand Work': MAC's Women's Course of Study," *Michigan History* (May/June 1996): 44.

178. "Miniatures," *Holcad* 3, no. 32 (29 May 1911): 4; Gilchrist, *The First Three Decades,* 32–34.

179. "Alumni Notes," *M.A.C. Record* 20, no. 3 (13 October 1914): 8.

180. "Miniatures," *Holcad* 3, no. 31 (22 May 1911): 5; "News and Comment," *M.A.C. Record* 22, no. 28 (8 May 1917): 8.

181. "About the Campus," *M.A.C. Record* 18, no. 3 (8 Oct. 1912): 4.

182. Snyder to Honorable State Board of Agriculture, 14 September 1910, Snyder Papers, UA 2.1.7, Box 815, Folder 121; Ward Giltner to Snyder, 13 September 1912, Snyder Papers, UA 2.1.7, Box 818, Folder 84.

183.   Ernst Bessey to Snyder, 16 April 1912, Snyder Papers, UA 2.1.7, Box, 818, Folder 9.

184.   Thomas Blaisdell to Snyder, 22 April 1910, Snyder Papers, UA 2.1.7, Box 815, Folder 114.

185.   "Men and Women Who Have Left M. A. C. for More Attractive Fields," [1919,] Kedzie Papers, UA 2.1.8, Box 893, Folder 32.

186.   John W. Beaumont to Kedzie, 17 May 1916, Kedzie Papers, UA 2.1.8, Box 892, Folder 19.

187.   Gilchrist, *The First Three Decades*, 50; Burness G. Wenberg, "News release," 20 March, 1996, Michigan Women's Hall of Fame-Marie Dye Award, 1998, Dean's Office-internal file, College of Human Ecology, Michigan State University.

188.   Beal, *History*, 143; J. L. Snyder, Memorandum, 17 December 1900, Snyder Papers, UA 2.1.7, Box 804, Folder 144.

189.   *Catalogue: Officers and Students of the Michigan State Agricultural College for the Year 1901–1902* (Agricultural College: The College, 1902), 35.

190.   *Catalogue: Officers and Students of the Michigan State Agricultural College for the Year 1902–1903* (Agricultural College: The College, 1903), 53–55; "Forestry at the Michigan Agricultural College," [1907,] Snyder Papers, UA 2.1.7, Box 810, Folder 123.

191.   Charles A. Davis to Snyder, 20 June 1902, Snyder Papers, UA 2.1.7, Box 806, Folder 40.

192.   Gifford Pinchot to Snyder, 11 October 1907, Snyder Papers, UA 2.1.7, Box 811, Folder 124. Concern over taxpayer opposition to the two programs is discussed in an exchange of letters in May 1912 between Snyder and H. B. Hutchins, president of the University of Michigan: Snyder to Hutchins, 4 May 1912 and Hutchins to Snyder, 31 May 1912, Snyder Papers, UA 2.1.7, Box 818, Folder 97.

193.   Snyder to Pinchot, 22 August 1907, Snyder Papers, UA 2.1.7, Box 811, Folder 122; Snyder to C. A. Scott, 23 August 1907, Snyder Papers, UA 2.1.7, Box 811, Folder 148; Snyder to Scott, 11 September 1907, Snyder Papers, UA 2.1.7, Box 811, Folder 149.

194.   *Catalog of the Michigan State College . . . for the Year 1925–1926*, 74.

195.   Morrill, *Veterinary Medicine in Michigan*, 49–54.

196.   Robert Gibbons to G. W. Dunphy, 3 March 1903, Snyder Papers, UA 2.1.7, Box 807, Folder 22.

197.   See Morrill, *Veterinary Medicine in Michigan*, 62–76, for a good discussion of the Veterinary School at Michigan State from 1910 until 1930.

198.   Richard P. Lyman, "Report of the Dean of Veterinary Science," *Fiftieth Annual Report of the Secretary of the State Board of Agriculture of the State of Michigan* (Lansing: Wynkoop Hallenbeck Crawford Co., 1911), 75–76.

199.   Snyder to S. H. Roginos, 17 July 1912, Snyder Papers, UA 2.1.7, Box 819, Folder 82.

200.   Max Wershow to Frank S. Kedzie, 2 October 1913, Kedzie Papers, UA 2.1.8, Box 892, Folder 16.

201.   Lyman to Snyder, 7 June 1913, Snyder Papers, UA 2.1.7, Box 820, Folder 71; Lyman to Snyder, 12 December 1912, Snyder Papers, Box 818, Folder 155; and Lyman to Snyder, 13 May 1912, Snyder Papers, Box 818, Folder 147.

202.   Lyman, "Report of the Dean of the Veterinary Division," 30 June 1915, *Fifty-Fourth Annual Report of the Secretary of the State Board of Agriculture of the State of Michigan. . . from July 1, 1914, to June 30, 1915* (Lansing: Wynkoop Hallenbeck Crawford Co., 1915), 122–14.

203.   Ward Giltner to Snyder, 20 May 1914, Snyder Papers, UA 2.1.7, Box 855, Folder 16.

204.   Lyman to Snyder, 9 April 1914, Snyder Papers, UA 2.1.7, Box 855, Folder 55.

205.   Lyman to Snyder, 26 October 1914, Snyder Papers, UA 2.1.7, Box 855, Folder 21.

206.   Giltner to Snyder, 20 May 1914.

207.   Snyder to The Hon. State Board of Agriculture, [June 1915,] Snyder Papers, UA 2.1.7, Box 866, Folder 26.

208.   Ward Giltner, "Report of the Dean of Veterinary Medicine," *Sixty-Fourth Annual Report of the Secretary of the State Board of Agriculture of the State of Michigan* (Lansing: [Michigan State College,] 1926), 72–73.

209.   E. H. Ryder, "Facts on the New Course Condensed for Alumni," *M.A.C. Record* 26, no 28 (6 May 1921): 8.

210.   *Catalog of the Michigan State College . . . for the Year 1925–1926*, 80–87.

211.   Ibid, 80. In the mid 1920s, Michigan State College of Agriculture and Applied Science issued a pamphlet titled Vocational Opportunities for the College Graduate that outlined the many potential careers available to its graduates. This publication demonstrated the college's commitment to training students to find productive work in the growing American economy of the 1920s. A copy of this brochure is in Kuhn Collection, UA 17.107, Box 1142, Folder 93.

212.   "Applied Science has Varied Courses," *M.A.C. Record* 28, no. 29 (14 May 1923): 8–9; "Training in Applied Science: Announcement of the New Applied Science Course, Michigan Agricultural College, Effective Fall Term, 1921," Kuhn Collection, UA 17.107, Box 1142, Folder 89.

213.   "Board Authorizes A.B. Degree Course," and "Views and Comments," *M.A.C. Record* 29, no. 20 (25 February 1924), 3, 6; "New Courses Offer Liberal Choice," M.A.C. Record 29, no. 29 (12 May 1924), 12;

214.   For a complete listing of courses offered in 1925–26 see *Catalog of the Michigan State College . . . for the Year 1925-1926*.

215.   *Catalog of the Michigan State College . . . for the Year 1925–1926*, 90–92.

216.   *Michigan State Agricultural College: General Catalogue of Officers and Students 1857–1900* (Agricultural College: The College, 1900), 251–55; *Twenty-Third Annual Catalogue of the Officers and Students of the State Agricultural College of Michigan 1879–80* (Lansing: W. S. George and Co., State Printers, 1879), 37; *Catalogue*, 1886–1887, 29.

217.   Extract from report of M.S. Degree Committee made to the Faculty 7 May 1894, Kuhn Collection, UA 17.107, Box 1137, Folder 123.

218.   Snyder to I. A. Williams, 2 December 1908, Snyder Papers, UA 2.1.7, Box 813, Folder 140.

219.   Charles E. Marshall to Snyder, 13 March 1909, Snyder Papers, UA 2.1., Box 814, Folder 26.

220. Snyder to R. S. Shaw, 2 June 1909, Snyder Papers, UA 2.1.7, Box 814, Folder 54.

221. Ward Giltner to Snyder, 15 May 1915, Snyder Papers, UA 2.1.7, Box 866, Folder 4.

222. Minutes of the Committee on Advanced Degrees, 1 February 1915, Snyder Papers, UA 2.1.7, Box 866, Folder 14.

223. "The Library and the Student," *M.A.C. Record* 6, no. 5 (16 October 1900), 1.

224. Lewis R. Fisk to State Board of Agriculture, 4 April 1861, Lewis R. Fisk Papers, UA 2.1.2, Box 862, Folder 4, MSU Archives.

225. F. S. Kedzie, "The Library and the Librarian Honored Among Present Day Founders," *The Michigan State College Record*, May 1931, 7.

226. "An Urgent Need," *M.A.C. Record* 20, no. 16 (26 January 1915): 4.

227. Linda E. Landon, "The Library," *M.A.C. Record* 2, no. 22 (8 June 1897): 1; "How to Use the Library," *Holcad* 3, no. 15 (16 January 1911), 3–4.

228. Landon, "Report of the Librarian," *Sixty-Fourth Annual Report of the Secretary of the State Board of Agriculture,* 99.

229. [Charles Edward] Marshall to Members of the Library Committee, 15 November 1906, Kuhn Collection, UA 17.107, Box 1138, Folder 19; Kuhn, *Michigan State,* 184–85.

230. Landon, "The Library," 1.

231. Requests for materials in 1914, 1915, 1916, Linda Landon file, Kuhn Collection, UA 17.107, Box 1139, Folder 74; *Sixty-Fourth Annual Report of the Secretary of the State Board of Agriculture,* 99.

232. Walter B. Barrows, "Report of the Department of Zoology and Physiology," 30 June 1916, *Fifty-Fifth Annual Report of the Secretary of the State Board of Agriculture of the State of Michigan, . . . July 1, 1915 to June 30, 1916* (Lansing: Wynkoop Hallenbeck Crawford Co., 1916). Also see Beal, *History,* 249.

233. Beal, *History,* 251; "History of the Michigan State University Herbarium"; http://www.bpp.msu.edu/herbarium/history.htm

234. Beal, *History,* 251–54; "W. J. Beal Botanical Garden"; http://www.cpp.msu.edu/beal/

CHAPTER 4. REACHING THE PEOPLE

1. R. S. Shaw to Heads of Departments, 23 August 1912, J. L. Snyder Papers, UA 2.1.7, Box 819, Folder 100.

2. Beal, *History,* 174.

3. W. L. Webber to Lewis G. Gorton, 27 October 1894 and 1 November 1894, Lewis Griffin Gorton Papers, UA 2.1.6, Box 873, Folder 37, MSU Archives.

4. M. Elva Worden to President Agricultural College, 22 March 1894, and Gorton to Worden, 24 March 1894, Gorton Papers, UA 2.1.6, Box 873, Folder 37.

5. Mrs. C. H. Strand to Proprietor Agricultural College, 22 August 1894, Gorton Papers, UA 2.1.6, Box 873, Folder 33.

6. W. L. Steele to Agricultural College, 21 June 1894, Gorton Papers, UA 2.1.6, Box 873, Folder 33.

7. F. M. Briggs to Gorton, 13 August 1895, Gorton Papers, UA 2.1.6, Box 873, Folder 39.

8. J. W. Sharp to Mr President, 7 April 1893, Oscar Clute Papers, UA 2.1.5, Box 863, Folder 90, MSU Archives.

9. Jno. L. Buell to Gorton, 21 November 1894 and 27 November 1894, Gorton Papers, UA 2.1.6, Box 873, Folder 17.

10. Belle Tanner to Supt. of Agricultural College, Gorton Papers, 30 March 1894, UA 2.1.6, Box 873, Folder 34.

11. Chas. Ziem to Superintendent Mich. Agricultural College Farm, 12 June 1897, Kuhn Collection, UA 17.107, Folder 38.

12. Edward James Dies, *Titans of the Soil: Great Builders of Agriculture* (Chapel Hill: University of North Carolina Press, 1949), 33–38. See also Julie A. Avery, ed., *Agricultural Fairs in America: Tradition, Education, Celebration* (East Lansing: Michigan State University Museum, 2000).

13. P. R. L. Pierce to H. P. Baldwin, 8 August 1871, Kuhn Collection, UA 17.107, Box 1140, Folder 32.

14. Duncan Buchanan to President Agricultural College, Clute Papers, UA 2.1.5, Box 863, Folder 77.

15. J. L. Snyder to I. H. Butterfield, Snyder Papers, UA 2.1.7, Box 806, Folder 138.

16. Winton U. Solberg, *The University of Illinois 1867–1894: An Intellectual and Cultural History* (Urbana: University of Illinois Press, 1968), 139.

17. "Farmers' Institutes," *Fourteenth Annual Report of the Secretary of the State Board of Agriculture of the State of Michigan for the Year 1875* (Lansing: W. S. George and Co., State Printers, 1876), 73.

18. Ibid., 74–76.

19. Ibid., 72–314.

20. Annie Hall, "Home Culture," *Fourteenth Annual Report of the Secretary of the State Board of Agriculture,* 181–82.

21. Mrs. Keeler, "Social Culture for Farmers," *Fourteenth Annual Report of the Secretary of the State Board of Agriculture,* 183–84.

22. Mary Mayo, "Does Education Lead to Extravagance," 21 January 1881, *Nineteenth Annual Report of the Secretary of the State Board of Agriculture,* 196.

23. Beal, *History,* 160, 162.

24. Mrs. Albert Granger, "Management of the Dairy," *First Biennial Report of the Secretary of the State Board of Agriculture of the State of Michigan, from Sept. 1, 1880, to Sept. 30, 1882* (Lansing: W. S. George and Co., State Printers, 1882), 283.

25. Margaret Sill to Lewis G. Gorton, 8 January 1894, Gorton Papers, UA 2.1.6, Box 873, Folder 22.

26. Nellie H. Mayo to Gorton, 3 April 1894, Gorton Papers, UA 2.1.6, Box 873, Folder 28.

27. Kuhn, *Michigan State,* 239; Beal, *History,* 160–162.

28. "Farmers' Institutes Held in 1895–6," *Thirty-Fifth Annual Report of the Secretary of the State Board of Agriculture,* 477–80.

29. Beal, *History,* 163.

30. *Forty-Fifth Annual Report of the Secretary of the State Board of Agriculture,* 37.

31. *Fifty-Fifth Annual Report of the Secretary of the State Board of Agriculture,* 112–28.

32. "Report of Board of Visitors," *Thirty-Fifth Annual Report of the Secretary of the State Board of Agriculture,* 76.

33. "Master's Annual Address," *Proceedings of the Twenty-Eighth Annual Session of the Michigan State Grange or Patrons of Husbandry Held in Armory Hall, Lansing, Michigan, December 11, 12, 13, and 14, 1900,* 19–20, Kuhn Collection, UA 17.107, Box 1138, Folder 94.

34. "First Annual Michigan Round-Up Farmers' Institute," *Thirty-Fifth Annual Report of the Secretary of the State Board of Agriculture,* 483–85.

35. Beal, *History,* 160–61.

36. Harry J. Eustace to Snyder, 10 March 1914, Snyder Papers, UA 2.1.7, Box 854, Folder 156.

37. Jennie Buell to Snyder, 17 March 1914, Snyder Papers, UA 2.1.7, Box 854, Folder 115.

38. *M.A.C. Record* 27, no. 15 (27 January 1922): 5.

39. *M.A.C. Record* 27, no. 19 (24 February 1922): 3.

40. Buell to Snyder, 17 March 1914.

41. Circular No. 1, "The Farm-Home Reading Circle of Michigan," Kuhn Collection, UA 17.107, Box 1138, Folder 3.

42. Herbert W. Mumford, "Farm Home Reading Circle," *Thirty-Fifth Annual Report of the Secretary of the State Board of Agriculture,* 56.

43. Beal, *History,* 175–76.

44. Michigan Agricultural College, "College Extension Reading Course," 1912-13, Kuhn Collection, UA 17.107, Box 1136, Folder 31.

45. Michigan State College of Agriculture and Applied Science, "Home Reading Courses," March 1926, 4, Kuhn Collection, UA 17.107, Box 1136, Folder 71.

46. Ibid., 4–6.

47. A good survey of the demonstration trains is LeRoy Barnett, "Putting Michigan Farmers on the Right Track," *Michigan History* (January/February 2000): 48–55. Snyder to E. E. Balcomb, 9 December 1911, Snyder Papers, UA 2.1.7, Box 815, Folder 135.

48. "The Institute Train," *M.A.C. Record* 16, no. 29 (11 April 1911): 2.

49. "Programme of the Sheep Shearing Exhibition," 23 & 24 May 1866, Kuhn Collection, UA 17.107, Box 1137, Folder 39; "Catalogue: Fourth Public Sale of Short-Horn and Hereford Cattle," 18 April 1888, Kuhn Collection, UA 17. 107, Box 1137, Folder 41.

50. "First Poultry Institute," 13–18 February 1911, Kuhn Collection, UA 17.107, Box 1137, Folder 21.

51. *Michigan Improved Live Stock Breeders' and Feeders' Association: Twenty-Fifth Annual Meeting, January 18, 19, 20, 1916,* Kuhn Collection, UA 17.107, Box 1138, Folder 118.

52. Snyder to "My dear Sir," [1911,] Snyder Papers, UA 2.1.7, Box 817, Folder 36.

53. *Michigan Agricultural College: Fifth Annual Summer Conference for Ministers and Laymen, July 7 to 12, 1919 & First Annual Summer Short Course for Ministers, July 7 to 19, 1919,* Kuhn Collection, UA 17.107, Box 1137, Folder 13.

54. "Excursion to the Agricultural College," [August 1899,] Kenyon L. Butterfield Papers, UA 2.1.10, Box 860, Folder 8, MSU Archives.

55. "At the College," *M.A.C. Record* 3, no. 2 (21 September 1897): 3.

56. Snyder to James Wilson, 28 January 1905, Snyder Papers, UA 2.1.7, Box 809, Folder 1; Beal, *History,* 256–58.

57. "Teaching Sewing in Detroit," *M.A.C. Record* 7, no. 9 (12 November 1901): 1.

58. Maude Gilchrist, *The First Three Decades of Home Economics at Michigan State College* ([East Lansing]: The School of Home Economics, 1947), 40.

59. E. E. Ferguson to Snyder, 15 July 1903, and Snyder to Ferguson, 17 July 1903, Snyder Papers, UA 2.1.7, Box 807, Folder 14.

60. E. W. Blackhurst to Snyder, 13 July 1903, and Snyder to Blackhurst, 17 July 1903, Snyder Papers, UA 2.1.7, Box 806, Folder 136.

61. "Alumni," *M.A.C. Record* 9, no. 11 (1 December 1903): 2.

62. "Alumni," *M.A.C. Record* 13, no. 11 (3 December 1907): 1; *M.A.C. Record* 17, no. 10 (28 November 1911): 2.

63. "Alumni," *M.A.C. Record* 16, no. 29 (11 April 1911): 1.

64. Kenyon L. Butterfield to Snyder, 12 October 1901, Snyder Papers, UA 2.1.7, Box 805, Folder 31.

65. Snyder to I. R. Waterbury, 21 June 1907, Snyder Papers, UA 2.1.7, Box 812, Folder 29.

66. Snyder to Chas. P. Craig, 7 November 1907, Snyder Papers, UA 2.1.7, Box 810, Folder 141; Snyder to L. D. Harvey, 5 May 1904, Snyder Papers, UA 2.1.7, Box 808, Folder 8; Charles R. Starring and James O. Knauss, *The Michigan Search for Educational Standards* (Lansing: Michigan Historical Commission, 1969), 156–57.

67. Jesse Hubbard to Snyder, 10 February 1908, Snyder Papers, UA 2.1.7, Box 812, Folder 127; Snyder to J. F. Wojta, 18 May 1908, Snyder Papers, UA 2.1.7, Box 813, Folder 3.

68. Snyder to Thomas R. Easterday, 19 July 1911, Snyder Papers, UA 2.1.7, Box 816, Folder 32; *Dunbar School of Agriculture and Domestic Science,* Bulletin Number One, [1911,] Snyder Papers, UA 2.1.7, Box 816, Folder 25.

69. Snyder to B. M. Davis, 8 June 1908, Snyder Papers, UA 2.1.7, Box 812, Folder 94.

70. W. H. French to County Normal Training Class Teachers, 4 April 1908, Snyder Papers, UA 2.1.7, Box 812, Folder 107.

71. "A Course in Agriculture for the Public Schools of Michigan," 1908, Snyder Papers, UA 2.1.7, Box 812, Folder 108. In 1907, the State Superintendent of Public Instruction published "An Elementary Laboratory Study in Crops For the Schools of Michigan," by Joseph A. Jeffery, Professor of Agronomy at M.A.C. Superintendent L. L. Wright said that he hoped that this work would serve "as an elementary study in crops," and that "this bulletin is published for the purpose of putting into the hands of our teachers some simple and definite work in the subject of agriculture." Kuhn Collection, UA 17.107, Box 1139, Folder 69.

72. "Department of Agricultural Education," in Beal, *History,* 177; Kuhn, *Michigan State,* 234.

73. French to Snyder, 11 April 1908, Snyder Papers, UA 2.1.7, Box 812, Folder 107.

74. Teachers of Agriculture in High Schools, 1912–1913, Snyder Papers, UA 2.1.7, Box 818, Folder 81.

75. "Report of the Department of Agricultural Education," *Fifty-Seventh Annual Report of the Secretary of the State Board of Agriculture of the State of Michigan . . . from July 1, 1917, to June 30, 1918* ([East Lansing]: The College, 1919), 70-71.

76. Wayne D. Rasmussen, *Taking the University to the People: Seventy-Five Years of Cooperative Extension* (Ames: Iowa State University Press, 1989), 86.

77. *Fifty-Seventh Annual Report of the Secretary of the State Board of Agriculture,* 70.

78. *Catalog of the Michigan State College . . . for the Year 1925–1926,* 149–50.

79. *Sixty-Fourth Annual Report of the Secretary of the State Board of Agriculture,* 147; For an excellent survey of the beginning and early years of teacher training at M.A.C., see Victor H. Noll, *The Preparation of Teachers at Michigan State University* (East Lansing: College of Education, 1968), 3–70.

80. "Alumni," *M.A.C. Record* 10, no. 39 (20 June 1905): 1.

81. Kuhn, *Michigan State,* 162–63.

82. "The Hatch Bill," *Twenty-Sixth Annual Report of the Secretary of the State Board of Agriculture of the State of Michigan, from October 1, 1886, to June 30, 1887* (Lansing: Thorp and Godfrey, State Printers, 1887), 28.

83. Kuhn, *Michigan State,* 161.

84. Beal, *History,* 300–301.

85. Stanley Johnston, "T. T. Lyon, South Haven (Michigan) Experiment Station," unpublished mss. Kuhn Collection, UA 17.107, Box 1137, Folder 138.

86. Kuhn, *Michigan State,* 164–65; I. H. Butterfield, undated handwritten account of Grayling sub-station, Oscar Clute Papers, UA 2.1.5, Box 863, Folder 81.

87. C. D. Smith to Snyder, 14 June 1901, Snyder Papers, UA 2.1.7, Box 805, Folder 176.

88. A. C. True to Snyder, 28 May 1904, Snyder Papers, UA 2.1.7, Box 808, Folder 87.

89. Snyder, [1908,] Snyder Papers, UA 2.1.7, Box 813, Folder 65.

90. Snyder to L. H. Bailey, 6 July 1905, Snyder Papers, UA 2.1.7, Box 809, Folder 65.

91. G. C. Davis, Plan for Experimental Work in Entomology for the Season of 1893, Kuhn Collection, UA 17.107, Box 1137, Folder 37.

92. "Experiment Station Reports," *Forty-Fifth Annual Report of the Secretary of the State Board of Agriculture,* 98.

93. "Report of the Secretary and Treasurer," *Fifty-Fifth Annual Report of the Secretary of the State Board of Agriculture,* 243.

94. Clinton D. Smith, "Dairy Records," Bulletin No. 127—Farm Department, *Thirty-Fifth Annual Report of the Secretary of the State Board of Agriculture,* 223–66.

95. Charles E. Marshall and Bell Farrand, *Bacterial Association in the*

Souring of Milk, Special Bulletin No. 42, *Forty-Seventh Annual Report of the Secretary of the State Board of Agriculture of the State of Michigan . . . from July 1, 1907, to June 30, 1908* (Lansing: Wynkoop Hallenbeck Crawford Co., 1908), 302–62.

96.   C. P. Halligan, *Onion Culture on Muck Lands,* Special Bulletin No. 67, *Fifty-Third Annual Report of the Secretary of the State Board of Agriculture,* 410–25.

97.   C. P. Halligan, *Trees, Shrubs and Plants for Farm and Home Planting,* Bulletin No. 281, *Fifty-Seventh Annual Report of the Secretary of the State Board of Agriculture,* 420–69.

98.   "Report of the Director and Agriculturalist," *Forty-Fifth Annual Report of the Secretary of the State Board of Agriculture,* 98.

99.   Leo M. Geismar to Snyder, 3 February 1912, Snyder Papers, UA 2.1.7, Box 818, Folder 90.

100. L. C. Holden to Snyder, 10 February 1912, Snyder Papers, UA 2.1.7, Box 818, Folder 91.

101. Thomas B. Wyman to Snyder, 21 February 1912, Snyder Papers, UA 2.1.7, Box 818, Folder 87.

102. Thornton A. Green to I. R. Waterbury, 11 March 1912, Snyder Papers, UA 2.1.7, Box 818, Folder 88.

103. Snyder to Holden, 12 February 1912, Snyder Papers, UA 2.1.7, Box 818, Folder 91.

104. Geismar to Snyder, 23 February 1912, Snyder Papers, UA 2.1.7, Box 818, Folder 90.

105. Geismar to Snyder, 1 April and 11 April 1912, Snyder Papers, UA 2.1.7, Box 818, Folder 89; E. W. MacPherran to Snyder, 19 April 1912 and Snyder to MacPherran, 29 April 1912, Snyder Papers, UA 2.1.7, Box 819, Folder 10.

106. Geismar to Snyder, 4 May 1912, Snyder Papers, UA 2.1.7, Box 818, Folder 89.

107. Wayne D. Rasmussen, *Taking the University to the People,* 32.

108. Curti and Carstensen, *The University of Wisconsin,* 2:563–64.

109. Earle D. Ross, *A History of the Iowa State College of Agriculture and Mechanic Arts* (Ames: The Iowa State College Press, 1942), 287–88.

110. Morris Bishop, *A History of Cornell* (Cornell University Press, 1962), 381.

111. C. R. Barns to Snyder, 17 June 1912, Snyder Papers, UA 2.1.7, Box 818, Folder 15.

112. R. S. Shaw, "Report of the Dean of Agriculture," *Fifty-Second Annual Report of the Secretary of the State Board of Agriculture of the State of Michigan. . . from July 1, 1912 to June 30, 1913* (Lansing: Wynkoop Hallenbeck Crawford Co., 1913), 45–46.

113. H. J. Eustace, "Report," *Fiftieth Annual Report of the Secretary of the State Board of Agriculture,* 44.

114. Eustace, "Report," *Fifty-First Annual Report of the Secretary of the State Board of Agriculture of the State of Michigan . . . from July 1, 1911, to June 30, 1912* (Lansing: Wynkoop Hallenbeck Crawford Co., 1912), 40–41.

115. Memorandum of Understanding Between the Alpena, Michigan, Chamber of Commerce, the Board of Supervisors of Presque Isle County, Michigan, the Board of Supervisors of Montmorency County,

Michigan, the Board of Supervisors of Alpena County, Michigan, the Michigan Agricultural College, and the Bureau of Plant Industry, U.S. Department of Agriculture, Relative to Investigations and Demonstrations in Farm Management, 1 July 1912, Kuhn Collection, UA 17.107, Box 1136, Folder 59.

116. Memorandum of Understanding Between the Michigan Agricultural College and the Bureau of Plant Industry, U.S. Department of Agriculture, Relative to Farm Management Field Studies and Demonstrations in the State of Michigan, 1 July 1912, Kuhn Collection, UA 17.107, Box 1136, Folder 59.

117. Snyder to W. J. Spillman, 13 September 1912, Snyder Papers, UA 2.1.7, Box 819, Folder 101.

118. H. F. Williams to C. B. Smith, 30 June 1913, Cooperative Extension Service Records, UA 16.34, Box 1:1913–17, Folder 1, MSU Archives. Eben Mumford, "Report," *Fifty-Second Annual Report of the Secretary of the State Board of Agriculture,* 64.

119. James C. McLaughlin to Snyder, 4 March 1914, Snyder Papers, UA 2.1.7, Box 855, Folder 176.

120. "Smith-Lever Act, 1914," in Rasmussen, *Agriculture in the United States,* 2:1384.

121. Rasmussen, *Taking the University to the People,* 42–48.

122. "College Extension Work," *Fifty-Third Annual Report of the Secretary of the State Board of Agriculture,* 47.

123. C. B. Smith to Frank S. Kedzie, 16 October 1915, . Kedzie Papers, UA 2.1.8, Box 892, Folder 2.

124. Mumford, "Report," *Fifty-Second Annual Report of the Secretary of the State Board of Agriculture,* 64.

125. C. B. Cook, Semi-Annual Report of County Agent, Allegan County, 30 June 1914, Cooperative Extension Service Records, UA 16.34, Box 1, Folder 12.

126. Eben Mumford, Report of County Agricultural Agent Work in Michigan for the Fiscal Year Ending June 30, 1918, Cooperative Extension Service Records, UA 16.34, Box 2, Folder 56.

127. Ibid.

128. Michigan Crop Improvement Association: Cooperative Demonstrations—1918, Kuhn Collection, UA 17.107, Box 873, Folder 114.

129. "Report of Seed Analyst," *Fifty-Seventh Annual Report of the Secretary of the State Board of Agriculture,* 105–6.

130. May M. Parson, Report of Extension Work in Home Economics, [1918,] Cooperative Extension Service Records, UA 16.34, Box 3, Folder 44.

131. "Extension Work," *Catalog: Officers and Students of the Michigan Agricultural College for the Year 1917–1918* (East Lansing: The College, 1918), 177–80.

132. F. Marguerite Erikson, Monthly Report, February 1918, Cooperative Extension Service Records, UA 16.34, Box 3, Folder 3; Erikson, Monthly Report, May 1918, Cooperative Extension Service Records, UA 16.34, Box 3, Folder 3; Erikson, Monthly Report, March 1918, Cooperative Extension Service Records, UA 16.34, Box 3, Folder 3.

133. Clara Rogers, Annual Report, 5 December 1917 to 1 December 1918, Cooperative Extension Service Records, UA 16.34, Box 3, Folder 3.

134. Edna V. Smith, Monthly Report of State Home Demonstration Leader, September, 1918, Cooperative Extension Service Records, UA 16.34, Box 3, Folder 46.

135. "Extension Work," *Sixty-Fourth Annual Report of the Secretary of the State Board of Agriculture,* 19–21; Robert J. Baldwin and Karl H. McDonel, *History of Cooperative Extension Work in Michigan, 1914–1939* (East Lansing: Michigan State College, Extension Division, 1941), 26.

136. Baldwin and McDonel, *History of Cooperative Extension,* 6.

137. Michigan Agricultural College, Extension Division, Club Bulletin No. 4, January 1915, 5, University Serials 310, MSU Archives.

138. Walter H. French, "Report of the Department of Agricultural Education," *Fifty-Third Annual Report of the Secretary of the State Board of Agriculture,* 78.

139. H. G. Smith to R. J. Baldwin, 6 April 1917, Kuhn Collection, UA 17.107, Box 1140, Folder 18.

140. Eben Mumford, "Report of County Agricultural Agent Work in Michigan for the Fiscal Year Ending June 30, 1918," Cooperative Extension Service Records, UA 16.34, Box 2, Folder 56.

141. Michigan Agricultural College, Extension Division, Club Bulletin No. 1, January 1915, and Bulletin Nos. 4, 8–9, University Serials, 310, MSU Archives.

142. Mumford, "Report, . . . 1918," 16.

143. Michigan Agricultural College, Extension Division, Club Bulletin No. 9a (revised), January 1923, 1, University Serials, 310.

144. Anna Bryant Cowles, *Home Canning By the One-Period Cold-Pack Method,* Michigan Agricultural College, Club Bulletin, No. 10, May 1917, University Serials, 310.

145. J. F. Cox, *How to Carry Out the Bean-Growing Project,* Michigan Agricultural College, Club Bulletin No. 3, May, 1915, University Serials, 310.

146. G. A. Brown, *Pork Production,* Michigan Agricultural College, Club Bulletin No. 5, January 1916, University Serial, 310.

147. "Report on Extension Schools and Miscellaneous Farmers' Meetings," *Fifty-Seventh Annual Report of the Secretary of the State Board of Agriculture,* 166–71.

148. Program: Agricultural Extension School, Bad Axe, 19–23 January [1914,] Snyder Papers, UA 2.1.7, Box 854, Folder 112.

149. M. C. Wilson and R. J. Baldwin, "Extension Results as Influenced by Various Factors: A Study of 451 Farms and Farm Homes in Jackson and Menominee Counties, Mich., 1928," Kuhn Collection, UA 17.107, Box 1140, Folder 121.

150. "College Radio Steady Development," *M.A.C. Record* 29, no. 20 (25 February 1924): 8; J. B. Hasselman to Kenyon L. Butterfield, 12 February 1926, Butterfield Papers, UA 2.1.10, Box 860H, Folder 3.

151. "Equip Radio Station for 2000 Mile Radius," *M.A.C. Record* 29, no. 6 (29 October 1923): 4; "Radio Programs Started Jan. 23," *M.A.C. Record* 29, no. 16 (28 January 1924): 10.

152. "Radio Message Unites Alumni on Founders' Day, *M.A.C. Record* 27, no. 30 (19 May 1922): 4–5; *M.A.C. Record* 28, no. 13 (18 December 1922): 7.

153. "Opera Songs Are Sent Out by Radio," *M.A.C. Record* 28, no. 22 (12 March 1923): 4.

154. "Radio Programs to be Continued," *M.A.C. Record* 28, no. 25 (16 April 1923): 14.

155. "Many Reached by Radio Station," *M.A.C. Record* 29, no. 19 (18 February 1924): 5.

156. "Use Radio Station for Short Courses," *M.A.C. Record* 29, no. 23 (17 March 1924): 9.

157. Hasselman to Butterfield, 12 February 1926; Curti and Carstensen, *The University of Wisconsin* 2:347; Carey, *Kansas State University*, 135.

158. "An Act to Reorganize the Agricultural College of the State of Michigan," 1179–80.

CHAPTER 5. THE ALUMNI AT THE TURN OF THE TWENTIETH CENTURY

1. *College Speculum* 1, no. 4 (1 June 1882): 53–54.

2. *General Catalogue 1857–1900.*

3. "Farm Population, Farms, Land in Farms, and Value of Farm Property and Farm Products Sold, by State, 1850 to 1969," *Historical Statistics of the United States: Colonial Times to 1970,* part 1 (Washington, D.C.: U.S. Department of Commerce, Bureau of the Census, 1975), 458–59.

4. Romanzo Adams, "Agriculture in Michigan," *Publications of the Michigan Political Science Association* 3, no. 7 (March 1899): 22–36. See William Cronon, *Nature's Metropolis: Chicago and the Great West* (New York: W. W. Norton, 1991), for a thorough discussion of the growth of Chicago as the center of the Midwest.

5. *Census of the State of Michigan, 1904: Agriculture, Manufactures and Mines,* (Lansing: Wynkoop Hallenbeck Crawford Co., 1905), 2: 64–67, 126–29, 188–91, 204–7, 258–61, 314–17, 368–71, 418–21, 470–73, 516–19, 610–13.

6. Dunbar, *Michigan,* 469–84. See also: Rolland H. Maybee, *Michigan's White Pine Era, 1840–1900* (Lansing: Michigan Historical Commission, 1960).

7. Dunbar, *Michigan,* 484. For a good overview Michigan railroads see Dunbar, *All Aboard! A History of Railroads in Michigan* (Grand Rapids: Eerdmans, 1969); Frank N. Elliott, *When the Railroad was King: The Nineteenth-Century Era in Michigan* (Lansing: Michigan Historical Commission, 1966). For a listing of railroads serving Michigan, in 1900, with the names of stations and miles between them see *Michigan Official Directory and Legislative Manual for the Years 1901–1902* (Lansing: Wynkoop Hallenbeck Crawford Co., 1901), 239–57 (hereafter *Michigan Manual*).

8. Dunbar, *Michigan,* 497–506.

9. F. Clever Bald, *Michigan in Four Centuries* (New York: Harper and Row, 1961), 298.

10. "Comparative Summary, Urban and Rural 1904 and 1900," *Census of the State of Michigan, 1904,* 2: 618–19.

11. John R. Swanton, *The Indian Tribes of North America* (Washington, D.C.: Smithsonian Institution Press, 1969), 246, 249, 263.

12. Dennis Badaczewski, *Poles in Michigan* (East Lansing: Michigan State University Press, 2002), 4.

13. Lewis Walker, Benjamin C. Wilson, and Linwood Cousins, *African Americans in Michigan* (East Lansing: Michigan State University Press, 2001), 15–17; Joe T. Darden, "Patterns of Residential Segregation in Michigan Cities in the Nineteenth Century," in *Blacks and Chicanos in Urban Michigan,* ed. Homer C. Hawkins and Richard W. Thomas ([Lansing:] Michigan Department of State, Michigan History Division, n.d.), 22.

14. Jeremy W. Kilar, *Germans in Michigan* (East Lansing: Michigan State University Press, 2002), 4–5.

15. Russell M. Magnaghi, *Italians in Michigan* (East Lansing: Michigan State University Press, 2001), 1–2.

16. C. Warren Vander Hill, *Settling the Great Lakes Frontier: Immigration to Michigan, 1837–1924* (Lansing: Michigan Historical Commission, 1970), 45–56.

17. Judith Levin Cantor, *Jews in Michigan* (East Lansing: Michigan State University Press, 2001), 6–17.

18. Vander Hill, *Settling the Great Lakes,* 35–44.

19. John P. DuLong, *French Canadians in Michigan* (East Lansing: Michigan State University Press, 2001), 11–25.

20. Vander Hill, *Settling the Great Lakes,* 27–34, 57–67.

21. Jack Glazier and Arthur W. Helweg, *Ethnicity in Michigan: Issues and People* (East Lansing: Michigan State University Press, 2001), 74–77.

22. "Public School System of Michigan," *Michigan Manual,* 258–59; "1900 State Level Census Data—Sorted by State/County," United States Historical Census Data Browser (http://fisher.lib.virginia.edu/census/).

23. For a complete list of newspapers published in Michigan in 1900 see "List of Publications in Michigan," *Michigan Manual,* 223–38.

24. Russell Allen Clark, "The Alumnus as a Citizen," in *Semi-Centennial Celebration of Michigan State Agricultural College,* ed. Thomas C. Blaisdell, 157–63.

25. Lucius Whitney Watkins, "Address for the Farmers' Clubs of Michigan," in *Semi-Centennial Celebration,* ed. Blaisdell, 37.

26. For an informative account of the development of public education in Michigan during later half of the nineteenth century and the early part of the twentieth century see Charles R. Starring and James O. Knauss, *The Michigan Search for Educational Standards* (Lansing: Michigan Historical Commission, 1969).

27. For discussions of efforts put forth by women to gain admission to higher education in Michigan, see Ruth Bordin *Women at Michigan: The "Dangerous Experiment," 1870s to the Present* (Ann Arbor: University of Michigan Press, 1999); and Gail B. Griffin, "Lucinda Hinsdale Stone: Champion of Women's Education," in *Historic Women of Michigan: A Sesquicentennial Celebration,* ed. Rosalie Riegle Troester (Lansing: Michigan Women's Studies Association, 1987), 41–60.

28. "Orlando John Root," Kuhn Collection, UA 17.107, Box 1140, Folder 22.

29. W. H. Jordan, "The Promotion of Agricultural Science," *Proceedings of the Twenty-Third Annual Meeting of the Society for the Promotion of Agricultural Science* ([Published by Authority of the Society, [1902]), 23.

30. "Membership List," *Proceedings of the Twenty-Third Annual Meeting*, 8–14; extracts from an editorial written by Dr. H. L. Knight, August 1924, in *An American Pioneer in Science: The Life and Service of William James Beal* [Ray Stannard Baker and Jessie Beal Baker] (Amherst, Mass.: Printed privately, 1925), 87.

31. Robert W. Topping, *A Century and Beyond: The History of Purdue University* (West Lafayette: Purdue University Press, 1988), 133–34.

32. Douglas D. Martin, *The Lamp in the Desert: The Story of the University of Arizona* (Tucson: The University of Arizona Press, 1960), 36–46.

33. "Alumni," *M.A.C. Record* 9, no. 36 (31 May 1904): 2.

34. E. M. Shelton to Oscar Clute, 21 September 1890, Oscar Clute Papers, UA 2.1.5, Box 863, Folder 33, MSU Archives.

35. Charles E. Bessey, "Industrial Education," speech printed in the *Annual Report of the Nebraska State Board of Agriculture, 1885*, 81, quoted in Thomas R. Walsh, "Charles E. Bessey and the Transformation of the Industrial College," *Nebraska History* 52, no. 4 (1971), 391.

36. Walsh, "Charles E. Bessey," 390–406.

37. Richard A. Overfield, "Charles E. Bessey: The Impact of the 'New' Botany on American Agriculture, 1880–1910," *Technology and Culture* 16, no. 2 (1975): 164–65.

38. Charles E. Bessey, *Botany for High Schools and Colleges* (New York: H. Holt and Co., 1880), iii.

39. Charles E. Bessey, *The Essentials of Botany* (New York: H. Holt and Co., 1884), iii. In 1914 Bessey and his son, Ernst, rewrote the book and issued it as *The Essentials of College Botany* in 1914. The first versions saw seven editions (Walsh, "Charles E. Bessey," 390).

40. Overfield, "Charles E. Bessey," 162–81. For an extensive treatment of Bessey's career see Richard A. Overfield, *Science with Practice: Charles E. Bessey and the Maturing of American Botany* (Ames: Iowa State University Press, 1993).

41. Barrett Willoughby, "Dr. Georgeson: The Man Who Put the Midnight Sun to Work," *The American Magazine* (August 1928): 36–38.

42. Andrew Denny Rodgers III, *Liberty Hyde Bailey: A Story of American Plant Sciences* (Princeton: Princeton University Press, 1949), 6–8.

43. George H. M. Lawrence, "Liberty Hyde Bailey, 1858–1954: An Appreciation," *Baileya,* 3 (March 1955): 28.

44. Rodgers, *Liberty Hyde Bailey,* 87.

45. Morris Bishop, *Earl Cornell, 1865–1900: The First Part of a History of Cornell* (Ithaca, N.Y.: Cornell University Press, 1962), 282, 290–92; Lawrence, "Liberty Hyde Bailey," 31.

46. Rodgers, *Liberty Hyde Bailey,* 121–81.

47. Liberty Hyde Bailey, *Field Notes on Apple Culture* (New York: O. Judd Co., 1886), 6.

48. Liberty Hyde Bailey, *American Grape Training* (New York: Rural Publishing Co., 1893), 5–6.

49. Liberty Hyde Bailey, *The Principles of Fruit-Growing* (New York: Macmillan, 1897).

50. Liberty Hyde Bailey, *The Pruning-Book: A Monograph of the Pruning and Training of Plants as Applied to American Conditions* (New York: Macmillan, 1898).

51. Liberty Hyde Bailey, *The Forcing-Book: A Manual of the Cultivation of Vegetables in Glass Houses* (New York: Macmillan, 1897).

52. Liberty Hyde Bailey, ed., *The Principles of Agriculture: A Text-Book for Schools and Rural Societies* (New York: Macmillan, 1898), vii.

53. Elizabeth Childe [pseud.], "'The Price of a Home,'" *The New Republic* 13, no. 158 (17 November 1917): 79.

54. France Wright Saunders, *Katharine and Isabel: Mother's Light, Daughter's Journey* (Palo Alto: Consulting Psychologists Press, 1991), 10–14.

55. Elizabeth Childe, "Personal Experiences of Mothers: Why I Find Children Slow in Their School Work," and "Why I Believe the Home is the Best School," *Ladies Home Journal,* 29, no. 10 (October 1912): 42, 68. She expressed more of her educational philosophy in "Teaching Concentration," *The Outlook* 109 (20 January 1915):155–59, and "Parents and Education," *The Outlook,* 109 (3 March 1915): 539–41.

56. Saunders, *Katharine and Isabel,* 56.

57. Katharine Cook Briggs, "Meet Yourself: How to Use the Personality Paint Box," *The New Republic* 49, no. 629 (22 December 1926): 129–33.

58. Saunders, *Katharine and Isabel,* 81, 107.

59. Winton U. Solberg, *The University of Illinois, 1894–1904: The Shaping of the University* (Urbana: University of Illinois Press, 2000), 121–22; Eugene Davenport, "A Son of the Timberlands: In the Land of Amanha," *The Country Gentleman* (26 January 1926): 32, 132–33. Davenport wrote "Experience in Brazil," *Speculum* 11, no. 8 (11 July 1892), 121–25, in which he shared his experience with the college family.

60. Solberg, *University of Illinois,* 122. Davenport's first ten years at Illinois are covered in detail in chapter 5 of Solberg's book.

61. Eugene Davenport to Elida Yakeley, 23 October 1939, Kuhn Collection, UA 17.107, Box 1139, Folder 39.

62. Eugene Davenport, *Education for Efficiency: A Discussion of Certain Phases of the Problem of Universal Education with Special Reference to Academic Ideals and Methods* (Boston: D.C. Heath and Co., 1914), 11.

63. Eugene Davenport, *Principles of Breeding: A Treatise on Thremmatology or the Principles and Practices Involved in the Economic Improvement of Domesticated Animals and Plants* (Boston: Ginn and Company, 1907), v.

64. Eugene Davenport, *Domesticated Animals and Plants: A Brief Treatise upon the Origin and Development of Domesticated Races, with Special Reference to the Methods of Improvement* (Boston: Ginn and Company, 1910), iii.

65. Kolia S. Thabue to Oscar Clute, 18 Jan. 1893, Oscar Clute Papers, UA 2.1.5, Box 863, Folder 91, MSU Archives.

66. Seneca Taylor to President of Agricultural College, 22 March 1894, Lewis Griffin Gorton Papers, UA 2.1.6, Box 873, Folder 34, MSU Archives.

67. J. H. Brown to Lewis G. Gorton, 29 January 1894, Gorton Papers, UA

2.1.6, Box 873, Folder 17.

68. P. D. Warner to Fred Warner, 19 October 1880, quoted in Jean M. Fox, *Fred M. Warner: Progressive Governor* (Farmington Hills, Michigan: Farmington Hills Historical Commission, 1988), 27.

69. Fox, *Fred M. Warner*, 122–30, 274–79.

70. Booker T. Washington to President, Michigan State Agricultural College, 28 April 1893, Oscar Clute Papers, UA 2.1.5, Box 863, Folder 92.

71. Clute to Washington, 2 May 1893, Clute Papers, UA 2.1.5, Box 863, Folder 92.

72. Washington to Lewis Gorton, 12 May 1894 and 22 May 1894, Gorton Papers, UA 2.1.6, Box 873, Folder 37.

73. Thos. Calloway to Gorton, 5 November 1895, Gorton Papers, UA 2.1.6, Box 873, Folder 40.

74. Washington to President, Michigan State Agriculture College, 28 March 1900, and Snyder to Washington, 31 March [1900], J. L. Snyder Papers, UA 2.1.7, Box 894, Folder 163.

75. *Catalogue: Officers and Students of the Michigan State Agricultural College for the Year 1899–1900* (Agricultural College: The College, 1900); *Michigan Agricultural College: Catalogue of Officers and Graduates, 1857–1911* (East Lansing: [The College], 1911), 80; "Alumni," *M.A.C. Record* 11, no. 14 (19 December 1905): 2.

76. "A Senior Goes to Tuskegee," *M.A.C. Record* 5, no. 8 (31 October 1899): 1; *General Catalogue 1857–1900*, 239.

77. Snyder to Washington, 8 February 1908 and Washington to Snyder, 3 February 1908, Snyder Papers, UA 2.1.7, Box 813, Folder 117.

78. Washington to Snyder, 28 February 1908 and Snyder to Washington, 4 March 1908, Snyder Papers, UA 2.1.7, Box 813, Folder 117.

79. T. B. Keogh to Gorton, 18 July 1895, Gorton Papers, UA 2.1.6, Box 873, Folder 47.

80. Keogh to Gorton, 10 August 1895, Gorton Papers, UA 2.1.6, Box 873, Folder 47.

81. Gorton to Keogh, 25 July 1895, Gorton Papers, UA 2.1.6, Box 873, Folder 47.

82. Keogh to Gorton, 29 July 1895, Gorton Papers, UA 2.1.6, Box 873, Folder 47.

83. A. T. Stevens to Gorton, 6 November 1895, Gorton Papers, UA 2.1.6, Box 873, Folder 54.

84. John W. Robinson to [Snyder], 7 June 1899, Snyder Papers, UA 2.1.7, Box 804, Folder 94.

85. Snyder to Robinson, 12 June 1899, Snyder Papers, UA 2.1.7, Box 804, Folder 94.

86. W. G. Beasley, *The Rise of Modern Japan* (London: Weidenfeld and Nicolson, 1990), 88.

87. Ishizuki Minoru, "Overseas Study by Japanese in the Early Meiji Period," in *The Modernizer: Overseas Students, Foreign Employees, and Meiji Japan,* ed. Ardath W. Burks (Boulder, Col.: Westview Press, 1985), 179.

88. Dairoku Kikuchi, *Japanese Education; Lectures Delivered in the University of London by Baron Dairoku Kiluchi* (London: J. Murray, 1909), 346.

89. Shoichi Yebina to Maurice G. Kains, 19 February 1915, Papers of Shoichi Yebina, UA 10.3.31, F. D., Folder 1, MSU Archives.

90. "Personals," *College Speculum* 5, no. 1 (1 August 1885): 17.

91. Takekazu Ogura, ed., *Agricultural Development in Modern Japan* (Tokyo: Fuji Publishing Co., 1967), 15–16.

92. V. A. Sobennikoff, "My Trip from Siberia to Japan—A few Pages from My Diary," *Speculum,* 12, no. 11 (10 October 1893): 173–74.

93. Wahey Matsura, "Buddhism and Its Influence," *Speculum* 13, no. 3 (15 October 1894): 36–37.

CHAPTER 6. STUDENT LIFE

1. Henry Haigh, Diary, 9 March 1871, note after entry, Kuhn Collection, UA 17.107, Box 1140, Folder 10.

2. Theodore F. Garvin to mother, 21 October 1859, Kuhn Collection, UA 17.107, Box 1140, Folder 3.

3. James H. Gunnison, "The Dawn of the Michigan Agricultural College," student reminiscences (after 1900), Kuhn Collection, UA 17.107, Box 1140, Folder 8; S. O. Greene, Accounts, 22 April 1861, Kuhn Collection, UA 17.107, Box 1142, Folder 81.

4. T. C. Abbot to Henry P. Baldwin, 1 December 1870, T. C. Abbot Papers, UA 2.1.3, Box 861, Folder 26.

5. Beal, *History*, 150.

6. Maude Gilchrist, *The First Three Decades of Home Economics at Michigan State College* ([East Lansing]: School of Home Economics, 1947), 9, 15.

7. Edwards, Smith, and Kedzie, "Attendance at Agricultural College," *Thirty-Fifth Annual Report of the Secretary of the State Board of Agriculture*, 68.

8. "Improvements at the College," *M.A.C. Record* 3, no. 15 (21 December 1897): 1.

9. Snyder to C. J. Monroe, 15 April 1905, J. L. Snyder Papers, UA 2.1.7, Box 809, Folder 19.

10. Snyder to Mrs. G. B. Kitchen, 21 August 1908, Snyder Papers, UA 2.1.7, Box 812, Folder 47.

11. Snyder to [faculty], 14 June 1905, Snyder Papers, UA 2.1.7, Box 809, Folder 77a.

12. Snyder, untitled document, [1907,] Kuhn Collection, UA 17.107, Box 1138, Folder 39.

13. Henry G. Reynolds to Snyder, 31 March 1909, Snyder Papers, UA 2.1.7, Box 814, Folder 48.

14. "Report of the President," *Forty-Seventh Annual Report of the Secretary of State Board of Agriculture*, 33; Snyder to Robert D. Graham, 3 July 1911, Snyder Papers, UA 2.1.7, Box 816, Folder 58.

15. Snyder to Wm. J. Oberderffer, 17 July 1911, Snyder Papers, UA 2.1.7, Box 817, Folder 8; Snyder to [State Board], 9 July 1912, Snyder Papers, UA 2.1.7, Box 819, Folder 99; Snyder to William G. Merritt, 10 September 1912, Snyder Papers, UA 2.1.7, Box 819, Folder 23.

16. Ben Westrate, *Greek History: Michigan State College 1939* (East Lansing: Interfraternity Council, Michigan State College, 1939), [6–8]; [Snyder] to F. C. Kuhn, 20 July 1911, Snyder Papers, UA 2.1.7, Box 816, Folder 84; *Catalog of the Michigan State College . . . for the Year*

1925–1926, 25–26; "Views and Comments," *M.A.C. Record* 29, no. 7 (5 November 1923): 8

17. *M.A.C. Record* 29, no. 4 (15 October 1923): 7; *Catalog of the Michigan State College . . . for the Year 1925–1926,* 26.

18. *College Speculum* 1, no. 1 (1 August 1881): 7; a good account of the boarding club system appears in "How Students Board at M.A.C.," *M.A.C. Record* 15, no. 20 (15 February 1910): 3.

19. Leon Drake to mother and father and Hazel, 24 September 1899, Leon Drake Papers.

20. Snyder to C. P. Bramhill, 31 January 1906, Snyder Papers, UA 2.1.7, Box 809, Folder 107; Snyder to W. L. Roberts, 10 November 1906, Snyder Papers, UA 2.1.7, Box 810, Folder 28; Snyder to Reiley Calvert, 26 December 1907, Snyder Papers, UA 2.1.7, Box 810, Folder 146; *Catalog of the Michigan State College . . . for the Year 1925–1926,* 21, Beal, *History,* 216–17.

21. *M.A.C. Record* 26, no. 34 (29 July 1921): 8.

22. Geo. E. Breck, J. S. Porter and R. B. Norton to Abbot, 21 October 1877, Abbot Papers, UA 2.1.3, Box 861, Folder 47.

23. Clarence J. Judson to C. B. Judson, 25 March 1883, Clarence Herbert Judson Papers, UA 10.3.126, Folder 1, MSU Archives; Lyster H. Dewey, "Michigan Agricultural College, 1885–'88," May 1940, 14, Kuhn Collection, UA 17.107, Box 1139, Folder 120; J. E. Coulter to Abbot, 23 August 1880, Abbot Papers, UA 2.1.3, Box 861, Folder 53.

24. Edward G. Granger, diary, 17 December 1858, Edward G. Granger Papers, UA 10.3.56, F.D., Folder 1, MSU Archives.

25. "The Junior Exhibition," *Bubble,* no. 5 (29 August 1866): [4]; "The First Publication at M.A.C.," *M.A.C. Record* 20, no. 28 (27 April 1915): 1.

26. "The Social," *Bubble,* no. 6 (19 September 1868): [3].

27. "Public Discussion Between the Senior and Junior Classes of the State Agricultural College, Friday Evening, June 18, 1869," Kuhn Collection, UA 17.107, Box 1137, Folder 23.

28. Henry Haigh, diary, 8 May 1873, Kuhn Collection, UA 17.107, Box 1140, Folder 10.

29. Dewey, "Michigan Agricultural College, 1885–'88," 18; Haigh, diary, Note after entry for 28 February 1872."

30. *College Speculum* 1, no. 1 (1 August 1881): 6–7.

31. A collection of the *Union Lit* is in the collections of Michigan State University Museum.

32. Kuhn, *Michigan State,* 186.

33. Edwards, Smith, and Kedzie, "Attendance at Agricultural College," 63.

34. "The Student and Community," *Holcad* 1, no. 1 (10 March 1909): 10.

35. Westrate, *Greek History.*

36. Irma Thompson Ireland, "Something About Themian," unpublished manuscript, Kuhn Collection, UA 17.107, Box 1138, Folder 59.

37. Helen M. Sheldon, page from notebook, n.d., Helen Sheldon Lundberg Papers, UA 10.3.38, F.D., Folder 3, MSU Archives. A fictionalized account of rushing is found in Helen Hull, *Quest* (New York: The Feminist Press, 1990), 175–81.

38. Snyder to Louise Goodbody, 13 December 1906, Snyder Papers, UA 2.1.7, Box 809, Folder 168.

39. "Union Literary Eleven O'clock," 14 March 1908, Scrapbook 44, Ruth

E. Carrel Papers, UA 10.3.16, MSU Archives. Helen Hull provides a fictional representation of the anxieties and awkwardness that going to a dance could produce for some students in *Quest* (181–88).

40. "Locals," *Holcad* 2, no. 4 (9 November 1909): 23.

41. "Social," *Holcad* 2, no. 5 (23 November 1909): 13.

42. "Military," *Holcad* 2, no. 5 (23 November 1909): 14.

43. "Soronian Fall Party," *Holcad* 2, no. 6 (7 December 1909): 12.

44. "Class of 1915 J Hop Proves to Be Most Brilliant Social Function," *Holcad* 6, no. 18 (16 February 1914): 10–11; Kuhn, *Michigan State*, 191, 252–53.

45. Louis A. Bregger to Snyder, 27 February 1914, Snyder Papers, UA 2.1.7, Box 854, Folder 113; Notes on M.A.C. Junior Hop and Mr. Helme's Alleged Criticisms, n. d., Snyder Papers, UA 2.1.7, Box 855, Folder 63.

46. Snyder to Bregger, 13 March 1914, Snyder Papers, UA 2.1.7, Box 854, Folder 114.

47. Constitution, Natural History Society, 1877, Kuhn Collection, UA 17.107, Box 1139, Folder 1; Proceedings of Meeting, Natural History Society, 8 July 1880, Kuhn Collection, UA 17.107, Box 1139, Folder 2; "History of the Natural History Society," *College Speculum* 1, no. 1 (1 August 1881): 4; "Notes from the Natural History Society," *College Speculum* 3, no. 1 (1 August 1883): 11.

48. Beal, *History*, 210–11.

49. R. H. Gulley, 1877, G. Stewart, n.d., Kuhn Collection, UA 17.107, Box 1137, Folder 89.

50. Snyder to Thomas Yuill, 28 January 1907, Snyder Papers, UA 2.1.7, Box 812, Folder 36.

51. Snyder to I. J. Cortright, 10 March 1914, Snyder Papers, UA 2.1.7, Box 854, Folder 134; Beal, *History,* 217–18.

52. "M.A.C. 62 Olivet 0," *M.A.C. Record* 16, no. 10 (22 November 1910): 1.

53. Kuhn, *Michigan State,* 249; "Oratorical Contest," *M.A.C. Record* 9, no. 24 (8 March 1904): 1; "Thirteenth Annual Men's Contest and Third Annual Women's Contest of the Michigan Oratorical League, Michigan Agricultural College, . . . March 4, 1910," Kuhn Collection, UA 17.107, Box 1137, Folder 17.

54. Beal, *History,* 207; *M.A.C. Record* 8, no. 18 (20 January 1903): 2.

55. *M.A.C. Record* 10, no. 10 (22 November 1904): 2.

56. "Debating Club," *M.A.C. Record* 12, no. 12 (4 December 1906): 4.

57. "Public Speaking Association," *Holcad* 5, no. 7 (11 November 1912): 5.

58. "M.A.C. and Ypsilanti Divide Honors," *M.A.C. Record* 15, no. 35 (7 June 1910): 1.

59. "Iowa Defeats M.A.C.," *M.A.C. Record* 19, no. 23 (17 March 1914): 7.

60. *M.A.C. Record* 27, no. 17 (10 February 1922): 3.

61. *M.A.C. Record* 28, no. 20 (26 February 1923): 6.

62. *M.A.C. Record* 28, no. 31 (28 May 1923): 7.

63. "Liberal Arts Union," *M.A.C. Record* 14, no. 15 (12 January 1909): 1.

64. "Entertainment Course," *Holcad* 2, no. 1 (30 September 1909): 7.

65. Handbill, "The Liberal Arts Council," 1914–1915, Kuhn Collection, UA 17.107, Box 1138, Folder 17.

66. "M.A.C. Chorus," *M.A.C. Record* 13, no. 3 (8 October 1907): 1.

67. "Fourth Annual May Festival Concert," 29 May 1907; *Charlotte Leader,* 4 June 1908; "Eighth Annual May Festival," 19 May 1911, Kuhn Collection, UA 17.107, Box 1137, Folder 6.

68. "Tenth Annual May Festival a Success," *Holcad* 5, no. 30 (26 May 1913): 12.

69. "Glee Club for M.A.C.," *M.A.C. Record* 14, no. 25 (23 March 1909): 2; "Glee Club to go on Extensive Tour," *Holcad* 6, no. 22 (16 March 1914): 8–9.

70. "Girls' Glee Club Concert," *Holcad* 6, no. 27 (4 May 1914): 1, 19.

71. "Notice," *M.A.C. Record* 9, no. 17 (19 January 1904): 1.

72. "Piano Recital," *M.A.C. Record* 16, no. 21 (14 February 1911): 1.

73. "Music and Story," *M.A.C. Record* 18, no. 18 (28 January 1913): 2.

74. "Shakespearian Drama This Week," *Holcad* 6, no. 32 (8 June 1914): 1; "'Midsummer Night's Dream' Ably Given By College Dramatic Club," *M.A.C. Record* 19, no. 35 (16 June 1914): 7.

75. For a discussion of the "sacred space" see Linda O. Stanford and C. Kurt Dewhurst, *MSU Campus; Buildings, Places, Spaces: Architecture and the Campus Park of Michigan State University* (East Lansing: Michigan State University Press, 2002), 8–15.

76. Edward G. Granger, diary, 8 and 10 February 1859, Edward G. Granger Papers, UA 10.3.56, F.D., Folder 1.

77. J. S. Tibbits to T. C. Abbot, 7 March 1863, Abbot Papers, UA 2.1.3, Box 861, Folder 8.

78. Beal, *History,* 77.

79. H. E. Thomas, editorial, *College Speculum* 4, no. 2 (15 October 1884): 4–5.

80. Jonathan L. Snyder to T. C. Blaisdell, 30 June 1911, Snyder Papers, UA 2.1.7, Box 815, Folder 131; and Snyder to O. D. Baker, 16 March 1909, Snyder Papers, UA 2.1.7, Box 814, Folder 3.

81. Snyder to H. S. Holdridge and Snyder to O. D. Fisher, 24 June 1901 and Snyder to F. T. Ranney, 25 June 1901, Snyder Papers, UA 2.1.7, Box 805, Folders 92, 72, and 170.

82. Snyder to William Schwaderer, 27 October 1902, Snyder Papers, UA 2.1.7, Box 806, Folder 97; Snyder to Andrew Anderson, 27 October 1902, Snyder Papers, UA 2.1.7, Box 806, Folder 18; Snyder to Robert Gibbons, 3 November 1902, Snyder Papers, UA 2.1.7, Box 806, Folder 48.

83. Snyder to A. C. Gongwer, 27 April 1906, Snyder Papers, UA 2.1.7, Box 809, Folder 161.

84. *College Speculum* 3, no. 2 (1 October 1883): 10–11.

85. Snyder to unaddressed, 1 November 1906, Snyder Papers, UA 2.1.7, Box 809, Folder 169.

86. Snyder to L. J. Lupinsky, 29 October 1908, Snyder Papers UA 2.1.7, Box 813, Folder 6.

87. "Information to new Students—Sept. 28 '08," Snyder Papers, UA 2.1.7, Box 1813, Folder 32.

88. Snyder to Charles E. Temple, 19 February 1913, Snyder Papers, UA 2.1.7, Box 854, Folder 31; Temple to Snyder, 7 January 1913, Snyder Papers, UA 2.1.7, Box 854, Folder 30.

89. Frank S. Kedzie to E. H. Halliday, 23 December 1916, Kedzie Papers, UA 2.1.8, Box 892, Folder 25; H. C. King to Kedzie, 1 February 1916, Kedzie Papers, UA 2.1.8, Box 892, Folder 28.

90. Beal, *History,* 218–19.

91. Student Council to the Faculty, 9 April 1912, Snyder Papers, UA 2.1.7, Box 818, Folder 40.

92. *Holcad* 6, no. 2 (6 October 1913): 9.

93. Snyder to G. A. Kelley, 20 September 1898, Snyder Papers, UA 2.1.7, Box 804, Folder 59; Snyder to R. J. Boynton, 12 February 1912, Snyder Papers, UA 2.1.7, Box 818, Folder 4.

94. *Holcad* 5, no. 15 (27 January 1913): 14.

95. *Holcad* 5, no. 23 (24 March 1913): 19.

96. "History," 1879, Kuhn Collection, UA 17.107, Box 1137, Folder 100.

97. Snyder to Mr. Gibson, 2 February 1898, Snyder Papers, UA 2.1.7, Box 804, Folder 54.

98. Snyder to Jacob Gansley, 5 February 1898, Snyder Papers, UA 2.1.7, Box 804, Folder 54.

99. I. Stiefel to Snyder, 18 May 1909; J.E.S., report, 22 May 1909, Snyder Papers, UA 2.1.7, Box 814, Folder 42. There is a considerable volume of correspondence pertaining to this investigation in folders 42 and 43 of Box 814. Of particular interest are Spencer reports of 27, 28, 29, 30 May 1909 and 1, 3, 9 June 1909.

100. J.E.S., report, 9 June 1909, Snyder Papers, UA 2.1.7, Box 814, Folder 43.

101. Stiefel to Snyder, 12 June 1909, Snyder Papers, UA 2.1.7, Box 814, Folder 43.

102. William J. Orr to W. H. Wallace, 17 June 1909, Snyder Papers, UA 2.1.7, Box 814, Folder 38.

103. Snyder to C. W. Liken, 28 January 1905, Snyder Papers, UA 2.1.7, Box 809, Folder 7.

104. Snyder to E. E. Peters, 22 January 1903 and Snyder to Gertrude Peters, 19 January 1903, Snyder Papers, UA 2.1.7, Box 807, Folder 93; *Alumni Catalogue Number Michigan State College Bulletin: List of Graduates, Officers, and Professors of the Faculties, 1857–1930,* [East Lansing: Michigan State College], 1931: 44.

105. Snyder to C. W. Liken, 28 January 1905, Snyder Papers, UA 2.1.7, Box 809, Folder 7.

106. Kittie Huckins to Maude Gilchrist, 1 September 1903, Snyder Papers, UA 2.1.7, Box 807, Folder 42.

107. Snyder to J. D. Huckins, 29 September 1903, Snyder Papers, UA 2.1.7, Box 807, Folder 43.

108. Snyder to Mrs. Melissa Adams, 13 April 1904, Snyder Papers, UA 2.1.7, Box 807, Folder 157.

109. Snyder to C. H. Hall, 6 March 1901, Snyder Papers, UA 2.1.7, Box 805, Folder 87.

110. Snyder to Will Felker, 23 January 1907, Snyder Papers, UA 2.1.7, Box 811, Folder 25.

111. [Committee on Doubtful Cases] to faculty, 13 January 1908, Snyder Papers UA 2.1.7, Box 812, Folder 71; Report of Committee on Doubtful Cases on Elgin Mifflin, Jr., 5 April 1904, Snyder Papers, UA 2.1.7, Box 808, Folder 61; Elida Yakeley to Snyder, 24 December 1910, Snyder Papers, UA 2.1.7, Box 817, Folder 111; Yakeley to Snyder, 1 April 1911, Snyder Papers, UA 2.1.7, Box 817, Folder 111.

112. Snyder to Ernest Vaughn, 24 December 1907, Snyder Papers, UA 2.1.7, Box 813, Folder 115.

113. Snyder to Louis Vatz, 12 April 1911, Snyder Papers, UA 2.1.7, Box 817, Folder 82.

114. Snyder to H. Edwards, 2 May 1905, Snyder Papers, UA 2.1.7, Box 808, Folder 149.

115. Minnie Hendrick to Committee on Doubtful Cases, 27 March 1908, Snyder Papers, UA 2.1.7, Box 812, Folder 128.

116. Kedzie to Snyder, 23 November 1908, Snyder Papers, UA 2.1.7, Box 812, Folder 151.

117. Kedzie to Snyder, 27 December 1911, Snyder Papers, UA 2.1.7, Box 816, Folder 89.

118. Mark A. Noll, *A History of Christianity in the United States and Canada* (Grand Rapids: Eerdmans, 1922), 163–422; Sidney E. Ahlstrom, *A Religious History of the American People* (New Haven: Yale University Press, 1972), 731–872; George M. Marsden, *The Soul of the American University: From Protestant Establishment to Established Nonbelief* (New York: Oxford University Press, 1994); Walter Rauschenbusch, *Christianity and the Social Crisis*, ed. Robert D. Cross (New York: Harper and Row, 1964); Charles Darwin, *The Origin of Species* (1859; reprint, New York: Mentor Books, 1958).

119. See Frank Kedzie's unpublished papers on the history of Michigan State College, Kedzie Papers, UA 2.1.8, Box 1196, Folder 33.

120. Granger, Diary, 11 January 1859.

121. *College Speculum* 5, no. 1 (1 August 1885): 7–8.

122. "College Sunday School," *Speculum* 13, no. 3 (15 October 1894): 47.

123. "Church Membership," *Forty-Fifth Annual Report of the Secretary of the State Board of Agriculture*, 28; "Church Membership," *Fifty-Fourth Annual Report of the Secretary of the State Board of Agriculture*, 150.

124. Snyder to John J. Farrell, 12 March 1907, Snyder Papers UA 2.1.7, Box 811, Folder 27.

125. "News and Community," *M.A.C. Record* 20, no. 28 (27 April 1915): 5.

126. Dewey, "Michigan Agricultural College, 1885–88," 19–20.

127. *M.A.C. Record* 6, no. 17 (15 January 1901): 3.

128. "Missionary Notes," *M.A.C. Record* 8, no. 18 (20 January 1903): 2.

129. *M.A.C. Record* 6, no. 5 (16 October 1900): 3; *Holcad* 2, no. 7 (13 January 1910): 8

130. *M.A.C. Record* 15, no. 5 (19 October 1909): 1–2; *M.A.C. Record* 20, no. 23 (16 March 1915): 8; *M.A.C. Record* 26, no. 2 (1 October 1920): 12; *M.A.C. Record* 28, no.18 (12 February 1923): 12–13; *M.A.C. Record* 29, no. 5 (22 October 1923): 13.

131. *M.A.C. Record* 21, no. 26 (11 April 1916): 3; *M.A.C. Record* 24, no. 20 (28 February 1919):7; *M.A.C. Record* 26, no. 31 (27 May 1921): 12. See Maude Nason Powell Papers, UA 10.3.59, MSU Archives.

132. *M.A.C. Record* 27, no. 34 (16 June 1922): 8.

133. *M.A.C. Record* 27, no. 4 (21 October 1921): 13.

134. *M.A.C. Record* 28, no. 7 (6 November 1922): 5.

135. Program, 8 December 1907, Kuhn Collection, UA 17.107, Box 1141, Folder 42; Baker, *American Chronicle*, 216–17.

136. *M.A.C. Record* 17, no. 5 (24 October 1911): 1.

137. *M.A.C. Record* 18, no. 18 (28 January 1913): 3.

138. *M.A.C. Record* 17, no. 16 (16 January 1912): 2; *M.A.C. Record* 18, no. 23 (4 March 1913): 4.

139. "Student Pastor for Community Church," *M.A.C. Record* 24, no. 25 (11 April 1919): 5; "Plan of Activities for the Christian Agencies of Michigan Agricultural College," n.d., Kuhn Collection, UA 17.107, Box 1142, Folder 63; Invitation to Dedication of People's Church, 11–16 May 1926, Kuhn Collection UA 17.107, Box 1141, Folder 52.

140. See Frank S. Kedzie, "History of the Health Department at Michigan State College," unpublished manuscript, for a quick introduction into the early medical history of the college, Kedzie Papers, UA 2.1.8, Box 1196, Folder 33.

141. *Detroit Journal,* 27 April 1892; *Lansing Republican,* April 1892; Clute Papers, UA 2.1.5, Box 863, Folder 57.

142. "Sickness at the College," *Speculum* 11, no. 6 (10 May 1892): 101.

143. Beal, *History,* 105, 283.

144. Snyder to Henry B. Baker, 26 October 1901, Snyder Papers UA 2.1.7, Box 805, Folder 32.

145. Snyder to Baker, 13 September 1904 and Baker to Snyder, 15 September 1904, Snyder Papers, UA 2.1.7, Box 807, Folder 172.

146. "College Water Supply Found to Be Impure," *M.A.C. Record* 19, no. 2 (16 December 1913): 3.

147. Snyder to Kilian Klement, 12 April 1911, Snyder Papers, UA 2.1.7, Box 816, Folder 81.

148. Snyder to James Knowles, 15 April 1911, Snyder Papers, UA 2.1.7, Box 816, Folder 81.

149. Snyder to H. A. Post, 15 April 1911, Snyder Papers, UA 2.1.7, Box 817, Folder 4.

150. F. Kedzie to Snyder, 6 April 1911, Snyder Papers, UA 2.1.7, Box 816, Folder 81.

151. Baker to Snyder, 23 January 1902, Snyder Papers, UA 2.1.7, Box 806, Folder 21.

152. Snyder, Faculty Action, 12 November 1910, Kuhn Collection, UA 17.107, Box 1141, Folder 131.

153. F. Kedzie, Scarlet Fever Bulletin, [21 March 1917,] Kedzie Papers, UA 2.1.8, Box 892, Folder 56.

154. "Report of the President," *Fifty-Eighth Annual Report of the Secretary of the State Board of Agriculture of the State of Michigan . . . from July 1, 1918, to June 30, 1919* (Fort Wayne, Ind.: The College, 1920), 32–34; "H. E. Girls Dieticians for Camp Hospital," *M.A.C. Record* 24, no. 4 (25 October 1918): 4.

155. *M.A.C. Record* 24, no. 11 (13 December 1918): 3–4.

156. *M.A.C. Record* 24, no. 12 (20 December 1918): 3.

157. *M.A.C. Record* 24, no. 4 (25 October 1918): 3.

158. Ward Giltner to F. S. Kedzie, 29 November 1915, Kedzie Papers, UA 2.1.8, Box 892, Folder 7; Memorandum of Agreement Respecting the Future Administration of the College Hospitals, [1918,] Kedzie Papers, UA 2.1.8, Box 893, Folder 26.

159. Health Service Committee to the President and Faculty, 19 May 1919, Kedzie Papers, UA 2.1.8, Box 893, Folder 30.

160. "Student Health Service Established," *M.A.C. Record* 26, no. 2 (1 October 1920): 7–8; "Report of the Health Service" and "Report of the Department of Physical Training," *Sixtieth Annual Report of the Secretary of the State Board of Agriculture of the State of Michigan . . . from*

July 1, 1920, to June 30, 1921 (Lansing: Wynkoop Hallenbeck Crawford Co., 1922), 111, 147.

161.   Beal, *History*, 214–15.

162.   "Announcement," *M.A.C. Record* 19, no. 1 (30 September 1913): 1.

163.   George C. Sheffield to State Board of Agriculture, 30 June 1914, Snyder Papers, UA 2.1.7, Box 855, Folder 147.

164.   "M.A.C. Association of Washington, D.C.," *M.A.C. Record* 7, no. 33 (13 May 1902): 1–2.

165.   H. A. Taft to Members of the Chicago M.A.C. Association, 18 February 1914, Snyder Papers UA 2.1.7, Box 855, Folder 156.

166.   Sheffield to Board, 30 June 1914.

167.   "Detroiters' Doings," *M.A.C. Record* 29, no. 20 (25 February 1924): 4.

168.   "Prudden Turns Sod for New Union," *M.A.C. Record* 28, no. 34 (16 July 1923): 14–15; Kuhn, *Michigan State,* 263.

169.   "Holcad Compiles Excavation Figures," *M.A.C. Record* 29, no. 12 (10 December 1923): 10.

170.   Fred Honhart, "The MSU Union's 75th Birthday," *MSU Alumni Magazine* 17, no. 2 (2000): 26–30; Kuhn, *Michigan State,* 264–66.

CHAPTER 7. AFRICAN-AMERICAN STUDENTS

1.   For an excellent synthesis of scholarship interpreting the power and extent of racial prejudice in the United States, see David A. Hollinger, "Amalgamation and Hypodescent: The Question of Ethnoracial Mixture in the History of the United States," *The American Historical Review* 108, no. 5 (December 2003): 1363–1390.

2.   Jonathan L. Snyder to S. L. Maddox, 13 April 1905, J. L. Snyder Papers, UA 2.1.7, Box 809, Folder 17.

3.   Snyder to Maude Gilchrist, 30 August 1906, Snyder Papers, UA 2.1.7, Box 819, Folder 164.

4.   Richard Thomas, *Life for Us Is What We Make It: Building Black Community in Detroit, 1915–1945* (Bloomington: Indiana University Press, 1992), 1–19.

5.   Joseph W. Jarvis, *Golden Jubilee: African Methodist Episcopal Church: Lansing, Michigan, 1867–1917* (N.p., 1917), 8–9.

6.   Fourteenth Census of the United States: 1920—Population, Michigan, Ingham County, City of Lansing, Enumeration District 91, Sheet 16; *Catalogue . . . for the Year 1899–1900* (Agricultural College: The College, 1900); *Wolverine,* 1900, 76; Claiie '05, Glüd=Uuf M.A.C.

7.   "Alumni," *M.A.C. Record* 11, no. 14 (19 December 1905): 2.

8.   *Catalogue of Officers and Graduates, 1857–1911,* 80; *Lansing City Directory 1911–12,* vol. 11 (Lansing: Chilson, McKinley and Co., 1911), 495.

9.   Fourteenth Census of the United States, Sheet 16; *Lansing City Directory 1916,* vol. 16 (Lansing: Chilson, McKinley and Co., 1916), 622.

10.   Lisa Fine, *The Story of Reo Joe: Work, Kin, and Community in Autotown, U.S.A.* (Philadelphia: Temple University Press, 2004), 1–37.

11.   *Lansing City Directory 1912,* vol. 12 (Lansing: Chilson, McKinley and Co., 1912), 510; *Lansing City Directory 1913,* vol. 13 (Lansing: Chilson, McKinley and Co., 1913), 544; and *Lansing City Directory 1914,* vol. 14 (Lansing: Chilson, McKinley and Co., 1914), 535.

12. Jarvis, *Golden Jubilee,* 24, 25, 33; "Thomson [sic] Funeral," *Lansing State Journal,* 8 February 1923, 2.

13. Jarvis, *Golden Jubilee,* 41; *Lansing City Directory 1916,* 471.

14. Joe T. Darden, "Patterns of Residential Segregation in Michigan Cities in the Nineteenth Century," in *Blacks and Chicanos in Urban Michigan,* eds. Homer C. Hawkins and Richard W. Thomas ([Lansing]: Michigan History Division, Michigan Department of State, n.d., 20–27; Douglas K. Meyer, "The Changing Negro Residential Patterns in Lansing, Michigan, 1850–1969" (Ph.D. diss., Michigan State University, 1970), 74–101.

15. After 1920 the U.S. census did not use "mulatto" as a racial designation. Hollinger, "Amalgamation and Hypodescent," 1369.

16. William O. Thompson, '04," *M.A.C. Record* 28, no. 19 (19 February 1923): 12.

17. Snyder to Booker T. Washington, 12 April 1907, Snyder Papers, UA 2.1.7, Box 810, Folder 129.

18. Cindi Steinway, "Myrtle Mowbray, '07: MSU's First Black Graduate, Remembers M.A.C.," *MSU Alumni Magazine* 18, no. 2 (November 1972): 15; *Lansing City Directory 1904,* vol. 7 (Lansing: Chilson, McKinley and Co., 1904), 123.

19. Jarvis, *Golden Jubilee,* 30.

20. "Conferring of Degrees," *Semi-Centennial Celebration of Michigan State Agricultural College,* ed. Thomas C. Blaisdell, 256–57.

21. Snyder to W. T. Vernon, 7 May 1907, Snyder Papers, UA 2.1.7, Box 810, Folder 130.

22. "Alumni," *M.A.C. Record* 13, no. 11 (3 December 1907): 1; Thomas Jesse Jones, ed., *Negro Education: A Study of the Private and Higher Schools for Colored People in the United States* (New York: Arno Press, 1969), 2:679–80.

23. Jones, ed., *Negro Education* 2:381–83.

24. *M.A.C. Record* 17, no. 10 (28 November 1911): 2.

25. "Praise Exemplary Life of Mrs. Myrtle C. Mowbray," *The [Kansas City, Missouri] Call,* 15 November to 21 November 1974.

26. David Thomas, "MSU Black is a Pioneer in American Football," *Lansing City Magazine* 6, no. 2 (February 1992): 27; George S. Alderton, "The Sport Grist," *Lansing State Journal,* 11 April 1947, 21.

27. "Cosmopolitanism a Movement," *M.A.C. Record* 15, no. 22 (1 March 1910): 1.

28. *Wolverine,* 1915, 108, 191, 241; *Wolverine,* 1916, 82, 214, 291.

29. Alderton, "The Sport Grist," 21.

30. Jones, ed., *Negro Education* 2:325–27, 623–24, 672–74.

31. News release, 7 May 1968, Hampton Institute.

32. "Band Slated for Junket," *M.A.C. Record* 21, no. 2 (28 September 1915): 4.

33. *Wolverine,* 1916, 88, 191, 278, 299; *Wolverine,* 1914, 120, 258, 259.

34. Boston School Department, Teacher Qualifications: Certificates of Qualification, Book "X," 248, City of Boston, Archives and Records Management Division.

35. It is puzzling as to why Thompson has disappeared from MSU's institutional memory. Perhaps it is because there are no known photographs of him in college publications, nor does it appear that he par-

ticipated in any extracurricular activities that might have drawn attention to him. When I was reading the 19 December 1905 issue of the *M.A.C. Record,* I noticed in the Alumni News column that W. O. Thompson, '04, had "been elected to a position with Booker T. Washington's school at Tuskegee." This stimulated my interest, since I thought that only African Americans taught at Tuskegee Institute at that time. Cynthia Beavers Wilson, archivist at Tuskegee University confirmed this for me. Unfortunately, faculty and staff records for the time when Thompson was at Tuskegee are not available (e-mail from Cynthia Beavers Wilson, 6 September 2002).

36. "The Cosmopolitan Club," *Wolverine,* 1915, 241.

37. "Inter-Class Tennis," *Wolverine,* 1917, 182.

38. *Alumni Catalogue Number: List of Graduates, Officers and Professors of the Faculties 1857–1930* ([East Lansing: Michigan State College], 1931), 75; Jones, ed., *Negro Education* 2, 674–75.

39. *Lansing City Directory 1920,* vol. 20 (Lansing: Chilson, McKinley and Co., 1920), 367; *Lansing City Directory 1924,* vol. 24 (Lansing: McKinley-Reynolds Co., 1924), 374; Fourteenth Census of the United States: 1920—Population, Michigan, Ingham County, City of Lansing, Enumeration District 91, Sheet 26.

40. *Lansing City Directory 1913,* 193; *Lansing City Directory 1914,* 201; *Lansing City Directory 1920,* 339.

41. Jarvis, *Golden Jubilee,* 25–27.

42. *Wolverine,* 1924, 34.

43. *Lansing City Directory 1925,* vol. 25 (Lansing: McKinley-Reynolds Co., 1925): 265; *Alumni Catalogue . . . 1857–1930,* 100.

44. *Lansing City Directory 1923,* vol. 23 (Lansing: Chilson, McKinley and Co., 1923), 631–32.

45. *Lansing City Directory 1900,* vol. 5 (Lansing: Chilson and McKinley, 1900), 190; *Lansing City Directory 1912,* 362; *Lansing City Directory 1920,* 609.

46. Jarvis, *Golden Jubilee* 24.

47. Mabel J. Lucas to F. S. Kedzie, 9 January 1930, Kedzie Papers, UA 2.1.7, Box 894, Folder 26; Jones, ed., *Negro Education* 2:598–600.

48. Lucas to Kedzie, 4 October 1930, Kedzie Papers, UA 2.1.8, Box 894, Folder 26.

49. *Wolverine,* 1925, 40.

50. *Wolverine,* 1924, 153; *Wolverine,* 1925, 161; and *Wolverine,* 1926, 141, 148.

51. *Wolverine,* 1923, 348; *Wolverine,* 1925, 42, 343, 345, 348.

52. *Alumni Catalogue . . . 1857–1930,* 105, 109; Jones, ed., *Negro Education* 2:682–83.

53. *Wolverine,* 1926, 64.

54. *Wolverine,* 1927, 89.

55. *Lansing and East Lansing City Directory 1928,* vol. 28 (Lansing: McKinley-Reynolds Co., 1928), 455, 575.

NOTES TO PAGES 355–362

CHAPTER 8. INTERNATIONAL STUDENTS

1. W. Reginald Wheeler, Henry H. King, and Alexander B. Davidson, eds. *The Foreign Student in America: A Study by the Commission on Survey of Foreign Students in the United States, under the Auspices of the Friendly Relations Committees of the Young Men's Christian Association and the Young Women's Christian Association* (New York: Association Press, 1925), 11–12.

2. Faculty Meeting Minutes, 1887–1889, 23 April 1888, Kuhn Collection, UA 17.107, Box 1143, Folder 75.

3. K. G. Minakata to Edwin Willits, 25 April 1888, Kuhn Collection, UA 17.107, Box 1143, Folder 75.

4. Leon Drake to Mother, Father, and Sister, 23 March 1900, Leon Drake Papers.

5. S. L. Maddox to Charles J. Mourse, February 1905, Jonathan L. Snyder Papers, UA 2.1.7, Box 809, Folder 17.

6. Snyder to Maddox, 13 April 1905, Snyder Papers, UA 2.1.7, Box 809, Folder 17.

7. Snyder to V. L. Tissera, 27 June 1908, Snyder Papers, UA 2.1.7, Box 813, Folder 97.

8. W. J. McGee to Snyder, 4 August 1908, Snyder Papers, UA 2.1.7, Box 813, Folder 18.

9. Snyder to McGee, 8 August 1908, Snyder Papers, UA 2.1.7, Box 813, Folder 18.

10. H. Caramanian to Snyder, 2 April 1906, Snyder Papers, UA 2.1.7, Box 890, Folder 129.

11. D. A. W. Lacklaw to Director of the Michigan Agricultural College, 21 April 1897, Snyder Papers, UA 2.1.7, Box 804, Folder 34.

12. T. H. Pardo de Tavera to Snyder, 18 April 1903, and Snyder to Pardo de Tavera, 18 May 1903, Snyder Papers UA 2.1.7, Box 807, Folder 118; Snyder to Alex Sutherland, 8 September 1904, Snyder Papers, UA 2.1.7, Box 808, Folder 85; *M.A.C. Record* 10, no. 1 (20 September 1904): 1, *M.A.C. Record* 10, no. 3 (4 October 1904): 4; *M.A.C. Record* 12, no. 1 (18 September 1906): 4.

13. Eastella C. Long to Snyder, 14 May 1908, Snyder Papers, UA 2.1.7, Box 813, Folder 122.

14. J. R. Maurer to University of Michigan, 26 May 1910, Snyder Papers, UA 2.1.7, Box 815, Folder 44.

15. A. S. Onischick to Snyder, 14 November 1908, Snyder Papers, UA 2.1.7, Box 813, Folder 40.

16. Ricardo Anibal Tello to Snyder, 1 August 1907, Snyder Papers, UA 2.1.7, Box 810, Folder 49; Snyder to Tello, 7 November 1907, Snyder Papers, UA 2.1.7, Box 810, Folder 47.

17. Snyder to Zakir Abdullin, 6 July 1912, Snyder Papers, UA 2.1.7, Box 817, Folder 119; Snyder to Dikran Erdoghlian, 27 June 1910, Snyder Papers, UA 2.1.7, Box 814, Folder 98.

18. H. V. Jackson to Snyder, 25 November 1903, Snyder Papers, UA 2.1.7, Box 807, Folder 56.

19. Snyder to Mrs. H. H. Jackson, 24 November 1903, Snyder Papers, UA 2.1.7, Box 807, Folder 56.

20. Snyder to Jose Susano Sanjurjo, 17 October 1906, Snyder Papers, UA 2.1.7, Box 810, Folder 40.

21. Snyder to Concul General of Peru, 10 October 1903, Snyder Papers, UA 2.1.7, Box 807, Folder 32.

22. "News and Comment," *M.A.C. Record* 20, no. 12 (16 December 1914): 7; Ivan Zwetan to Snyder, 5 May 1915, Snyder Papers, UA 2.1.7, Box 866, Folder 24.

23. "Cosmopolitan Club," *Holcad* 2, no. 8 (27 January 1910): 6; typescripts of Cosmopolitan Club constitution, Kuhn Collection, UA 17.107, Box 1137, Folder 109.

24. "Cosmopolitanism a Movement," *M.A.C. Record* 15, no. 22 (1 March 1910): 1.

25. "Cosmopolitan Club," *Holcad* 3, no. 17 (30 January 1911): 6.

26. "The Cosmopolitan Club," *M.A.C. Record* 16, no. 21 (14 February 1911): 1.

27. "Cosmopolitan Club," *Holcad* 3, no. 28 (1 May 1911): 12.

28. "Cosmopolitan Club," *M.A.C. Record* 18, no. 10 (26 November 1912): 2.

29. *M.A.C. Record* 27, no. 28 (5 May 1922): 3.

30. "Cosmopolitan Club" and "Program for the Fall Quarter, 1925," Kuhn Collection, UA 17.107, Box 1137, Folder 114.

31. Vadim Sobennikoff, "Russian Peasant Life," *M.A.C. Record* 1, no. 4 (1 February 1896) and "Occupations and Systems of Land Ownership," *M.A.C. Record* 1, no. 5 (8 February 1896).

32. F. Yebina, "Agriculture in Japan," *M.A.C. Record* 1, no. 11 (24 March 1896).

33. Gikan Fujimura, "Relations between Japan and the United States," *Holcad* 1, no. 2 (23 March 1909): 3.

34. Osman Abdel Razik, "Agriculture in Egypt," *Holcad* 2, no. 3 (26 October 1909): 3–4, and "Co-Eds," *Holcad* 2, no. 5 (23 November 1909): 10.

35. Morris C. Ellman, "The Wandering Jew," *Holcad* 2, no. 10 (24 February 1910): 4.

36. "Cosmopolitan Club," *M.A.C. Record* 16, no. 26 (21 March 1911): 2. For a full discussion of Chinese students in America, see Stacey Bieler, *"Patriots" or "Traitors"?: A History of American-Educated Chinese Students* (Armonk, New York: M. E. Sharpe, 2004).

37. *M.A.C. Record* 27, no. 10 (2 December 1921): 5; "Alumni Opinion," *M.A.C. Record* 29, no. 2 (1 October 1923): 10.

38. "The New Class," *M.A.C. Record* 17, no. 2 (3 October 1911): 1.

39. *Alumni Catalogue . . . 1857–1930,* , 65.

40. *Alumni Catalogue . . . 1857–1930,* 105.

41. "Itano, '11 Accepts Position in Japan," *M.A.C. Record* 29, no. 19 (18 February 1924): 10.

42. "Alumni Notes," *M.A.C. Record* 21, no. 13 (14 December 1915): 11; "Class Notes," *M.A.C. Record* 27, no. 2 (7 October 1921): 14.

43. "Henry W. Geller," *M.A.C. Record* 9, no. 39 (21 June 1904): 1.

44. Henry W. Geller to Snyder, 3 September 1907, Snyder Papers, UA 2.1.7, Box 811, Folder 45.

45. "Alumni," *M.A.C. Record* 13, no. 29 (21 April 1908): 1.

46. Moses N. Levine to Snyder, 24 August 1910, Snyder Papers, UA 2.1.7, Box 815, Folder 26.

47. "Demonstration Train," *Holcad* 5, no. 2 (7 October 1912): 5.

48. Dana G. Dalrymple, "Russian Studies of American Agriculture: 1908–1914," unpublished manuscript, Kuhn Collection, UA 17.107, Box 1139, Folder 38; "Rosen's Russian Rye," *The Michigan State University Magazine* 8, no. 1 (September 1962): 4–6.

49. Joseph A. Rosen to Frank Kedzie, 24 September 1921, *M.A.C. Record* 27, no. 5 (28 October 1921): 13.

50. Dalrymple, "Russian Studies of American Agriculture," 8–9; "Rosen Introduces Corn into Russia," *M.A.C. Record* 28, no. 5 (23 October 1922): 5.

51. Dalrymple, "Russian Studies of American Agriculture," 10.

52. "Rosen, '08 Has Hand in Sugar Beet Seed Situation," *M.A.C. Record* 22, no. 24 (27 March 1917): 3.

53. *The New International Encyclopedia,* 20 vols. (New York: Dodd, Mead and Company, 1905).

54. M. G. Kains, ed., *Making Horticulture Pay: Experiences in Gardening and Fruit Growing* (New York: Orange Judd Company, 1913), v-vi.

55. M. G. Kains to Snyder, 26 February 1913, Snyder Papers, UA 2.1.7, Box 820, Folder 46; "M.A.C. Graduate to Join Faculty at Penn State," *M.A.C. Record* 19, no. 29 (5 May 1914): 5; "Kains, '95, Gives up Teaching for Larger Field," *M.A.C. Record* 22, no. 4 (17 October 1916): 3. M. G. Kains, *Plant Propagation: Greenhouse and Nursery Practice* (New York: Orange Judd Company, 1925). Another book edited by Kains is *Culinary Herbs* (New York: Orange Judd Company, 1912).

CHAPTER 9. ATHLETICS

1. D. P. Yerkes, "Athletics in Colleges," *College Speculum* 6, no. 4 (1 June 1887): 4–5.

2. "Athletics," *Speculum* 12, no. 4 (10 November 1892): 66.

3. Curti and Carstensen, *The University of Wisconsin*, 2:536–38; John Sayle Watterson, *College Football: History, Spectacle, Controversy* (Baltimore: Johns Hopkins University Press, 2000), 9–140; Ronald A. Smith, *Sports and Freedom: The Rise of Big-Time College Athletics* (New York: Oxford University Press, 1988), 67–98.

4. "Notes from Alfred G. Gulley," typescript, Kuhn Collection, UA 17.107, Box 1140, Folder 7.

5. William Fraser to John C. Fraser, 9 September 1866, Kuhn Collection, UA 17.107, Box 1140, Folder 1; "Tracy, '68, Writes of Baseball at M.A.C. 50 Years Ago," *M.A.C. Record* 21, no. 32 (23 May 1916): 6.

6. "Athletics," *College Speculum* 8, no. 6 (10 May 1889): 96; *College Speculum* 5, no. 4 (1 June 1886): 10.

7. "New Baseball Uniforms," *M.A.C. Record* 4, no. 35 (16 May 1899): 1.

8. *College Speculum* 1, no. 3 (1 April 1882): 39–40.

9. *College Speculum* 6, no. 2 (1 October 1886): 13.

10. Beal, *History*, 220.

11. "At the College," *M.A.C. Record* 3, no. 2 (21 September 1897): 3.

12. *M.A.C. Record* 4, no. 2 (27 September 1898): 1.

13. "The Week in Athletics," *M.A.C. Record* 4, no. 31 (18 April 1899): 1.

14. Harry J. Driskel to Jonathan L. Snyder, 13 August 1901, J. L. Snyder Papers, UA 2.1.7, Box 895, Folder 58.

15. "Expenses," *Catalogue: Officers and Students of the Michigan Agricultural College for the Year 1912–1913* (East Lansing: The College, 1913), 51; G. E. Gauthier, "Report of Department of Athletics," *Fifty-Fifth Annual Report of the Secretary of the State Board of Agriculture,* 193.

16. "Attendance at Agricultural College," *Thirty-Fifth Annual Report of the Secretary of the State Board of Agriculture,* chart C, 70–71.

17. "Our Field Day," *College Speculum* 5, no. 4 (1 June 1886): 12–13; *Lansing State Republican,* 7 June 1886: 4; Beal, *History,* 220.

18. "Constitution of Inter-Collegiate Athletic Association," *College Speculum* 7, no. 5 (10 June 1888): 9; Todd E. Harburn and Gerald E. Harburn, *MIAA Football: The Illustrated Gridiron History of the Michigan Inter-collegiate Athletic Association* (Chelsea, Michigan: Bookcrafters, 1986), 1–4.

19. "Field Day," *College Speculum* 7, no. 5 (10 June 1888): 7–9; U. P. Hedrick, *The Land of the Crooked Tree* (New York: Oxford University Press, 1948), 32–34.

20. *College Speculum* 3, no. 4 (1 June 1884): 12; *College Speculum* 5, no. 4 (1 June 1886): 13.

21. *College Speculum* 7, no. 3 (10 April 1888): 8; *College Speculum* 13, no. 3 (15 October 1894): 51.

22. "Foot Ball," *M.A.C. Record* 1, no. 44 (8 December 1896): 1; Fred W. Stabley, *The Spartans: A Story of Michigan State Football* (Huntsville, Ala.: Strobe, 1975), 283.

23. Stabley, *The Spartans,* 14.

24. Charles O. Bemies, "Report of the Department of Physical Culture," 30 June 1900, *Thirty-Ninth Annual Report of the Secretary of the State Board of Agriculture of the State of Michigan . . . from July 1, 1899, to June 30, 1900* (Lansing: Wynkoop Hallenbeck Crawford Co., 1900), 52.

25. Leon Drake to Mother, Father and Sister, 4 November 1900, Leon Drake Papers.

26. Bemies, "Report of the Department of Physical Culture," 30 June 1901, *Fortieth Annual Report of the Secretary of the State Board of Agriculture of the State of Michigan . . . from July 1, 1900, to June 30, 1901* (Lansing: Wynkoop Hallenbeck Crawford Co., 1901), 83; "Intercollegiate Field Day," *M.A.C. Record* 6, no. 36 (11 June 1901): 1.

27. Chas. H. Gurney to Snyder, 24 October 1901, and Snyder to Gurney, 26 October 1901, Snyder Papers UA 2.1.7, Box 805, Folder 84.

28. Snyder to W. M. Burke, 9 June 1905, Snyder Papers, UA 2.1.7, Box 808, Folder 122.

29. Snyder to W. M. Burke, 9 June 1905, Snyder Papers, UA 2.1.7, Box 808, Folder 122; Snyder to C. L. Brewer, 28 February 1910, Snyder Papers, UA 2.1.7, Box 814, Folder 84.

30. "Suggested Rules for Government of Athletics at M.A.C.," adopted by faculty, 2 May 1910, Snyder Papers, UA 2.1.7, Box 814, Folder 87.

31. Snyder to J. F. Macklin, 20 June 1913, Snyder Papers, UA 2.1.7, Box 820, Folder 91.

32. Snyder to Macklin, 12 December 1911, Snyder Papers, UA 2.1.7, Box 816, Folder 123.

33. Snyder to Macklin, 17 November 1914, Snyder Papers, UA 2.1.7, Box 855, Folder 72.

34. George Denman to Snyder, 13 June 1902, and 25 August 1902, Snyder Papers, UA 2.1.7, Box 806, Folder 42 and 31 July 1902, Box 806, Folder 41.

35. Stabley, *The Spartans,* 14; Beal, *History,* 220.

36. Jack Seibold, *Spartan Sports Encyclopedia* (Champaign: Sports Publishing, 2003), xvi-xvii.

37. "M.A.C. Out of M.I.A.A.," *M.A.C. Record* 13, no. 19 (4 February 1908): 1.

38. C. L. Brewer, "Report of the Department of Physical Culture," *Forty-Seventh Annual Report of the Secretary of the State Board of Agriculture,* 99–100.

39. Snyder to C. R. Van Hise, 6 December 1904, Snyder Papers, UA 2.1.7, Box 808, Folder 95.

40. Snyder to W. H. Wallace, 27 May 1908, Snyder Papers, UA 2.1.7, Box 813, Folder 123.

41. Snyder to A. J. Anderson, 28 April 1905, Snyder Papers, UA 2.1.7, Box 809, Folder 59.

42. Snyder to Charles H. Edwards, 5 November 1910, Snyder Papers, UA 2.1.7, Box 814, Folder 98; Snyder to Brewer, 16 December 1910, Snyder Papers, UA 2.1.7, Box 814, Folder 94.

43. *M.A.C. Record* 8, no. 7 (28 October 1902): 2; "Athletic Notes," *M.A.C. Record* 8, no. 6 (21 October 1902): 4.

44. "Hold Michigan to a Scoreless Tie," *M.A.C. Record* 14, no. 4 (13 October 1908): 2; *M.A.C. Record* 14, no. 7 (3 November 1908): 4.

45. Snyder to John W. Beaumont, 26 June 1915, Snyder Papers, UA 2.1.7, Box 855, Folder 191.

46. "The Greatest Game Ever Played Here," *M.A.C. Record* 15, no. 9 (16 November 1909): 1.

47. "Notre Dame is Badly Beaten," *M.A.C. Record* 16, no. 7 (1 November 1910): 1.

48. "M.A.C. 62 Olivet 0," *M.A.C. Record* 16, no. 10 (22 November 1910): 1.

49. "M.A.C. 12; U. of M. 7," *Holcad* 6, no. 4 (20 October 1913): 4; "M.A.C. Humbles Michigan," *M.A.C. Record* 19, no. 4 (21 October 1913): 1; Stabley, *The Spartans,* 35–39.

50. "M.A.C. Did It," *Holcad* 6, no. 5 (27 October 1913): 8, 10–11; "Badgers Bow to Aggies," *M.A.C. Record* 19, no. 5 (28 October 1913): 1, 3; Stabley, *The Spartans,* 38–39.

51. Snyder to Macklin, 28 October 1913, Snyder Papers, UA 2.1.7, Box 819, Folder 117A.

52. "Faculty Celebration of Wisconsin Victory," *Holcad* 6, no. 6 (3 November 1913): 1.

53. Snyder to Maude Gilchrist, 27 October 1913, Snyder Papers, UA 2.1.7, Box 819, Folder 140.

54. Snyder to E. D. Partridge, 14 October 1914, Snyder Papers, UA 2.1.7, Box 855, Folder 114; Snyder to Macklin, 3 October 1914, Snyder Papers, UA 2.1.7, Box 855, Folder 71.

55. Snyder to George Sheffield, 14 October 1914, Snyder Papers, UA 2.1.7, Box 855, Folder 71.

56. Snyder to Partridge, 14 October 1914, Snyder Papers, UA 2.1.7, Box 819, Folder 140.

57. "Rules Governing the Board in Control of Athletics at Michigan Agri-

cultural College," [1915,] Snyder Papers, UA 2.1.7, Box 866, Folder 11; "Athletic Board of Control," *Wolverine*, 1917, 145; "Athletic Board of Control," *Wolverine*, 1912, n.p.; *Catalog of the Michigan State College . . . for the Year 1925–1926*, 27.

58. Frank Kedzie to J. W. Beaumont, 1 December 1916 and Beaumont to Kedzie, 28 November 1916, Kedzie Papers, UA 2.1.8, Box 892, Folder 21; *Detroit Free Press*, 28 November 1916, 14; Seibold, *Spartan Sports Encyclopedia*, 55.

59. E. A. Burnett to Snyder, 21 September 1914 and Snyder to Burnett, 30 September 1914, Snyder Papers, UA 2.1.7, Box 854, Folder 125.

60. Stabley, *The Spartans*, 39–48.

61. Ibid., 57; "Stadium Going South of River," *M.A.C. Record* 28, no. 29 (14 May 1923): 3.

62. "Stadium Has Many Modern Features," *M.A.C. Record* 29, no. 1 (24 September 1923): 5.

63. *Wolverine*, 1901, 182.

64. "The U. of M. Game," *M.A.C. Record* 8, no. 30 (21 April 1903): 1.

65. "1911 Baseball," *Wolverine*, 1912, n.p.

66. "Baseball," *Wolverine*, 1917, 160–61.

67. "Baseball," *Wolverine*, 1926, 134; "'Spartan'—How It Came to Be," 2 July 1951, Kuhn Collection, UA 17.107, Box 1142, Folder 75.

68. Seibold, *Spartan Sports Encyclopedia*, 125–26.

69. "Basketball," *M.A.C. Record* 3, no. 20 (1 February 1898): 2–3.

70. "Basketball," *M.A.C. Record* 5, no. 24 (6 March 1900): 1; "Basketball," *M.A.C. Record* 8, no. 27 (24 March 1903): 1.

71. "Basketball," *M.A.C. Record* 6, no. 20 (5 February 1901): 4.

72. "Co-ed Basketball Teams," *M.A.C. Record* 5, no. 14 (12 December 1899): 1.

73. "Basketball," *M.A.C. Record* 7, no. 12 (3 December 1901): 2.

74. "Reds 37-Blacks 21," *M.A.C. Record* 7, no. 27 (25 March 1902): 4.

75. "Senior Girls to Have a Good Basketball Team," *M.A.C. Record* 19, no. 10 (2 December 1913): 3.

76. "Basket-ball—Olivet vs M.A.C." *M.A.C. Record* 4, no. 24 (7 March 1899): 4.

77. "Basket-ball," *M.A.C. Record* 5, no. 24 (6 March 1900): 1; "Athletics," *Wolverine*, 1901, 182.

78. "Basketball Season Opens," *M.A.C. Record* 8, no. 20 (3 February 1903): 1.

79. "Review of the 1916–17 Athletic Season," *Wolverine*, 1917, 146.

80. "Basketball," *Wolverine*, 1921, 193.

81. *College Speculum* 7, no. 3 (10 April 1888): 8.

82. "Intercollegiate Field Day," *M.A.C. Record* 5, no. 37 (5 June 1900): 1–2; "Field Day," *College Speculum* 7, no. 5 (10 June 1888): 8–9; Seibold, *Spartan Sports Encyclopedia*, 49.

83. "M.A.C. Easily Wins in Tennis Over Olivet," *M.A.C. Record* 14, no. 33 (25 May 1909): 2.

84. Seibold, *Spartan Sports Encyclopedia*, 60.

85. *Wolverine*, 1920, 192.

86. *Wolverine*, 1917, 183.

87. *Wolverine*, 1926, 146.

88. "Indoor Athletic Meets," *M.A.C. Record* 5, no. 26 (20 March 1900): 1.

89. "Intercollegiate Field Day," *M.A.C. Record* 5, no. 37 (5 June 1900): 1.

90. "M.A.C. Out of M.I.A.A.," *M.A.C. Record* 13, no. 19 (4 February 1908): 1.

91. *Wolverine,* 1917, 171.

92. Seibold, *Spartan Sports Encyclopedia,* 19.

93. *Wolverine,* 1926, 141–42.

94. "Athletics," *M.A.C. Record* 12, no. 29 (9 April 1907): 1; "Waite Wins," *M.A.C. Record* 12, no. 30 (16 April 1907): 2; Lyman L. Frimodig and Fred W. Stabley, *Spartan Saga: A History of Michigan State Athletics* (East Lansing: Michigan State University, 1971), 27.

95. Seibold, *Spartan Sports Encyclopedia,* 28.

96. *Wolverine,* 1912, n.p.

97. *Wolverine,* 1917, 171.

98. *Wolverine,* 1921, 219.

99. *Wolverine,* 1926, 148.

100. *M.A.C. Record* 24, no. 31 (23 May 1919): 3.

101. *M.A.C. Record* 27, no. 5 (28 October 1921): 6.

102. *M.A.C. Record* 28, no. 19 (19 February 1923): 5, 7; Frimodig and Stabley, *Spartan Saga,* 131.

103. "Varsity Swimmers Defeat Ypsilanti," *M.A.C. Record* 29, no. 20 (25 February 1924): 11.

104. *Wolverine,* 1921, 240.

105. *Wolverine,* 1917, 195.

106. *M.A.C. Record* 27, no. 16 (3 February 1922): 3; *M.A.C. Record* 29, no. 23 (17 March 1924): 8.

107. *Wolverine,* 1926, 150, 161.

108. "Hockey at M.A.C.," *M.A.C. Record* 11, no. 17 (16 January 1906): 1; "Lansing H. S. Takes First Game by Close Score," *M.A.C. Record* 11, no. 21 (13 February 1906): 1; and "Turn Tables on High School," *M.A.C. Record* 11, no. 22 (11 February 1906): 1.

109. "Ice Hockey Arrives," *M.A.C. Record* 27, no. 13 (13 January 1922): 11; "Hockey Team Busy," *M.A.C. Record* 27, no. 14 (20 January 1922): 12.

110. "Rink for Hockey is Ordered Built," *M.A.C. Record* 29, no. 12 (10 December 1923): 11; Frimodig and Stabley, *Spartan Saga,* 121; Seibold, *Spartan Sports Encyclopedia,* 117, 130.

111. "Varsity Hockey," *Wolverine,* 1926, 149.

112. "Minor Sports Teams Lose," *M.A.C. Record* 27, no. 19 (24 February 1922): 5; "Wrestlers Defeat Michigan University," *M.A.C. Record* 27, no. 20, 3 March 1922): 5; Frimodig and Stabley, *Spartan Saga,* 143–48, 103–4.

113. "Varsity Wrestling," *Wolverine,* 1926, 124.

114. "Fencing," *Wolverine,* 1926, 147; Frimodig and Stabley, *Spartan Saga* 103.

115. *Catalog of the Michigan State College . . . for the Year 1925–1926,* 221–23.

116. *M.A.C. Record* 24, no. 27 (25 April 1919): 3.

117. *Wolverine,* 1926, 158–68.

118. "Report of the Department of Physical Education," *Sixty-Fourth Annual Report of the Secretary of the State Board of Agriculture,* 139–40.

119. "Aggies Will Seek Place in Big Ten," *Lansing Capital Times* 4, no. 34 (9 February 1923): 2.

CHAPTER 10. THE FIRST WORLD WAR
TURNS M.A.C. INTO A MILITARY CAMP

1. E. H. Ryder, "The World Conflict with Militarism," *M.A.C. Record* 23, no. 27 (5 April 1918): 5–6, no. 30 (26 April 1918): 7–8, and no. 31 (3 May 1918): 6–7.

2. Ryder, "The World Conflict with Militarism," *M.A.C. Record* 23, no. 31 (3 May 1918): 7.

3. "The President's Definition of Germanism," *Holcad* 10, no. 11 (8 January 1918): 5.

4. Frank S. Kedzie to Ellwood C. Perisho, 26 October 1917, Kedzie Papers, UA 2.1.8, Box 892, Folder 64.

5. "Report of Division of Extension Work," *Fifty-Seventh Annual Report of the Secretary of the State Board of Agriculture*, 159–229.

6. "Report of the Dean of Engineering," *Fifty-Seventh Annual Report*, 73; "Radio Course Begun," *M.A.C. Record* 23, no. 14 (21 December 1917): 3.

7. "New Farm Tractor Course," *M.A.C. Record*, 23, no. 22 (22 February 1918): 3.

8. "Soldiers for Auto Mechanics Course Arrive This Week," *M.A.C. Record* 23, no. 32 (17 May 1918): 5; "Report of the Dean of Engineering," *Fifty-Seventh Annual Report of the Secretary of the State Board of Agriculture*, 74.

9. "Report of the President," *Fifty-Eighth Annual Report of the Secretary of the Agriculture*, 32–33; "College on Full-Time War Basis," *M.A.C. Record* 24, no. 1 (30 September 1918), 6.

10. Kedzie to students, 26 November 1917, Kedzie Papers, UA 2.1.8, Box 892, Folder 58; "Campus Guarded by Constabulary," *M.A.C. Record* 23, no. 12 (7 December 1917): 3.

11. *M.A.C. Record* 24, no. 2 (11 October 1918): 3.

12. "News and Comment," *M.A.C. Record* 22, no. 28 (8 May 1917): 7; "H. E. Girls Dieticians for Camp Hospital," *M.A.C. Record* 24, no. 4 (25 October 1918): 4.

13. "Report of the President," *Fifty-Seventh Annual Report of the Secretary of the State Board of Agriculture*, 49.

14. "Forestry Dept. Serves Fuel Administrator," *M.A.C. Record* 23, no. 18 (25 January 1918): 3; "Students Complete Fuel Wood Survey," *M.A.C. Record* 23, no. 22 (22 February 1918): 7.

15. "Eustace Loaned to Hoover," *M.A.C. Record* 23, no. 1 (14 September 1917): 6.

16. "Report of the Department of Bacteriology," *Fifty-Eighth Annual Report of the Secretary of the State Board of Agriculture*, 91–93.

17. "Faculty Men Entering War Work," *M.A.C. Record* 23, no. 35 (30 August 1918): 3.

18. "Two M.A.C. Girls Enter Red Cross," *M.A.C. Record* 24, no. 1 (30 September 1918): 3.

19. *M.A.C. Record* 23, no. 12 (7 December 1917): 4.

20. "Twenty Engineers Enter Technical R. C.," *M.A.C. Record* 23, no. 18 (25 January 1918): 4.

21. "Abbot Hall to Remain Closed," *M.A.C. Record* 23, no. 25 (15 March 1918): 3.

22. "Report of the President," *Fifty-Seventh Annual Report of the Secretary of the State Board of Agriculture,* 52–53.

23. "The First," *M.A.C. Record* 23, no. 1 (14 September 1917): 9.

24. "The Casualty Lists," *M.A.C. Record* 23, no. 21 (15 February 1918): 4–5; "Hodgkins, '17, Survivor of Tuscania," and "Wm. R. Johnson, '12," *M.A.C. Record* 23, no. 23 (1 March 1918): 3, 8.

25. "Gordon Webster Cooper, '18," *M.A.C. Record* 23, no. 34 (8 July 1918): 4.

26. *M.A.C. Record* 24, no. 1 (30 September 1918): 4–5.

27. R. J. Johnson to A. K. Chittenden, in *M.A.C. Record* 23, no. 33 (1 June 1918): 10.

28. Russel A. Warner to Dean Bissell, 26 January 1918, *M.A.C. Record* 23, no. 25 (15 March 1918): 3.

29. Roger W. Billings to Dr. Hibbard, 11 February 1918, *M.A.C. Record* 23, no. 28 (12 April 1918): 3.

30. O. A. Taylor to Editor, 16 December 1917, *M.A.C. Record* 23, no. 16 (11 January 1918): 5.

31. W. J. MacKenzie to Kedzie, 2 February 1916, Kedzie Papers, UA 2.1.8, Box 892, Folder 33.

32. "M.A.C. Girl with Red Cross in France," *M.A.C. Record* 24, no. 23 (21 March 1919): 5.

33. "War Figures Worth Pondering," *M.A.C. Record* 24, no. 20 (28 Feb. 1919): 5.

34. Unsigned letter to Kedzie, 14 May 1919, Kedzie Papers, UA 2.1.8, Box 893, Folder 30.

35. "Ex-President Taft at M.A.C.," *M.A.C. Record* 24, no. 24 (4 April 1919): 7.

EPILOGUE

1. Patricia McClelland Miller, "Afterword," in *Quest,* by Helen R. Hull (New York: The Feminist Press, 1990), 355–83; "Helen Hull Wins Prize List Place," *M.A.C. Record* 28, no. 29 (14 May 1923): 3.

2. "Alumni," *M.A.C. Record* 16, no. 17 (17 January 1911): 1. For a full account of Michael's experience in Russia, see Louis Guy Michael, *More Corn for Bessarabia: Russian Experience 1910–1917* (East Lansing: Michigan State University Press, 1983).

3. Louis G. Michael to F. S. Kedzie, 17 October 1932, Kedzie Papers, UA 2.1.8, Box 894, Folder 51. For a more complete biographical sketch of Michael see the "Biographical Sketch of Louis G. Michael," prepared by him, that is in the Kuhn Collection, UA 17.107, Box 1140, Folder 17.

4. "Mary Allen's Story Romantic," *M.A.C. Record* 28, no. 28 (7 May 1923): 9.

5. "Allen Concert Triumph for Singer," *M.A.C. Record* 28, no. 31 (28 May 1923): 8.

6. "New Extension Department," *M.A.C. Record* 20, no. 16 (26 January 1915): 5.

7. "News and Comment," *M.A.C. Record* 20, no. 16 (26 January 1915): 5; "Home Economics Extension Work," *Fifty-Fourth Annual Report of the*

Secretary of the State Board of Agriculture, 71; "Alumni Notes," *M.A.C. Record* 22, no. 26 (24 April 1917): 8.

8. Paulina Raven Morse to F. S. Kedzie, Kedzie Papers, UA 2.1.8, Box 894, Folder 14.

9. Don Francisco to F. S. Kedzie, 1 August 1928, and Francisco, "The Advertising of Agricultural Specialties," Kedzie Papers, UA 2.1.8, Box 894, Folder 12; "Prominent Alumni of M.S.C.," *Wolverine,* 1926, 305.

10. Baker, *American Chronicle,* 195.

11. Ibid., 297.

12. Ray Stannard Baker, *Woodrow Wilson: Life and Letters,* 8 vols. (Garden City, N.Y.: Doubleday, Page and Co., 1927–39).

Select Bibliography

MICHIGAN STATE UNIVERSITY ARCHIVES
AND HISTORICAL COLLECTIONS: MANUSCRIPTS

UA 1: State Board of Agriculture Papers

Correspondence, 1855–.
Trustee Minutes.

UA 2: Office of the President.

Abbot, Theophilus Capen. Papers. 1848–1893. UA 2.1.3.
Butterfield, Kenyon L. Papers. 1924–1928, UA 2.1.10.
Clute, Oscar. Papers. 1889–1891. UA 2.1.5.
Fiske, Lewis Ransom. Papers. 1855–1862. UA 2.1.2.
Friday, David L. Papers, 1913, 1921–1923, UA 2.1.9.
Gorton, Lewis Griffin. Papers. 1893–1895. UA 2.1.6.
Kedzie. Frank S. . Papers. 1877–1947. UA 2.1.8.
Snyder, Jonathan L. Papers. 1887–1918. UA 2.1.7.
Williams, Joseph R. Papers. 1827–1867. UA 2.1.1.
Willits, Edwin. Papers. 1884–1889. UA 2.1.4.

*UA 10: Vice President for University Development, Records:
Papers of Individual Alumni*

Alvord, Charles H. Papers. 1894–1901. UA 10.3.20. Scrapbook 55.
Bregger, Louis A. Papers.1884–1909. UA 10.3.93.
Bulkeley, Joseph. Papers. 1897–1898. UA 10.3.98.
Carr, Roswell G. Papers. 1892–1908, 1905–1908. UA 10.3.13. Scrapbook 45.
Carrel, Ruth E. Papers. 1905–1909. UA 10.3.16, 1 vol. Scrapbook 44.
Coryell, Eva Diann. Collection. 1877–1879. UA 10.3.10, F.D., Folder 1.
Garcinava, Alfonso. Papers. 1906–1909. UA 10.3.14. Scrapbook 52.
Glazier, Hugh. Papers. 1903–1907. UA 10.3.144.
Granger, Edward G. Papers. 1858–1860. UA 10.3.56, F.D., Folders 1–2.
Jewell, Charles A., II. Papers. 1860–1979. UA 10.3.5, Folders 1–3.
Judson, Clarence Herbert. Papers. 1874–1988. UA 10.3.126, Folders 1–8.
Kilbourne, Emily L. Papers. 1899–1900. UA 10.3.12, Folders 1–5.
Koch, Carol R. Papers. 1977. UA 10.3.55, F.D., Folder 1.
Lundberg, Helen Sheldon. Papers. 1908–1916. UA 10.3.38, F.D., Folders 1–5.
Nevins, Bartlett A. Papers. 1875–1878. UA 10.3.11, F.D., Folder 1.
Pratt, Alvin C. Papers. 1902–1909. UA 10.3.72.

Sessions, Sidney S. Papers. 1859–1860. UA 10.3.9, F.D., Folder 1.
Snyder, Le Moyne. Papers. 1900–1920. UA 10.3.97, transcripts, 24 July 1972.
Thompson, Irma. Papers. 1835–1964, 1896–1904. UA 10.3.35.
Wellington, James. Papers. 1860. UA 10.3.6, F.D., Folder 1.
Yebina, Shoichi. Papers. 1915. UA 10.3.31, F. D., Folder 1.

UA 16.7: Agricultural Experiment Station Records

UA 16.34: Cooperative Extension Service Records

UA 17.4: Beal, William J. Papers. 1859–1921

UA 17.14: Dye, Marie. Papers. 1896–1973

UA 17.107: Kuhn, Madison. Collection. 1827–1966

Land Grant Collection. Morrill Justin S. Papers. Microfilm copy. Library of Congress.

MICHIGAN STATE UNIVERSITY ARCHIVES AND HISTORICAL COLLECTIONS: PERIODICALS

The Bubble. No. 2 (20 June 1868)—No. 7 (24 October 1868).
College Speculum. Vols. 1–7, 1 August 1881–10 July 1888. Serial no. 662.
Holcad. Vols. 1–6, 10 March 1909–8 June 1914. Serial no. 669.
M.A.C. Record. Vols. 1–15, 14 January 1896–28 June 1910.
Speculum. Vols. 8–15, 10 August 1888–15 November 1895. Serial no. 662.

MICHIGAN STATE UNIVERSITY LIBRARIES– SPECIAL COLLECTIONS

Holcad. Vols. 1–17, 1909–1925.
M.A.C. Record. Vols. 1–30, 1896–1926.
Bessey, Charles Edwin. Papers. 1865–1915. Microfilm copy. University of Nebraska Archives, University of Nebraska, Lincoln.

MICHIGAN STATE UNIVERSITY MUSEUM COLLECTIONS

Drake, Leon. Papers.
The Union Lit.

BENTLEY LIBRARY, UNIVERSITY OF MICHIGAN

Michigan University. Board of Regents, Records.
Tappen, Henry Philip. Papers. 1840–1881.

CITY OF BOSTON, ARCHIVES AND RECORDS MANAGEMENT DIVISION

Boston School Department Records.

LIBRARY OF MICHIGAN

Reynolds, Henry G. Diaries. 1867 and 1872–1874.

PRINTED WORKS

Abbot, Theophilus C. "The Michigan State Agricultural College," *Michigan Pioneer and Historical Collections,* Vol. 6. Lansing: Wynkoop Hallenbeck Crawford Co., 1907.

Adams, Romanzo. "Agriculture in Michigan." *Publications of the Michigan Political Science Association* 3, no. 7 (March 1899): 1–40.

Ainsworth, G. C. *Introduction to the History of Plant Pathology.* Cambridge: Cambridge University Press, 1981.

Alumni Catalogue Number: Michigan State College Bulletin: List of Graduates, Officers and Professors of the Faculties 1857–1930. [East Lansing: Michigan State College], 1931.

Alvord, Henry E. "Dairy Development in the United States." In *Yearbook of the United States Department of Agriculture, 1899.* Washington, D.C.: Government Printing Office, 1900.

Avery, Julie A., ed. *Agricultural Fairs in America: Tradition, Education, Celebration.* East Lansing: Michigan State University Museum, 2000.

Bailey, Liberty Hyde. *American Grape Training.* New York: Rural Publishing Co., 1893.

———. *Field Notes on Apple Culture.* New York: O. Judd Co., 1886.

———. *The Forcing-Book: A Manual of the Cultivation of Vegetables in Glass Houses.* New York: Macmillan, 1897.

———. *The Principles of Agriculture: A Text-Book for Schools and Rural Societies.* New York: Macmillan, 1898.

———. *The Principles of Fruit-Growing.* New York: Macmillan, 1897.

———. *The Pruning-Book: A Monograph of the Pruning and Training of Plants as Applied to American Conditions.* New York: Macmillan, 1898.

Bailyn, Bernard. *Education in the Forming of American Society: Needs and Opportunities for Study.* New York: Vintage Books, 1960.

Baker, Gladys L., Wayne Rasmussen, Vivian Wiser, and Jane M. Porter. *Century of Service: The First 100 Years of the United States Department of Agriculture.* [Washington, D.C.]: U.S. Department of Agriculture, 1963.

Baker, Ray Stannard. *American Chronicle.* New York: Charles Scribner's Sons, 1945.

———. *Native American: The Book of My Youth.* New York: Charles Scribner's Sons, 1941.

Barber, Mary I., ed. *History of the American Dietetic Association, 1917–1959.* Philadelphia: J. B. Lippincott, 1959.

Barnett, LeRoy. "Putting Michigan Farmers on the Right Track." *Michigan History* (January/February 2000): 48–55.

Basler, Roy P., Marion Dolores Pratt, and Lloyd A. Dunlap, eds. *The Collected Works of Abraham Lincoln.* 8 vols. New Brunswick: Rutgers University Press, 1953.

Beal, W. J. *History of the Michigan Agricultural College and Biographical Sketches of Trustees and Professors.* East Lansing: The Agricultural College, 1915.

Beasley, W. G. *The Rise of Modern Japan.* London: Weidenfeld and Nicolson, 1990.

Beecher, Catharine E. "An Address on Female Seminary." In *Pioneers of Women's Education in the United States: Emma Willard, Catherine Beecher, Mary Lyon,* ed. by Willystine Goodsell. New York: McGraw-Hill, 1931.

————. *A Treatise on Domestic Economy, for the Use of Young Ladies at School.* Rev. ed. New York: Harper and Brothers, 1859.

Beecher, Catharine E., and Harriet Beecher Stowe. *The American Woman's Home: Or Principles of Domestic Science; Being a Guide to the Formation and Maintenance of Economical, Healthful, Beautiful, and Christian Homes.* New York: J. B. Ford and Company, 1869.

Bernstein, Richard J., ed. *John Dewey on Experience, Nature, and Freedom: Representative Selections.* New York: Liberal Arts Press, 1960.

Bessey, Charles E., *Botany for High Schools and Colleges.* New York: Henry Holt and Company, 1880.

————. *The Essentials of Botany.* New York: Henry Holt and Company, 1884.

Betters, Paul V. *The Bureau of Home Economics: Its History, Activities and Organization.* Washington, D.C.: Brookings Institution, 1930.

Bevier, Isabel. *Home Economics in Education.* Philadelphia: J. B. Lippincott, 1928.

Bezilla, Michael. *The College of Agriculture at Penn State: A Tradition of Excellence.* University Park: Pennsylvania State University Press, 1987.

Bieler, Stacey. *"Patriots" or "Traitors"?: A History of American-Educated Chinese Students.* Armonk, N.Y.: M.E. Sharpe, 2004.

Bierer, B. W. *A Short History of Veterinary Medicine.* East Lansing: Michigan State University Press, 1955.

Bingham, Kinsley S. "Address to the Board of Education." In *The Agricultural College of the State of Michigan.* Lansing, Mich.: Hosmer and Fitch, 1857.

Bishop, Morris. *Early Cornell, 1865–1900: The First Part of a History of Cornell.* Ithaca, N.Y.: Cornell University Press, 1962.

————. *A History of Cornell.* Ithaca, N.Y.: Cornell University Press, 1962.

Blaisdell, Thomas C., ed. *Semi-Centennial Celebration of Michigan State Agricultural College: May Twenty-Sixth, Twenty-Ninth, Thirtieth and Thirty-First, Nineteen Hundred Seven.* Chicago: University of Chicago Press, 1908.

Bordin, Ruth. *Women at Michigan: The "Dangerous Experiment," 1870s to the Present.* Ann Arbor: University of Michigan Press, 1999.

Bowling, G. A. *History of Dairy Development at Michigan State College.* East Lansing: Dairy Department[, 1929].

Briggs, Katharine Cook. "Meet Yourself: How to Use the Personality Paint Box." *The New Republic* 49, no. 629 (22 December 1926): 129–33.

Brody, Clark L. *In the Service of the Farmer: My Life in the Michigan Farm Bureau.* East Lansing: Michigan State University Press, 1959.

Browne, Charles A. "Agricultural Chemistry." In *A Half-Century of Chemistry in America, 1876–1926,* ed. Charles A. Browne. Easton, Penn.: American Chemical Society, 1926.

Buck, Solon Justus. *The Granger Movement: A Study of Agricultural Organization and Its Political Economic and Social Manifestations, 1870–1880.* Cambridge: Harvard University Press, 1913.

Byerly, T. C. "Changes in Animal Science." *Agricultural History* 50, no. 1 (January 1976): 258–74.

Caldwell, G. C. "The More Notable Events in the Progress in Agricultural Chemistry, Since 1870." *The Journal of the American Chemical Society* 14 (1892): 83–111.

Cameron, Jenks. *The Bureau of Dairy Industry: Its History, Activities and Organizations.* Baltimore: Johns Hopkins University Press, 1929.

Carey, James C. *Kansas State University: The Quest for Identity.* Lawrence: The Regents Press of Kansas, 1977.

Catalogue of Officers and Graduates, 1857–1911. East Lansing, Michigan, 1911.

Catalogue of Officers and Graduates of the Michigan State Agricultural College up to 1873. Lansing: W. S. George and Co., 1873.

Ceasar, Ford Stevens. *The Bicentennial History of Ingham County, Michigan.* Lansing: Ingham County, 1976.

Census of the State of Michigan, 1904: Agriculture, Manufacture and Mines. Vol. 2. Lansing: Wynkoop Hallenbeck Crawford Co., 1905.

Chamberlain, Joseph, ed. *Chemistry in Agriculture: A Cooperative Work Intended to Give Examples of the Contributions Made to Agriculture by Chemistry.* New York: The Chemical Foundation, 1926.

Childe, Elizabeth [pseud.]. "Personal Experiences of Mothers: Why I Find Children Slow in Their School Work." *Ladies Home Journal* 29, no. 10 (October 1912): 42

———. "'The Price of a Home.'" *The New Republic* 13, no. 158 (10 November 1917): 78–79.

———. "Why I Believe the Home is the Best School." *Ladies Home Journal* 29, no. 10 (October 1912): 68.

Clarke, Robert. *Ellen Swallow: The Woman Who Founded Ecology.* Chicago: Follett, 1973.

Clemen, Rudolf Alexander. *The American Livestock and Meat Industry.* New York: Ronald Press, 1923.

Cochrane, Willard W. *The Development of American Agriculture: A Historical Analysis.* Minneapolis: University of Minnesota Press, 1979.

Conable, Charlotte Williams. *Women at Cornell: The Myth of Equal Education.* Ithaca, N.Y.: Cornell University Press, 1977.

Craig, Lee A. "Constrained Resource Allocation and the Investment in the Education of Black Americans: The 1890 Land-Grant Colleges." *Agricultural History* 65, no. 2 (1991): 73–84.

Crèvecoeur, J. Hector St. John, de. *Letters from an American Farmer and Sketches of Eighteenth-Century America.* New York: New American Library, 1963.

Cronon, William. *Nature's Metropolis: Chicago and the Great West.* New York: W. W. Norton, 1991.

Cullen, Maurice Raymond, Jr. "The Presidency of Jonathan Lemoyne Snyder at Michigan Agricultural College, 1896–1915." Ph.D. diss., Michigan State University, 1965.

Culpepper, Marilyn. "Mary Mayo: Leader of Rural Women." In *Historic Women of Michigan: A Sesquicentennial Celebration,* ed. Rosalie Riegle Troester. Lansing: Michigan Women's Studies Association, 1987.

Curti, Merle, and Vernon Carstensen. *The University of Wisconsin: A History, 1848–1925.* 2 vols. Madison: University of Wisconsin Press, 1949.

Darby, William J. *Nutrition Science: An Overview of American Genius.* Washington, D.C.: Agricultural Research Service, USDA, 1976.

Darden, Joe T. "Patterns of Residential Segregation in Michigan Cities in the Nineteenth Century." In *Blacks and Chicanos in Urban Michigan,* ed. Homer C. Hawkins and Richard W. Thomas. [Lansing:] Michigan History Division, Michigan Department of State, n.d.

Davenport, Eugene. *Domesticated Animals and Plants: A Brief Treatise upon the Origin and Development of Domesticated Races with Special Reference to the Methods of Improvement.* Boston: Ginn and Company, 1910.

———. *Education for Efficiency: A Discussion of Certain Phases of the Problem of Universal Education with Special Reference to Academic Ideals and Methods.* Rev. ed. Boston: D.C. Heath and Co., 1914.

———. *Principles of Breeding: A Treatise on Thremmatology or the Principles and Practices Involved in the Economic Improvement of Domesticated Animals and Plants.* Boston: Ginn and Company, 1907.

———. "A Son of the Timberlands: In the Land of Amanha." *The Country Gentleman* (January 1926): 32, 132–34.

———. *Timberland Times.* Urbana: University of Illinois Press, 1950.

De'Sigmond, Alexius A. J. "Development of Soil Science." *Soil Science* 40 (June-December 1935): 77–86.

Dewey, James S., comp. *The Compiled Laws of the State of Michigan.* 2 vols. Lansing: W.S. George and Co., 1872.

Dewey, John. *Lectures in the Philosophy of Education, 1899,* ed. Reginald D. Archambault. New York: Random House, 1966.

Dies, Edward Jerome. *Titans of the Soil: Great Builders of Agriculture.* Chapel Hill: University of North Carolina Press, 1949.

Dressel, Paul L. *College to University: The Hannah Years at Michigan State, 1935–1969.* East Lansing: Michigan State University, University Publications, 1987.

Dunbar, Willis F. *All Aboard!: A History of Railroads in Michigan.* Grand Rapids: Eerdmans, 1969.

———. *The Michigan Record in Higher Education.* Detroit: Wayne State University Press, 1963.

Dunlap, Thomas R. "Farmers, Scientists, and Insects." *Agricultural History* 54, no. 1 (January 1980): 93–107.

Eddy, Edward Danforth, Jr. *Colleges for Our Land and Time: The Land-Grant Idea in American Education.* New York: Harper and Brothers, 1957.

Edmond. J. B. *The Magnificent Charter: The Origin and Role of the Morrill Land-Grant Colleges and Universities.* Hicksville, N.Y.: Exposition Press, 1978.

Ellis, Robert. *Michigan State Baseball: A Concerned Look.* Grand Rapids, Mich.: McGriff and Bell, 1992.

Farrand, Elizabeth M. *History of the University of Michigan.* Ann Arbor: Register Publishing Houses, 1885.

Ferleger, Louis and William Lazonick. "Higher Education for an Innovative Economy: Land-Grant Colleges and the Managerial Revolution in America." *Business and Economic History* 23, no. 1 (1994): 116–28.

Finch, James Kip. *The Story of Engineering.* Garden City, N.Y.: Doubleday, 1960.

"First Morrill Act, 1862, The." In *Agriculture in the United States: A Documentary History,* ed. Wayne D. Rasmussen. 4 vols. New York: Random House, 1975.

Florer, John H. "Major Issues in the Congressional Debate of the Morrill Act of 1862." *History of Education Quarterly* 8, no. 4 (1968): 459-78.

Foner, Eric. "Free Labor and Nineteenth-Century Ideology." In *The Market Revolution in America: Social, Political, and Religious Expressions, 1800–1880,* ed. Melvyn Stokes and Stephen Conway. Charlottesville: University Press of Virginia, 1996.

———. *The Story of American Freedom.* New York: Norton, 1998.

Fox, Jean M. *Fred M. Warner: Progressive Governor.* Farmington Hills, Mich.: Farmington Hills Historical Commission, 1988.

Frimodig, Lyman L., and Fred W. Stabley. *Spartan Saga: A History of Michigan State Athletics.* East Lansing: Michigan State University, 1971.

Frykman, George A. *Creating the People's University: Washington State University, 1890–1990.* Pullman: Washington State University Press, 1990.

Galloway, B. T. "Progress in the Treatment of Plant Diseases in the United States." In *Yearbook of the United State Department of Agriculture, 1899.* Washington, D.C.: Government Printing Office, 1900.

———. "Twenty Years Progress in Plant Pathology." In *Proceedings of the Twenty-First Annual Meeting of the Society for the Promotion of Agricultural Science, 1900.* Syracuse, N.Y.: The Society, 1900.

Gates, Paul W., "Western Opposition to the Agricultural College Act." *Indiana Magazine of History* 37, no. 1 (March 1941): 103–36.

Geiger, Roger, ed. *The American College in the Nineteenth Century.* Nashville: Vanderbilt University Press, 2000.

General Catalogue of the Officers and Students, 1857–1900. Agricultural College, Mich.: Published by the College, 1900.

Gilchrist, Maude. *The First Three Decades of Home Economics at Michigan State College.* [East Lansing:] School of Home Economics, 1947.

Gilman, Daniel Coit, "Report on the National Schools of Science," in "Report of the Commissioner of Education." In *Report of The Secretary of the Interior; Being Part of the Message and Documents Communicated to the Two Houses of Congress at the Beginning of the Second Session of the Forty-Second Congress.* Vol. 2. Washington, D.C.: Government Printing Office, 1872.

Glover, W. H. *Farm and College: The College of Agriculture of the University of Wisconsin, A History.* Madison: University of Wisconsin Press, 1952.

Goodsell, Willystine, ed. *Pioneers of Women's Education in the United States: Emma Willard, Catharine Beecher, Mary Lyon.* New York: McGraw-Hill, 1931.

Goodwyn, Lawrence. *The Populist Moment: A Short History of the Agrarian Revolt in America.* Oxford: Oxford University Press, 1981.

Gough, Robert. *Farming the Cutover: A Social History of Northern Wisconsin, 1900–1940.* Lawrence: University Press of Kansas, 1997.

Griffin, Gail B. "Lucinda Hinsdale Stone: Champion of Women's Education." In *Historic Women of Michigan: A Sesquicentennial Celebration,* ed. Rosalie Riegle Troester. Lansing: Michigan Women's Studies Association, 1987.

Hamilton, Thomas H., and Edward Blackman, eds. *The Basic College of Michigan State.* East Lansing: Michigan State College Press, 1955.

Hannah, John A. *A Memoir.* East Lansing: Michigan State University Press, 1980.

Harburn, Todd E., and Gerald E. Harburn. *MIAA Football: The Illustrated Gridiron History of the Michigan Intercollegiate Athletic Association.* Chelsea, Mich.: BookCrafters, 1986.

Harding, T. Swann. *Two Blades of Grass: A History of Scientific Development in the U. S. Department of Agriculture.* Norman: University of Oklahoma Press, 1947.

Harris, Leslie J. "The Discovery of Vitamins." In *The Chemistry of Life: Eight Lectures on the History of Biochemistry,* ed. Joseph Needham. Cambridge: Cambridge University Press, 1970.

Hedrick, U. P. *The Land of the Crooked Tree.* New York: Oxford University Press, 1948.

Helvenston, Sally I., and Margaret M. Bubolz. "Home Economics and Home Sewing in the United States 1870–1940." In *The Culture of Sewing: Gender, Consumption and Home Dressmaking,* ed. Barbara Burman. Oxford: Berg, 1999.

Herbst, Jergen. "The Development of Public Universities in the Old Northwest." In *The Northwest Ordinance: Essays on Its Formulation, Provisions, and Legacy,* ed. Frederick D. Williams. East Lansing: Michigan State University Press, 1989.

Hodgman, Frank. *The Wandering Singer and His Songs and Other Poems.* Climax, Mich.: F. Hodgman, 1898.

Hofstadter, Richard. *The Age of Reform from Bryan to F.D.R.* New York: Vintage, 1955.

Holland, W. J. "The Development of Entomology in North America." *Annals of the Entomological Society of America* 13, no. 1 (March 1920): 1–15.

Hollinger, David A., "Amalgamation and Hypodescent: The Question of Ethnoracial Mixture in the History of the United States," *The American Historical Review* 108, no. 5 (December 2003): 1363–90.

Holmes, Frederic L. "The Transformation of the Science of Nutrition." *Journal of the History of Biology* 8, no. 1 (1975): 135–44.

Honhart, Fred. "The MSU Union's 75th Birthday." *MSU Alumni Magazine* 17, no. 2 (2000): 26–30.

Houck, U. G. *The Bureau of Animal Industry of the United States Department of Agriculture: Its Establishment, Achievements and Current Activities.* Washington, D.C.: The Author, 1924.

Howard, L. O. "Progress in Economic Entomology in the United States." In *Yearbook of the United States Department of Agriculture, 1899.* Washington, D.C.: Government Printing Office, 1900.

———. "The Rise of Applied Entomology in the United States." *Agricultural History* 3, no. 3 (July 1929): 131–39.

Hull, Helen. *Quest.* New York: The Feminist Press, 1990.

Humphries, Frederick S., "1890 Land-Grant Institutions: Their Struggle for Survival and Equality." *Agricultural History* 65, no. 2 (1991): 3-11.

Hurt, R. Douglas. *Agricultural Technology in the Twentieth Century.* Manhattan, Kan.: Sunflower University Press, 1991.

———. *American Agriculture: A Brief History.* Ames: Iowa State University Press, 1994.

James, Edmund J. *The Origin of the Land Grant Act of 1862 (The So-called Morrill Act) and Some Account of its Author Jonathan B. Turner.* Urbana-Champaign: University Press, 1910.

Jarvis, Joseph W. *Golden Jubilee: African Methodist Episcopal Church, Lansing, Michigan, 1867–1917.* N.p., 1917.

Jefferson, Thomas. *Notes on the State of Virginia,* ed. William Peden. New York: Norton, 1982.

Jenkins, Robert L. "The Black Land-Grant Colleges in Their Formative Years, 1890–1920." *Agricultural History* 65, no. 2 (1991): 63–72.

Jones, Thomas Jesse, ed. *Negro Education: A Study of the Private and Higher Schools for Colored People in the United States.* 2 vols. New York: Arno Press, 1969.

Jordan, W. H. "The Promotion of Agricultural Science." *Proceedings of the Twenty-Third Annual Meeting of the Society for the Promotion of Agricultural Science.* Published by Authority of the Society[, 1902].

Kaestle, Carl F. "The Development of Common School Systems in the States of the Old Northwest." In *" . . . Schools and the Means of Education Shall Forever Be Encouraged,"* ed. Paul H. Mattingly and Edward W. Stevens Jr.. Athens: Ohio University Libraries, 1987.

Kains, Maurice G. *Culinary Herbs: Their Cultivation, Harvesting, Curing and Uses.* New York: Orange Judd Company, 1912.

———. *Making Horticulture Pay: Experiences in Gardening and Fruit Growing.* New York: Orange Judd Company, 1913.

———. *Plant Propagation: Greenhouse and Nursery Practice.* New York: Orange Judd Publishing Company, 1925.

Kestenbaum, Justin L., ed. *At the Campus Gate: A History of East Lansing.* East Lansing: East Lansing Bicentennial Committee, 1976.

———. *Out of a Wilderness: An Illustrated History of Greater Lansing.* Woodland Hills, Calif.: Windsor Publications, 1981.

Key, Scott. "Economics or Education: The Establishment of American Land-Grant Universities." *Journal of Higher Education* 67, no. 2 (March/April 1996): 196–220.

Kikuchi, Dairoku. *Japanese Education: Lectures Delivered in the University of London by Baron Dairoku Kikuchi.* London: J. Murray, 1909.

Kirkland, Edward C. "'Scientific Eating': New Englanders Prepare and Promote a Reform, 1873–1907." In *Proceedings of the Massachusetts Historical Society* 86. Boston: The Society, 1975.

Kuhn. Madison. *Michigan State: The First Hundred Years.* East Lansing: Michigan State University Press, 1955.

Lang, Daniel W. "The People's College: The Mechanics' Mutual Protection and the Agricultural College Act." *History of Education Quarterly* 18, no. 3 (1978): 295–321.

Lautner, Harold W. "*From an Oak Opening:* A Record of the Development of the Campus Park of Michigan State University, 1855–1969," 2 vols.

Lee, Jeanette A., Katherine M. Hart, and Rosalind B. Mentzer. *From Home Economics to Human Ecology.* East Lansing: Michigan State University, 1972.

Lewis, Kenneth E. *West to Far Michigan: Settling the Lower Peninsula, 1815-1860.* East Lansing: Michigan State University Press, 2002.

McCain, James A. "Designed for Relevance: The Land-Grant Universities." *The Centennial Review* 14, no. 1 (1970): 91–107.

McCall, A. G. "The Development of Soil Science." *Agricultural History* 5, no. 2 (April 1931): 43–56.

MacLean, James, and Craig A. Whitford. *Lansing City on the Grand 1836–1939.* Charleston, S.C.: Arcadia, 2003.

Mann, Horace. *Lectures on Education.* New York: Arno Press, 1969.

Manual, Containing the Rules of the Senate and House of Representatives of the State of Michigan, with the Joint Rules of the Two Houses, and Other Matter. Lansing: By Authority, 1861.

Marcus, Alan I. *Agricultural Science and the Quest of Legitimacy: Farmers, Agricultural Colleges, and Experiment Stations, 1870–1890.* Ames: Iowa State University Press, 1985.

Marsden, George M. *The Soul of the American University: From Protestant Establishment to Established Nonbelief.* New York: Oxford University Press, 1994.

Martin, Douglas D. *The Lamp in the Desert: The Story of the University of Arizona.* Tucson: University of Arizona Press, 1960.

Marx, Leo. *The Machine in the Garden: Technology and the Pastoral Ideal in America.* Oxford: Oxford University Press, 2000.

Mayberry, B. D. *A Century of Agriculture in the 1890 Land-Grant Institutions and Tuskegee University—1890–1990.* New York: Vantage Press, 1991.

Meyer, Douglas K. "The Changing Negro Residential Patterns in Lansing, Michigan, 1850–1969." Ph.D. diss., Michigan State University, 1970.

Michael, Louis Guy. *More Corn for Bessarabia: Russian Experience 1910–1917.* East Lansing: Michigan State University Press, 1983.

Miller, Patricia McClelland. "Afterword." In *Quest,* by Helen R. Hull. New York: The Feminist Press, 1990.

Miller, Whitney. *East Lansing: Collegeville Revisited.* Chicago: Arcadia, 2002.

Minoru, Ishizuki. "Overseas Study by Japanese in the Early Meiji Period." In *The Modernizers: Overseas Students, Foreign Employees, and Meiji Japan,* ed. Ardath W. Burks. Boulder: Westview Press, 1985.

Morrill, Charles Cleon. *Veterinary Medicine in Michigan: An Illustrated History.* East Lansing: College of Veterinary Medicine, Michigan State University, 1979.

Mortimer, Jeff. *Pigeons, Bloody Noses and Little Skinny Kids: A Story of Wolverine Basketball.* Dexter, Mich.: Thomson-Store Publishers, 1978.

Nevins, Allan. *The Origins of the Land-Grant Colleges and State Universities: A Brief Account of the Morrill Act of 1862 and Its Results.* Washington, D.C.: Civil War Centennial Commission, 1962.

———. *The State University and Democracy.* Urbana: University of Illinois Press, 1962.

Noll, Victor H. *The Preparation of Teachers at Michigan State University.* East Lansing: College of Education, 1968.

Ogura, Takekazu, ed. *Agricultural Development in Modern Japan.* Tokyo: Fuji Publishing Co., 1967.

Overfield, Richard. "Charles E. Bessey: The Impact of the 'New' Botany on American Agriculture, 1880–1910." *Technology and Culture* 16, no. 2 (April 1975): 162–81.

———. *Science with Practice: Charles E. Bessey and the Maturing of American Botany.* Ames: Iowa State University Press, 1993.

Peckham, Howard H. *The Making of The University of Michigan, 1817–1992.* Ann Arbor: University of Michigan, Bentley Historical Library, 1994.

Pirtle, T. R. *History of the Dairy Industry.* Chicago: Mojonnier Bros. Co., 1926.

Pollack, Norman. *The Humane Economy: Populism, Capitalism, and Democracy.* New Brunswick: Rutgers University Press, 1990.

———. *The Populist Response to Industrial America: Midwestern Populist Thought.* Cambridge: Harvard University Press, 1962.

Pollard, James E. *History of the Ohio State University.* Columbus: Ohio State University Press, 1952.

Rasmussen, Wayne D., ed. *Agriculture in the United States: A Documentary History.* 4 vols. New York: Random House, 1975.

———. "The 1890 Land-Grant Colleges and Universities: A Centennial Overview." *Agricultural History* 65, no. 2 (1991): 168–72.

———. *Taking the University to the People: Seventy-Five Years of Cooperative Extension.* Ames: Iowa State University Press, 1989.

Ratermann, Dale. *The Big Ten: A Century of Excellence.* Champaign: Sagamore, 1996.

Renne, Roland R. "Land-Grant Institutions, the Public, and the Public Interest." *The Annals of the American Academy of Political and Social Science* 331 (September 1960): 46–51.

Report of the Commission on Country Life. New York: Sturgis and Walton, 1911.

Riis, Jacob. *How the Other Half Lives: Studies Among the Tenements of New York.* New York: Scribner's and Sons, 1890.

Rodgers, Andrew Denny, III. *Liberty Hyde Bailey: A Story of America Plant Sciences.* Princeton: Princeton University Press, 1949.

Romig, Walter. *Michigan Place Names: The History of the Founding and the Naming of More Than Five Thousand Past and Present Michigan Communities.* Detroit: Wayne State University Press, 1986.

Rosenberg, Charles E. "The Adams Act: Politics and the Cause of Scientific Research." *Agricultural History* 38, no. 1 (January 1964): 3–12.

———. "Rationalization and Reality in the Shaping of American Agricultural Research, 1875–1914." *Social Studies of Science* 7, no. 4 (November 1977): 401–22.

Ross, Earle Dudley, *Democracy's College: The Land-Grant Movement in the Formative Stage.* Ames: Iowa State College Press, 1942.

———. *A History of the Iowa State College of Agriculture and Mechanic Arts.* Ames: Iowa State College Press, 1942.

———. *The Land-Grant Idea at Iowa State College: A Centennial Trial Balance, 1858–1958.* Ames: Iowa State College Press, 1958.

Rossiter, Margaret W. *The Emergence of Agricultural Science: Justus Liebig and the Americans, 1840–1880.* New Haven: Yale University Press, 1975.

———. *Women Scientists in America: Struggles and Strategies to 1940.* Baltimore: Johns Hopkins University Press, 1982.

Sack, Saul. *History of Higher Education in Pennsylvania.* 2 vols. Harrisburg: Pennsylvania Historical and Museum Commission, 1963.

Salmon, D. E. "Some Examples of the Development of Knowledge Concerning Animal Diseases." In *Yearbook of the United States Department of Agriculture, 1899.* Washington, D.C.: Government Printing Office, 1900.

Saunders, Frances Wright. *Katharine and Isabel: Mother's Light, Daughter's Journey.* Palo Alto, Calif.: Consulting Psychologists Press, 1991.

Schlebecker, John T. *A History of American Dairying.* Chicago: Rand McNally, 1967.

Sellers, Charles. *The Market Revolution: Jacksonian America, 1815–1846.* New York: Oxford University Press, 1991.

Shapiro, Laura. *Perfection Salad: Women and Cooking at the Turn of the Century* New York: Farrar, Straus and Giroux, 1986.

Sklar, Kathryn Kish. *Catharine Beecher: A Study in American Domesticity.* New York: W. W. Norton, 1973.

Smith, Eliza C. "History of the Michigan Female College, and a Sketch of the Life and Work of Miss A. C. Rogers," in *Michigan Pioneer and Historical Collections.* Vol. 6. Lansing: Michigan Pioneer and Historical Society, 1907.

Smith, Henry Nash. *Virgin Land: The American West as Symbol and Myth.* New York: Vintage, 1950.

Smith, Ronald A. *Sports and Freedom: The Rise of Big-Time College Athletics.* New York: Oxford University Press, 1988.

Solberg, Winton U. *The University of Illinois 1867–1894: An Intellectual and Cultural History.* Urbana: University of Illinois Press, 1968.

————. *The University of Illinois 1894–1904: The Shaping of the University.* Urbana: University of Illinois Press, 2000.

Spargo, John. *The Bitter Cry of the Children.* New York: Macmillan, 1968.

Stabley, Fred W. *The Spartans: A Story of Michigan State Football.* Huntsville, Ala.: Strode, 1975.

Stage, Sarah. "Ellen Richards and the Social Significance of the Home Economics Movement." In *Rethinking Home Economics: Women and the History of a Profession,* ed. Sarah Stage and Virginia B. Vincenti. Ithaca, N.Y.: Cornell University Press, 1997,

Steinway, Cindi. "Myrtle Mowbray, '07: MSU's first black graduate, remembers M.A.C." *MSU Alumni Magazine* 18, no. 2 (November 1972): 15.

Stevenson, John A. "The Beginnings of Plant Pathology in North America." In *Plant Pathology: Problems and Progress, 1908–1958,* ed. C. S. Holton. G. W. Fischer, R. W. Fulton, Helen Hart, and S. E. A. McCallan. Madison: University of Wisconsin Press, 1959.

Thackrey, Russell, and Jay Richter. "The Land-Grant Colleges and Universities, 1862–1962: An American Institution." *Higher Education* 16, no. 3 (November 1959): 3–7.

Thomas, David. "A Fortnight in Michigan's Forests." *Michigan History* 72, no. 4 (July/August 1988): 36–43.

————. "MSU Black is a Pioneer in American Football." *Lansing City Magazine* 6, no. 2 (February 1992): 27.

Thomas, David, and Marc Thomas. "Leander Burnett: Saga of an Athlete." *Michigan History* 70, no. 1 (January/February 1986): 12–15.

Thomas, Richard W. *Life for Us is What We Make It: Building Black Community in Detroit, 1915–1945.* Bloomington: Indiana University Press, 1992.

Topping, Robert W. *A Century and Beyond: The History of Purdue University.* West Lafayette: Purdue University Press, 1988.

Transactions of the Michigan State Agricultural Society with Reports of County Agricultural Societies [1849]. [Lansing: N.p., 1850.]

Transactions of the State Agricultural Society with Reports of County Agricultural Societies for 1850. Lansing: Published by Order of the Legislature, 1851.

Trump, Fred. *The Grange in Michigan: An Agricultural History of Michigan over the past Ninety Years.* Grand Rapids: Fred Trump, 1963.

Tyler, Alice Felt. *Freedom's Ferment: Phases of American Social History from the Colonial Period to the Outbreak of the Civil War.* New York: Harper and Row, 1944.

Walker, John Charles. *Plant Pathology.* New York: McGraw-Hill, 1969.

Walsh, Christy, ed. *Intercollegiate Football: A Complete Pictorial and Statistical Review from 1869 to 1934.* St. Paul: Doubleday, Doran and Company, 1934.

Walsh, Thomas R. "Charles E. Bessey and the Transformation of the Industrial College." *Nebraska History* 52, no. 4 (1971): 383–409.

Watterson, John Sayle. *College Football: History, Spectacle, Controversy.* Baltimore: Johns Hopkins University Press, 2000.

Weber, Gustavus A. *The Bureau of Chemistry and Soils: Its History, Activities and Organization.* Baltimore: Johns Hopkins Press, 1928.

———. *The Bureau of Entomology: Its History, Activities and Organization.* Washington, D.C.: Brookings Institution, 1930.

Wheeler, W. Reginald, Henry H. King, and Alexander B. Davidson, eds. *The Foreign Student in America: A Study by the Commission on Survey of Foreign Students in the United States of America, under the Auspices of the Friendly Relations Committees of the Young Men's Christian Association and the Young Women's Christian Association,* ed.. New York: Association Press, 1925.

Whitaker, Adelynne H. "Pesticide Use in Early Twentieth Century Animal Disease Control." *Agricultural History* 54, no. 1 (January 1980): 71–81.

Whitney, Milton. "Soil Investigations in the United States." In *Yearbook of the United States Department of Agriculture, 1899.* Washington D. C.: Government Printing Office, 1900.

Wik, Reynold M. "Science and American Agriculture." In *Science and Society in the United States,* ed. David D. Van Tassel and Michael G. Hall. Homewood, Ill.: Dorsey Press, 1966.

Wiley, H. W. "The Relation of Chemistry to the Progress of Agriculture." In *Yearbook of the United States Department of Agriculture, 1899.* Washington, D.C.: Government Printing Office, 1900.

Willard, Emma. "An Address to the Public Particularly to the Members of the Legislature of New-York proposing a Plan for Improving Female Education." In *Pioneers of Women's Education in the United States: Emma Willard, Catharine Beecher, Mary Lyon,* ed. Willystine Goodsell. New York: McGraw-Hill, 1931.

Williams, Brian A. *Thought and Action: John Dewey at the University of Michigan.* Ann Arbor: Bentley Historical Library, 1998.

Williams, Joseph R., "Address to the Board of Education." In *The Agricultural College of the State of Michigan.* Lansing, Mich.: Hosmer and Fitch, 1857.

Willoughby, Barrett. "Dr. Georgeson: The Man Who Put the Midnight Sun to Work." *The American Magazine* (August 1928): 36–38.

Winters, Robert K., ed. *Fifty Years of Forestry in the U.S.A.* Washington, D.C.: Society of American Foresters, 1950.

Woody, Thomas. *A History of Women's Education in the United States.* Vol. 2. New York: Octagon Books, 1966.

Index

Entries in **boldface** refer to illustrations and captions.